TRANSLATIONS
OF
MATHEMATICAL
MONOGRAPHS

Volume 24

THEORY AND APPLICATIONS OF VOLTERRA OPERATORS IN HILBERT SPACE

by
I. C. GOHBERG
M. G. KREǏN

AMERICAN MATHEMATICAL SOCIETY

Providence, Rhode Island 02904

1970

ТЕОРИЯ ВОЛЬТЕРРОВЫХ ОПЕРАТОРОВ
В ГИЛЬБЕРТОВОМ ПРОСТРАНСТВЕ И ЕЕ ПРИЛОЖЕНИЯ

И. Ц. ГОХБЕРГ
М. Г. КРЕЙН

Издательство „Наука"
Главная Редакция
Физико-Математической Литературы
Москва 1967

Translated from the Russian by
A. Feinstein

International Standard Book Number 0-8218-1574-1
Library of Congress Card Number 71-120134

Printed in the United States

AUTHORS' PREFACE TO THE ENGLISH EDITION

This book is in some ways a continuation of our first book "Introduction to the theory of linear nonselfadjoint operators", which has already been published in English translation by the American Mathematical Society. Nevertheless, it can be read independently of the former.

We thank the American Mathematical Society for the interest which they have manifested in this book.

We wish to express our sincere thanks to Professor A. Feinstein for his considerable efforts in the translation of this book.

Odessa, Arcadia
June 1, 1969

I. C. GOHBERG
M. G. KREĬN

PREFACE

This book may be read independently of our earlier book "Introduction to the theory of linear nonselfadjoint operators" (henceforth cited as [GK]). At the same time, the appearance of this book is inseparable from that of the first book, if only because in first getting down to work we had intended to write a review paper, and later, in the course of developing the theme, we had attempted to present it in one compact monograph.

In this initial period (Spring 1959) the abstract theory of triangular integration—the basis of this book—could not have figured in our plans; we had not the slightest inkling of it. In fact, this theory did not then exist.

But in roughly a year this theory had already assumed a leading position in our review paper, which was proposed for the journal Uspehi Matematičeskih Nauk. Although this paper was not published (cf. the preface of [GK]), copies of the manuscript made their way to a number of scientific centers. Evidently this circumstance raised some interest in the embryonic theory. At any rate, in 1960 the number of investigators working on this theory increased noticeably—from three... to four.

In the summer of 1963 we submitted to "Fizmatgiz" a manuscript totalling 37 printer's sheets. For reasons which we shall not go into here, in about a year, by means of a certain amount of "vivisection", we separated out from our manuscript our first book—[GK]. The second volume of our book, however, was delayed. A certain amount of time was needed for its regeneration. But this was only one of the reasons. During the time which had passed considerable progress had been made in the theory of nonselfadjoint operators with continuous spectra. We wanted to show the results of these methods and the ideas in these new investigations. To this end we had decided to write another chapter. When the size of this chapter (after much review) began to endanger the size of the book, we realized that we had actually written several chapters of a new, third book, which ought to be titled "The spectral analysis of operators, close to selfadjoint or unitary ones". At the same time it became clear that the second part of the "trilogy" was completed.

In this book we discuss a new theory. To master it, as any new theory, one has to overcome a certain psychological barrier. In this connection Academician N. N. Luzin wrote, in a letter to a young scientist: "One has to coax scientists from their attitudes in the same way that a brisk May morning coaxes a man from his bed... But such coaxing should be maturely deliberate. And it goes into your scientific work...".[1)]

We should frankly admit that we tried to follow this advice; to this end we adopted certain measures. Let us point out the two most important: 1) we chose the proper editor for the book and secured his consent, 2) we did not miss opportunities to show unexpected and fruitful connections of the new theory with questions from classical analysis.

Thus we have fulfilled our promise, made in the preface to [GK], to continue our account of how these first two books came to be written. Moreover, we hope that this account will be continued and—we would hope—concluded in our third book.

It is pleasent for us to mention that the group of colleagues who assisted us (M. S. Brodskiĭ, S. G. Kreĭn, B. Ja. Levin, V. B. Lidskiĭ, Ju. I. Ljubič, A. S. Markus, V. I. Macaev, L. A. Šahnovič), to whom we have already expressed our profound thanks, which we again confirm, has increased. We are very grateful for the valuable assistance of M. Š. Birman and M. Z. Solomjak, Ju. P. Ginzburg, G. È. Kisilevskiĭ, E. M. Semenov and Ju. L. Šmul′jan.

As in the first volume, our editor F. V. Širokov vigilantly and firmly stood guard over the reader's interests. He frequently made us build platforms on steep slopes to allow the reader a rest, not overburden his memory, and illuminate the path ahead with the pleasant light of heuristics. More generally, in taking upon himself the right to speak in the reader's name, our editor with his characteristic conscientiousness found for himself and for us varied and at times hard work.

However we must admit that ultimately the reader's interests coincide with our own. We wish to express to F. V. Širokov our deep and sincere gratitude for his considerable and friendly assistance.

We are also very much obliged to A. Z. Ryvkin for his considerable work in preparing this book for publication.

Odessa, Arcadia I. C. GOHBERG
July 1966 M. G. KREĬN

[1)] Cf. "Letters of Academician N. N. Luzin" in the collection *Paths in the unknown*, "Sovetskiĭ pisatel′", Moscow, 1960, pp. 522-525.

TABLE OF CONTENTS

INTRODUCTION

This book discusses a new theory—the theory of abstract Volterra operators. This theory arose very recently in connection with ideas and results of the general theory of nonselfadjoint operators. The problems studied in the theory of abstract Volterra operators were nowhere formulated in the traditional theory of linear Volterra integral equations, whose treatment is usually found at the beginning of courses on integral equations. Nevertheless, the Volterra integral operator

$$(*) \qquad (Vf)(t) = \int_t^b \mathscr{V}(t,s)f(s)\,ds$$

is a "model" (see Chapter V) for this book. We shall consider the operator V in the Hilbert space $L_2(a,b)$ of scalar functions or (for the general case of a matrix kernel $\mathscr{V}(t,s)$) in the space $L_2^{(r)}(a,b)$ of r-dimensional vector-functions ($r \leqq \infty$).

As is known from the theory of integral equations, under very general assumptions regarding the kernel $\mathscr{V}(t,s)$ ($a \leqq t \leqq s \leqq b$) the operator $I - \mu V$ is invertible for all values of the complex parameter μ; its inverse is given by the Neumann series

$$(**) \qquad (I - \mu V)^{-1} = \sum_{j=0}^{\infty} \mu^j V^j.$$

We also mention that the usual restrictions on the kernel $\mathscr{V}(t,s)$ guarantee not only the continuity of the operator V, but also its complete continuity.

In the language of the abstract theory of operators in a Hilbert space, the convergence of the Neumann series (**) for every complex μ means that the spectrum of the operator V reduces to a single point $\lambda = 0$ ($\lambda = 1/\mu$).

And so, singling out the "abstract power" of this concept, we shall call a linear operator in a Hilbert space a *Volterra* operator, if it is completely continuous and has the one-point spectrum $\lambda = 0$.[1)]

[1)] In the foreign literature Volterra operators are called, correctly but cumbersomely, *completely continuous quasi-nilpotent* operators.

The question naturally arises: How justified is this attempt to extend the content of the notion of a "Volterra operator"?

It turns out that under very general assumptions every abstract Volterra operator is unitarily equivalent to the traditional Volterra integral operator defined by (*) in the space $L_2^{(r)}(0,1)$ (see Chapter V).

This fundamental fact follows from more general results of the theory of nonselfadjoint operators.

The Volterra integral operator V (in contrast with the Fredholm integral operator) has an explicitly defined continuous monotonically increasing chain of invariant subspaces, namely the chain of subspaces L_τ $(a \leqq \tau \leqq b)$, where $L_\tau = \{f : f \in L_2(a,b),\ f(t) \equiv 0 \text{ for } \tau < t \leqq b\}$. It turns out that every Volterra operator acting in a Hilbert space $\mathfrak{H}$ has, following an *inessential extension* (see Chapter I, §3), a monotonically increasing continuous chain $\mathfrak{H}_\tau$ $(0 \leqq \tau \leqq 1)$ of invariant subspaces. By means of such a chain one can now obtain an abstract triangular representation of the Volterra operator—an abstract analog of the representation (*) (see Chapter I).

These results were obtained rather recently; they belong to the new trend in the theory of nonselfadjoint operators which was mentioned above.

Very diverse problems of the spectral theory of nonselfadjoint operators lead to abstract Volterra operators; Volterra operators are important constituents in the structure of operators of a very general type.

It was not expected that the theory of Volterra operators would also clarify many deep facts of the theory of the extensions of Hermitian operators.

Let us clarify, by means of two examples, how abstract Volterra operators arise in investigations in the theory of nonselfadjoint and even selfadjoint operators.

Let A be a completely continuous linear operator in a Hilbert space $\mathfrak{H}$. Will the system of its root vectors be complete in $\mathfrak{H}$?

The solution of this problem, as a rule, proceeds as follows (see [GK], Chapter V).

We denote by $\mathfrak{E}$ the closed linear hull of all root vectors of the operator A. Obviously $\mathfrak{E}$ lies in the closure of the range $\mathfrak{R}(A)$ of A, and therefore completeness can occur only if $\mathfrak{R}(A)$ is dense in the space $\mathfrak{H}$, which is equivalent to the requirement that A^* not annihilate any nonzero vector. Suppose that this condition is fulfilled. Let us suppose that $\mathfrak{E} \neq \mathfrak{H}$ and consider the orthogonal complement $\mathfrak{E}^\perp$ of $\mathfrak{E}$. It is easily shown that $\mathfrak{E}^{\perp}$ is invariant with respect to A^* and that the restriction of A^* to $\mathfrak{E}^{\perp}$ is a *Volterra* operator. It now remains to clarify under what conditions this restriction equals zero. In many cases this last question has achieved solution only on the basis of deep and fairly recently discovered properties of Volterra operators. Sometimes a decisive role is played here by various nontrivial relations between the spectra of the Hermitian components $A_{\mathscr{R}} = (A + A^*)/2$ and $A_{\mathscr{I}} = (A - A^*)/2i$ of the Volterra operator.

These same relations also play a role in totally different questions. Let H_0 be a Hermitian operator with a dense domain and with finite and equal deficiency indices. Let us assume that H_0 can be extended to a nonselfadjoint operator H_v *which has no spectrum* (i.e., such that H_v^{-1} is a Volterra operator). Then we can assert that any selfadjoint extension H of H_0 will have a spectrum with completely determined asymptotic properties. This fact follows from a general asymptotic theorem (Theorem VII.1.1) concerning the spectrum of the real component of a Volterra operator whose imaginary component is nuclear.

The situation just described for the operator H is encountered, for example, in the theory of boundary value problems for ordinary differential equations, since in inverting a differential operator with given initial conditions a Volterra integral operator arises.

Among other results, it has become clear fairly recently that *any* selfadjoint boundary value problem (with parameters which enter linearly) for an ordinary differential equation or for a system of such equations can be reduced to a boundary value problem for a so-called *canonical* differential equation.

With the development of the theory of canonical differential equations, many problems of the theory of the spectral functions of differential operators, related with their generalized Fourier transforms, as well as other problems, have become tractable. The theory of canonical differential equations with periodic coefficients has also developed.

It is noteworthy that the abstract theory of Volterra operators dictated the necessity of studying just such canonical differential equations: the theory of triangular models of Volterra operators leads to these equations.

Considerable attention is given over in this book to those questions in the theory of abstract Volterra operators and the theory of canonical differential equations where this connection turns out to be fruitful on both sides.

This circle of questions has widened so much in recent years that it is difficult to encompass them all in a monograph of this size. Therefore we have been forced to merely give a survey of some of them, despite their importance and beauty, in the Appendix.

Lastly, the so-called inverse problem of the spectral theory of regular and singular differential operators leads naturally to canonical differential equations. This extensive field has been completely ignored in this book, if we do not count certain questions taken up in the Appendix.[2)]

Thus, the theory of Volterra operators—in former times an elementary chapter in the general theory of integral equations—has undergone a remarkable evolution. The profundity of this evolution is in part characterized by the appearance in the theory of Volterra operators of theorems with "multiparagraph" proofs. One theorem of this sort is the aforementioned asymptotic Theorem VII.1.1. That it appears in the last, seventh chapter is no accident; its proof required preparatory "artillery" which was called up throughout almost all the preceding chapters of the book and even in the preceding book [GK]. An analysis of the proof of this theorem in all detail requires patience and persistence of the reader.

A second theorem of this kind is Theorem A.4.1, whose proof we have been forced to omit.

Of course, the Appendix "conceals" other theorems with "imposing" proofs.

We have discussed above certain qualitative characteristics of the material contained in this book. The quantitative characteristics are given in the table of contents. However, we hope that this information will not be redundant if we add a brief survey of the content of this book, taken chapter by chapter.

The theory of the triangular integral (and more generally of integration with respect to a chain of projectors)—the foundation of this book—is discussed in two locales, the first of these being Chapter I, and the second, Chapter III. Further results of this theory are contained in the fourth chapter, which studies the problem of the factorization of operators. If the abstract triangular representation of an operator serves as a continuous analog of the triangular form of a square matrix, then the problem of the factorization of an operator along a chain turns out to be the continuous analog of the representation of a square matrix in the form of a product of left and right triangular matrices.

[2)] A number of the basic results of the theory of spectral functions for canonical equations (direct problems) are discussed in the recent book by F. V. Atkinson [**2**].

In the third and fourth chapters a large role is played by certain *transformators*. A transformator is a linear operator in a Banach space whose elements, in turn, are themselves operators. General results on transformators are discussed in Chapter II.

In the fifth chapter we construct analytic models of abstract Volterra operators; in other words, these operators are *realized* in the form of concrete Volterra operators of the type (*). The fifth chapter is at the same time a bridge for passing from the theory of abstract Volterra operators to the theory of canonical differential equations, which is discussed in the sixth chapter. It will probably be surprising to many readers that investigations in the abstract theory of Volterra operators have made it possible, for example, to establish new and precise tests for the strong stability of the solutions of canonical differential equations with periodic coefficients.

The seventh chapter is devoted, basically, to the proof of the already thrice mentioned asymptotic theorem for Volterra operators with nuclear imaginary component.

Finally, in the Appendix we give a brief survey (with some commentary) of recent results in the theory of unicellular operators. Here are clarified the connections of this theory with inverse problems of the spectral theory of differential operators and with the multiplicative representation of entire matrix-functions.

We now speak frankly to the reader whom, taking into account the variety of his tastes, we are trying to entice (see the Preface).

For specialists in the theory of differential equations we point out Chapter VI on canonical equations, material which appears for the first time on the pages of a book. This chapter contains a number of new results.

For those who are attracted to the constructive theory of functions we present an "instantaneous" derivation of the deep analytic properties of the Riesz-Titchmarsh transform—the discrete analog of the Hilbert transform (Chapter III, §10).

Specialists in the theory of analytic functions have already been able to learn from the book [GK], we hope with satisfaction, of the large role played by the methods and results of the theory of functions in the theory of operators. Here, in the Appendix, they can become acquainted with certain nontrivial generalizations of theorems on scalar entire functions to entire matrix-functions.

The reader with a taste for applied analysis will find in Chapter IV some methods for the effective solution of integral equations. Although these methods have already proved themselves in contact problems of elasticity theory, their justification is presented here for the first time (rigorously only for integral equations of the second kind).

As for lovers of functional analysis, it is basically for them that this book is intended.

In conclusion let us say a few words regarding our future plans. The investigations discussed here in the theory of Volterra operators play an essential role in the theory of nonselfadjoint operators of a substantially more general type. We hope to show this rather fully in our third book (in preparation) on the theory of nonselfadjoint operators: "Spectral analysis of operators, close to selfadjoint or unitary ones".

To define the place of the present book in our "trilogy", we explain that from the spectral point of view the first book, [GK], is basically devoted to linear operators with a "discrete spectrum", the second—to linear operators having a one-point spectrum, and the third—being prepared—to operators with a continuous and discrete spectrum.

It is also necessary to emphasize that in the present book the characteristic operator-function of a nonselfadjoint operator—a fundamental concept introduced by M. S. Livšic—hardly figures. This function first appears in Chapter V, but only in a modest role as an invariant of the triangular model of an operator. The reader may form an idea of its more profound value from the Appendix.

The fundamentals of the general theory of the characteristic operator-function, which is a satellite of nonselfadjoint operators, will be discussed in the third book. We shall attempt there to further present the remarkable evolution of this theory and its applications.

To do this we have to discuss tests for the existence of chains of invariant subspaces for operators of a considerably more complicated structure than Volterra operators, and to extend the theory of triangular representations to such operators. The problem of factorization appears in a new role. In particular, it proves to be closely related with the problem of the multiplicative representation of characteristic operator-functions. We obtain the opportunity of discussing a number of important results from the series of outstanding papers by B. Sz.-Nagy and C. Foiaş [**1—10**], and also the recent papers by L. de Branges [**1—11**] (unfortunately, for some reason or another these papers did not get into the brief historical survey[3] which was given in the introduction of the book [GK]).

We have not spoken anywhere here (avoiding excessive promises) on the possibilities of applying these new investigations in the abstract theory of operators to the theory of partial differential equations and, in particular, to scattering problems (see P. Lax and R. S. Phillips [**1**]), which we also hope to touch upon.

[3] Despite its brevity, this survey has a direct relation to the material of this book, and, in part, to that of the following (third) book.

CHAPTER I

ABSTRACT TRIANGULAR REPRESENTATION OF COMPLETELY CONTINUOUS OPERATORS

Starting with the well-known work of M. S. Livšic [2] (cf. also M. S. Brodskiĭ and M. S. Livšic [1]), the algebraic idea of I. Schur of reducing a matrix to triangular form by means of a unitary transformation has undergone profound development in the theory of linear operators.

Considerable progress in this direction has been achieved in the last decade. The starting point of these new investigations is a theorem of J. von Neumann and N. Aronszajn, according to which every completely continuous linear operator has a nontrivial invariant subspace. From this result it is easy to deduce (using, however, the lemma of Kuratowski—Zorn) that for every completely continuous linear operator in a Hilbert space there exists an infinite maximal chain of invariant subspaces which are linearly ordered by inclusion.

L. A. Sahnovič was the first to show (in 1958) that this geometric fact makes it possible to obtain essentially new results in the theory of triangular models for nonselfadjoint operators.

At about the same time, there appeared in papers by M. S. Brodskiĭ [**3**—**5**] an integral giving an abstract triangular representation for Volterra operators. In a short time the theory of the new integral and its applications was developed considerably (cf. the review paper by M. S. Brodskiĭ, I. C. Gohberg, M. G. Kreĭn and V. I. Macaev [**1**]). This theory called into being new analytic investigations of the spectral properties of nonselfadjoint as well as selfadjoint operators.

In this chapter we discuss the general aspects of the abstract triangular representation of Volterra operators.

The role of this representation in the construction of various models for Volterra operators will be clarified in Chapter V.

The aforementioned analytic investigations, connected with the triangular representation, are discussed in Chapters VI and VII.

§1. Notation and some well-known results

In this section we introduce basic notation and recall a number of well-known concepts and results of the theory of operators in Hilbert space, which are systematically used in the sequel. All details relating to this material are presented in the first chapter of the authors' book [GK].

We denote by the symbol $\mathfrak{H}$ a separable Hilbert space.

The domain of a linear operator A, acting in $\mathfrak{H}$, is denoted by $\mathfrak{D}(A)$, and its range, by $\mathfrak{R}(A)$ or $A\mathfrak{H}$.

Everywhere henceforth we shall be concerned, basically, with *bounded* linear operators. For such operators, always $\mathfrak{D}(A) = \mathfrak{H}$.

We denote the set of all bounded linear operators in $\mathfrak{H}$ by $\mathfrak{R} = \mathfrak{R}(\mathfrak{H})$. By the *uniform* norm of an operator $A \in \mathfrak{R}$ is meant the number

$$|A| = \sup_{|\phi|=1} |A\phi|.$$

With this definition of the norm, $\mathfrak{R}$ becomes a complete normed ring.

A subspace $\mathfrak{L} \subseteq \mathfrak{H}$ is said to be an *invariant subspace* of the operator $A \in \mathfrak{R}$, if for any vector $f \in \mathfrak{L}$ the vector Af likewise belongs to $\mathfrak{L}$.

Let P $(\in \mathfrak{R})$ be a projector $(P^2 = P)$.

Obviously, the subspace $P\mathfrak{H}$ will be invariant with respect to an operator A $(\in \mathfrak{R})$ if and only if

$$PAP = AP. \tag{1.1}$$

If, moreover, the subspace $Q\mathfrak{H}$ $(Q = I - P)$ is invariant with respect to A, then (and only then) the stronger condition

$$PA = AP \tag{1.2}$$

is fulfilled.

However, if the operator A is selfadjoint and P is an orthoprojector (orthogonal projector, $P^* = P$), the conditions (1.1) and (1.2) are equivalent.

It is easily deduced ([GK], p. 2) that *if a subspace $\mathfrak{L}$ is invariant with respect to the operator A $(\in \mathfrak{R})$, then its orthogonal complement $\mathfrak{L}^{\perp} = \mathfrak{H} \ominus \mathfrak{L}$ is invariant with respect to the adjoint A^*.*

By the *dimension of an operator* A $(\in \mathfrak{R})$ is meant the number $r(A) = \dim A$ $(\leqq \infty)$, the dimension of the subspace $\overline{A\mathfrak{H}}$. An operator $A \in \mathfrak{R}$ is said to be *finite-dimensional* if $r(A) < \infty$.

Obviously, every operator K defined by

$$Kf = \sum_{j=1}^{n} (f, \phi_j)\psi_j \quad (f \in \mathfrak{H}), \tag{1.3}$$

where $\{\phi_j\}_1^n$ and $\{\psi_j\}_1^n$ are some systems of vectors from $\mathfrak{H}$, is finite-dimensional, and $r(K) \leqq n$.

It is easily seen that, conversely, every finite-dimensional operator K can be represented in the form (1.3) with $n = r(K)$.

Henceforth we shall write the operator K, defined by (1.3), in the form

$$K = \sum_{j=1}^{n} (\cdot, \phi_j)\psi_j.$$

For an operator A $(\in \mathfrak{R})$, the symbol $\mathfrak{Z}(A)$ will denote the subspace consisting of all solutions $\phi \in \mathfrak{H}$ of the equation $A\phi = 0$. This subspace is obviously closed.

An operator U $(\in \mathfrak{R})$ is said to be *partially isometric* if it maps the subspace $\mathfrak{H} \ominus \mathfrak{Z}(U)$ isometrically onto $U\mathfrak{H}$. If U is a partially isometric operator, the same is true of the operator U^*.

Every operator $A \in \mathfrak{R}$ admits a *polar representation* in the form

$$A = UH, \tag{1.4}$$

where $H = (A^*A)^{1/2}$, and U is a partially isometric operator which maps the subspace $\overline{A^*\mathfrak{H}}$ isometrically onto $\overline{A\mathfrak{H}}$.

Every operator $A \in \mathfrak{R}$ can be represented in the form

$$A = A_{\mathscr{R}} + iA_{\mathscr{I}};$$

$$A_{\mathscr{R}} = (A + A^*)/2 \quad \text{and} \quad A_{\mathscr{I}} = (A - A^*)/2i$$

are selfadjoint operators, called the *real* and *imaginary components* of the operator A.

An operator $A \in \mathfrak{R}$ is said to be *dissipative* if its imaginary component is nonnegative: $A_{\mathscr{I}} \geqq 0$.

One distinguishes in the ring $\mathfrak{R}$ the set of all completely continuous linear operators; we shall denote it by $\mathfrak{S}_\infty$. As is known, $\mathfrak{S}_\infty$ is a closed two-sided ideal in the ring $\mathfrak{R}$.

For reasons which will be clarified later, for $X \in \mathfrak{S}_\infty$ the uniform norm $|X|$ will at times be denoted by $|X|_\infty$.

If a selfadjoint operator A belongs to $\mathfrak{S}_\infty$, we shall denote by $\{\lambda_j(A)\}$ the complete sequence of its eigenvalues, arbitrarily enumerated in the order of decreasing absolute value, and in this enumeration every nonzero eigenvalue will occur as many times as its multiplicity.

It should be clarified that the term 'complete' has a special meaning here: if the operator A has infinitely many nonzero eigenvalues, then the sequence $\{\lambda_j(A)\}$ will consist only of them, and will not contain any zero eigenvalues of A, if there are such. If A has only a finite number

ν of nonzero eigenvalues, then we shall put $\lambda_{\nu+j}(A) = 0$ $(j = 1, 2, \cdots)$.

An operator $A \in \mathfrak{S}_\infty$ is said to be a *Volterra* operator, if the point $\lambda = 0$ is the only spectral point of A.

As is known (cf. [GK], p. 17), *the limit in the uniform norm of a sequence of Volterra operators is a Volterra operator.*

§2. Theorem on the existence of a proper invariant subspace for any completely continuous operator

The following theorem holds.

THEOREM 2.1 (J. VON NEUMANN, N. ARONSZAJN). *Every completely continuous linear operator A in $\mathfrak{H}$ has a proper*[1] *invariant subspace.*

PROOF. Let f $(\neq 0)$ be an arbitrary vector of the space $\mathfrak{H}$. The closed linear hull $\mathfrak{M}$ of the sequence $f, Af, \cdots, A^n f, \cdots$ is an invariant subspace of the operator A. Therefore to prove the theorem it is sufficient to restrict attention to the case $\mathfrak{M} = \mathfrak{H}$.

In this case the vectors $A^n f$ $(n = 0, 1, \cdots)$ are linearly independent, i.e., any finite subset of them is linearly independent. Let us denote by $\mathfrak{L}_k$ $(k = 1, 2, \cdots)$ the linear hull of the vectors $A^n f$ $(n = 0, 1, \cdots, k-1)$ and by P_k the projector which projects $\mathfrak{H}$ orthogonally onto $\mathfrak{L}_k$. Obviously, for every $\phi \in \mathfrak{H}$ we have

$$\lim_{k \to \infty} | P_k \phi - \phi | = 0,$$

i.e., the sequence of orthoprojectors P_k converges strongly to I.

We define an operator A_k in the subspace $\mathfrak{L}_k$ $(k = 1, 2, \cdots)$ by

$$A_k \phi = P_k A \phi \quad (\phi \in \mathfrak{L}_k).$$

Since A_k is a linear operator in the k-dimensional space $\mathfrak{L}_k$, with respect to an appropriate basis of $\mathfrak{L}_k$ it can be represented by a triangular matrix. There follows the existence of an increasing sequence of subspaces $\{0\} = \mathfrak{L}_{k0} \subset \mathfrak{L}_{k1} \subset \cdots \subset \mathfrak{L}_{kk} = \mathfrak{L}_k$, where $\mathfrak{L}_{kj}$ is a j-dimensional invariant subspace of the operator A_k.

Let us denote by P_{kj} the operator of orthogonal projection of the space $\mathfrak{H}$ onto $\mathfrak{L}_{kj}$.

If $f \in \mathfrak{L}_k$, then obviously

$$|f| = |f - P_{k0} f| \geq |f - P_{k1} f| \geq \cdots \geq |f - P_{kk} f| = 0.$$

---

[1] A subspace $\mathfrak{H}_1$ $(\subseteq \mathfrak{H})$ is called a *proper subspace* if $\mathfrak{H}_1 \neq \mathfrak{H}$ and $\mathfrak{H}_1 \neq \{0\}$.

Let α be a positive number less than unity $(0 < \alpha < 1)$. Then for any positive integer k there exists a unique positive integer $j(k)$, $0 \leqq j(k) < k$, such that

$$|f - P_{kj(k)} f| \geqq \alpha |f| > |f - P_{kj(k)+1} f|. \tag{2.1}$$

Any set $\{P\}$ of orthoprojectors is weakly compact, since on the unit sphere of the separable space $\mathfrak{H}$ the functions $(P\phi, \psi)$ $(|\phi| = |\psi| = 1)$ are uniformly bounded and equicontinuous. Hence there exists a sequence of indices k_m $(m = 1, 2, \cdots)$ such that the sequences $\{P_{k_m j(k_m)}\}$ and $\{P_{k_m j(k_m)+1}\}$ will be weakly convergent. We denote their limits by Q and Q_1.

The operators Q and Q_1 are bounded, and their norms do not exceed unity: $|Q|$, $|Q_1| \leqq 1$ (however, generally speaking, Q and Q_1 are not projectors).

The sequence $\{P_{k_m j(k_m)} A P_{k_m j(k_m)}\}$ has a weak limit which equals QAQ.

In fact, since A can be approximated in the uniform norm to any degree of accuracy by finite-dimensional operators, which are made up of a finite number of one-dimensional operators, it suffices to prove the stated result for a one-dimensional operator $A = (\cdot, \phi)\psi$. In this case the relation

$$\lim_{m \to \infty} (\chi, P_{k_m j(k_m)} \phi)(P_{k_m j(k_m)} \psi, \omega) = (\chi, Q\phi)(Q\psi, \omega) \quad (\chi, \omega \in \mathfrak{H})$$

is equivalent to the relation

$$\lim_{m \to \infty} (P_{k_m j(k_m)} A P_{k_m j(k_m)} \chi, \omega) = (QAQ\chi, \omega) \quad (\chi, \ \omega \in \mathfrak{H}).$$

Let us now consider the equality

$$P_{k_m j(k_m)} A P_{k_m j(k_m)} = P_{k_m} A P_{k_m j(k_m)};$$

passing to the limit $m \to \infty$, we obtain, as is easily seen, AQ on the right-hand side, i.e., we obtain the equality

$$QAQ = AQ.^{2)} \tag{2.2}$$

Let us denote by $\mathfrak{N}$ the closed subspace of all vectors ϕ which are invariant under Q $(Q\phi = \phi)$. According to (2.2) we have $QA\phi = A\phi$ for all $\phi \in \mathfrak{N}$, i.e., $\mathfrak{N}$ is an invariant subspace of the operator A.

If $\{0\} \neq \mathfrak{N} \neq \mathfrak{H}$, the theorem is proved.

---

2) If the operator Q were a projector, this would complete the proof.

We first consider the case in which $\mathfrak{N} = \{0\}$ but $Q \neq 0$. Since $\mathfrak{R}(AQ) \subseteq \mathfrak{N}$, in this case $AQ = 0$, and so any one-dimensional subspace of $\mathfrak{R}(Q)$ is a proper invariant subspace of A.

Since the equality $\mathfrak{N} = \mathfrak{H}$ means that $Q = I$, it remains to consider two cases: $Q = 0$ and $Q = I$.

The second case, however, is not possible.

Indeed, if $Q = I$, then

$$\lim_{m\to\infty} |P_{k_m j(k_m)} f| = |f| \quad (f \in \mathfrak{H})$$

and so

$$\lim_{m\to\infty} |P_{k_m j(k_m)} f - f| = 0 \quad (f \in \mathfrak{H}).$$

This contradicts (2.1).

By arguments similar to the preceding, we can establish that in all cases where $0 \neq Q_1 \neq I$, the operator A has a proper invariant subspace. The equality $Q_1 = 0$ cannot hold, since this would imply that the projectors $P_{k_m j(k_m)+1}$ converge strongly to zero, and this again contradicts the inequality (2.1).

Thus, we will complete the proof of the theorem by showing that the equalities $Q = 0$, $Q_1 = I$ cannot hold simultaneously. If $Q = 0$ and $Q_1 = I$, the unit operator $I = Q - Q_1$ would be the weak limit of the one-dimensional operators $P_{k_m j(k_m)+1} - P_{k_m j(k_m)}$, which is impossible.

The theorem is proved.

A simple corollary of this theorem is the following important theorem.

THEOREM 2.2. *If two subspaces $\mathfrak{N}$ and $\mathfrak{M}$ are invariant with respect to a completely continuous linear operator A, where $\mathfrak{N} \subset \mathfrak{M}$ and* $\dim(\mathfrak{M} \ominus \mathfrak{N}) > 1$, *then there exists an invariant subspace $\mathfrak{E}$ of A which is intermediate between $\mathfrak{N}$ and $\mathfrak{M}$:*

$$\mathfrak{N} \subseteq \mathfrak{E} \subseteq \mathfrak{M}, \quad \mathfrak{N} \neq \mathfrak{E} \neq \mathfrak{M}. \tag{2.3}$$

PROOF. Indeed, let us denote by A_0 the operator which A induces in $\mathfrak{M}$. Then the subspace $\mathfrak{N}_1 = \mathfrak{M} \ominus \mathfrak{N}$ is invariant with respect to A_0^* (cf. p. 9). According to Theorem 2.1, there is a nontrivial subspace $\mathfrak{N}_2 \subseteq \mathfrak{N}_1$ ($\mathfrak{N}_2 \neq \mathfrak{N}_1$) which is invariant with respect to A_0^*. The subspace $\mathfrak{E} = \mathfrak{M} \ominus \mathfrak{N}_2$ obviously satisfies the relations (2.3) and is invariant with respect to A.

Theorem 2.1, with the proof given here, was published in a joint paper by N. Aronszajn and K. Smith [1], in which its generalization to completely continuous operators in a Banach space was also obtained. In this paper it was stated that for the case of Hilbert space the theorem was already known to J. von Neumann in 1935. Later it was proved independently by N. Aronszajn.

§3. Maximal chains of orthoprojectors for a completely continuous operator[3)]

A set $\mathfrak{P}$ of orthoprojectors is called a *chain*, if for any pair $P_1, P_2 \in \mathfrak{P}$ either[4)] $P_1 < P_2$ or $P_2 < P_1$.

A chain $\mathfrak{P}$ is said to be *bordered*, if it contains the operators 0 or I.

By the *closure* $\overline{\mathfrak{P}}$ of a chain $\mathfrak{P}$ is meant the set of all operators which are the strong limits of sequences of projectors from $\mathfrak{P}$.

It is obvious that the closure of every chain is a chain.

If $\overline{\mathfrak{P}} = \mathfrak{P}$, the chain $\mathfrak{P}$ is said to be closed.

A pair (P^-, P^+) $(P^- < P^+)$ of orthoprojectors belonging to a closed chain $\mathfrak{P}$ is called a *gap* of this chain, if for any $P \in \mathfrak{P}$ either $P \leqq P^-$ or $P \geqq P^+$.

The dimension of the orthoprojector $P^+ - P^-$ is called the *dimension* of the gap (P^-, P^+).

It follows from the separability of the space $\mathfrak{H}$ that each chain has no more than a countable set of gaps.

The relation of inclusion, $\mathfrak{P}_1 \subset \mathfrak{P}_2$, having the property of transitivity, makes it possible to regard the set Z of all chains as partially ordered.

We shall say that a chain $\mathfrak{P}$ is *maximal* if it cannot be enlarged, i.e., if it is larger than any chain $\mathfrak{P}_1$ $(\in Z)$ which can be compared with it.

Obviously, every maximal chain is bordered and closed, and its gaps are one-dimensional.

The above properties characterize a maximal chain. Indeed, if we suppose that a chain $\mathfrak{P}$ has these properties and is not maximal, there exists a chain $\mathfrak{P}_1 \supset \mathfrak{P}$ which contains some projector Q which does not belong to $\mathfrak{P}$, and since $\mathfrak{P}$ is closed, the projectors

$$P^- = \max_{P \in \mathfrak{P}\,;\, P < Q} P \quad \text{and} \quad P^+ = \min_{P \in \mathfrak{P}\,;\, P > Q} P$$

[3)] The assertions of this section, as well as Theorem 2.2, appear in papers by L. A. Sahnovič [4] and M. S. Brodskiĭ [5].

[4)] For two orthoprojectors P_1, P_2 we write $P_1 < P_2$ $(P_2 > P_1)$ if the subspace $P_1\mathfrak{H}$ is a proper subset of the subspace $P_2\mathfrak{H}$. The relation $P_1 \leqq P_2$ $(P_2 \geqq P_1)$ means that either $P_1 < P_2$ or $P_1 = P_2$.

exist and belong to $\mathfrak{P}$. Clearly the pair of projectors (P^-, P^+) forms a gap of the chain $\mathfrak{P}$ of dimension greater than unity, which is not possible.

1. We shall say that a chain $\mathfrak{P}$ is an *eigenchain* of the operator A, if each of the subspaces $\mathfrak{R}(P)$ $(P \in \mathfrak{P})$ is invariant with respect to A, in other words, if $PAP = AP$ $(P \in \mathfrak{P})$.

THEOREM 3.1. *Every completely continuous linear operator has a maximal eigenchain.*

PROOF. Let us denote by $\mathfrak{A}$ the set of all bordered eigenchains of the operator A, which as a subset of Z is a partially ordered set.

If some set Z_0 of chains is linearly ordered, then, taking the union of all projectors, each of which belongs to at least one chain from Z_0, we obtain a chain which is the supremum of Z_0. Therefore the lemma of Kuratowski—Zorn (see, e.g., Dunford and Schwartz [1], p. 6) is applicable to the chains from $\mathfrak{A}$, in virtue of which there exists at least one maximal element in $\mathfrak{A}$.

Let $\mathfrak{P}_0$ be a fixed maximal element of the set $\mathfrak{A}$. The chain $\mathfrak{P}_0$ is closed. Indeed, the chain $\overline{\mathfrak{P}}_0$, the closure of $\mathfrak{P}_0$, belongs to $\mathfrak{A}$ and $\mathfrak{P}_0 \subseteq \overline{\mathfrak{P}}_0$; hence, $\mathfrak{P}_0 = \overline{\mathfrak{P}}_0$.

Since the chain $\mathfrak{P}_0$ is bordered, to complete the proof of the theorem it remains to show that all the gaps of $\mathfrak{P}_0$ are one-dimensional. Let us assume the contrary, i.e., that some pair of projectors (P^-, P^+) forms a gap of $\mathfrak{P}_0$ of dimension greater than unity. But then, by Theorem 2.2, there exists an orthoprojector Q such that $P^- < Q < P^+$ and $QAQ = AQ$.

Obviously, the set of projectors $\mathfrak{P}$, consisting of the chain $\mathfrak{P}_0$ and the projector Q, is a chain which belongs to $\mathfrak{A}$, and $\mathfrak{P} \supset \mathfrak{P}_0$. This is not possible. The theorem is proved.

2. We note two properties of Volterra operators.

Let $P_1 < P_2$ be orthoprojectors, and let $A \in \mathfrak{R}$. We shall call the operator $A_0 = (P_2 - P_1)A(P_2 - P_1)$ the *block* of the operator A on the pair of orthoprojectors (P_1, P_2). If this pair of orthoprojectors forms a gap of some chain $\mathfrak{P}$, the operator A_0 will be called the *block of the operator* A *on the gap* (P_1, P_2).

1. *If the chain* $\mathfrak{P}$ *is a maximal eigenchain of a Volterra operator* A, *then for any gap* (P^-, P^+) *of* $\mathfrak{P}$ *one has*

$$(P^+ - P^-)A(P^+ - P^-) = 0, \tag{3.1}$$

i.e., each block of a Volterra operator A *on the gaps of a maximal eigenchain of* A *equals* 0.

Indeed, let us consider the Volterra operator A_0 induced by A in the subspace $\mathfrak{R}(P^+)$;

$$A_0 = P^+AP^+ \quad \text{on } \mathfrak{R}(P^+).$$

The subspace $\mathfrak{R}(P^-)$ is invariant with respect to A_0; consequently, the one-dimensional subspace $\mathfrak{N} = \mathfrak{R}(P^+) \ominus \mathfrak{R}(P^-)$ is invariant with respect to the operator A_0^*, the adjoint of A_0;

$$A_0^* = P^+A^*P^+ \quad \text{on } \mathfrak{R}(P^+).$$

As A_0^* is a Volterra operator, it follows that $A_0^*\mathfrak{N} = 0$. This means that the range of A_0 is orthogonal to $\mathfrak{N}$, i.e., (3.1) holds.

A closed chain $\mathfrak{P}$ is said to be *continuous* if it has no gaps. A continuous chain is characterized by the property that for any pair of projectors $P_1, P_2 \in \mathfrak{P}$ there exists at least one projector $P_3 \in \mathfrak{P}$ such that

$$P_1 < P_3 < P_2.$$

Obviously, every bordered continuous chain is maximal. We shall call a bordered continuous chain a *complete* chain.

2. *If a completely continuous dissipative operator A does not have any eigenvectors,*[5] *then each of its maximal eigenchains is continuous.*[6]

By hypothesis, A is, in particular, a Volterra operator. Let $\mathfrak{P}$ be one of its maximal eigenchains. Suppose that $\mathfrak{P}$ has a gap (P^-, P^+). Then the operator $P^+ - P^-$ is one-dimensional, and it follows from $(P^+ - P^-)A(P^+ - P^-) = 0$ that

$$(P^+ - P^-)A_{\mathscr{J}}(P^+ - P^-) = 0. \tag{3.2}$$

Since $A_{\mathscr{J}}$ is nonnegative, it follows from (3.2) that

$$A_{\mathscr{J}}(P^+ - P^-) = (P^+ - P^-)A_{\mathscr{J}} = 0,$$

as $|A_{\mathscr{J}}^{1/2}(P^+ - P^-)\phi|^2 = ((P^+ - P^-)A_{\mathscr{J}}(P^+ - P^-)\phi, \phi) = 0$. Hence $A_{\mathscr{J}}P^+ = A_{\mathscr{J}}P^-$ and $P^+A_{\mathscr{J}} = P^-A_{\mathscr{J}}$; therefore

$$P^+A_{\mathscr{J}}P^+ = P^-A_{\mathscr{J}}P^-. \tag{3.3}$$

Let ψ be a unit vector from the one-dimensional subspace $\mathfrak{R}(P^+ - P^-)$; then $P^+ - P^- = (\cdot, \psi)\psi$, $|\psi| = 1$. Since $P^\pm AP^\pm = AP^\pm$, we have

---

[5] This condition means that A is a Volterra operator and that it vanishes only at the zero vector.

[6] This result was communicated to us by M. S. Brodskiĭ.

$$P^+AP^+f = P^-AP^-f + (f,\psi)A\psi \quad (f \in \mathfrak{H}), \tag{3.4}$$

and (3.2) is equivalent to the equality $(A\psi,\psi) = 0$. Comparing (3.3) and (3.4), we obtain

$$P^+A_{\mathscr{R}}P^+f = P^-A_{\mathscr{R}}P^-f + (f,\psi)A\psi. \tag{3.5}$$

Since the operators $P^{\pm}A_{\mathscr{R}}P^{\pm}$ are selfadjoint, it follows from (3.5) that the operator $(\cdot,\psi)A\psi$ is selfadjoint and, consequently, the vector $A\psi$ is colinear with the vector ψ. Since $(A\psi,\psi) = 0$, it follows that $A\psi = 0$. This contradicts the hypothesis that A does not annihilate any nonzero vector.

3. Henceforth we shall assume that all chains are closed. We need the following concept.

A linear operator $\widetilde{A}$, acting in a Hilbert space $\widetilde{\mathfrak{H}} \supset \mathfrak{H}$, will be called an inessential extension of the operator A $(\in \mathfrak{R})$, if

$$\widetilde{A}\phi = A\phi \ (\phi \in \mathfrak{H}), \quad \widetilde{A}\phi = 0 \ (\phi \in \widetilde{\mathfrak{H}} \ominus \mathfrak{H}).$$

THEOREM 3.2. *Let the operator $A \in \mathfrak{R}$ have a bordered eigenchain $\mathfrak{P}$. If the block of A on each gap (P^-,P^+) of $\mathfrak{P}$ equals zero;*

$$(P^+ - P^-)A(P^+ - P^-) = 0, \tag{3.6}$$

then A has an inessential extension $\widetilde{A}$ which has a complete (i.e., bordered continuous) eigenchain $\widetilde{\mathfrak{P}}$.

PROOF. Let us denote by (P_j^-,P_j^+) $(j = 1,2,\cdots,\omega;\ 0 \leqq \omega \leqq \infty)$ all the gaps of the chain $\mathfrak{P}$. We first consider the case in which all the gaps of $\mathfrak{P}$ are infinite-dimensional;

$$\dim(P_j^+ - P_j^-) = \infty \quad (j = 1,2,\cdots,\omega).$$

We shall show that in this case A *itself* has a complete (i.e., bordered continuous) eigenchain. Let us denote by $\mathfrak{Q}_j$ $(j = 1,2,\cdots,\omega)$ an arbitrary complete chain in the space $\mathfrak{H}_j = (P_j^+ - P_j^-)\mathfrak{H}$, and extend the operators of these chains to all of $\mathfrak{H}$ by putting

$$Qf = 0 \quad (f \in \mathfrak{H}_j^{\perp},\ Q \in \mathfrak{Q}_j).$$

We consider the chain $\widetilde{\mathfrak{P}}$, consisting of the union of the chains

$$\mathfrak{P}, \{P_j^- + Q\}_{Q \in \mathfrak{Q}_j} \quad (j = 1,2,\cdots,\omega).$$

It is easily seen that $\widetilde{\mathfrak{P}}$ is a complete chain. Moreover, since $Q \in \mathfrak{Q}_j$ implies $QP_j^- = 0$, $QP_j^+ = Q$, we have

$$
\begin{aligned}
&(P_j^- + Q)A(P_j^- + Q)\\
&\quad = AP_j^- + P_j^- AQ + QAQ\\
&\quad = AP_j^- + P_j^- A(P_j^+ - P_j^-)Q + Q(P_j^+ - P_j^-)A(P_j^+ - P_j^-)Q\\
&\quad = AP_j^- + [P_j^+ - (P_j^+ - P_j^-)]A(P_j^+ - P_j^-)Q\\
&\quad = AP_j^- + P_j^+ AQ = A(P_j^- + Q) \quad (Q \in \mathfrak{Q}_j),
\end{aligned}
$$

from which it follows that $\widetilde{\mathfrak{P}}$ is an eigenchain of A.

Let us pass to the general case, in which some of the gaps of the chain $\mathfrak{P}$ are finite-dimensional. Let $\mathfrak{H}_j$ $(j = 1, 2, \cdots, \omega)$ be a separable infinite-dimensional Hilbert space. We form the orthogonal sum of these spaces and the space $\mathfrak{H}$:

$$\widetilde{\mathfrak{H}} = \mathfrak{H} \oplus \mathfrak{H}_1 \oplus \mathfrak{H}_2 \oplus \cdots.$$

We extend all the projectors $P \in \mathfrak{P}$ to all of $\widetilde{\mathfrak{H}}$, putting

$$\widetilde{P}f = \widetilde{P}(f_0 + f_1 + \cdots) = Pf_0 + \sum_j f_j \quad (f_0 \in \mathfrak{H}, f_j \in \mathfrak{H}_j),$$

where the summation is extended over all j for which $P_j^+ \leqq P$. Obviously $\{\widetilde{P}\}$ is a bordered closed chain, and

$$\dim(\widetilde{P}_j^+ - \widetilde{P}_j^-) = \infty \quad (j = 1, 2, \cdots, \omega).$$

It is not hard to verify that $\{\widetilde{P}\}$ constitutes an eigenchain for the inessential extension $\widetilde{A}$ of the operator A to the space $\widetilde{\mathfrak{H}}$, and that

$$(\widetilde{P}_j^+ - \widetilde{P}_j^-)\widetilde{A}(\widetilde{P}_j^+ - \widetilde{P}_j^-) = 0 \quad (j = 1, 2, \cdots, \omega).$$

Thus, by what has been proved earlier the operator $\widetilde{A}$ has a bordered continuous eigenchain.

The theorem is proved.

§4. The general notion of an integral with respect to a chain

Let $\mathfrak{P}$ be an arbitrary chain.[7] We shall call a chain $\mathfrak{z}$ a *partition* of $\mathfrak{P}$, if it consists of a finite number of projectors from $\mathfrak{P}$ and if

$$\min_{P \in \mathfrak{z}} P = \min_{P \in \mathfrak{P}} P\,(\geqq 0) \quad \text{and} \quad \max_{P \in \mathfrak{z}} P = \max_{P \in \mathfrak{P}} P\,(\leqq I).$$

By the union of the partitions $\mathfrak{z}_1$ and $\mathfrak{z}_2$ of $\mathfrak{P}$ we shall mean the partition $\mathfrak{z} = \mathfrak{z}_1 \cup \mathfrak{z}_2$, consisting of all projectors which belong to at least one of the partitions $\mathfrak{z}_1, \mathfrak{z}_2$. Obviously $\mathfrak{z}_1 \cup \mathfrak{z}_2 \supseteq \mathfrak{z}_1$ and $\mathfrak{z}_1 \cup \mathfrak{z}_2 \supseteq \mathfrak{z}_2$.

---

[7] Recall that we have agreed to consider only closed chains.

Suppose that we are given an arbitrary, in general many-valued operator-function $\Phi(\mathfrak{z})$, defined on the set $\mathfrak{z}_{\mathfrak{P}}$ of all partitions $\mathfrak{z}$ of the chain $\mathfrak{P}$, and having as values linear operators from $\mathfrak{R}$.

We need the following definition of the limit of the operator-function $\Phi(\mathfrak{z})$, which makes use of the general notion of the Šatunovskiĭ limit [1]. An operator A is called the *limit* in norm of the function $\Phi(\mathfrak{z})$, written

$$A = \lim \Phi(\mathfrak{z}),$$

if for any $\epsilon > 0$ there exists a partition $\mathfrak{z}_\epsilon$ of the chain $\mathfrak{P}$ such that for all partitions $\mathfrak{z}$ of this chain for which $\mathfrak{z} \supset \mathfrak{z}_\epsilon$, every value of $\Phi(\mathfrak{z})$ satisfies the inequality

$$|\Phi(\mathfrak{z}) - A| < \epsilon.$$

Suppose that we are given an arbitrary operator-function $F(P)$, defined on the chain $\mathfrak{P}$ and having as its values operators from $\mathfrak{R}$. With this function we can form a many-valued function $S(\mathfrak{z})$, putting, for each partition $\mathfrak{z} = \{P_0 < P_1 < \cdots < P_{n-1} < P_n\}$ of $\mathfrak{P}$,

$$S(\mathfrak{z}) = S_Q(\mathfrak{z}) = \sum_{j=1}^{n} F(Q_j)(P_j - P_{j-1}),$$

where Q denotes any family of projectors $Q_j \in \mathfrak{P}$ $(j = 1, 2, \cdots, n)$ such that

$$P_{j-1} \leqq Q_j \leqq P_j \quad (j = 1, 2, \cdots, n).$$

If $S_Q(\mathfrak{z})$ has as its limit in norm some operator A $(\in \mathfrak{R})$, we shall say that the integral

$$\int_{\mathfrak{P}} F(P)\, dP \tag{4.1}$$

converges and equals A. The integral

$$B = \int_{\mathfrak{P}} dPF(P)$$

is defined analogously.

Let us note certain properties of the integral (4.1).

1. *If the integral* (4.1) *converges, then the integral*

$$A_1 = \int_{\mathfrak{P}_1} F(P)\, dP \tag{4.2}$$

converges, where $\mathfrak{P}_1$ is any segment of the chain $\mathfrak{P}$, i.e., a chain consisting *of all projectors* $P \in \mathfrak{P}$ *such that*

$$P_1 \leqq P \leqq P_2 \quad (P_1, P_2 \in \mathfrak{P}).$$

In fact, the convergence of the integral (4.1) is equivalent to the following property for its Riemann sums: for any $\epsilon > 0$ there exists a partition $\mathfrak{z}_\epsilon$ of the chain $\mathfrak{P}$ such that

$$| S_{Q_1}(\mathfrak{z}_1) - S_{Q_2}(\mathfrak{z}_2) | < \epsilon$$

as long as $\mathfrak{z}_j \supset \mathfrak{z}$ $(j = 1, 2)$.

We can obviously assume that the end projectors P_1, P_2 belong to $\mathfrak{z}_\epsilon$.

Let us denote by $\mathfrak{z}_\epsilon^{(1)}$ the 'portion' of the partition $\mathfrak{z}_\epsilon$ belonging to the chain $\mathfrak{P}_1$; $\mathfrak{z}_\epsilon^{(1)}$ is a partition of the chain $\mathfrak{P}_1$. Let $\mathfrak{z}_1^{(1)}$ and $\mathfrak{z}_2^{(1)}$ be any partitions of $\mathfrak{P}_1$ for which $\mathfrak{z}_j^{(1)} \supset \mathfrak{z}_\epsilon^{(1)}$ $(j = 1, 2)$. Denoting by $S_Q^{(1)}(\mathfrak{z})$ a Riemann sum of the integral (4.2), we note that the difference $S_{Q_1}^{(1)}(\mathfrak{z}_1^{(1)}) - S_{Q_2}^{(1)}(\mathfrak{z}_2^{(1)})$ can be represented as some difference $S_{Q_1}(\mathfrak{z}_1) - S_{Q_2}(\mathfrak{z}_2)$, with $\mathfrak{z}_1, \mathfrak{z}_2 \supset \mathfrak{z}_\epsilon$. Consequently,

$$| S_{Q_1}^{(1)}(\mathfrak{z}_1^{(1)}) - S_{Q_2}^{(1)}(\mathfrak{z}_2^{(1)}) | < \epsilon.$$

Thus the integral (4.2) converges.

From this result there follows at once

2. *If the integral* (4.1) *converges, then*

$$\int_{\mathfrak{P}} F(P)\,dP = \int_{P \leqq P_1} F(P)\,dP + \int_{P \geqq P_1} F(P)\,dP,$$

where the integrals on the right side are taken with respect to the chains consisting, respectively, of all projectors $P \in \mathfrak{P}$ *for which* $P \leqq P_1$, $P \geqq P_1$.

The following result holds.

3. *For the convergence of the integral* (4.1) *it is necessary that for any gap* (P^-, P^+) *of the chain* $\mathfrak{P}$

$$(F(P^+) - F(P^-))(P^+ - P^-) = 0. \tag{4.3}$$

In fact, let us consider the part $\mathfrak{P}_1$ of $\mathfrak{P}$ consisting of the projectors P^-, P^+. According to 1, the integral

$$\int_{\mathfrak{P}_1} F(P)\,dP$$

converges. The Riemann sums corresponding to this integral can take on only two values:

$$F(P^-)(P^+ - P^-) \quad \text{and} \quad F(P^+)(P^+ - P^-).$$

Therefore these Riemann sums are equal, and (4.3) holds.

We remark that if the condition (4.3) is not fulfilled, one can consider the integrals

$$\int_{\mathfrak{P}}^{m} F(P)\,dP \quad \text{and} \quad \int_{\mathfrak{P}}^{M} F(P)\,dP;$$

the first of these is defined as the (Šatunovskiĭ) limit of sums of the form

$$\sum_{j=1}^{n} F(P_{j-1})\,\Delta P_j,$$

and the second as the limit of sums of the form

$$\sum_{j=1}^{n} F(P_j)\,\Delta P_j.$$

We will encounter such integrals in Chapter IV.

§5. Abstract triangular representation of an operator by means of its chains

1. One can form an initial idea of this kind of representation by considering operators in a finite-dimensional unitary space.

A linear operator A in n-dimensional unitary space R_n, having only zero as an eigenvalue, and which is therefore nilpotent ($A^n = 0$), should be considered the analog of a Volterra operator.

By Schur's theorem, for every linear operator A in R_n there exists an orthonormal basis $\{e_j\}$ $(j = 1, 2, \cdots, n)$ in which A reduces to triangular form, i.e.,

$$Ae_j = a_{1j}e_1 + a_{2j}e_2 + \cdots + a_{jj}e_j \quad (j = 1, 2, \cdots, n). \tag{5.1}$$

In the case of a nilpotent operator, obviously,

$$a_{jj} = 0 \quad (j = 1, 2, \cdots, n).$$

Let us denote by E_j $(j = 1, 2, \cdots, n)$ the operator of orthogonal projection onto the one-dimensional subspace generated by the vector e_j. Then obviously

$$E_j A E_k = 0 \quad \text{for } j \geqq k, \tag{5.2}$$

and consequently

$$A = \sum_{j,k=1}^{n} E_j A E_k = \sum_{j<k} E_j A E_k - \sum_{j<k} E_j A^* E_k = 2i \sum_{j<k} E_j A_{\mathscr{I}} E_k, \tag{5.3}$$

since

$$\sum_{j<k} E_j A^* E_k = \Big(\sum_{j<k} E_k A E_j\Big)^* = 0.$$

The projectors

$$P_0 = 0 \quad \text{and} \quad P_k = \sum_{j=1}^{k} E_j \quad (k = 1, 2, \cdots, n) \tag{5.4}$$

form a maximal eigenchain $\mathfrak{P}$ of the operator A. The equality (5.3) can now be written in the form

$$A = 2i \sum_{j=1}^{n} P_j A_{\mathscr{J}} \Delta P_j \quad (\Delta P_j = P_j - P_{j-1};\ j = 1, 2, \cdots, n).$$

It follows from (5.2) that

$$\Delta P_j A_{\mathscr{J}} \Delta P_j = 0 \quad \text{or} \quad P_j A_{\mathscr{J}} \Delta P_j = P_{j-1} A_{\mathscr{J}} \Delta P_j \quad (j = 1, 2, \cdots, n),$$

and therefore (5.3) can be rewritten in the following way:

$$A = 2i \sum_{j=1}^{n} \hat{P}_j A_{\mathscr{J}} \Delta P_j, \tag{5.5}$$

where P_j $(j = 1, 2, \cdots, n)$ is either of the projectors P_j, P_{j-1}.

It is natural to write the representation (5.5) of the operator A in the form of an integral representation

$$A = 2i \int_{\mathfrak{P}} P A_{\mathscr{J}} dP.$$

This, obviously, is equivalent to a triangular representation of A.

Now let H be some selfadjoint operator in R_n, whose matrix relative to some orthonormal basis $\{e_j\}$ $(j = 1, 2, \cdots, n)$ has zero diagonal elements. Defining the maximal chain $\mathfrak{P}$ by (5.4), we can form the integral

$$A = 2i \int_{\mathfrak{P}} P H dP.$$

It is easily seen that the operator A thus obtained is nilpotent, reduces in the basis $\{e_j\}_1^n$ to triangular form, and that H is its imaginary Hermitean component. One can say that the integral in question 'truncates' the matrix of the operator H, leaving only the elements above the main diagonal.

2. We precede the main theorem of this section by the following result.

LEMMA 5.1. *Let $A \in \mathfrak{S}_\infty$ and let $\mathfrak{P}$ be a bordered chain. Then the operator-function $\Psi_A(\mathfrak{z})$, defined on the set $\mathfrak{z}_\mathfrak{P}$ of all partitions $\mathfrak{z} = \{P_j\}_0^n$ of the chain $\mathfrak{P}$ by*

$$\Psi_A(\mathfrak{z}) = \sum_{j=1}^{n} \Delta P_j A \Delta P_j, \tag{5.6}$$

tends on $\mathfrak{z}_\mathfrak{P}$, in the uniform norm, to the operator

$$\widetilde{A} = \sum_{j=1}^{\omega} (P_j^+ - P_j^-) A (P_j^+ - P_j^-),$$

where (P_j^-, P_j^+) $(j = 1, 2, \cdots, \omega;\ 0 \leqq \omega \leqq \infty)$ is the sequence of all gaps of $\mathfrak{P}$.

In particular, if $\mathfrak{P}$ is continuous or the blocks of the operator A on its gaps equal zero:

$$(P_j^+ - P_j^-) A (P_j^+ - P_j^-) = 0 \quad (j = 1, 2, \cdots, \omega),$$

then

$$\lim |\Psi_A(\mathfrak{z})| = 0.$$

PROOF. Let us first consider the simple case in which A is a one-dimensional selfadjoint operator $A = A_\phi = (\cdot, \phi)\phi$ $(\phi \in \mathfrak{H},\ |\phi| = 1)$. Let ϵ be an arbitrary positive number, and N a positive integer such that

$$\left| \sum_{j=N+1}^{\omega} Q_j \phi \right| < \frac{\epsilon}{6} \quad (Q_j = P_j^+ - P_j^-).$$

For the vector

$$\psi = \phi - \sum_{j=N+1}^{\omega} Q_j \phi$$

the quantity $|P\psi|^2$ $(P \in \mathfrak{P})$ assumes all values from zero to $|\psi|^2$ $(\leqq 1)$, excluding the values from the intervals $(|P_j^- \psi|^2, |P_j^+ \psi|^2)$ $(j = 1, 2, \cdots, N)$.

Let $\mathfrak{z}_\epsilon = \{P_k\}_0^n$ be a partition of the chain $\mathfrak{P}$ which contains all of the projectors $P_j^\pm$ $(j = 1, 2, \cdots, N)$ and has the property that for any pair of projectors P_{k-1}, P_k $(\in \mathfrak{z}_\epsilon)$, of which at least one is not contained among the $P_j^\pm$ $(j = 1, 2, \cdots, N)$,

$$|\Delta P_k \psi|^2 < \epsilon/3. \tag{5.7}$$

Then

$$\left| \sum_{k=1}^{n} \Delta P_k A_\psi \Delta P_k - \sum_{j=1}^{N} Q_j A_\psi Q_j \right| = \max_k |\Delta P_k A_\psi \Delta P_k|, \tag{5.8}$$

where the maximum is taken over all k for which the pair (P_{k-1}, P_k) does not coincide with any gap (P_j^-, P_j^+) $(j = 1, 2, \cdots, N)$.

Since

$$|\Delta P_j A_\psi \Delta P_j| = |\Delta P_j \psi|^2,$$

it follows from (5.7) and (5.8) that

$$\left| \sum_{j=1}^{n} \Delta P_j A_\psi \Delta P_j - \sum_{j=1}^{\omega} Q_j A_\psi Q_j \right| < \frac{\epsilon}{3}. \tag{5.9}$$

Obviously the relation (5.9) holds for any partition $\mathfrak{z} \supset \mathfrak{z}_\epsilon$. Since for the one-dimensional operators A_ϕ and A_ψ

$$|A_\phi - A_\psi| \leqq 2|\phi - \psi| < \epsilon/3,$$

it follows from (5.9) that for any partition $\mathfrak{z} \supset \mathfrak{z}_\epsilon$

$$\begin{aligned} &\left| \sum_{j=1}^{n} \Delta P_j A_\phi \Delta P_j - \sum_{j=1}^{\omega} Q_j A_\phi Q_j \right| \\ &\qquad \leqq \left| \sum_{j=1}^{n} \Delta P_j (A_\phi - A_\psi) \Delta P_j \right| + \left| \sum_{j=1}^{\omega} Q_j (A_\phi - A_\psi) Q_j \right| \\ &\qquad + \left| \sum_{j=1}^{n} \Delta P_j A_\psi \Delta P_j - \sum_{j=1}^{\omega} Q_j A_\psi Q_j \right| < \epsilon. \end{aligned}$$

Any finite-dimensional operator can be represented as a linear combination of selfadjoint one-dimensional operators of the form A_ϕ; consequently, for the case of a finite-dimensional operator A the lemma is proved.

Finally, let A be any completely continuous operator, and K a finite-dimensional operator. Then for any partition $\mathfrak{z} = \{P_j\}_0^n$ of the chain $\mathfrak{P}$ we have

$$\begin{aligned} &\left| \sum_{j=1}^{n} \Delta P_j A \Delta P_j - \sum_{j=1}^{\omega} Q_j A Q_j \right| \\ &\qquad \leqq 2|A - K| + \left| \sum_{j=1}^{n} \Delta P_j K \Delta P_j - \sum_{j=1}^{\omega} Q_j K Q_j \right|. \end{aligned}$$

Since the set of finite-dimensional operators is dense in $\mathfrak{S}_\infty$, the validity of the lemma for the general case follows at once.

The lemma is proved. We now proceed to the basic theorem of this section, the theorem on the integral triangular representation of a Volterra operator.

THEOREM 5.1 (M. S. BRODSKIĬ [**5**]).[8)] *Let* $A \in \mathfrak{S}_\infty$ *and let* $\mathfrak{P}$ *be a bordered eigenchain of the operator* A. *If for every gap* (P^-, P^+) *of* $\mathfrak{P}$ *the block of* A *equals zero*:

$$(P^+ - P^-)A(P^+ - P^-) = 0,$$

then A *has a triangular representation*

$$A = 2i \int_{\mathfrak{P}} PA_{\mathscr{I}}\,dP$$

which converges in the uniform norm.

Recall that according to the result 1 of §3, the conditions of the theorem will be fulfilled if A is a Volterra operator and $\mathfrak{P}$ is a maximal eigenchain of A.

PROOF. Let $\mathfrak{z} = \{P_j\}_0^n$ be a partition of the chain $\mathfrak{P}$. Denoting by ΔP_k the difference $P_k - P_{k-1}$ $(k = 1, 2, \cdots, n)$ and taking into account that for $j > k$

$$\Delta P_j A \Delta P_k = 0 \quad \text{and} \quad \Delta P_k A^* \Delta P_j = 0,$$

we obtain

$$\begin{aligned} A &= \sum_{j,k=1}^{n} \Delta P_j A \Delta P_k = \sum_{j \leqq k} \Delta P_j A \Delta P_k \\ &= \sum_{j=1}^{n} \Delta P_j A \Delta P_j + 2i \sum_{j<k} \Delta P_j A_{\mathscr{I}} \Delta P_k \end{aligned}$$

or

$$A = \sum_{j=1}^{n} \Delta P_j A \Delta P_j + 2i \sum_{j=1}^{n} P_{j-1} A_{\mathscr{I}} \Delta P_j.$$

Let $Q_j \in \mathfrak{P}$ $(j = 1, 2, \cdots, n)$ be any set of projectors satisfying the inequalities $P_{j-1} \leqq Q_j \leqq P_j$ $(j = 1, 2, \cdots, n)$.

Since

$$\begin{aligned} &\left| A - 2i \sum_{j=1}^{n} Q_j A_{\mathscr{I}} \Delta P_j \right| \\ &\qquad \leqq \left| A - 2i \sum_{j=1}^{n} P_{j-1} A_{\mathscr{I}} \Delta P_j \right| + 2 \left| \sum_{j=1}^{n} (Q_j - P_{j-1}) A_{\mathscr{I}} \Delta P_j \right| \end{aligned}$$

[8)] Our formulation of this theorem differs from that of the corresponding theorem of M. S. Brodskiĭ in that we do not assume in advance that A is a Volterra operator.

and for any $f \in \mathfrak{H}$

$$\left| \sum_{j=1}^{n} (Q_j - P_{j-1}) A_{\mathscr{I}} \Delta P_j f \right|^2 = \sum_{j=1}^{n} | (Q_j - P_{j-1}) A_{\mathscr{I}} \Delta P_j f |^2$$

$$\leqq \sum_{j=1}^{n} | \Delta P_j A_{\mathscr{I}} \Delta P_j f |^2 = \left| \sum_{j=1}^{n} \Delta P_j A_{\mathscr{I}} \Delta P_j f \right|^2,$$

it follows that

$$\left| A - 2i \sum_{j=1}^{n} Q_j A_{\mathscr{I}} \Delta P_j \right| \leqq 3 \left| \sum_{j=1}^{n} \Delta P_j A_{\mathscr{I}} \Delta P_j \right|.$$

Thus, the proof of the theorem reduces to proving that the operator-function

$$\Phi(\mathfrak{z}) = \sum_{j=1}^{n} \Delta P_j A \Delta P_j$$

tends to zero in the uniform norm, and this is so according to Lemma 5.1.

The theorem is proved.

§6. Inversion of the theorem on the abstract triangular representation of operators

Theorem 5.1 of the preceding section admits an inverse. Before formulating this inverse, we prove an auxiliary result.

LEMMA 6.1. *Let $X \in \mathfrak{S}_\infty$ and let $\mathfrak{P}$ be a bordered chain. If the integral*

$$Y = \int_{\mathfrak{P}} PXdP \tag{6.1}$$

converges in the uniform norm, then the integral

$$\int_{\mathfrak{P}} dP\, XP \tag{6.1'}$$

converges in the same norm, and

$$\int_{\mathfrak{P}} dPXP = X - \int_{\mathfrak{P}} PXdP.$$

It is natural to call this formula *the formula for integration by parts.*

PROOF. In fact, for any partition $\mathfrak{z} = \{P_j\}_0^n$ of the chain $\mathfrak{P}$

$$X = \sum_{j,k=1}^{n} \Delta P_j X \Delta P_k = \sum_{j \leq k} \Delta P_j X \Delta P_k + \sum_{j > k} \Delta P_j X \Delta P_k$$

or

$$X = \sum_{k=1}^{n} P_k X \Delta P_k + \sum_{j=1}^{n} \Delta P_j X P_{j-1}. \tag{6.2}$$

Since the integral (6.1) converges in the uniform norm, it follows from (6.2) that the sums

$$S = \sum_{j=1}^{n} \Delta P_j X P_{j-1}$$

have the limit $X - Y$ in this norm. Moreover, for any Q_j $(P_{j-1} \leqq Q_j \leqq P_j)$

$$\left| \sum_{j=1}^{n} \Delta P_j X (Q_j - P_{j-1}) \right| \leqq \left| \sum_{j=1}^{n} \Delta P_j X \Delta P_j \right|.$$

According to Lemma 5.1 and result 3 of §4,

$$\lim \left| \sum_{j=1}^{n} \Delta P_j X \Delta P_j \right| = 0;$$

consequently, the integral (6.1′) converges in the uniform norm and equals $X - Y$.

We shall now prove the theorem which is the inverse of Theorem 5.1.

THEOREM 6.1. *Let* $\mathfrak{P}$ *be a bordered chain, and let* X *be a completely continuous linear operator for which the integral*

$$Y = 2i \int_{\mathfrak{P}} P X \, dP \tag{6.3}$$

converges in the uniform norm.

Then the operator Y *has the following properties:*

a) *each of the subspaces* $\mathfrak{R}(P)$ $(P \in \mathfrak{P})$ *is an invariant subspace of* Y,

b) Y *is a Volterra operator,*

c) *for any gap* (P^-, P^+) *of the chain* $\mathfrak{P}$ *the block of* Y *equals zero*:

$$(P^+ - P^-) Y (P^+ - P^-) = 0. \tag{6.4}$$

If X *is a selfadjoint operator, then, moreover,*

d) $Y_{\mathcal{J}} = X$,

and the integral (6.3) *yields the unique completely continuous operator* Y *having the properties* a), c), d).[9)]

[9)] This theorem was first obtained by the authors in [3] for the case in which X is a Hilbert-Schmidt operator. Then M. S. Brodskii [5] proved properties a), b), d) for the operator Y under the hypotheses of Theorem 6.1 for a maximal chain $\mathfrak{P}$.

PROOF. Let Q be any projector from $\mathfrak{P}$. Then it is easily seen that

$$YQ = 2i\int_{\mathfrak{P}} PXdPQ = 2i\int_{P\leq Q} PXdP = 2iQ\int_{P\leq Q} PXdP = QYQ;$$

hence property a) is proved.

The operator Y is completely continuous, since it is the limit in the uniform norm of the completely continuous operators

$$S(\mathfrak{z}) = \sum_{j=1}^{n} P_{j-1}X\Delta P_j, \tag{6.5}$$

where $\mathfrak{z} = \{P_j\}_0^n$ is some partition of the chain $\mathfrak{P}$.

From the representation of the operator $S(\mathfrak{z})$ in the form

$$S(\mathfrak{z}) = \sum_{j<k} \Delta P_j X \Delta P_k$$

it follows that $S^n(\mathfrak{z}) = 0$. It follows that the operators of the form (6.5) are Volterra operators. On the other hand, according to Theorem I.4.1 of [GK], a limit of Volterra operators is again a Volterra operator.

Assuming, for a given gap (P^-, P^+) of the chain $\mathfrak{P}$, that each of the partitions $\mathfrak{z}$ contains the pair P^-, P^+, we obtain

$$(P^+ - P^-)S(\mathfrak{z})(P^+ - P^-) = 0.$$

Hence (6.4) holds for the limit operator Y. Thus, the properties b) and c) are proved. Property d) follows at once from the formula for integration by parts, since for a selfadjoint operator X this formula assumes the form

$$X = (Y - Y^*)/2i = Y_{\mathcal{J}}.$$

Finally, let Z be a completely continuous operator which has the properties a), c), d). Then according to Theorem 5.1, Z has a uniformly convergent integral representation

$$Z = 2i\int_{\mathfrak{P}} PZ_{\mathcal{J}}\,dP = 2i\int_{\mathfrak{P}} PXdP,$$

from which it follows that $Z = Y$.

The theorem is proved.

COROLLARY 6.1. *If the hypotheses of Theorem* 6.1 *are fulfilled, and if, moreover,* $\mathfrak{P}$ *is a maximal chain, then* Y *is the unique operator having the properties* a), b), d).

§7. Simple operators, the rank of a chain and the rank of an operator

1. An operator $A \in \mathfrak{N}$ is said to be *simple* if A and A^* do not have a common invariant subspace on which they coincide.

If A $(\in \mathfrak{N})$ is a nonsimple operator, then, obviously, there exists a maximal subspace $\mathfrak{L}_T = \mathfrak{L}_T(A)$, invariant with respect to A and A^*, on which these operators coincide.

The subspace $\mathfrak{L}_T$ is called the *trivial* subspace of the nonselfadjoint operator A.

If $A = A^*$, then $\mathfrak{L}_T(A) = \mathfrak{H}$, and if A is a simple operator, then we put $\mathfrak{L}_T = \{0\}$.

In the general case, as is easily seen, the subspace $\mathfrak{L}_T(A)$ consists of precisely those vectors $f \in \mathfrak{H}$ for which

$$A^n f = (A^*)^n f \quad (n = 1, 2, \cdots).$$

It follows from the definition of the trivial subspace that A and A^* have the same trivial subspace:

$$\mathfrak{L}_T(A) = \mathfrak{L}_T(A^*).$$

Thus, the operators A and A^* can only be simple or nonsimple simultaneously.

The following result holds (cf. [GK], p. 223).

1. *To every nonsimple operator $A \in \mathfrak{N}$ there corresponds a unique decomposition of the space $\mathfrak{H}$ into the orthogonal sum of two subspaces, invariant with respect to A and A^*,*

$$\mathfrak{H} = \mathfrak{H}_1 \oplus \mathfrak{H}_2,$$

which have the following properties: 1) *the operators A and A^* coincide on $\mathfrak{H}_1$, and* 2) *A induces a simple operator in $\mathfrak{H}_2$.*

In this decomposition $\mathfrak{H}_1 = \mathfrak{L}_T(A)$.

The restriction $\hat{A}$ of A to the subspace $\hat{\mathfrak{H}} = \mathfrak{L}_T^{\perp}(A)$ will be called the *simple part* of the operator A.

The trivial subspace $\mathfrak{L}_T(A)$ admits yet another characterization.

2. *Let $A \in \mathfrak{N}$ and let $\mathfrak{L}$ be the linear hull of all vectors of the form $A^n A_{\mathscr{J}} f$ $(f \in \mathfrak{H};\ n = 0, 1, \cdots)$. Then*

$$\mathfrak{L}_T(A) = \mathfrak{L}^{\perp}.$$

In fact, for any vector $\phi \in \mathfrak{L}_T(A)$ one obviously has

$$(A^n A_{\mathscr{J}} f, \phi) = (f, A_{\mathscr{J}} (A^*)^n \phi) = 0 \quad (f \in \mathfrak{H};\ n = 0, 1, \cdots);$$

hence $\mathfrak{L}_T(A) \subseteq \mathfrak{L}^{\perp}$.

Since $\mathfrak{L}$ is the linear hull of the ranges of the operators $A^n A_{\mathscr{I}}$ $(n = 0, 1, \cdots)$, the subspace $\mathfrak{L}^\perp$ consists of all the common solutions of the equations $A_{\mathscr{I}}(A^*)^n \phi = 0$ $(n = 0, 1, \cdots)$. It follows that $\mathfrak{L}^\perp$ is invariant with respect to A^*, and that $A_{\mathscr{I}} \mathfrak{L}^\perp = 0$.

Consequently, $\mathfrak{L}^\perp$ is invariant with respect to the operator $A = A^* + 2iA_{\mathscr{I}}$, and the restriction of A to $\mathfrak{L}^\perp$ is a selfadjoint operator. Thus $\mathfrak{L}^\perp \subseteq \mathfrak{L}_T(A)$.

We have thus established that $\mathfrak{L}^\perp = \mathfrak{L}_T(A)$.

As a corollary of result 2 we obtain

3. *In order that an operator A $(\in \mathfrak{R})$ be simple, it is necessary and sufficient that the linear hull $\mathfrak{L}$ of all vectors of the form $A^n A_{\mathscr{I}} f$ $(f \in \mathfrak{H}$; $n = 0, 1, \cdots)$ be dense in $\mathfrak{H}$.*

2. For a dissipative operator A the trivial subspace has a simpler characterization.

4. *The trivial subspace $\mathfrak{L}_T(A)$ of a dissipative operator A coincides with the maximal invariant subspace of A in which A induces a selfadjoint operator.*

In fact, the operators A and A^* coincide on $\mathfrak{L}_T(A)$, so that A induces a selfadjoint operator in $\mathfrak{L}_T$.

Let us consider any invariant subspace $\mathfrak{L}_1$ of A in which A induces a selfadjoint operator. For $f \in \mathfrak{L}_1$ we have

$$(7.1) \qquad |A_{\mathscr{I}}^{1/2} f|^2 = (A_{\mathscr{I}} f, f) = \operatorname{Im}(Af, f) = 0,$$

and thus $A_{\mathscr{I}}$ is identically zero on $\mathfrak{L}_1$. Thus $Af = A^* f$ for $f \in \mathfrak{L}_1$, and so $\mathfrak{L}_1 \subseteq \mathfrak{L}_T$, which was to be proved.

As a simple corollary of result 4 one obtains

5. *A simple dissipative operator A $(\in \mathfrak{R})$ induces a simple dissipative operator in any one of its invariant subspaces.*

3. Let A be a Volterra operator. Then the restriction $\hat{A}$ of A to the invariant subspace $\mathfrak{L}_T(A)$ is a selfadjoint Volterra operator. This is possible if and only if $A\mathfrak{L}_T = \hat{A}\mathfrak{L}_T = 0$. It follows that for a Volterra operator A the subspace $\mathfrak{L}_T(A)$ consists of all the common solutions of the equations $A\phi = 0$, $A^*\phi = 0$.

Thus, in order that a Volterra operator A be simple, it is necessary and sufficient that the equations $A\phi = 0$, $A^*\phi = 0$ not have any nontrivial common solution, or, in other words, that A not be an inessential extension of any completely continuous operator.

If A is a dissipative Volterra operator, then by virtue of (7.1) the equality $Af = 0$ implies $A_{\mathscr{J}} f = 0$, and consequently $A^*f = 0$. Thus, a *dissipative Volterra operator* A *will be simple if and only if the equation* $A\phi = 0$ *has only the zero solution.*

THEOREM 7.1.[10] *Let* $\mathfrak{P}$ *be a bordered chain, and suppose that for some* $X = X^* \in \mathfrak{S}_\infty$ *the integral*

$$Y = 2i \int_{\mathfrak{P}} PX\,dP$$

converges. Then in order that the operator Y *be simple, it is necessary that the linear hull of all the vectors of the form* PXf $(P \in \mathfrak{P}, f \in \mathfrak{H})$ *be dense in* $\mathfrak{H}$. *If* Y *is a dissipative* operator (*i.e.*, $X \geqq 0$), *this condition is also sufficient.*

PROOF. Let us denote by $\mathfrak{H}_0$ the set of all vectors $f \in \mathfrak{H}$ such that $XPf = 0$ for every $P \in \mathfrak{P}$. For $f \in \mathfrak{H}_0$ we obviously have

$$Yf = 2i \int_{\mathfrak{P}} PX\,dPf = 0.$$

The subspace $\mathfrak{H}_1 = \mathfrak{H} \ominus \mathfrak{H}_0$ is the closed linear hull of all vectors of the form PXf $(P \in \mathfrak{P}, f \in \mathfrak{H})$, and so $\mathfrak{R}(Y) \subseteq \mathfrak{H}_1$. Thus, $\mathfrak{H}_1$ is invariant with respect to Y, and Y is an inessential extension of the operator which it induces in $\mathfrak{H}_1$. It follows that for Y to be simple, it is necessary that $\mathfrak{H}_1 = \mathfrak{H}$.

For a dissipative operator Y this last condition is sufficient. In fact, if $Yf = 0$ for some $f \in \mathfrak{H}$, then $Y^*f = 0$. Thus, for any $P \in \mathfrak{P}$

$$(PXPf,f) = \frac{1}{2i}[(PYPf,f) - (PY^*Pf,f)]$$

$$= \frac{1}{2i}[(Pf, Y^*f) - (PY^*f,f)] = 0,$$

whence $XPf = 0$ $(P \in \mathfrak{P})$ or $f \in \mathfrak{H}_1^\perp$.

Since $\mathfrak{H}_1 = \mathfrak{H}$, we have $f = 0$, and the result is proved.

4. The concept of the rank of a chain enables us to strengthen a number of theorems proved earlier. It plays an important role in the construction of analytic models of Volterra operators.

A system of vectors $\{\chi_j\}_1^r$ $(1 \leqq r \leqq \infty)$ is called a *reproducing* system

[10] The second assertion of the theorem was communicated to the authors by M. S. Brodskiĭ.

for the chain $\mathfrak{P}$, if the closed linear hull of the set of vectors $P\chi_j$ $(j = 1, 2, \cdots, r;\ P \in \mathfrak{P})$ coincides with the entire space $\mathfrak{H}$. The smallest of the cardinalities of all possible reproducing systems for the chain $\mathfrak{P}$ is called the *rank* of this chain, and is denoted by $r(\mathfrak{P})$. It is easy to construct examples of chains of any rank $r(\mathfrak{P}) \leqq \infty$.

COROLLARY 7.1. *The rank $r(\mathfrak{P})$ of any maximal eigenchain of a simple Volterra operator A does not exceed the dimension of its imaginary component*:

$$r(\mathfrak{P}) \leqq r(A_{\mathscr{J}}). \tag{7.2}$$

Let A be a simple Volterra operator, and $\mathfrak{P}$ a maximal eigenchain of A; then

$$A = 2i \int_{\mathfrak{P}} PA_{\mathscr{J}}\, dP,$$

and according to Theorem 7.1 the linear hull of all vectors of the form $PA_{\mathscr{J}} f$ $(f \in \mathfrak{H},\ P \in \mathfrak{P})$ is dense in $\mathfrak{H}$. It follows that an orthonormal system of eigenvectors ϕ_j $(j = 1, 2, \cdots, r(A_{\mathscr{J}}))$ of $A_{\mathscr{J}}$ which is complete in $\mathfrak{R}(A_{\mathscr{J}})$ will be a reproducing system for the chain $\mathfrak{P}$, and so the inequality (7.2) holds.

THEOREM 7.2. *If the simple part of a Volterra operator A has a maximal eigenchain of rank r, then the operator A itself has a maximal eigenchain of rank r.*

PROOF. Let us denote by $\hat{\mathfrak{P}}$ a maximal eigenchain of rank r of the operator $\hat{A}$. Let $\hat{\mathfrak{H}}$ be the subspace in which $\hat{A}$ acts, and let $\mathfrak{Q}$ be any maximal chain of rank 1 in the subspace $\mathfrak{H} \ominus \hat{\mathfrak{H}}$.

We lengthen the chain $\hat{\mathfrak{P}}$ by adding to it all orthoprojectors of the form $\hat{P} + Q$ $(Q \in \mathfrak{Q})$, where $\hat{P}$ is the orthoprojector which projects $\mathfrak{H}$ orthogonally onto the subspace $\hat{\mathfrak{H}}$. The chain $\mathfrak{P}$ thus obtained is obviously a maximal eigenchain of the operator A. If $\{\chi_j\}_1^r$ is a reproducing system for the chain $\hat{\mathfrak{P}}$ and y is a reproducing vector for the chain $\mathfrak{Q}$, then obviously the system $\{\chi_j + y\}_1^r$ is a reproducing system for the chain $\mathfrak{P}$.

The theorem is proved.

5. The examples considered above suggest the concept of the rank of an operator.

The smallest of the ranks of all maximal eigenchains of a completely continuous operator A will be called the *rank* of A.

It is easily seen that a completely continuous operator with a complete system of root vectors, in particular a completely continuous self-adjoint operator, has rank equal to unity.

We remark that the rank of an operator A always coincides with the rank of its adjoint A^*. According to Corollary 7.1, the rank of a Volterra operator is not greater than the dimension of its imaginary component $A_{\mathcal{J}}$.

The concept of the rank of an operator will be valuable, of course, only under the condition that there exist completely continuous operators with rank greater than unity. However, the authors do not have available a proof of the existence of such operators.

If such operators do exist, then, possibly, each of them has an inessential extension of rank 1.

Recently G. È. Kisilevskiĭ (see the Appendix, §§1—3) established the following important result.

THEOREM 7.3 (G. È. KISILEVSKIĬ [5]). *The rank of every dissipative Volterra operator with nuclear imaginary component is equal to unity.*

§8. Volterra operators with a one-dimensional imaginary component

This section is devoted to the following important result of M. S. Livšic.[11)]

THEOREM 8.1. *Every simple dissipative operator A which has a one-dimensional imaginary component $A_{\mathcal{J}} = (\cdot, e)e$ is unitarily equivalent to the operator*

$$Jf = 2i\int_x^1 f(s)\,ds \quad (f \in L_2(0,1)).$$

We preceed the proof of this theorem by a lemma.[12)]

LEMMA 8.1. *Let A be a simple dissipative operator with a one-dimensional imaginary component $A_{\mathcal{J}} = (\cdot, e)e$ $(e \in \mathfrak{H}, |e| = 1)$, and let $\mathfrak{P}$ be a maximal eigenchain of A. Then the function*

$$t(P) = (Pe, e) \quad (P \in \mathfrak{P})$$

[11)] This result is not formulated explicitly in the work of M. S. Livšic. However, it is contained in the results of his paper [2], and can also be deduced from Theorem 14 of his old paper [1]. In explicit form, this result was stated in a paper by M. S. Brodskiĭ [2]. The proof given below, based on the triangular integral, was presented by the authors [3].

[12)] A general result on the parametrization of chains will be proved in Theorem V.1.1. We suggest that the reader compare Lemma 8.1 with this theorem.

is continuous and monotonically increasing, and its values fill the entire interval $[0,1]$.

PROOF. It is easily seen that the function $t(P)$ is continuous on $\mathfrak{P}$ and does not decrease. By virtue of the result 2 of §3, the chain $\mathfrak{P}$ is continuous, from which it follows at once that $t(P)$ assumes all values from the interval $[0,1]$. It remains to establish that $t(P)$ is strictly increasing. Let us assume the contrary, i.e., that there exist projectors $P_1, P_2 \in \mathfrak{P}$ $(P_1 < P_2, P_1 \neq 0)$ for which $t(P_1) = t(P_2)$. Then obviously $P_1 e = P_2 e$.

Let us denote by $\hat{A}$ the restriction of the operator A to the subspace $P_2\mathfrak{H}$. The subspace $P_1\mathfrak{H}$ is invariant with respect to $\hat{A}$. Since

$$\hat{A} = P_2 A_{\mathscr{R}} P_2 + i(\cdot, h)h,$$

where $h = P_2 e$ and $P_1 h = h$, the subspace $P_1\mathfrak{H}$ is invariant with respect to $\hat{A}_{\mathscr{J}}$, and hence with respect to $\hat{A}_{\mathscr{R}}$. But then the subspace $(P_2 - P_1)\mathfrak{H}$ is invariant with respect to $\hat{A}_{\mathscr{R}}$ and, moreover, $A_{\mathscr{J}}(P_2 - P_1)\mathfrak{H} = 0$. Consequently, the subspace $(P_2 - P_1)\mathfrak{H}$ is invariant with respect to A, and A induces a selfadjoint operator in it. This contradicts the simplicity of A. The lemma is proved.

PROOF OF THE THEOREM. Let $\mathfrak{P} = \{P\}$ be a maximal eigenchain of operator A, which is assumed to satisfy the hypothesis of the theorem. According to the result 2 of §3, the chain $\mathfrak{P}$ is continuous. By Lemma 8.1, one can construct the continuous monotone operator-function $P(t)$ $(0 \leqq t \leqq 1)$, the inverse of the function $t(P) = (Pe, e)$ $(P \in \mathfrak{P})$.

We consider the operator U, defined on the set $\mathfrak{E}$ of all piecewise constant functions $f(t)$ $(0 \leqq t \leqq 1)$ by

$$Uf = \int_0^1 f(t)\, dP(t)e \qquad (\in \mathfrak{H}). \tag{8.1}$$

The operator U has the property

$$(Uf, Ug) = \int_0^1 f(t)\overline{g(t)}\, d(P(t)e, e) = \int_0^1 f(t)\overline{g(t)}\, dt \qquad (f, g \in \mathfrak{E}).$$

Since the set $U\mathfrak{E}$ is the linear hull of all the vectors $P(t)e$ $(0 \leqq t \leqq 1)$, according to Theorem 7.1 $U\mathfrak{E}$ is dense in $\mathfrak{H}$. Extending the operator U by continuity to all of the space $L_2(0,1)$, we obtain a unitary operator which maps $L_2(0,1)$ onto $\mathfrak{H}$ according to the formula (8.1).

By Theorem 5.1, the operator A admits a continuous triangular representation

$$Af = 2i \int_0^1 (f, dP(t)e)P(t)e \quad (f \in \mathfrak{H}).$$

It follows that for any functions $f, g \in L_2(0,1)$

$$(U^*AUf, g) = (AUf, Ug) = 2i \int_0^1 (e, P(t)Ug)d(P(t)Uf, e).$$

Since

$$P(x)Ug = \int_0^x g(t)dP(t)e \quad \text{and} \quad (P(x)Ug, e) = \int_0^x g(t)dt,$$

we have

$$(U^*AUf, g) = 2i \int_0^1 \int_0^x \overline{g(s)}dsf(x)dx = 2i \int_0^1 \int_x^1 f(s)ds\overline{g(x)}dx.$$

Thus,

$$U^*AU = J.$$

The theorem is proved.

§9. Unicellular operators

As is known, in a finite-dimensional space every linear operator can be decomposed into the direct sum of "simplest" operators, each of which acts in a suitably chosen subspace and has, in a certain basis of this subspace, a normal Jordan form consisting of only one cell. It is natural to call the simplest operators unicellular operators.

A unicellular operator is characterized geometrically by the fact that of any two distinct invariant subspaces of this operator, one is always contained in the other. This last property is taken by M. S. Brodskiĭ [1][13] as the definition of the unicellularity of an operator in the case of an infinite-dimensional space.

Thus, a bounded linear operator A, acting in $\mathfrak{H}$, is said to be *unicellular* if for any two of its invariant subspaces $\mathfrak{L}_1$, $\mathfrak{L}_2$ $(\mathfrak{L}_1 \neq \mathfrak{L}_2)$, either $\mathfrak{L}_1 \subset \mathfrak{L}_2$ or $\mathfrak{L}_2 \subset \mathfrak{L}_1$.

It is easily seen (cf. [GK], Chapter I, §1.3) that the spectrum of a unicellular operator is a connected set. It follows that a completely continuous unicellular operator is always a simple Volterra operator. In other words, a completely continuous operator is unicellular if and only if it has a unique maximal eigenchain.

---

[13] Brodskiĭ applied this definition of unicellularity only to Volterra operators.

If $\mathfrak{L}$ is an invariant subspace of an operator $A \in \mathfrak{R}$, then $\mathfrak{L}^{\perp}$ is an invariant subspace of the operator A^*. It follows at once that *if the operator A is unicellular, then A^* will be unicellular.*

If the operators A_1 and A_2, acting in the Hilbert spaces $\mathfrak{H}_1$ and $\mathfrak{H}_2$ respectively, are unitarily equivalent, then the unicellularity of one of them obviously implies the unicellularity of the other. Equally obvious is the following more general result.

If the operators A_1 and A_2 are similar, then the unicellularity of one of them implies the unicellularity of the other.

Let us clarify that two operators A_1 and A_2 are said to be *similar*, if there exists a continuous linear and invertible mapping B of the space $\mathfrak{H}_1$ onto the space $\mathfrak{H}_2$ which transforms A_1 into A_2, i.e.,

$$A_2 = BA_1B^{-1}.$$

We remark that in a finite-dimensional space all unicellular operators having the unique eigenvalue zero are similar to one another.

In contrast with this, it will be shown in the Appendix, §6, that in any separable infinite-dimensional Hilbert space there exists a continuum of pairwise nonsimilar unicellular Volterra operators.

1. A vector $f \in \mathfrak{H}$ is said to be a *cyclic vector* for a bounded linear operator A, acting in $\mathfrak{H}$, if the closed linear hull $\mathfrak{L}_A(f)$ of the sequence $f, Af, \cdots, A^nf, \cdots$ coincides with $\mathfrak{H}$.

If a bounded linear operator A has at least one cyclic vector, then A is said to be *cyclic*. In the case of a finite-dimensional space the unicellular operators are completely characterized by the fact that they are cyclic and their spectrum consists of one point.

In the case of an infinite-dimensional space the situation is otherwise. It will be shown in the Appendix (§1.1) that there exist cyclic Volterra operators which are not unicellular. Nevertheless, one has

THEOREM 9.1. *Every unicellular operator A ($\in \mathfrak{R}$) is cyclic.*

PROOF. If the operator A does not have any proper invariant subspace, they any vector $f \neq 0$ is a cyclic vector for A, since the closed linear hull $\mathfrak{L}_A(f)$ of the sequence $f, Af, A^2f, \cdots$ is obviously an invariant subspace of A.

Let us consider the case in which A has a proper invariant subspace. The orthoprojectors P onto all of these subspaces, together with 0 and I, form a bordered closed eigenchain $\mathfrak{P}_A$ of A.

We shall say that a vector f *covers* a chain $\mathfrak{P}$ if it does not belong to any of the subspaces $P\mathfrak{H}$ ($P \in \mathfrak{P}$; $P \neq I$). It is easily seen that an arbitrary chain $\mathfrak{P}$ always has a covering vector. In fact, it suffices to consider the case in which $\mathfrak{P}$ is closed and contains I.

If $\mathfrak{P}$ contains a largest orthoprojector P_∞ among those satisfying $P < I$, then obviously any vector $f \neq 0$ from the subspace $(I - P_\infty)\mathfrak{H}$ will be a covering vector. If P_∞ does not exist, then there exists a sequence $P_1 < P_2 < \cdots$ of orthoprojectors in $\mathfrak{P}$ which converges strongly to I. In this case every vector f of the form

$$f = \sum_{j=1}^{\infty} f_j,$$

where $f_j \in (P_{j+1} - P_j)\mathfrak{H}$, $f_j \neq 0$ and $\sum_j |f_j|^2 < \infty$, is a covering vector for $\mathfrak{P}$.

Therefore the theorem will be proved as soon as the following general result is proved.

In order that a vector f be a cyclic vector for a unicellular operator A with maximal eigenchain $\mathfrak{P}_A$, it is necessary and sufficient that f be a covering vector for $\mathfrak{P}_A$.

Indeed, if f does not cover $\mathfrak{P}_A$, there exists $P \in \mathfrak{P}_A$ ($P \neq I$), such that $f \in P\mathfrak{H}$; but then $\mathfrak{L}_A(f) \subseteq P\mathfrak{H}$ and f is not a cyclic vector for A.

If f covers $\mathfrak{P}_A$, then $\mathfrak{L}_A(f)$, being an invariant subspace of A, cannot coincide with any subspace $\mathfrak{R}(P)$ ($P \in \mathfrak{P}_A$; $P \neq I$), since it contains the vector f. Consequently $\mathfrak{L}_A(f) = \mathfrak{H}$.

The theorem is proved.

We have already mentioned that in the general case Theorem 9.1 does not admit a converse. Nevertheless, G. È. Kisilevskiĭ has been able to show that for one important class of dissipative Volterra operators the converse of Theorem 9.1 is also valid (cf. Theorem 1.1 of the Appendix).

2. The following theorem distinguishes the simplest unicellular operators.

THEOREM 9.2. *Every simple Volterra operator with a one-dimensional imaginary component is unicellular.*

According to a theorem of M. S. Livšic (Theorem 8.1), every such operator becomes unitarily equivalent to the operator

$$(Jf)(t) = 2i \int_t^1 f(s)\, ds,$$

acting in the space $L_2(0,1)$, upon multiplication by a real number. Thus Theorem 9.2 is equivalent to the following.

THEOREM 9.2′. *The operator of integration is a unicellular operator. The set of all its invariant subspaces coincides with the set of subspaces* $\{\mathfrak{L}_\sigma\}_{0 \leq \sigma \leq 1}$, *where* $\mathfrak{L}_\sigma$ *is the set of all functions* $f(t) \in L_2(0,1)$ *which vanish almost everywhere on the interval* $[\sigma, 1]$.

This last theorem was discovered almost simultaneously by several authors[14] (cf. M. S. Brodskiĭ [2], L. A. Sahnovič [2], G. K. Kalisch [1]). All of these proofs of Theorem 9.2 were based on various analytical results: on Titchmarsh's theorem concerning the vanishing of a convolution or on the theory of characteristic matrix-functions of nonselfadjoint operators. Recently M. S. Brodskiĭ [9] gave an elementary proof of Theorem 9.2 based on simple results of the theory of Volterra operators. We reproduce his proof here.

PROOF OF THEOREM 9.2. Let A be a simple Volterra operator with a one-dimensional imaginary component. We may assume without loss of generality that $A_{\mathcal{J}} = (\cdot, e)e$ $(e \in \mathfrak{H},\ |e| = 1)$. Let us suppose that $\mathfrak{P} = \{P\}$ and $\mathfrak{P}' = \{P'\}$ are two maximal eigenchains of A.

By Lemma 8.1 the functions $t(P) = (Pe, e)$ $(P \in \mathfrak{P})$ and $t(P') = (P'e, e)$ $(P' \in \mathfrak{P}')$ are continuous and strictly increasing, and take on all values in $[0,1]$. Thus we can establish a one-to-one correspondence between the elements of the chains $\mathfrak{P}$ and $\mathfrak{P}'$, setting $P \leftrightarrow P'$ if

$$(Pe, e) = (P'e, e). \tag{9.1}$$

From now on we shall denote by $P' \in \mathfrak{P}'$ the projector corresponding to the projector $P \in \mathfrak{P}$.

We define a linear operator U, to begin with, on the vectors of the form Pe by

$$UPe = P'e \quad (P \in \mathfrak{P}).$$

According to Theorem 7.1, the linear hull of each of the systems of vectors $\{Pe\}_{P \in \mathfrak{P}}$, $\{P'e\}_{P' \in \mathfrak{P}'}$ is dense in $\mathfrak{H}$. It follows from (9.1) that U can be extended to a unitary operator on the entire space.

For every pair of projectors $P_1, P_2 \in \mathfrak{P}$ one has

$$UP_1P_2e = P_1'P_2'e = P_1'UP_2e.$$

Considering the last equality for a fixed projector $P_1 = P$ and arbitrary

[14] The first proof of this theorem was apparently obtained by B. Ja. Levin (by methods of the theory of entire functions): this proof is unpublished.

$P_2 \in \mathfrak{P}$, we conclude that

$$UP = P'U \quad (P \in \mathfrak{P}). \tag{9.2}$$

We further note that from $Ue = e$ follows

$$UA_{\mathcal{J}} = A_{\mathcal{J}}U. \tag{9.3}$$

According to Theorem 5.1, the operator A has a triangular representation

$$A = 2i \int_{\mathfrak{P}} PA_{\mathcal{J}}\, dP.$$

It follows, by virtue of (9.2) and (9.3), that

$$UA = 2i \int_{\mathfrak{P}} UPA_{\mathcal{J}}\, dP = 2i \int_{\mathfrak{P}'} P'A_{\mathcal{J}}\, dP'U = AU,$$

and consequently

$$UA^n e = A^n Ue = A^n e \quad (n = 0, 1, \cdots). \tag{9.4}$$

By the result 3 of §7, the linear hull of the vectors $A^n e$ $(n = 0, 1, \cdots)$ is dense in $\mathfrak{H}$. It follows from (9.4) that $U = I$. Thus $P = P'$ $(P \in \mathfrak{P})$ and the operator A has a unique maximal eigenchain.

The theorem is proved.

As was mentioned, Theorem 9.2 can be deduced from the following theorem of Titchmarsh concerning convolutions.

THEOREM 9.3 (E. C. TITCHMARSH [1]). *Let $F(t)$, $G(t) \in L_1(0,1)$. If the convolution*

$$(F * G)(t) = \int_0^t F(t-s)\, G(s)\, ds \tag{9.5}$$

equals zero almost everywhere, there exists a number α $(0 \leqq \alpha \leqq 1)$ such that the functions $F(t)$ and $G(t)$ equal zero almost everywhere on the intervals $[0, \alpha]$ and $[0, 1-\alpha]$, respectively.

A direct elementary proof of this theorem has been obtained by I. Mikusinski [1].

G. K. Kalisch [1] discovered that the opposite result holds, i.e., Titchmarsh's theorem follows from Theorem 9.2. We present a proof of this fact, following a paper by M. S. Brodskii [9].

PROOF OF THEOREM 9.3. Suppose that the convolution (9.5) equals zero almost everywhere. We consider the functions

$$f(t) = 1 * F(t) = \int_0^t F(s)\,ds, \quad g(t) = 1 * G(t) = \int_0^t G(s)\,ds.$$

Since the operation of convolution is commutative and associative (cf. I. M. Gel′fand, D. A. Raĭkov and G. E. Šilov [1]), the function

$$\begin{aligned}&\int_0^t f(t-s_n)\int_0^{s_n}\cdots\int_0^{s_1} g(s)\,ds\,ds_1\cdots ds_n\\ &\qquad = f*(1*\cdots*(1*(1*g)\cdots)\end{aligned} \tag{9.6}$$

equals zero for all $t \in [0,1]$. In particular,

$$\int_0^1 f(1-s_n)\int_0^{s_n}\cdots\int_0^{s_1} g(s)\,ds\,ds_1\cdots ds_n = 0 \quad (n = 0,1,2,\cdots). \tag{9.7}$$

The operator J^* has the form

$$(J^*g)(t) = -2i\int_0^t g(s)\,ds;$$

hence (9.7) can be rewritten in the form

$$\int_0^1 f(1-s)\,[(J^*)^n g](s)\,ds = (f(1-s), \overline{[(J^*)^n g](s)}) = 0 \quad (n = 0,1,\cdots)$$

or in the form

$$(J^n(f(1-t)), \overline{g(t)}) = 0 \quad (n = 0,1,\cdots).$$

The closure of the linear hull of the functions $J^n(f(1-t))$ $(n = 0,1,\cdots)$ is an invariant subspace of the operator J, and so by Theorem 9.2 this subspace coincides with some subspace $\mathfrak{L}_\sigma$ $(0 \leqq \sigma \leqq 1)$. Thus, $f(1-t) \in \mathfrak{L}_\sigma$, $g(t) \in \mathfrak{L}_\sigma^\perp$, and so the functions $f(t)$ and $g(t)$, and with them the functions $F(t)$ and $G(t)$, equal zero almost everywhere in the intervals $[0,\alpha]$ and $[0,1-\alpha]$, respectively, where $\alpha = 1-\sigma$.

The theorem is proved.

§10. The integral of triangular truncation in the class of Hilbert-Schmidt operators[15)]

1. In this section we shall consider the integral

$$Y = \int_{\mathfrak{P}} PX\,dP \tag{10.1}$$

[15)] A brief report of the results of this section was published by the authors in [5].

as an operator acting in the space $\mathfrak{S}_2$ of Hilbert-Schmidt operators.

Let us give the definition of the space $\mathfrak{S}_2$ (for details concerning this space see [GK], Chapter III, §9). An operator $A \in \mathfrak{S}_\infty$ is called a *Hilbert-Schmidt operator* if $\operatorname{sp} A^*A < \infty$.[16] The set of all Hilbert-Schmidt operators forms a linear set, denoted by $\mathfrak{S}_2$ (the meaning of this notation will be clarified later).

If we introduce a scalar product in $\mathfrak{S}_2$ by the rule

$$(X, Y) = \operatorname{sp}(XY^*) \quad (X, Y \in \mathfrak{S}_2),$$

then $\mathfrak{S}_2$ becomes a Hilbert space with the norm

$$|X|_2 = (\operatorname{sp} XX^*)^{1/2}.$$

In the sequel, operators acting in spaces of operators (in particular, in $\mathfrak{S}_2$) will for convenience be called *transformators*.

2. We recall that a necessary condition for the convergence in the uniform norm of the integral (10.1) is the vanishing of the blocks of the operator X on the gaps of the chain $\mathfrak{P}$:

$$(P^+ - P^-)X(P^+ - P^-) = 0, \tag{10.2}$$

where (P^-, P^+) is an arbitrary gap of $\mathfrak{P}$.

Let us denote by $\mathfrak{S}_2[\mathfrak{P}]$ the set of all operators $X \in \mathfrak{S}_2$ which satisfy the condition (10.2). Obviously $\mathfrak{S}_2[\mathfrak{P}]$ is a subspace of the Hilbert space $\mathfrak{S}_2$.

We note that for the case of a bordered continuous chain $\mathfrak{S}_2[\mathfrak{P}]$ coincides with $\mathfrak{S}_2$.

The basic result of the section is the following theorem.

THEOREM 10.1. *Let $\mathfrak{P}$ be an arbitrary bordered chain. Then for every $X \in \mathfrak{S}_2[\mathfrak{P}]$ the integral* (10.1) *converges in the norm of the space $\mathfrak{S}_2$ and represents an orthogonal projector which projects $\mathfrak{S}_2[\mathfrak{P}]$ onto the set of all operators from $\mathfrak{S}_2[\mathfrak{P}]$ which have $\mathfrak{P}$ as an eigenchain.*

PROOF. We note first that for any operator $A \in \mathfrak{S}_2$ and any partition $\mathfrak{z} = \{P_j\}_0^n$ of the chain $\mathfrak{P}$

$$|A|_2^2 = \sum_{k,j=1}^{n} |\Delta P_k A \Delta P_j|_2^2 \quad (\Delta P_j = P_j - P_{j-1}). \tag{10.3}$$

In fact, A can be represented in the form

---

[16] For the concept of the trace $\operatorname{sp} A$ of an operator see [GK], Chapter III, §8. The basic properties of this concept will be recalled below in Chapter III, §1.

$$A = \sum_{k,j=1}^{n} \Delta P_k A \Delta P_j;$$

hence

$$A^*A = \sum_{k,j,r=1}^{n} \Delta P_k A^* \Delta P_j A \Delta P_r$$

and

$$|A|_2^2 = \operatorname{sp}(A^*A) = \sum_{k,j,r=1}^{n} \operatorname{sp}(\Delta P_k A^* \Delta P_j A \Delta P_r)$$
$$= \sum_{k,j=1}^{n} \operatorname{sp}(\Delta P_k A^* \Delta P_j A \Delta P_k) = \sum_{k,j=1}^{n} |\Delta P_k A \Delta P_j|_2^2.$$

We consider the Riemann sum

$$S_{\mathfrak{z}} = \sum_{j=1}^{n} P_{j-1} X \Delta P_j = \sum_{j<k} \Delta P_j X \Delta P_k. \tag{10.4}$$

The set of sums of the form (10.4) is bounded. In fact, by (10.3) we have the inequality

$$|S_{\mathfrak{z}}|_2^2 = \sum_{j<k} |\Delta P_j X \Delta P_k|_2^2 \leqq \sum_{j,k=1}^{n} |\Delta P_j X \Delta P_k|_2^2 = |X|_2^2. \tag{10.5}$$

Moreover, for any two partitions $\mathfrak{z}_1 \subset \mathfrak{z}_2$ of $\mathfrak{P}$, by (10.3),

$$|S_{\mathfrak{z}_2} - S_{\mathfrak{z}_1}|_2^2 = |S_{\mathfrak{z}_2}|_2^2 - |S_{\mathfrak{z}_1}|_2^2,$$

and so

$$|S_{\mathfrak{z}_2}|_2 \geqq |S_{\mathfrak{z}_1}|_2.$$

Let M be the supremum of the set of all norms $|S_{\mathfrak{z}}|_2$. For any $\epsilon > 0$ there exists a partition $\mathfrak{z}_\epsilon$ of $\mathfrak{P}$ such that

$$|S_{\mathfrak{z}_\epsilon}|_2^2 > M^2 - \epsilon^2/4.$$

Then if $\mathfrak{z}_1, \mathfrak{z}_2$ are any two partitions of $\mathfrak{P}$ containing $\mathfrak{z}_\epsilon$, we have

$$|S_{\mathfrak{z}_1} - S_{\mathfrak{z}_2}|_2^2 \leqq 2(|S_{\mathfrak{z}_1} - S_{\mathfrak{z}_\epsilon}|_2^2 + |S_{\mathfrak{z}_2} - S_{\mathfrak{z}_\epsilon}|_2^2)$$
$$= 2(|S_{\mathfrak{z}_1}|_2^2 + |S_{\mathfrak{z}_2}|_2^2 - 2|S_{\mathfrak{z}_\epsilon}|_2^2),$$

and so

$$|S_{\mathfrak{z}_1} - S_{\mathfrak{z}_2}|_2^2 \leqq 4(M^2 - |S_{\mathfrak{z}_\epsilon}|_2^2) < \epsilon^2.$$

Thus the sums $S_{\mathfrak{z}}$ converge in the norm of $\mathfrak{S}_2$ to some operator $Y \in \mathfrak{S}_2$.

For any partition $\mathfrak{z}$ of $\mathfrak{P}$ satisfying the condition $\mathfrak{z} \supset \mathfrak{z}_\epsilon$, we will have

$$\left| Y - \sum_{j=1}^{n} Q_j X \Delta P_j \right|_2 \leqq | Y - S_{\mathfrak{z}} |_2 + \left| \sum_{j=1}^{n} (P_{j-1} - Q_j) X \Delta P_j \right|_2,$$

$$P_{j-1} \leqq Q_j \leqq P_j \quad (j = 1, 2, \cdots, n).$$

Since

$$\left| \sum_{j=1}^{n} (P_{j-1} - Q_j) X \Delta P_j \right|_2 \leqq \left| \sum_{j=1}^{n} \Delta P_j X \Delta P_j \right|_2,$$

to complete the proof of the convergence of the integral (10.1) it suffices to establish that

$$\lim \left| \sum_{j=1}^{n} \Delta P_j X \Delta P_j \right|_2 = 0.$$

The proof of this relation can be carried out in the same way as the proof of Lemma 5.1. Incidentally, in Chapter III, §7 a lemma will be proved which contains, in particular, the relation in question.

According to what has been proved and Theorem 6.1, the operator Y defined by (10.1) belongs to $\mathfrak{S}_2[\mathfrak{P}]$ and has $\mathfrak{P}$ as an eigenchain. Consequently, for any partition $\mathfrak{z} = \{P_j\}_0^n$ of $\mathfrak{P}$

$$Y = \sum_{j=1}^{n} P_j Y \Delta P_j.$$

It follows that

$$\mathscr{C}^2(X) = \mathscr{C}(X) \quad (X \in \mathfrak{S}_2[\mathfrak{P}]),$$

where $\mathscr{C}(X) = \int_{\mathfrak{P}} PX\, dP.$

The transformator $\mathscr{C}$, acting in $\mathfrak{S}_2[\mathfrak{P}]$, is selfadjoint. Indeed, for any $X, Z \in \mathfrak{S}_2[\mathfrak{P}]$

$$\begin{aligned} \left(\sum_{j=1}^{n} P_j X \Delta P_j, Z \right) &= \operatorname{sp} \left(\sum_{j=1}^{n} P_j X \Delta P_j Z^* \right) \\ &= \sum_{j=1}^{n} \operatorname{sp}(P_j X \Delta P_j Z^*) = \sum_{j=1}^{n} \operatorname{sp}(X \Delta P_j Z^* P_j) \\ &= \operatorname{sp} \left(X \left[\sum_{j=1}^{n} P_j Z \Delta P_j \right]^* \right) = \left(X, \sum_{j=1}^{n} P_j Z \Delta P_j \right). \end{aligned}$$

Passing to the limit in the relation

$$\left(\sum_{j=1}^{n} P_j X \Delta P_j, Z\right) = \left(X, \sum_{j=1}^{n} P_j Z \Delta P_j\right),$$

we obtain

$$(\mathscr{C}(X), Z) = (X, \mathscr{C}(Z)),$$

or $\mathscr{C} = \mathscr{C}^*$. The relations $\mathscr{C}^2 = \mathscr{C}$ and $\mathscr{C} = \mathscr{C}^*$ show that $\mathscr{C}$ is an orthoprojector. The theorem is proved.

From this theorem there follows at once

THEOREM 10.2 (L. A. SAHNOVIČ [5]). *If the imaginary component $A_{\mathscr{I}}$ of a Volterra operator A belongs to $\mathfrak{S}_2$, then its real component $A_{\mathscr{R}}$ also belongs to $\mathfrak{S}_2$, and*

$$|A_{\mathscr{I}}|_2^2 = |A_{\mathscr{R}}|_2^2 = |A|_2^2/2. \tag{10.6}$$

PROOF. In fact, if $\mathfrak{P}$ is a maximal eigenchain of A, then by Theorem 5.1

$$A = 2i \int_{\mathfrak{P}} P A_{\mathscr{I}} \, dP.$$

It follows by Theorem 10.1 that the operators A and $A_{\mathscr{R}}$ belong to $\mathfrak{S}_2$. Let $\mathfrak{z} = \{P_j\}_0^n$ be any partition of the chain $\mathfrak{P}$. In the course of proving Theorem 5.1 it was shown that

$$A = \sum_{j=1}^{n} \Delta P_j A \Delta P_j + 2i \sum_{j<k} \Delta P_j A_{\mathscr{I}} \Delta P_k;$$

hence

$$A_{\mathscr{R}} = \frac{1}{2}(A + A^*) = \sum_{j=1}^{n} \Delta P_j A_{\mathscr{R}} \Delta P_j + i \sum_{j<k} (\Delta P_j A_{\mathscr{I}} \Delta P_k - \Delta P_k A_{\mathscr{I}} \Delta P_j).$$

Further calculation yields

$$|A_{\mathscr{R}}|_2^2 = \sum_{j=1}^{n} |\Delta P_j A_{\mathscr{R}} \Delta P_j|_2^2 + |A_{\mathscr{I}}|_2^2 - \sum_{j=1}^{n} |\Delta P_j A_{\mathscr{I}} \Delta P_j|_2^2.$$

Passing to the limit along the chain $\mathfrak{P}$ in this relation, we obtain $|A_{\mathscr{R}}| = |A_{\mathscr{I}}|$.

Finally, for any operator $A \in \mathfrak{S}_2$ one has

$$\begin{aligned} |A|_2^2 &= \operatorname{sp}(AA^*) = \operatorname{sp}(A_{\mathscr{I}}^2 + A_{\mathscr{R}}^2 + i(A_{\mathscr{I}} A_{\mathscr{R}} - A_{\mathscr{R}} A_{\mathscr{I}})) \\ &= \operatorname{sp} A_{\mathscr{I}}^2 + \operatorname{sp} A_{\mathscr{R}}^2 = |A_{\mathscr{I}}|_2^2 + |A_{\mathscr{R}}|_2^2. \end{aligned}$$

Here we have used the fact that $\operatorname{sp} XY = \operatorname{sp} YX$ for any $X, Y \in \mathfrak{S}_2$. The theorem is proved.

3. With each bordered continuous chain $\mathfrak{P}$ we associate the transformator $\mathscr{C}(X: \mathfrak{P})$, acting in $\mathfrak{S}_2$ according to the formula

$$\mathscr{C}(X; \mathfrak{P}) = \int_{\mathfrak{P}} PX\,dP. \tag{10.7}$$

We shall call the transformator $\mathscr{C}(X; \mathfrak{P})$ the transformator of *triangular truncation* with respect to the chain $\mathfrak{P}$. In those statements where the individual properties of the chain $\mathfrak{P}$ play no role, we shall omit reference to it and use the notation $\mathscr{C}$.

According to Theorem 10.1, $\mathscr{C}$ is an orthogonal projector which projects $\mathfrak{S}_2$ onto the subspace of all Volterra operators in $\mathfrak{S}_2$ which have $\mathfrak{P}$ as an eigenchain.

We note that the projector $\mathscr{C}^{\perp} = I - \mathscr{C}$ has the form

$$\mathscr{C}^{\perp}(X; \mathfrak{P}) = \int_{\mathfrak{P}} dP\,XP.$$

This follows directly from the formula for integration by parts (see Lemma 6.1).

It is easily verified that

$$\mathscr{C}^{\perp}(X; \mathfrak{P}) = \mathscr{C}(X; \mathfrak{P}^{\perp}) \quad (X \in \mathfrak{S}_2),$$

where $\mathfrak{P}^{\perp}$ is the chain of all projectors $P^{\perp} = I - P$ $(P \in \mathfrak{P})$.

Along with the transformators $\mathscr{C}$ and $\mathscr{C}^{\perp}$ we shall consider the transformator $\mathscr{S}'$, defined by

$$\mathscr{S}'(X; \mathfrak{P}) = \int_{\mathfrak{P}} (PX\,dP - dP\,XP),$$

so that $\mathscr{S}' = \mathscr{C} - \mathscr{C}^{\perp}$.

The transformator $\mathscr{S}'$ is obviously *selfadjoint* and *unitary*:

$$(\mathscr{S}')^*\mathscr{S}' = \mathscr{S}'(\mathscr{S}')^* = (\mathscr{S}')^2 = \mathscr{C} + \mathscr{C}^{\perp} = I.$$

Thus, $\mathscr{S}'$ is the operator of reflection with respect to the subspace $\mathfrak{R}(\mathscr{C})$; it leaves the projection $\mathscr{C}(X)$ of any operator $X \in \mathfrak{S}_2$ unchanged, and changes the sign of the orthogonal component $\mathscr{C}^{\perp}(X)$.

Further on we shall also need the transformator $\mathscr{S} = i\mathscr{S}'$. The transformator $\mathscr{S} = \mathscr{S}(\cdot\,; \mathfrak{P})$ is real with respect to the involutive operation of Hermitian conjugation, i.e. $[\mathscr{S}(X)]^* = \mathscr{S}(X^*)$. In fact, since for any operator $X \in \mathfrak{S}_2$

$$[\mathscr{C}(X)]^* = \mathscr{C}^{\perp}(X^*) \quad \text{and} \quad [\mathscr{C}^{\perp}(X)]^* = \mathscr{C}(X^*),$$

we have

$$[\mathscr{S}(X)]^* = \mathscr{S}(X^*) \quad (X \in \mathfrak{S}_2).$$

It is easily seen that the transformator $\mathscr{S}(\cdot; \mathfrak{P})$ carries every selfadjoint operator $X \in \mathfrak{S}_2$ into a selfadjoint operator $Y = \mathscr{S}(X; \mathfrak{P})$ which is the real component of a Volterra operator having $\mathfrak{P}$ as an eigenchain and imaginary component equal to X. Obviously the transformator $\mathscr{S}$ is *skewsymmetric*.

CHAPTER II

GENERAL THEOREMS ON TRANSFORMATORS[1)]

In papers by the authors [**3-5**] and V. I. Macaev [**2, 3**] it was shown that the abstract triangular representation of Volterra operators leads naturally to a number of relations which connect the eigenvalues of the imaginary and real components of such operators. These results have been further essentially developed in a paper by V. I. Macaev [**3**]. Although in their derivation use was made of certain specific properties of the transformator of triangular truncation, it turned out that in principle these results can be generalized to any transformator. By such a generalization the nature of these relations becomes more transparent, and at the same time their derivation simplifies.

It turns out that if $\mathscr{T}$ is some transformator[2)] and $X \in \mathfrak{S}_2$, then there exists a series of dependencies, expressed by linear inequalities and determined solely by $\mathscr{T}$, between the s-numbers of the operators X and $Y = \mathscr{T}(X)$. Three general theorems concerning these dependencies will be established. These theorems find application in the general theory of the transformator of triangular truncation, studied in Chapter III. It is of no less interest, however, that in conjunction with certain general results of the theory of canonical systems of differential equations with periodic coefficients these theorems make it possible to obtain a number of exact criteria for the stable boundedness of solutions of such systems. These applications are discussed in Chapter VI.

The results discussed in the present chapter (except for §§1 and 6) are taken from the authors' paper [**6**].

[1)] The results of this chapter clarify the general aspects of different methods which are applicable for the study of triangular truncation. Nevertheless, this chapter can be omitted at first reading. The reading of Chapter III may serve to reawaken interest in the results of the present chapter.

[2)] We recall that a transformator is a linear operator acting in $\mathfrak{S}_2$.

§1. s-numbers of completely continuous operators and their properties

In this section we present without proof the basic results of the theory of the s-numbers of completely continuous operators, which will be used in the sequel. We owe the origin of this theory to the studies of E. Schmidt, H. Weyl and subsequent work by A. Horn, Ky Fan and others. Proofs of these results together with references to original sources may be found in the authors' book ([GK], Chapter II).

1. If $A \in \mathfrak{S}_\infty$, then $H = (A^*A)^{1/2} \in \mathfrak{S}_\infty$. The eigenvalues of the operator H are called the *s-numbers* of the operator A. We shall enumerate the nonzero s-numbers in decreasing order, taking their multiplicities into account, so that

$$s_j(A) = \lambda_j(H) \quad (j = 1, 2, \cdots, \dim H),$$

where $\dim H$ is the dimension of the operator H. If $r = \dim H < \infty$, then

$$s_j(A) = 0 \quad (j = r+1, \cdots).$$

If A is selfadjoint or, more generally, normal, then

$$s_j(A) = |\lambda_j(A)| \quad (j = 1, 2, \cdots).$$

For any operator $A \in \mathfrak{S}_\infty$

$$s_j(A) = s_j(A^*) \quad (j = 1, 2, \cdots),$$

and for any bounded operator B

$$s_j(BA) \leqq |B| s_j(A) \quad (j = 1, 2, \cdots),$$

$$s_j(AB) \leqq |B| s_j(A) \quad (j = 1, 2, \cdots).$$

Every operator $A \in \mathfrak{S}_\infty$ can be expanded in a Schmidt series

$$A = \sum_j s_j(A)(\cdot, \phi_j)\psi_j,$$

where $\{\phi_j\}$ and $\{\psi_j\}$ are certain orthonormal systems.

The s-numbers of an operator $A \in \mathfrak{S}_\infty$ have the following approximation properties:

$$s_{n+1}(A) = \min_{K \in \mathfrak{K}_n} |A - K|, \tag{1.1}$$

where $\mathfrak{K}_n$ $(n = 1, 2, \cdots)$ is the set of all finite-dimensional operators of dimension $\leqq n$.

From this property one deduces the relation

$$s_{n+r}(A) \leqq s_n(A+K) \leqq s_{n-r}(A)$$

for any operators $A \in \mathfrak{S}_\infty$ and $K \in \mathfrak{K}_r$, and also the relations

$$s_{m+n-1}(A+B) \leqq s_m(A) + s_n(B) \quad (m,n = 1,2,\cdots), \tag{1.2}$$

$$s_{m+n-1}(AB) \leqq s_m(a) s_n(B) \quad (m,n = 1,2,\cdots), \tag{1.3}$$

which hold for any operators $A, B \in \mathfrak{S}_\infty$.

We also need the following asymptotic property of s-numbers.

1. *Suppose that for the operators $A, B \in \mathfrak{S}_\infty$ and for some $r > 0$*

$$\lim_{n\to\infty} n^r s_n(A) = a \quad \textit{and} \quad \lim_{n\to\infty} n^r s_n(B) = 0. \tag{1.4}$$

Then

$$\lim_{n\to\infty} n^r s_n(A+B) = a\,. \tag{1.5}$$

2. The eigenvalues $\lambda_j(A)$ and the s-numbers $s_j(A)$ of an operator $A \in \mathfrak{S}_\infty$ are connected by the relations

$$|\lambda_1(A)\lambda_2(A)\cdots\lambda_n(A)| \leqq s_1(A)s_2(A)\cdots s_n(A) \quad (n = 1,2,\cdots), \tag{1.6}$$

where $\{\lambda_j(A)\}$ denotes the sequence consisting of the nonzero eigenvalues of A, arbitrarily enumerated according to decreasing modulus, where in this enumeration each eigenvalue is counted as many times as its algebraic multiplicity.[3)]

The sums of the first n s-numbers of an operator $A \in \mathfrak{S}_\infty$ have extremal properties which are expressed by the relations

$$\sum_{j=1}^n s_j(A) = \max \left| \sum_{j=1}^n (UA\phi_j, \phi_j) \right| \quad (n = 1,2,\cdots),$$

where the maximum is taken over all unitary operators U and orthonormal systems $\{\phi_j\}_1^n$.

This property enables one to deduce important relations for the s-numbers of sums and products of operators:

$$\sum_{j=1}^n s_j(A+B) \leqq \sum_{j=1}^n s_j(A) + \sum_{j=1}^n s_j(B) \quad (n = 1,2,\cdots), \tag{1.7}$$

[3)] We recall that the *algebraic multiplicity* of an eigenvalue $\lambda_0 \neq 0$ of an operator $A \in \mathfrak{S}_\infty$ is the dimension of the subspace of all root vectors of A corresponding to λ_0, i.e. the dimension of the subspace of all solutions of the equations $(A - \lambda_0 I)^n \phi = 0$ $(n = 1,2,\cdots)$. For details concerning this concept see the book [GK], Chapter I.

$$\prod_{j=1}^{n} s_j(AB) \leqq \prod_{j=1}^{n} s_j(A) \prod_{j=1}^{n} s_j(B) \quad (n = 1, 2, \cdots). \tag{1.8}$$

From (1.6), (1.7) and (1.8) one can immediately obtain a whole series of new relations, if one makes use of the following result.

2. *Let $\Phi(x)\,(-\infty < x < \infty)$ be a convex function which tends to zero as $x \to -\infty$, and let $\{a_j\}$ and $\{b_j\}$ be nonincreasing sequences of real numbers such that*

$$\sum_{j=1}^{n} a_j \leqq \sum_{j=1}^{n} b_j \quad (n = 1, 2, \cdots). \tag{1.9}$$

Then

$$\sum_{j=1}^{n} \Phi(a_j) \leqq \sum_{j=1}^{n} \Phi(b_j). \tag{1.10}$$

In particular, from this result and from (1.6) follows the relation

$$\sum_{j=1}^{n} |\lambda_j(A)| \leqq \sum_{j=1}^{n} s_j(A) \quad (A \in \mathfrak{S}_\infty; n = 1, 2, \cdots), \tag{1.11}$$

and also the relation

$$\sum_{j=1}^{n} s_j(AB) \leqq \sum_{j=1}^{n} s_j(A)\, s_j(B) \quad (A, B \in \mathfrak{S}_\infty;\ n = 1, 2, \cdots). \tag{1.12}$$

§2. Auxiliary lemmas

Let $\mathfrak{B}_m$ $(m = 1, 2, \cdots)$ denote the set of all partially isometric operators of dimension $\leqq m$, $\hat{\mathfrak{B}}_m$ the set of all selfadjoint operators[4] from $\mathfrak{B}_m$, and $\mathfrak{P}_m$ $(m = 1, 2, \cdots)$ the set of all orthoprojectors of dimension $\leqq m$.

LEMMA 2.1. *For any operator $A \in \mathfrak{S}_\infty$*

$$\sum_{j=1}^{m} s_j(A) = \max_{K \in \mathfrak{K}_m;\, |K| \leqq 1} |\mathrm{sp}(KA)| = \max_{V \in \mathfrak{B}_m} |\mathrm{sp}(VA)| \quad (m = 1, 2, \cdots). \tag{2.1}$$

If A is a selfadjoint operator, then

$$\sum_{j=1}^{m} s_j(A) = \max_{V \in \hat{\mathfrak{B}}_m} |\mathrm{sp}(VA)| \quad (m = 1, 2, \cdots).$$

If A is a nonnegative operator, then

---

[4] It is easily seen that if $X \in \hat{\mathfrak{B}}_m$, then X is either an orthoprojector of dimension $\leqq m$ or the difference of two orthoprojectors, the sum of whose dimensions is $\leqq m$.

$$\sum_{j=1}^{m} s_j(A) = \max_{P \in \mathfrak{P}_m} \operatorname{sp}(PA) \quad (m = 1, 2, \cdots).$$

PROOF. According to (1.11) and (1.12),

$$|\operatorname{sp}(KA)| \leqq \sum_{j=1}^{\infty} s_j(KA) \leqq \sum_{j=1}^{\infty} s_j(K) s_j(A) \quad (K \in \mathfrak{K}_m).$$

Taking into account that for $K \in \mathfrak{K}_m$ and $|K| \leqq 1$

$$s_j(K) \leqq 1 \ (j = 1, 2, \cdots, m) \quad \text{and} \quad s_j(K) = 0 \ (j = m+1, m+2, \cdots),$$

we obtain

$$|\operatorname{sp}(KA)| \leqq \sum_{j=1}^{m} s_j(A) \quad (K \in \mathfrak{K}_m;\ |K| \leqq 1;\ m = 1, 2, \cdots).$$

Let

$$A = \sum_j s_j(A)(\cdot, \phi_j)\psi_j$$

be the Schmidt expansion of A. We form the m-dimensional partially isometric operator

$$V = \sum_{j=1}^{m} (\cdot, \psi_j)\phi_j.$$

We have

$$VA = \sum_{j=1}^{m} s_j(A)(\cdot, \phi_j)\phi_j,$$

and so

$$\operatorname{sp}(VA) = \sum_{j=1}^{m} s_j(A).$$

If A is selfadjoint, then obviously $\psi_j = \pm \phi_j$, and hence V is selfadjoint. If A is nonnegative, then $\psi_j = \phi_j$, and V is consequently an orthoprojector.

The lemma is proved.

For the case of a selfadjoint operator A, Lemma 2.1 can be supplemented by the following.

LEMMA 2.2. *For any selfadjoint operator* $A \in \mathfrak{S}_\infty$

$$\max_{P\in\mathfrak{P}_m} \operatorname{sp}(PA) = \sum_{j=1}^{m} \lambda_j^+(A) \quad (m = 1, 2, \cdots), \tag{2.2}$$

$$\min_{P\in\mathfrak{P}_m} \operatorname{sp}(PA) = -\sum_{j=1}^{m} \lambda_j^-(A) \quad (m = 1, 2, \cdots). \tag{2.3}$$

PROOF. We represent the operator A in the form $A = A_+ - A_-$, where $A_\pm$ are mutually orthogonal nonnegative operators. We recall that $\lambda_j^\pm(A) = \lambda_j(A_\pm)$. Since the trace of the product of two nonnegative operators is nonnegative (see [GK], Theorem III.8.3), it follows that

$$\operatorname{sp}(PA) = \operatorname{sp}(PA_+) - \operatorname{sp}(PA_-) \leqq \operatorname{sp}(PA_+),$$

and so, using Lemma II.4.2 from [GK], we obtain

$$\operatorname{sp}(PA) \leqq \sum_{j=1}^{m} \lambda_j^+(A).$$

Let P_0 be the orthogonal projector which projects the space $\mathfrak{H}$ onto the subspace $\mathfrak{L}$ spanned by all the eigenvectors of A corresponding to the first m positive eigenvalues. If the number of positive eigenvalues of A is less than m, we take $\mathfrak{L}$ equal to the linear hull of all the eigenvectors of A corresponding to its positive eigenvalues. Obviously

$$\operatorname{sp}(P_0A) = \sum_{j=1}^{m} \lambda_j^+(A).$$

Thus (2.2) is proved. The proof of (2.3) is analogous.

§3. A theorem on transformators

As has already been indicated, we shall consider continuous linear operators $\mathscr{T}$ acting in the Hilbert space $\mathfrak{S}_2$, which, in contrast with operators $X \in \mathfrak{S}_2$, we shall call *transformators*.

As usual, we denote the norm of $\mathscr{T}$ by $|\mathscr{T}|$, so that

$$|\mathscr{T}| = \sup_{X\in\mathfrak{S}_2} (|\mathscr{T}(X)|_2/|X|_2).$$

For any $X, Y \in \mathfrak{S}_2$ we obviously have

$$|\operatorname{sp}(\mathscr{T}(X)Y)| \leqq |\mathscr{T}|\,|X|_2|Y|_2.$$

If $K \in \mathfrak{R}_n$ and $|K|_\infty \leqq 1$, then obviously

$$|K|_2 \leqq \sqrt{n},$$

from which it follows that for any $K \in \mathfrak{R}_n$ and $L \in \mathfrak{R}_m$ with $|K|_\infty \leqq 1$, $|L|_\infty \leqq 1$,[5)]

$$|\operatorname{sp}(\mathscr{T}(K)L)| \leqq \sqrt{mn}\, |\mathscr{T}|.$$

For a given transformator $\mathscr{T}$ we put

$$C_{mj} = C_{mj}(\mathscr{T}) = \sup_{\substack{L \in \mathfrak{R}_m;\, K \in \mathfrak{R}_j \\ |L| \leqq 1;\, |K| \leqq 1}} |\operatorname{sp}(L^* \mathscr{T}(K))| \quad (m, j = 1, 2, \cdots). \tag{3.1}$$

The definition (3.1) of the constants C_{mj} is equivalent to each of the following:

$$C_{mj}(\mathscr{T}) = \sup_{K \in \mathfrak{R}_j;\, |K| \leqq 1} \sum_{n=1}^{m} s_n[\mathscr{T}(K)] \quad (m, j = 1, 2, \cdots), \tag{3.2}$$

$$C_{mj}(\mathscr{T}) = \sup_{V \in \mathfrak{B}_m;\, W \in \mathfrak{B}_j} |\operatorname{sp}(V^* \mathscr{T}(W))| \quad (m, j = 1, 2, \cdots), \tag{3.3}$$

where $\mathfrak{B}_j$ $(\subset \mathfrak{R}_j)$ $(j = 1, 2, \cdots)$, as before, denotes the set of all partially isometric operators of dimension $\leqq j$.

The equivalence of the definitions (3.1) and (3.2) is a simple consequence of Lemma 2.1.

Let us prove the equivalence of the definitions (3.1) and (3.3). It is obviously sufficient to prove that the constant C_{mj} defined by (3.3) is not less than the constant C_{mj} defined by (3.1). According to the definition (3.1), for any $\epsilon > 0$ there exist operators $L_0 \in \mathfrak{R}_m$ and $K_0 \in \mathfrak{R}_j$ with norms $\leqq 1$ such that

$$|\operatorname{sp}(L_0^* \mathscr{T}(K_0))| \geqq C_{mj} - \epsilon.$$

By Lemma 2.1 there exists an operator $V \in \mathfrak{B}_m$ for which

$$\operatorname{sp}(V^* \mathscr{T}(K_0)) \geqq C_{mj} - \epsilon.$$

Since

$$\operatorname{sp}(V^* \mathscr{T}(K_0)) = \operatorname{sp}([\mathscr{T}^*(V)]^* K_0),$$

where $\mathscr{T}^*$ is the transformator adjoint to $\mathscr{T}$, by Lemma 2.1 again there exists $W \in \mathfrak{B}_j$ such that

$$\operatorname{sp}([\mathscr{T}^*(V)]^* W) \geqq \operatorname{sp}([\mathscr{T}^*(V)]^* K_0).$$

Thus

---

5) We recall that for the case of completely continuous operators the ordinary norm $|\cdot|$ is also denoted by $|\cdot|_\infty$.

$$\mathrm{sp}(V^* \mathcal{T}(W)) \geqq C_{mj} - \epsilon.$$

The assertion is proved.

If $\mathcal{T}^*$ is the adjoint of $\mathcal{T}$,

$$\mathrm{sp}(\mathcal{T}(X)Y^*) = \mathrm{sp}(X[\mathcal{T}^*(Y)]^*),$$

then obviously

$$C_{mj}(\mathcal{T}) = C_{jm}(\mathcal{T}^*) \quad (j, m = 1, 2, \cdots). \tag{3.4}$$

In particular, if $\mathcal{T}$ is selfadjoint, i.e.

$$\mathrm{sp}(\mathcal{T}(X)Y^*) = \mathrm{sp}(X[\mathcal{T}(Y)]^*)$$

for all $X, Y \in \mathfrak{S}_2$, then

$$C_{mj}(\mathcal{T}) = C_{jm}(\mathcal{T}) \quad (j, m = 1, 2, \cdots). \tag{3.5}$$

It is also clear from the definition of the constants C_{mj} that

$$C_{mj} \leqq C_{m+1,j}, \quad C_{mj} \leqq C_{m,j+1} \quad (j, m = 1, 2, \cdots). \tag{3.6}$$

We note that the C_{mj} are subadditive with respect to each index, i.e.

$$C_{m,p+q} \leqq C_{mp} + C_{mq} \quad (m, p, q = 1, 2, \cdots), \tag{3.7}$$

$$C_{p+q,j} \leqq C_{pj} + C_{qj} \quad (j, p, q = 1, 2, \cdots). \tag{3.8}$$

In fact, if $W \in \mathfrak{B}_{p+q}$, it can be represented in the form $W = W_1 + W_2$, where $W_1 \in \mathfrak{B}_p$ and $W_2 \in \mathfrak{B}_q$. For any $V \in \mathfrak{B}_m$ we will then have

$$|\mathrm{sp}(\mathcal{T}(V)W^*)| \leqq |\mathrm{sp}(\mathcal{T}(V)W_1^*)| + |\mathrm{sp}(\mathcal{T}(V)W_2^*)| \leqq C_{mp} + C_{mq},$$

which implies (3.7). The relation (3.8) follows from (3.7) and (3.4). Of course, it can just as easily be proved directly.

The relations (3.7) and (3.8) imply, in obvious fashion, the estimates

$$C_{mj} \leqq jC_{m1} \quad \text{and} \quad C_{mj} \leqq mC_{1j}. \tag{3.9}$$

THEOREM 3.1. *Let $\mathcal{T}$ be a continuous transformator acting in $\mathfrak{S}_2$. Then for any $A \in \mathfrak{S}_2$, with $B = \mathcal{T}(A)$, we have the inequalities*

$$\sum_{k=1}^{m} s_k(B) \leqq \sum_{j=1}^{\infty} [C_{mj}(\mathcal{T}) - C_{m,j-1}(\mathcal{T})]\, s_j(A) \quad (m = 1, 2, \cdots), \tag{3.10}$$

where we take $C_{m0}(\mathcal{T}) = 0$ $(m = 1, 2, \cdots)$. In particular,

$$|\mathcal{T}(A)| \leqq \sum_{j=1}^{\infty} [C_{1j}(\mathcal{T}) - C_{1,j-1}(\mathcal{T})]\, s_j(A).$$

PROOF. The Schmidt series of $A \in \mathfrak{S}_2$,

$$A = \sum_{j=1}^{\infty} s_j(A)(\cdot, \phi_j)\psi_j,$$

converges in the norm of $\mathfrak{S}_2$. This expansion of A generates the convergent in $\mathfrak{S}_2$ expansion

$$A = \sum_{j=1}^{\infty} [s_j(A) - s_{j+1}(A)] V_j, \tag{3.11}$$

where

$$V_j = \sum_{k=1}^{j} (\cdot, \phi_k)\psi_k \in \mathfrak{B}_j \quad (j = 1, 2, \cdots).$$

From (3.11) we obtain a convergent in $\mathfrak{S}_2$ expansion for the operator $B = \mathscr{T}(A)$, namely,

$$B = \sum_{j=1}^{\infty} [s_j(A) - s_{j+1}(A)] \mathscr{T}(V_j). \tag{3.12}$$

We now write the Schmidt expansion for B:

$$B = \sum_{j=1}^{\infty} s_j(B)(\cdot, \chi_j)\omega_j,$$

where $\{\chi_j\}_1^\infty$ and $\{\omega_j\}_1^\infty$ are certain orthonormal systems, and put

$$W_m = \sum_{j=1}^{m} (\cdot, \omega_j)\chi_j \ (\in \mathfrak{B}_m).$$

Obviously

$$\mathrm{sp}(BW_m) = \sum_{j=1}^{m} s_j(B);$$

comparing this with (3.12), we obtain

$$\sum_{k=1}^{m} s_k(B) = \sum_{j=1}^{\infty} [s_j(A) - s_{j+1}(A)] \mathrm{sp}(\mathscr{T}(V_j) W_m),$$

and consequently

$$\sum_{k=1}^{m} s_k(B) \leqq \sum_{j=1}^{\infty} [s_j(A) - s_{j+1}(A)] C_{mj}(\mathscr{T}). \tag{3.13}$$

Regrouping the terms in the series on the right, we arrive at the required relation.

In this regrouping—Abel's transformation—a 'remainder term' $s_{j+1}(A)C_{mj}(\mathscr{T})$ appears. This terms tends to 0 as $j\to\infty$, since for any $A\in\mathfrak{S}_2$ one has the estimate $s_j(A)=o(j^{-1/2})$, while for the constants $C_{mj}(\mathscr{T})$ one has the estimate $C_{mj}(\mathscr{T})\leqq\sqrt{mj}\,|\mathscr{T}|$.

The theorem is proved.

COROLLARY 3.1. *For $A\in\mathfrak{S}_2$ and $B=\mathscr{T}(A)$ one has the inequality*

$$\sum_{k=1}^{m}s_k(B)\leqq C_{m1}(\mathscr{T})\sum_{j=1}^{\infty}s_j(A)\quad(m=1,2,\cdots). \tag{3.14}$$

If, moreover, the transformator $\mathscr{T}$ is selfadjoint, then

$$\sum_{k=1}^{m}s_k(B)\leqq C_{m1}(\mathscr{T})\sum_{j=1}^{m}s_j(A)+m\sum_{j=m+1}^{\infty}(C_{j,1}(\mathscr{T})-C_{j-1,1}(\mathscr{T}))s_j(A)\quad(m=1,2,\cdots). \tag{3.15}$$

In fact, the relation (3.14) is obtained from (3.10) if one uses the subadditivity of the C_{mj} with respect to the index j.

The second relation, (3.15), is obtained from (3.13) if we replace the first m constants C_{mj} $(j=1,2,\cdots,m)$ by the numbers jC_{m1}, and the rest of the C_{mj} by the numbers mC_{1j}, taking into account that $C_{1j}=C_{j1}$.

REMARK 3.1. For a *normal* operator $A\in\mathfrak{S}_2$ the bound of Theorem 3.1 can be strengthened. Namely, it can be replaced by the following:

$$\sum_{k=1}^{m}s_k(B)\leqq\sum_{j=1}^{\infty}[C'_{mj}(\mathscr{T})-C'_{m,j-1}(\mathscr{T})]s_j(A)\quad(m=1,2,\cdots), \tag{3.16}$$

where the constants C'_{mj} are defined by

$$C'_{mj}(\mathscr{T})=\sup_{P\in\mathfrak{P}_j;\,V\in\mathfrak{B}_m}|\mathrm{sp}(V^*\mathscr{T}(P))|$$

or the equivalent relation

$$C'_{mj}(\mathscr{T})=\sup_{P\in\mathfrak{P}_j}\sum_{k=1}^{m}s_k(\mathscr{T}(P)).$$

In the above relations $\mathfrak{P}_j$ $(j=1,2,\cdots)$ denotes, as before, the set of all orthoprojectors of dimension $\leqq j$. Thus $C'_{mj}(\mathscr{T})\leqq C_{mj}(\mathscr{T})$.

The proof of the estimate (3.16) is carried out in the same way as that of the analogous relation in Theorem 3.1.

Let us also note the estimate

(3.17) $$\sum_{k=1}^{m} s_k(B) \leqq C'_{m1}(\mathscr{T}) \sum_{j=1}^{\infty} s_j(A),$$

which follows from (3.16).

§4. Two theorems on real transformators

A continuous transformator $\mathscr{T}$ acting in $\mathfrak{S}_2$ will be called *real* if

(4.1) $$[\mathscr{T}(X)]^* = \mathscr{T}(X^*)$$

for every $X \in \mathfrak{S}_2$. In other words, such a transformator is characterized by the fact that it carries every selfadjoint operator into a selfadjoint operator.

The transformator $\mathscr{T}^*$, the adjoint of a real transformator $\mathscr{T}$, is likewise real. In fact, for $X, Y \in \mathfrak{S}_2$

(4.2) $$\mathrm{sp}(\mathscr{T}^*(X^*)Y^*) = \mathrm{sp}(X^*[\mathscr{T}(Y)]^*),$$

and so, using (4.1),

$$\mathrm{sp}(\mathscr{T}^*(X^*)Y^*) = \mathrm{sp}(X^*\mathscr{T}(Y^*)).$$

Since

(4.3) $$\mathrm{sp}(X^*\mathscr{T}(Y^*)) = \mathrm{sp}([\mathscr{T}^*(X)]^*Y^*),$$

comparing (4.2) and (4.3) we obtain

$$\mathrm{sp}(\mathscr{T}^*(X^*)Y^*) = \mathrm{sp}([\mathscr{T}^*(X)]^*Y^*) \quad (X, Y \in \mathfrak{S}_2),$$

from which follows

$$\mathscr{T}^*(X^*) = [\mathscr{T}^*(X)]^*.$$

We shall call a real selfadjoint transformator a real *symmetric* transformator. For such a transformator we obviously have

$$\mathrm{sp}(\mathscr{T}(X)Y) = \mathrm{sp}(X\mathscr{T}(Y)) \quad (X, Y \in \mathfrak{S}_2).$$

We shall call a real transformator $\mathscr{T}$ real *skewsymmetric* if $\mathscr{T}^* = -\mathscr{T}$. Obviously, such a transformator is characterized by the relation

$$\mathrm{sp}(\mathscr{T}(X)Y) = -\,\mathrm{sp}(X\mathscr{T}(Y)) \quad (X, Y \in \mathfrak{S}_2).$$

For a given real transformator $\mathscr{T}$ we put

(4.4) $$C^+_{mj} = C^+_{mj}(\mathscr{T}) = \sup_{Q \in \mathfrak{P}_m;\, P \in \mathfrak{P}_j} \mathrm{sp}[Q\mathscr{T}(P)] \quad (m, j = 1, 2, \cdots)$$

and

$$C_{mj}^{-} = C_{mj}^{-}(\mathscr{T}) = -\inf_{Q\in\mathfrak{P}_m;\,P\in\mathfrak{P}_j} \operatorname{sp}[Q\mathscr{T}(P)] \quad (m,j = 1,2,\cdots), \tag{4.5}$$

where $\mathfrak{P}_m$ is the set of all orthoprojectors of dimension $\leqq m$.

According to Lemma 2.2, the definitions (4.4) and (4.5) of the constants $C_{mj}^{\pm}$ are equivalent to the following:

$$C_{mj}^{+} = \sup_{P\in\mathfrak{P}_j} \sum_{k=1}^{m} \lambda_k^{+}[\mathscr{T}(P)] \quad \text{and} \quad C_{mj}^{-} = \sup_{P\in\mathfrak{P}_j} \sum_{k=1}^{m} \lambda_k^{-}[\mathscr{T}(P)].$$

The constants $C_{mj}^{\pm}$, for a transformator of a special type (the skewsymmetric transformator $\mathscr{S}$) and for all m, j, were first introduced by V. I. Macaev [**3**], who established their basic properties. Estimates of these constants for the transformators $\mathscr{S}$ and $\mathscr{C}$ are given in §6.

For the transformator $\mathscr{T}^*$, the adjoint of $\mathscr{T}$,

$$C_{mj}^{\pm}(\mathscr{T}) = C_{jm}^{\pm}(\mathscr{T}^*) \quad (j,m = 1,2,\cdots).$$

In particular, if $\mathscr{T}$ is a symmetric transformator, then

$$C_{mj}^{\pm}(\mathscr{T}) = C_{jm}^{\pm}(\mathscr{T}) \quad (j,m = 1,2,\cdots). \tag{4.6}$$

If $\mathscr{T}$ is skewsymmetric, then

$$C_{mj}^{\pm}(\mathscr{T}) = C_{jm}^{\mp}(\mathscr{T}) \quad (j,m = 1,2,\cdots). \tag{4.7}$$

The constants $C_{mj}^{\pm}$ are monotone and subadditive with respect to each index:

I) $$C_{mj}^{\pm} \leqq C_{m+1,j}^{\pm};\ C_{mj}^{\pm} \leqq C_{m,j+1}^{\pm};$$

II) $$C_{m,p+q}^{\pm} \leqq C_{mp}^{\pm} + C_{mq}^{\pm};\ C_{p+q,j}^{\pm} \leqq C_{pj}^{\pm} + C_{qj}^{\pm},$$

where $p,q,m,j = 1,2,\cdots$.

The proof of these properties is analogous to the proof of the corresponding properties of the constants C_{mj}.

Property II implies, in evident fashion, the estimates

$$C_{mj}^{\pm} \leqq jC_{m1}^{\pm} \quad \text{and} \quad C_{mj}^{\pm} \leqq mC_{1j}^{\pm} \quad (j,m = 1,2,\cdots).$$

THEOREM 4.1. *Let $\mathscr{T}$ be a real transformator. Then for any selfadjoint operator $A \in \mathfrak{S}_2$ and $B = \mathscr{T}(A)$ we have the inequalities*

$$\begin{aligned}\sum_{k=1}^{m} \lambda_k^{+}(B) \leqq{}& \sum_{j=1}^{\infty} (C_{mj}^{+} - C_{m,j-1}^{+})\lambda_j^{+}(A)\\ &+ \sum_{j=1}^{\infty} (C_{mj}^{-} - C_{m,j-1}^{-})\lambda_j^{-}(A) \quad (m = 1,2,\cdots).\end{aligned} \tag{4.8}$$

Similarly,

$$\sum_{k=1}^{m}\lambda_k^-(B) \leqq \sum_{j=1}^{\infty}(C_{mj}^- - C_{m,j-1}^-)\lambda_j^+(A) + \sum_{j=1}^{\infty}(C_{mj}^+ - C_{m,j-1}^+)\lambda_j^-(A) \quad (m = 1,2,\cdots), \tag{4.9}$$

where

$$C_{mj}^{\pm} = C_{mj}^{\pm}(\mathscr{T}) \ (j,m = 1,2,\cdots) \quad \textit{and} \quad C_{m0} = 0 \ (m = 1,2,\cdots).$$

PROOF. Any selfadjoint operator $A \in \mathfrak{S}_2$ has a spectral decomposition

$$A = \sum_{j=1}^{\infty}\lambda_j^+(A)(\cdot,\phi_j^+)\phi_j^+ - \sum_{j=1}^{\infty}\lambda_j^-(A)(\cdot,\phi_j^-)\phi_j^-,$$

which converges in the norm of $\mathfrak{S}_2$. This decomposition gives rise to the convergent in $\mathfrak{S}_2$ expansion

$$A = \sum_{j=1}^{\infty}[\lambda_j^+(A) - \lambda_{j+1}^+(A)]P_j^+ - \sum_{j=1}^{\infty}[\lambda_j^-(A) - \lambda_{j+1}^-(A)]P_j^-, \tag{4.10}$$

where

$$P_j^{\pm} = \sum_{k=1}^{j}(\cdot,\phi_k^{\pm})\phi_k^{\pm} \in \mathfrak{P}_j \ (j = 1,2,\cdots).$$

Applying the transformator $\mathscr{T}$ to both sides of (4.10), we obtain

$$B = \mathscr{T}(A) = \sum_{j=1}^{\infty}[\lambda_j^+(A) - \lambda_{j+1}^+(A)]\mathscr{T}(P_j^+) - \sum_{j=1}^{\infty}[\lambda_j^-(A) - \lambda_{j+1}^-(A)]\mathscr{T}(P_j^-). \tag{4.11}$$

Let

$$B = \sum_{j=1}^{\infty}\lambda_j^+(B)(\cdot,\psi_j^+)\psi_j^+ - \sum_{j=1}^{\infty}\lambda_j^-(B)(\cdot,\psi_j^-)\psi_j^-$$

be the spectral decomposition of B. We put

$$Q_m^+ = \sum_{j=1}^{m}(\cdot,\psi_j^+)\psi_j^+ \quad (\in \mathfrak{P}_m).$$

Obviously

$$\mathrm{sp}(Q_m^+ B) = \sum_{j=1}^{m}\lambda_j^+(B) \quad (m = 1,2,\cdots).$$

Comparing this relation with (4.11), we obtain

$$\sum_{j=1}^{m}\lambda_j^+(B)=\sum_{j=1}^{\infty}[\lambda_j^+(A)-\lambda_{j+1}^+(A)]\mathrm{sp}(Q_m^+\mathscr{T}(P_j^+))$$

$$-\sum_{j=1}^{\infty}[\lambda_j^-(A)-\lambda_{j+1}^-(A)]\mathrm{sp}(Q_m^+\mathscr{T}(P_j^-)),$$

and consequently

$$\sum_{j=1}^{m}\lambda_j^+(B)\leqq\sum_{j=1}^{\infty}[\lambda_j^+(A)-\lambda_{j+1}^+(A)]C_{mj}^+ +\sum_{j=1}^{\infty}[\lambda_j^-(A)-\lambda_{j+1}^-(A)]C_{mj}^-.$$

Regrouping the terms in the sums on the right side, we obtain (4.8). As regards an estimate of the remainder term in this Abel transformation, see the remark at the end of the proof of Theorem 3.1.

The relation (4.9) can be proved similarly. However, it follows from (4.8) applied to the operator $-A$. The theorem is proved.

For the case at hand one can derive estimates analogous to (3.14) and (3.15).

COROLLARY 4.1. *Under the hypotheses of Theorem 4.1, the following relations hold:*

$$\sum_{k=1}^{m}\lambda_k^+(B)\leqq C_{m1}^+\sum_{j=1}^{m}\lambda_j^+(A)+C_{m1}^-\sum_{j=1}^{m}\lambda_j^-(A)$$

$$(4.12)\qquad +m\sum_{j=m+1}^{\infty}(C_{1j}^+-C_{1,j-1}^+)\lambda_j^+(A)$$

$$+m\sum_{j=m+1}^{\infty}(C_{1j}^--C_{1,j-1}^-)\lambda_j^-(A)\quad (m=1,2,\cdots)$$

and a similar relation for the sums $\sum_{k=1}^{m}\lambda_k^-(B)$.

The derivation of (4.12) is similar to that of the analogous relation (3.15).

We shall call a continuous transformator $\mathscr{T}$ $\mathscr{K}$-positive if it transforms every nonnegative operator $X\in\mathfrak{S}_2$ into a nonnegative operator. Obviously every $\mathscr{K}$-positive transformator is real.

By virtue of the definition of the constants C_{mj}^- we have

$$C_{mj}^-(\mathscr{T})=0\quad (m,j=1,2,\cdots)$$

for a $\mathscr{K}$-positive transformator $\mathscr{T}$. Thus, for such a transformator the estimates (4.8) and (4.9) simplify.

THEOREM 4.2. *Let $\mathscr{T}$ be a $\mathscr{K}$-positive transformator. Then for any selfadjoint operator $A \in \mathfrak{S}_2$ and $B = \mathscr{T}(A)$ we have the inequalities*

$$\sum_{k=1}^{m} \lambda_k^{\pm}(B) \leqq \sum_{j=1}^{\infty} (C_{mj}^{+} - C_{m,j-1}^{+})\lambda_j^{\pm}(A) \quad (m = 1, 2, \cdots). \tag{4.13}$$

In particular,

$$|\mathscr{T}(A)| \leqq \max\left\{\sum_{j=1}^{\infty} (C_{1j}^{+} - C_{1,j-1}^{+})\lambda_j^{+}(A),\ \sum_{j=1}^{\infty} (C_{1j}^{+} - C_{1,j-1}^{+})\lambda_j^{-}(A)\right\}.$$

§5. Examples of real symmetric transformators

1. Suppose that the space $\mathfrak{H}$ is realized in the form of the space $L_2^{(n)}(a, b)$ of n-dimensional vector-functions $f(t) = \{f_j(t)\}_1^n$ with coordinates $f_j(t) \in L_2(a, b)$ $(j = 1, 2, \cdots, n)$. Every Hilbert-Schmidt operator A in $L_2^{(n)}(a, b)$ is an integral operator of the form

$$(Af)(t) = \int_a^b \mathscr{A}(t, s) f(s)\, ds,$$

where $\mathscr{A}(t, s) = \| a_{jk}(t, s) \|_1^n$ $(a \leqq t,\ s \leqq b)$ is a matrix-function with elements from $L_2(Q)$, where Q is the square $a \leqq t,\ s \leqq b$.

In passing from such an operator A to its adjoint $B = A^*$ the matrix kernel will be replaced by its Hermitian conjugate with the variables interchanged:

$$\mathscr{B}(t, s) = \| b_{jk}(t, s) \|_1^n,$$

where

$$b_{jk}(t, s) = \overline{a_{kj}(s, t)} \quad (a \leqq t, s \leqq b;\ j, k = 1, 2, \cdots, n).$$

The quantity $\mathrm{sp}(AB^*)$ for two Hilbert-Schmidt operators A and B can be calculated from the formula

$$\mathrm{sp}(AB^*) = \int\!\!\int_a^b \mathrm{sp}(\mathscr{A}\mathscr{B}^*)\, ds\, dt = \int_a^b \int_a^b \left\{ \sum_{j,k=1}^{n} a_{jk}(t, s)\overline{b_{jk}(t, s)} \right\} dt\, ds.$$

2. Let $g(t, s)$ $(a \leqq t,\ s \leqq b)$ be a bounded scalar kernel. We associate with $g(t, s)$ the transformator $\mathscr{T}_g$, where by definition the relation $B = \mathscr{T}_g(A)$ means that

$$\mathscr{B}(t, s) = g(t, s)\mathscr{A}(t, s). \tag{5.1}$$

Since $|B|_2 = [\mathrm{sp}(B^*B)]^{1/2}$, according to (5.1)

$$|\mathscr{T}_g(A)|_2 \leqq M|A|_2,$$

where $M = \operatorname{ess\,sup}|g(t,s)|$.

It is not hard to see that

$$\mathscr{T}_g^* = \mathscr{T}_{\bar{g}},$$

where $\bar{g} = \overline{g(t,s)}$ is the conjugate of $g(t,s)$. It follows that

1. *The transformator $\mathscr{T}_g$ is selfadjoint if and only if $g(t,s)$ assumes only real values.*

It is easily seen that:

The transformator $\mathscr{T}_g$ is real (real symmetric) if and only if the kernel $g(t,s)$ is Hermitian (real and symmetric).

It is also easy to verify that

2. *The transformator $\mathscr{T}_g$ will be $\mathscr{K}$-positive if and only if the kernel $g(t,s)$ is Hermitian-positive, i.e.*

$$\int_a^b \int_a^b g(t,s)\phi(s)\overline{\phi(t)}\,dtds \geqq 0 \tag{5.2}$$

for every $\phi \in L_1(a,b)$.

In fact, the condition is necessary. Let $X = (\cdot, \chi)\chi$ be a nonnegative one-dimensional operator, i.e. $\mathscr{X}(t,s) = \chi(t)\chi^*(s)$ $(\chi \in L_2^{(n)}(a,b))$.[6] If $\mathscr{T}_g$ is positive, then the operator $Y = \mathscr{T}_g(X)$ with kernel $\mathscr{Y}(t,s) = g(t,s)\chi(t)\chi^*(s)$ will be nonnegative, i.e. for any $f \in L_2^{(n)}(a,b)$ we will have

$$\int_a^b \int_a^b f^*(t)\chi(t)g(t,s)\chi^*(s)f(s)\,dsdt \geqq 0.$$

Noting that any function $\phi \in L_1(a,b)$ can be represented in the form $\phi(t) = \chi^*(t)f(t)$, where χ, $f \in L_2^{(n)}(a,b)$, we see that (5.2) holds.

Conversely, suppose that (5.2) holds for every function $\phi \in L_1(a,b)$. Putting $\phi(t) = \chi^*(t)f(t)$ in (5.2), where f and χ are elements of $L_2^{(n)}(a,b)$, we see that the transformator $\mathscr{T}_g$ carries every nonnegative one-dimensional operator into a nonnegative operator. But then $\mathscr{T}_g$ carries every nonnegative operator $A \in \mathfrak{S}_2$ into a nonnegative operator, since A can be expanded in a convergent in $\mathfrak{S}_2$ sum of nonnegative one-dimensional operators. This proves the result 2.

3. Let us consider the integral equation corresponding to the operator $Y = \mathscr{T}_g(X)$, where $X = (\cdot,\chi)\chi$ is a one-dimensional orthoprojector $(|\chi| = 1)$.

---

[6] We regard vectors χ from $L_2^{(n)}(a,b)$ as column-vectors. The vector χ^* is the row-vector whose components are the complex conjugates of the components of χ.

In the case being considered the equation $\psi = \mu Y \psi$ will mean the following:

$$\psi(t) = \mu \int_a^b \chi(t) g(t,s) \chi^*(s) \psi(s)\, ds.$$

This is obviously equivalent to the system

$$\phi(t) = \mu \int_a^b g(t,s) \chi^*(s) \psi(s)\, ds,$$

$$\psi(t) = \phi(t) \chi(t).$$

Inserting the expression for $\psi(t)$ given by the second equation into the first, we obtain a scalar weighted integral equation

$$\phi(t) = \mu \int_a^b g(t,s) \phi(s) p(s)\, ds \tag{5.3}$$

with a nonnegative weight function $p(t) = \chi^*(t)\chi(t)$ which satisfies the condition

$$\int_a^b p(t)\, dt = 1.$$

Using equation (5.3), we can easily establish the result

3. *If the kernel* $g(t,s)$ $(a \leqq t,\ s \leqq b)$ *is continuous and Hermitian-positive, then for the transformator* $\mathscr{T}_g$ *we have*

$$C_{1m} = C_{m1} = g_{\max}$$

for every m, *where*[7)]

$$g_{\max} = \max_{a \leqq s \leqq b} g(s,s) = \max_{a \leqq t,s \leqq b} |g(t,s)|. \tag{5.4}$$

In fact, the numbers $\lambda_j^{-1}(Y)$ $(j = 1, 2, \cdots; \lambda_j \neq 0)$ are the characteristic numbers of the integral equation (5.3); therefore by (5.4)

$$\sum_{j=1}^{\infty} \lambda_j(Y) = \int_a^b g(s,s)\, p(s)\, ds \leqq g_{\max}.$$

In particular, for any $m = 1, 2, \cdots$ we have

$$\sum_{j=1}^{m} \lambda_j(Y) \leqq g_{\max}, \tag{5.5}$$

---

7) Let us recall that for a continuous Hermitian-positive kernel $g(t,s)$ one always has $|g(t,s)|^2 \leqq g(s,s) g(t,t)$ $(a \leqq t, s \leqq b)$.

where $Y = \mathscr{T}_g(X)$, and X is any one-dimensional orthoprojector.

On the other hand, according to the definition (3.2), for given m the supremum of the sum on the left is C_{1m}; hence $C_{1m} \leqq g_{\max}$ $(m = 1, 2, \cdots)$.

Let $s_0 \in [a, b]$ be a point for which $g_{\max} = g(s_0, s_0)$. Let $\chi_N(t)$ $(N = 1, 2, \cdots)$ be a unit vector from $L_2^{(n)}(a, b)$ such that $\chi_N(t) \equiv 0$ if $|t - s_0| > 1/N$. Then $p_N(t) = \chi_N^*(t)\chi_N(t) \equiv 0$ for $|t - s_0| > 1/N$. We put $Y_N = \mathscr{T}_g(X_N)$, where $X_N = (\cdot, \chi_N)\chi_N$ $(N = 1, 2, \cdots)$. Since in the limit $N \to \infty$ equation (5.3) with $p(t) = p_N(t)$ goes over into the equation

$$\phi(t) = \mu g(t, s_0)\phi(s_0),$$

which has one characteristic number $\mu = g^{-1}(s_0, s_0)$, it obviously follows that as $N \to \infty$

$$\lambda_1(Y_N) \to g(s_0, s_0) = g_{\max}.$$

Therefore the sum on the left side of (5.5) tends for $Y = Y_N$ to $g_{\max}$ as $N \to \infty$. Finally, bearing in mind that $C_{m1}(\mathscr{T}_g) = C_{1m}(\mathscr{T}_g)$ and that the kernal $\bar{g}(t, s)$ is Hermitian-positive, we obtain also $C_{m1} = g_{\max}$. This completes the proof of the result 3.

4. Let $a = 0$, $b = l$, and let

$$g(t, s) = g_\kappa(t - s),$$

$$g_\kappa(t) = \frac{1}{2\pi} \sum_{j=-\infty}^{\infty} \frac{e^{2\pi i(j+\kappa)t/l}}{j + \kappa} \qquad (0 < \kappa < 1).$$

It is not hard to see that

$$g_{1/2}(t) = (i/2)\operatorname{sign} t.$$

What is essential for us is that for any κ $(0 < \kappa < 1)$ the Hermitian kernel $g_\kappa(t - s)$ is the Green's function for the boundary-value problem

$$-i\frac{d}{dt}y = h(t), \quad y(l) = e^{i2\pi\kappa}y(0),$$

so that the solution of this boundary-value problem for any $h(t) \in L_1(0, l)$ is obtained from the formula

$$y(t) = \int_0^l g_\kappa(t - s)h(s)\,ds.$$

This circumstance enables us to obtain all the characteristic numbers of equation (5.3) very simply for the case at hand, in which this equation assumes the form

$$\phi(t) = \mu \int_0^l g_\kappa(t-s)\phi(s)p(s)\,ds.$$

In fact, this equation is equivalent to the homogeneous boundary-value problem

$$\phi'(t) - i\mu p(t)\phi(t) = 0, \quad \phi(l) - e^{i2\pi\kappa}\phi(0) = 0. \tag{5.6}$$

Since it follows from this differential equation that

$$\phi(t) = \exp\left[i\mu \int_0^t p(s)\,ds \right]\phi(0), \ \phi(l) = e^{i\mu}\phi(0),$$

the spectrum of the boundary-value problem and hence the spectrum of characteristic numbers of (5.6) coincide with the set of numbers

$$\mu_j = 2\pi(\kappa + j) \quad (j = 0, \pm 1, \pm 2, \cdots).$$

Thus, the spectrum of eigenvalues of the operator $Y = \mathscr{T}_\kappa(X)$, where $X = (\cdot, \chi)\chi$, is independent of the choice of the orthoprojector X, and consists of the numbers

$$\lambda_j^+ = \frac{1}{2\pi(j+\kappa-1)} \quad (j = 1, 2, \cdots),$$

$$-\lambda_j^- = -\frac{1}{2\pi(j-\kappa)} \quad (j = 1, 2, \cdots).$$

For brevity we have denoted the transformator $\mathscr{T}_{g_\kappa}$ by $\mathscr{T}_\kappa$.

At the same time we arrive at the following deduction.

4. *For the transformator* $\mathscr{T}_\kappa$ *the constants* $C_{1m}^{\pm}$ *have the following values:*

$$\left.\begin{aligned} C_{m1}^+ = C_{1m}^- &= \frac{1}{2\pi}\sum_{j=1}^m \frac{1}{j+\kappa-1}, \\ C_{m1}^- = C_{1m}^+ &= \frac{1}{2\pi}\sum_{j=1}^m \frac{1}{j-\kappa}. \end{aligned}\right\} \quad (m = 1, 2, \cdots),$$

In particular, for $\kappa = 1/2$

$$C_{m1}^{\pm} = C_{1m}^{\pm} = \frac{1}{\pi}\sum_{j=1}^m \frac{1}{2j-1} \quad (m = 1, 2, \cdots).$$

It remains to clarify the equality $C_{1m}^{\pm} = C_{m1}^{\mp}$. This equality follows from the fact that $C_{m1}^{\pm}(\mathscr{T}_{g_\kappa}) = C_{1m}^{\pm}(\mathscr{T}_{\bar{g}_\kappa})$, $\overline{g_\kappa(t)} = g_{1-\kappa}(t)$.

5. Recently, in connection with the perturbation theory of self-

adjoint operators, M. Š. Birman and M. Z. Solomjak [1-4] carried out an interesting study of transformators of the form

$$\mathscr{T}(X) = \int_{-\infty}^{\infty} \int_{-\infty}^{\infty} \phi(\lambda, \mu)\, dE_\lambda X dF_\mu, \tag{5.7}$$

where E_λ and F_μ are two resolutions of the identity, and $\phi(\lambda, \mu)$ is a bounded complex-valued function. These authors showed that the theory of such transformators finds important application in factor theory, the theory of multidimensional singular integrals, and other topics.

The first integrals of this form were apparently encountered in a paper by Ju. L. Daleckiĭ and S. G. Kreĭn [1] in connection with certain questions in perturbation theory.

One can obtain a more general class of transformators by considering transformators of the form

$$\mathscr{T}(X) = \int_{\mathfrak{P}} \int_{\mathfrak{Q}} \Phi(P, Q)\, dQ X dP, \tag{5.8}$$

where $\mathfrak{P}$ and $\mathfrak{Q}$ are two bordered closed chains, and $\Phi(P, Q)$ is an operator-function defined on the product $\mathfrak{P} \times \mathfrak{Q}$ with values in $\mathfrak{R}$.

The transformator $\mathscr{L}$ of triangular truncation is a special case of the transformator (5.7) and therefore, in particular, of the transformator (5.8), which is obtained by putting $\mathfrak{P} = \mathfrak{Q}$ and

$$\Phi(P, Q) = \begin{cases} I & \text{for } P < Q, \\ 0 & \text{for } P \geqq Q. \end{cases}$$

If we formally put

$$\int_{\mathfrak{Q}} \Phi(P, Q)\, dQ = F(P),$$

then the integral (5.8) can be written in the form

$$\mathscr{T}(X) = \int_{\mathfrak{P}} F(P) X\, dP.$$

Certain special integrals of this type will be studied in Chapter IV in connection with the problem of the factorization of operators.

We remark that for $E_\lambda = F_\lambda$ the transformator (5.7) is essentially of the same type as the transformator $\mathscr{T}_g$ already considered.

§6. The constants C_{mj} for the transformators $\mathscr{C}$ and $\mathscr{S}$

1. According to §§2 and 3 we can associate with the transformators $\mathscr{C}$ and $\mathscr{S}$ constants $C_{mj}(\mathscr{C})$, $C'_{mj}(\mathscr{C})$ and $C^{\pm}_{mj}(\mathscr{S})$, which are defined by

$$C_{mj}(\mathscr{C}) = \sup_{K\in\mathfrak{K}_j;\,|K|=1} \sum_{n=1}^{m} s_n(\mathscr{C}(K)) \quad (m,j = 1,2,\cdots), \tag{6.1}$$

$$C'_{mj}(\mathscr{C}) = \sup_{P\in\mathfrak{P}_j} \sum_{n=1}^{m} s_n(\mathscr{C}(P)) \quad (m,j = 1,2,\cdots), \tag{6.2}$$

$$C^{\pm}_{mj}(\mathscr{S}) = \sup_{P\in\mathfrak{P}_j} \sum_{n=1}^{m} \lambda^{\pm}_n(\mathscr{S}(P)) \quad (m,j = 1,2,\cdots), \tag{6.3}$$

where $\mathfrak{K}_j$ is the set of all finite-dimensional operators of dimension $\leqq j$, and $\mathfrak{P}_j$ is the set of all orthoprojectors from $\mathfrak{K}_j$.

In this section we shall determine the exact values of these constants for $j = 1$ and all m. Since for all the constants C_{mj} one has the estimate $C_{mj} \leqq jC_{m1}$, we obtain in this way certain estimates for these constants for all values of j and m. The exact values of the constants C_{mj} are not known. To obtain these estimates we need the following result.

LEMMA 6.1. *If A is a dissipative Volterra operator which has a one-dimensional imaginary component $A_{\mathscr{J}} = (\cdot, e)e$, $|e| = 1$, then*

$$\lambda_n(A_{\mathscr{R}}) = \frac{2}{(2n-1)\pi} \quad (n = 0, \pm 1, \cdots)$$

and

$$s_n(A) = \frac{4}{(2n-1)\pi} \quad (n = 1,2,\cdots).$$

PROOF. By Theorem 8.1 of Chapter I, the simple part of the operator A is unitarily equivalent to the operator of integration J, defined in $L_2(0,1)$ by

$$(Jf)(t) = 2i\int_t^1 f(s)\,ds.$$

Hence $\lambda_n(A_{\mathscr{R}}) = \lambda_n(J_{\mathscr{R}})$ and $s_n(A) = s_n(J)$. The operator J^* has the form

$$(J^*f)(t) = -2i\int_0^t f(s)\,ds;$$

thus the numbers $\mu_n = s_n^{-2}(J)$ $(n = 1, 2, \cdots)$ are the characteristic numbers of the integral equation

$$\phi(t) = 4\mu \int_0^t \left(\int_s^1 \phi(u)\, du \right) ds,$$

which is equivalent to the boundary-value problem

$$\phi'' + 4\mu\phi = 0, \qquad \phi(0) = \phi'(1) = 0.$$

It is easily seen that the sequence $\mu_n = \pi^2(2n-1)^2/16$ $(n = 1, 2, \cdots)$ constitutes the complete system of characteristic numbers of this problem, and therefore

$$s_n(J) = \frac{4}{(2n-1)\pi} \qquad (n = 1, 2, \cdots).$$

The operator $J_{\mathscr{R}}$ is defined by

$$(J_{\mathscr{R}} f)(t) = -i \int_0^1 \operatorname{sign}(t-s) f(s)\, ds.$$

Hence its characteristic numbers are the characteristic numbers of the integral equation

$$\phi(t) = -i\mu \int_0^1 \operatorname{sign}(t-s)\phi(s)\, ds,$$

which is equivalent to the boundary-value problem

$$\phi'(t) + 2i\mu\phi(t) = 0, \quad \phi(0) + \phi(1) = 0.$$

The numbers $\mu_n = (2n-1)\pi/2$ $(n = 0, \pm 1, \pm 2, \cdots)$ form the complete system of characteristic numbers of this boundary-value problem (the corresponding eigenfunctions are $\exp[-(2n-1)i\pi t]$). Thus

$$\lambda_n(J_{\mathscr{R}}) = \frac{2}{(2n-1)\pi} \qquad (n = 0, \pm 1, \cdots).$$

The lemma is proved.

2. Let $P_\phi = (\cdot, \phi)\phi$ $(\phi \in \mathfrak{H};\ |\phi| = 1)$ be a one-dimensional projector, and let $\mathfrak{P}$ be any complete chain. Then the operator $B = 2i\mathscr{C}(P_\phi; \mathfrak{P})$ is a Volterra operator with imaginary component $B_{\mathscr{I}} = P_\phi$. The real component $B_{\mathscr{R}}$ of B is equal to $\mathscr{S}(P_\phi; \mathfrak{P})$.

Thus, by Lemma 6.1

$$s_n(B) = \frac{4}{(2n-1)\pi} \quad (n = 1, 2, \cdots)$$

and

$$\lambda_n(B_{\mathscr{R}}) = \frac{2}{(2n-1)\pi} \quad (n = 0, \pm 1, \cdots).$$

By virtue of (6.2) and (6.3) we obtain

$$C'_{m1}(\mathscr{C}) = \frac{2}{\pi}\sigma_m, \quad C^{\pm}_{m1}(\mathscr{S}) = \frac{2}{\pi}\sigma_m \quad (m = 1, 2, \cdots), \tag{6.4}$$

where

$$\sigma_m = \sum_{k=1}^{m} \frac{1}{(2k-1)}.$$

Since the transformator $\mathscr{S}$ is skewsymmetric,

$$C^{\pm}_{1m}(\mathscr{S}) = C^{\mp}_{m1}(\mathscr{S}) = \frac{2}{\pi}\sigma_m \quad (m = 1, 2, \cdots). \tag{6.5}$$

We note that (6.4) implies the following estimates:

$$\left.\begin{matrix} C'_{mj}(\mathscr{C}) \\ C^{\pm}_{mj}(\mathscr{S}) \end{matrix}\right\} \leqq \frac{2}{\pi} j\sigma_m \quad (j, m = 1, 2, \cdots). \tag{6.6}$$

It also follows from (6.5) that

$$C^{\pm}_{mj}(\mathscr{S}) = C^{\mp}_{jm}(\mathscr{S}) \leqq mC_{1j} = \frac{2}{\pi} m\sigma_j \quad (j, m = 1, 2, \cdots). \tag{6.7}$$

3. The aim of this subsection is to prove the validity of the following estimates:

$$C_{m1}(\mathscr{C}) \leqq \sigma_m \quad (m = 1, 2, \cdots). \tag{6.8}$$

These estimates are simple consequences of the following lemma.

LEMMA 6.2. *Let $\mathfrak{P}$ be a complete chain. Then for any one-dimensional operator $X = (\cdot, \psi)\phi$ with $|\phi| = |\psi| = 1$, the s-numbers of the operator $Y = \mathscr{C}(X; \mathfrak{P})$ admit the exact estimates*

$$s_j(Y) \leqq \frac{1}{2j-1} \quad (j = 1, 2, \cdots). \tag{6.9}$$

PROOF. Let $\mathfrak{P} = \{P(t)\}$ $(0 \leqq t \leqq 1)$ be a parametrization[8)] of the chain $\mathfrak{P}$. For any $f \in \mathfrak{H}$

$$(6.10) \qquad \begin{aligned} Yf &= \int_0^1 P(t) X dP(t) = \int_0^1 (f, dP(t)\psi) P(t)\phi, \\ Y^*f &= \int_0^1 (f, P(t)\phi) dP(t)\psi. \end{aligned}$$

The numbers $\mu_j = s_j^{-2}(Y)$ $(j = 1, 2, \cdots)$ are the characteristic numbers of the equation

$$\chi = \mu Y^* Y \chi,$$

which according to (6.10) is equivalent to the equation

$$(6.11) \qquad \chi = \mu \int_0^1 \int_0^1 (\chi, dP(t)\psi)(P(t)\phi, P(s)\phi) dP(s)\psi.$$

Let us put

$$(6.12) \qquad \eta(t) = (\chi, \psi - P(t)\psi) \quad (0 \leqq t \leqq 1).$$

The integral equation (6.11) is obviously equivalent to the system consisting of equation (6.12) and the equation

$$\chi = -\mu \int_0^1 \int_0^1 (P(t)\phi, P(s)\phi) d\eta(t) dP(s)\psi.$$

Replacing the variable t by τ in (6.12) and inserting in it the value of χ from the last equation, we see that (6.12) is equivalent to the following:

$$(6.13) \qquad \eta(\tau) = -\mu \int_0^1 \int_0^1 (P(t)\phi, P(s)\phi) d\eta(t) d_s(P(s)\psi, \psi - P(\tau)\psi).$$

Since

$$(P(t)\phi, P(s)\phi) = (P(r)\phi, \phi), \quad r = \min(t, s),$$

and $\eta(1) = 0$, integration by parts yields

$$-\int_0^1 (P(t)\phi, P(s)\phi) d\eta(t) = \int_0^s \eta(t) d(P(t)\phi, \phi).$$

Putting

[8)] We shall show in §1, Chapter V that every complete chain $\mathfrak{P}$ can be regarded as the continuous monotonic image of the interval $[0, 1]$.

$$\sigma_1(t) = (P(t)\phi, \phi), \qquad \sigma_2(t) = (P(t)\psi, \psi),$$

we can therefore rewrite equation (6.13) in the form

$$\eta(\tau) = \mu \int_\tau^1 d\sigma_2(s) \int_0^s \eta(t)\, d\sigma_1(t) \quad (0 \leqq x \leqq 1). \tag{6.14}$$

From this last equation it is clear that if the function $\sigma_2(s)$ is constant on some interval (α, β) $(0 < \alpha < \beta < 1)$, then the function $\eta(\tau)$ will also be constant on this interval. Therefore there exists a continuous function $y(x)$ $(0 \leqq x \leqq 1)$ such that

$$y(\sigma_2(t)) = \eta(t).$$

Passing in (6.14) from the function $\eta(t)$ to the function $y(x)$ transforms it into the form

$$y(x) = \mu \int_x^1 d\xi \int_0^\xi y(\xi)\, dM(\xi), \tag{6.15}$$

where $M(\xi)$ $(M(0) = 0)$ is a nondecreasing function determined by the nondecreasing functions $\sigma_1(t)$ and $\sigma_2(t)$.

Since $\sigma_j(0) = 0$ and $\sigma_j(1) = 1$ $(j = 1, 2)$, obviously $M(1) = 1$.

The integral equation (6.15) is equivalent to the boundary-value problem

$$dy' + \mu y dM = 0, \tag{6.16}$$

$$y'(0) + M(+0)\mu y(0) = 0, \ y(1) = 0, \tag{6.17}$$

which determines the frequencies $p = \sqrt{\mu}$ and the amplitude functions $y(x)$ of a string S which is stretched between the points 0 and 1 by a unit force, whose right end is held fixed, and whose left end is fastened to a ringlet of mass $M(+0) > 0$ which is free to slide in a direction perpendicular to the equilibrium position of S. With this interpretation, $M(x + 0)$ denotes the mass of the segment $[0, x]$ of the string.

As is known (see subsection 4 below), for such a string the squares $\mu_j = p_j^2$ of the frequencies admit the precise lower bounds

$$\mu_j \geqq (2j - 1)^2 \quad (j = 1, 2, \cdots). \tag{6.18}$$

Since $s_j(Y) = \mu_j^{-2}$ $(j = 1, 2, \cdots)$, we have established the inequality (6.9).

For $j = 1$ equality will hold in (6.9) if the entire mass of the string is concentrated at its left end. In fact, in this case equation (6.15) reduces to the relation

$$y(x) = \mu(1 - x)y(0).$$

Putting $x = 0$, we find that $\mu_1 = 1$. All the other μ_j $(j > 1)$ may be regarded as being at infinity.

We arrive at the integral equation (6.15) with such a mass distribution if and only if the one-dimensional operator $X = (\cdot, \psi)\phi$ is so chosen that for some a $(0 < a < 1)$

$$P(a)\phi = \phi, \quad P(a)\psi = 0, \tag{6.19}$$

and so the entire variation of the function σ_1 takes place on the interval $[0, a]$ ($\sigma_1(t) = 1$ for $a \leqq t \leqq 1$), and that of the function $\sigma_2(t)$, on the interval $[a, 1]$ ($\sigma_2(t) = 0$ for $0 \leqq t \leqq a$).

It is easily seen that for the case (6.19) one has $Y = \mathscr{C}(X; \mathfrak{P}) = X$.

For $j > 1$ the equal sign in (6.9) is achieved for the string S which is a thread, weighted with masses $m = M/j$ at the points $x_k = 2k/(2j - 1)$ $(k = 0, 1, \cdots, j - 1)$ (see, in this connection, subsection 4 below).

All one-dimensional operators $X = (\cdot, \psi)\phi$, where ψ and ϕ are unit vectors, which satisfy certain conditions will lead to such a mass distribution. These conditions are that for some $0 = a_0 < a_1 < \cdots < a_{2j-1} < a_{2j} = 1$ the function $\sigma_1(t) = (P(t)\phi, \phi)$ is constant in each of the intervals (a_{2k-1}, a_{2k}) $(k = 1, 2, \cdots, j)$, and the function $\sigma_2(t) = (P(t)\psi, \psi)$ is constant in each of the complementary intervals (a_{2k-2}, a_{2k-1}) $(k = 1, 2, \cdots, j)$, and

$$\sigma_1(a_{2k-1}) - \sigma_1(a_{2k-2}) = \frac{2}{2j - 1} \quad (k = 1, 2, \cdots, j),$$

$$\sigma_2(a_{2k}) - \sigma_2(a_{2k-1}) = \frac{2}{2j - 1} \quad (k = 1, 2, \cdots, j - 1),$$

$$\sigma_2(1) - \sigma_2(a_{2j-1}) = \frac{1}{2j - 1}.$$

The lemma is proved.

4. Since the bound (6.18) is easily established, we shall derive it. To this end, let us consider a string S_1 which is stretched between the points $x = \pm 1$ by a unit force, and supports a total mass of two units.

Let us denote by $M_1(x)$ the mass of the segment $[-1, x]$ $(-1 \leqq x \leqq 1)$ of the string S_1 $(M_1(-1) = 0,\ M_1(1) = 2)$. The squares $\lambda_j = p_j^2$ $(j = 1, 2, \cdots)$ of the frequencies of the string S_1 are the eigenvalues of the boundary value problem

$$dy' + \lambda dM_1(x)y = 0, \quad y(-1) = y(1) = 0 \tag{6.20}$$

or, what is the same, the characteristic numbers of the integral equation

$$y(x) = \lambda \int_{-1}^{1} \mathcal{G}(x,s)y(s)\,dM_1(s), \tag{6.21}$$

where

$$\mathcal{G}(x,s) = \begin{cases} \frac{1}{2}(1+x)(1-s), & x \leqq s, \\ \frac{1}{2}(1-x)(1+s), & x \geqq s. \end{cases} \tag{6.22}$$

We shall show that the characteristic numbers $\lambda_1 < \lambda_2 < \cdots$ of the integral equation (6.21) admit the precise lower bounds

$$\lambda_k \geqq k^2 \quad (k = 1, 2, \cdots). \tag{6.23}$$

This will establish the bounds (6.18). In fact, if μ_j is an eigenvalue of the boundary-value problem (6.16), (6.17), and $y_j(x)$ ($\not\equiv 0$) is the corresponding solution of (6.16), then the corresponding even function $\tilde{y}_j(x) = \tilde{y}_j(-x) = y_j(x)$ $(0 \leqq x \leqq 1)$ will be a solution of (6.20) for $\lambda = \mu_j$, if the function $M_1(x)$ defines a mass distribution which is symmetric with respect to the origin $x = 0$ $(M_1(x) - M_1(+0) = M_1(-0) - M_1(-x);\ 0 < x \leqq 1)$, and which coincides on $[0,1]$ with the mass distribution defined by $M(x)$, i.e.

$$M(x) - M(-0) = M_1(x) - M_1(+0) \quad (0 \leqq x \leqq 1),$$
$$M_1(+0) - M_1(0) = M(+0) - M(0) = M(+0).$$

When these conditions relative to $M_1(x)$ and $M(x)$ are satisfied, the string S_1 can be regarded as consisting of S and the reflection of S about the y-axis.

As is known,[9] the eigenfunction $y_j(x)$ $(j = 1, 2, \cdots)$ has exactly $j-1$ nodes in the interval $(0,1)$, and $y_j(0) \neq 0$ (as otherwise, by the first boundary condition in (6.17), $y_j'(0) = 0$, and consequently $y_j(x) \equiv 0$). Therefore the function $\tilde{y}_j(x)$ will have exactly $2j-2$ nodes in the interval $(-1,1)$, and hence will correspond to the $(2j-2)$th overtone of the string S_1, i.e. $\mu_j = \lambda_{2j-1}$. Thus, the inequalities (6.18) are a consequence of the inequalities (6.23).

Inequality (6.23) for $j = 1$ was stated by H. E. Žukovskiĭ [1]. It can, for example, be obtained from the trace formula

[9] See, for example, the book by F. R. Gantmaher and M. G. Kreĭn [1], in which the theory of the nodes of the eigenvibrations of a string is discussed for any distribution of mass along the string.

$$\int_{-1}^{1} \mathscr{G}(x,x)\,dM_1(x) = \sum_j \frac{1}{\lambda_j}\left(\geqq \frac{1}{\lambda_1}\right).$$

Indeed, taking into account (6.22), we deduce that

$$1 = \frac{1}{2}\int_{-1}^{1} dM_1(x) \geqq \frac{1}{2}\int_{-1}^{1}(1-x^2)\,dM_1(x) \geqq \frac{1}{\lambda_1}.$$

Thus $\lambda_1 \geqq 1$, and equality holds when the entire mass of the string S_1 is concentrated at the point $x = 0$, i.e.

$$M_1(+0) - M_1(-0) = 2.$$

It is easy to ascertain that for a string with fixed end points and length L and any total mass M this bound assumes the form

$$\lambda_1 \geqq 4/LM. \tag{6.24}$$

As was noted by I. M. Rapoport [1] (see also M. G. Kreĭn [10]), from Žukovskiĭ's estimate it is not difficult to derive the general estimates

$$\lambda_k \geqq 4k^2/LM \quad (k = 1, 2, \cdots),$$

which go over into (6.23) for $L = M = 2$.

In fact, let $y = \phi_k(x)$ be the fundamental function of the boundary-value problem (6.20) for the string S_1, corresponding to λ_k. We denote by $(-1 <)\ x_1 < x_2 < \cdots < x_{k-1}\ (< 1)$ the nodes of ϕ_k. If the string S_1 is held fixed at the points x_j $(j = 1, 2, \cdots, k-1)$, it breaks up into k strings of lengths

$$L_j = x_j - x_{j-1} \quad (j = 1, 2, \cdots, k;\ x_0 = -1, x_k = 1)$$

and total masses

$$M_j = \int_{x_{j-1}+0}^{x_j - 0} dM_1(x) \quad (j = 1, 2, \cdots, k),$$

and for each of these pieces the frequency $p_k = \sqrt{\lambda_k}$ will now be the first (fundamental) frequency. Therefore, according to (6.24) we will have

$$\lambda_k \geqq 4/M_j L_j \quad (j = 1, 2, \cdots, k), \tag{6.25}$$

and so

$$\lambda_k \geqq \frac{1}{2}\sum_{j=1}^{k} M_j \lambda_k \geqq 2\sum_{j=1}^{k}\frac{1}{L_j}.$$

On the other hand, under the condition $\sum_{j=1}^{k} L_j = 2$ the minimum of the sum on the right side is achieved for $L_1 = L_2 = \cdots = L_k = 2/k$. Substituting these values for the L_j into (6.25), we obtain (6.23).

The bound (6.23) is obviously achieved for a string with the following mass distribution: the string is divided into k equal parts, and a mass $M/k = 2/k$ is suspended at the center of each part. The points at which the string is divided into k equal parts will be the nodes of the oscillation with frequency $p_k = \sqrt{\lambda_k} = k$ of a string with such a mass distribution. It is not difficult to show that this is the only mass distribution for which the bound (6.23) is achieved.

CHAPTER III

THE TRANSFORMATOR OF TRIANGULAR TRUNCATION. RELATIONS BETWEEN THE SPECTRA OF THE HERMITIAN COMPONENTS OF VOLTERRA OPERATORS

As we already know, in connection with the abstract triangular representation of a Volterra operator there naturally arises the concept of the transformator of triangular truncation generated by a given maximal chain.

In this chapter we shall study the action of this transformator in various symmetrically-normed ideals (see [GK], Chapter III).[1)] This study sometimes leads to unexpected and deep interrelations between the spectra of the Hermitian components of Volterra operators. In certain cases the clarification of these interrelations has required the use of a special analytic apparatus.

In the last section of the chapter, as one illustration of the application of the general results, we establish new analytic properties of the Riesz-Titchmarsh transform, the discrete analog of the classical integral transform of Hilbert.

§1. Symmetrically-normed ideals

In this section we review the basic concepts and results of the theory of symmetrically-normed ideals which will be needed further on. This theory was worked out by J. von Neumann and R. Schatten (see R. Schatten [**1**]) and was further developed in the authors' book [GK] (Chapter III), in which one can find a detailed account of the results discussed here.

1. Symmetrically-normed ideals. A *symmetrically-normed* ideal $\mathfrak{S}$ of the ring $\mathfrak{R}$ is a two-sided ideal in which is defined some norm $|X|_{\mathfrak{S}}$ which makes $\mathfrak{S}$ into a Banach space and has the following properties:

a) $|AXB|_{\mathfrak{S}} \leqq |A|\,|X|_{\mathfrak{S}}|B|$ $(A, B \in \mathfrak{R};\ X \in \mathfrak{S})$;

b) $|X|_{\mathfrak{S}} = |X|$ for every one-dimensional operator X.

[1)] The results of this chapter are, essentially, due to the authors [**3**–**5**] and to V. I. Macaev [**2, 3**] (see also the survey by Brodskiĭ, Gohberg, Kreĭn and Macaev [**1**]). To these results have recently been added results by E. M. Semenov, which are presented in §8.

For brevity we shall call a symmetrically-normed ideal an *s.n. ideal*.

It is known that every two-sided ideal $\mathfrak{S}$ ($\neq 0$) of the ring $\mathfrak{R}$ contains all finite-dimensional operators. On the finite-dimensional operators the norm $|X|_{\mathfrak{S}}$ is a function, with well-defined properties, which depends only on the s-numbers of the operator.

Let c_0 be the space of all sequences $\xi = \{\xi_j\}$ of real numbers which tend to zero. We denote by $\hat{c}$ the lineal in c_0 consisting of all sequences $\xi = \{\xi_j\}$ with a finite number of nonzero coordinates.

A real function $\Phi(\xi) = \Phi(\xi_1, \xi_2, \cdots)$ defined on $\hat{c}$ is called a *symmetric norming* (*s.n.*) function if it has the following properties:

1) $\Phi(\xi) > 0$ $(\xi \in \hat{c},\ \xi \neq 0)$;

2) $\Phi(\alpha\xi) = |\alpha|\Phi(\xi)$ ($\xi \in \hat{c}$; α a real number)

3) $\Phi(\xi + \eta) \leqq \Phi(\xi) + \Phi(\eta)$ $(\xi, \eta \in \hat{c})$;

4) $\Phi(1, 0, 0, \cdots) = 1$;

5) $\Phi(\xi_1, \xi_2, \cdots, \xi_n, 0, 0, \cdots) = \Phi(|\xi_{j_1}|, |\xi_{j_2}|, \cdots, |\xi_{j_n}|, 0, 0, \cdots)$,

where $\xi = \{\xi_j\}$ is any vector from $\hat{c}$, and $j_1, j_2, \cdots, j_n$ is any permutation of the integers $1, 2, \cdots, n$.

The following result holds.

Let $\xi = \{\xi_j\}$, $\eta = \{\eta_j\} \in \hat{c}$ *satisfy the relations*

$$\xi_1 \geqq \xi_2 \geqq \cdots \geqq 0, \quad \eta_1 \geqq \eta_2 \geqq \cdots \geqq 0,$$

$$\sum_{j=1}^{k} \xi_j \leqq \sum_{j=1}^{k} \eta_j \quad (k = 1, 2, \cdots).$$

Then for any s.n. function $\Phi(\xi)$ *one has*

$$\Phi(\xi) \leqq \Phi(\eta).$$

It is easy to show that the functions

$$\Phi_\infty(\xi) = \max_j |\xi_j|, \quad \Phi_1(\xi) = \sum_j |\xi_j|$$

are s.n. functions. For any s.n. function $\Phi(\xi)$ one has the inequalities

$$\Phi_\infty(\xi) \leqq \Phi(\xi) \leqq \Phi_1(\xi) \quad (\xi \in \hat{c}).$$

Therefore the functions $\Phi_\infty(\xi)$ and $\Phi_1(\xi)$ are called the *minimal* and *maximal* s.n. functions, respectively.

Two s.n. functions $\Phi_1(\xi)$ and $\Phi_2(\xi)$ are said to be *equivalent* if

$$\sup_{\xi \in \hat{c}} (\Phi_1(\xi)/\Phi_2(\xi)) < \infty \quad \text{and} \quad \sup_{\xi \in \hat{c}} (\Phi_2(\xi)/\Phi_1(\xi)) < \infty.$$

It turns out that an s.n. function $\Phi(\xi)$ is equivalent to the minimal one if and only if

$$\sup_n \Phi(\underbrace{1, 1, \cdots, 1}_{n}, 0, 0, \cdots) < \infty.$$

On the other hand, an s.n. function $\Phi(\xi)$ is equivalent to the maximal one if and only if

$$\sup_n (n/\Phi(\underbrace{1, 1, \cdots, 1}_{n}, 0, 0, \cdots)) < \infty.$$

The following theorem holds.

THEOREM 1.1. *Let $\Phi(\xi)$ be any s.n. function. Then the set $\mathfrak{S}_\Phi$, consisting of all operators $X \in \mathfrak{S}_\infty$ for which*

$$\sup_n \Phi(s_1(X), s_2(X), \cdots, s_n(X), 0, 0, \cdots) < \infty,$$

is an s.n. ideal with the norm

$$|X|_\Phi = \sup_n \Phi(s_1(X), s_2(X), \cdots, s_n(X), 0, 0, \cdots).$$

If $\Phi(\xi)$ is an s.n. function, we denote by $\mathfrak{S}_\Phi^{(0)}$ the subspace of $\mathfrak{S}_\Phi$ obtained by taking the closure with respect to the norm of $\mathfrak{S}_\Phi$ of the set of all finite-dimensional operators. One can show that $X \in \mathfrak{S}_\Phi^{(0)}$ if and only if

$$\lim_{n,p\to\infty} \Phi(s_{n+1}(X), s_{n+2}(X), \cdots, s_{n+p}(X), 0, 0, \cdots) = 0.$$

An s.n. function $\Phi(\xi)$ is said to be *mononormalizing* if for any vector $\xi = \{\xi_j\} \in c_0$ satisfying the condition

$$\sup_n \Phi(\xi_1, \xi_2, \cdots, \xi_n, 0, 0, \cdots) < \infty$$

one has

$$\lim_{n,p\to\infty} \Phi(\xi_{n+1}, \cdots, \xi_{n+2}, \cdots, \xi_{n+p}, 0, 0, \cdots) = 0.$$

One can show ([GK], Chapter III, §6.1) that the spaces $\mathfrak{S}_\Phi$ and $\mathfrak{S}_\Phi^{(0)}$ coincide if and only if the function $\Phi(\xi)$ is mononormalizing.

It is easily seen that the minimal and maximal s.n. functions are mononormalizing.

Every s.n. function which is not mononormalizing is said to be *binormalizing*.

THEOREM 1.2. *For any s.n. function $\Phi(\xi)$ the space $\mathfrak{S}_\Phi^{(0)}$ is a separable s.n. ideal of the ring $\mathfrak{R}$. Conversely, for each separable s.n. ideal $\mathfrak{S}$ of the ring $\mathfrak{R}$ there exists an s.n. function $\Phi(\xi)$ (and only one) such that $\mathfrak{S} = \mathfrak{S}_\Phi^{(0)}$.*

2. The symmetrically-normed ideals $\mathfrak{S}_p$ $(1 \leqq p \leqq \infty)$. Important examples of s.n. ideals are the s.n. ideals $\mathfrak{S}_p = \mathfrak{S}_{\Phi_p}$ $(1 \leqq p \leqq \infty)$, where $\Phi_p(\xi)$ is the mononormalizing s.n. function defined by

$$\Phi_p(\xi) = \left(\sum_j |\xi_j|^p\right)^{1/p}$$

for $1 \leqq p < \infty$.

Thus, *the s.n. ideal* $\mathfrak{S}_p$ $(1 \leqq p < \infty)$ *consists of all operators* $X \in \mathfrak{S}_\infty$ *for which*

$$(|X|_p)^p = \sum_{j=1}^{\infty} s_j^p(X) < \infty.$$

The norm in $\mathfrak{S}_p$ *is defined by* $|X|_{\mathfrak{S}_p} = |X|_p$.

The elements of the s.n. ideal $\mathfrak{S}_1$ are called *nuclear* operators. The nuclear operators X are precisely those having the property that for any orthonormal basis $\{\phi_j\}$ of the space $\mathfrak{H}$ the series

$$\sum_{j=1}^{\infty} (X\phi_j, \phi_j)$$

converges. The sum of this series does not depend upon the choice of basis; this sum is called the *trace* of the nuclear operator X and is denoted by $\operatorname{sp} X$. By a well-known theorem of V. B. Lidskiĭ (see [GK], p. 101)

$$\operatorname{sp} X = \sum_j \lambda_j(X) \quad (X \in \mathfrak{S}_1),$$

where $\{\lambda_j(X)\}$ is the complete sequence of eigenvalues of X. The trace $\operatorname{sp} X$ is a bounded linear functional on the space $\mathfrak{S}_1$ of nuclear operators. The following result holds.

If $X \in \mathfrak{S}_p$ *and* $Y \in \mathfrak{S}_q$, *where*

$$p^{-1} + q^{-1} = 1,$$

then $XY, YX \in \mathfrak{S}_1$ *and*

$$\operatorname{sp}(XY) = \operatorname{sp}(YX), \tag{1.1}$$

and

$$|XY|_1 \leqq |X|_p |Y|_q, \quad |YX|_1 \leqq |X|_p |Y|_q.$$

A special role is played by the ideal $\mathfrak{S}_2$, consisting of all Hilbert-Schmidt operators. As was already mentioned, $\mathfrak{S}_2$ forms a Hilbert space with the scalar product

$$(X, Y) = \operatorname{sp}(XY^*).$$

The s.n. ideal $\mathfrak{S}_\infty$ consisting of all completely continuous operators obviously corresponds to the minimal function

$$\Phi_\infty(\xi) = \lim_{p\to\infty} \Phi_p(\xi) = \max_j |\xi_j| \quad (\xi \in \mathbf{c}_0).$$

3. The symmetrically-normed ideals $\mathfrak{S}_\Pi^{(0)}, \mathfrak{S}_\pi, \mathfrak{S}_\Pi$. We give other important examples of s.n. ideals.

A nonincreasing sequence $\Pi = \{\pi_j\}_1^\infty$ of positive numbers is said to be *binormalizing* if it satisfies the following conditions:

$$p_1 = 1, \quad \lim_{n\to\infty} \pi_n = 0, \quad \sum_{n=1}^\infty \pi_n = \infty.$$

One can associate with such a sequence two functions $\Phi_\pi(\xi)$ and $\Phi_\Pi(\xi)$ $(\xi = \{\xi_j\} \in \mathbf{c}_0)$, defined by

$$\Phi_\pi(\xi) = \sum_{j=1}^\infty \pi_j \xi_j^*,$$

$$\Phi_\Pi(\xi) = \sup_n \left(\sum_{j=1}^n \xi_j^* \Big/ \sum_{j=1}^n \pi_j \right),$$

where $\xi_j^* = |\xi_{nj}|$ $(j = 1, 2, \cdots)$, and $\{n_j\}$ is a permutation of the positive integers such that the sequence $\{|\xi_{n_j}|\}$ is nonincreasing.

The function $\Phi_\pi(\xi)$ is a mononormalizing s.n. function, and $\Phi_\Pi(\xi)$ is a binormalizing s.n. function. Thus, with each sequence Π there is associated a triple of s.n. ideals

$$\mathfrak{S}_\pi = \mathfrak{S}_{\Phi_\pi}, \quad \mathfrak{S}_\Pi = \mathfrak{S}_{\Phi_\Pi} \quad \text{and} \quad \mathfrak{S}_\Pi^{(0)} = \mathfrak{S}_{\Phi_\Pi}^{(0)}.$$

The s.n. ideals $\mathfrak{S}_\Pi$ *and* $\mathfrak{S}_\pi$ *consist respectively of all operators* $X \in \mathfrak{S}_\infty$ *for which*

$$|X|_\Pi = \sup_n \left(\sum_{j=1}^n s_j(X) \Big/ \sum_{j=1}^n \pi_j \right) < \infty$$

or

$$|X|_\pi = \sum_{j=1}^\infty \pi_j s_j(X) < \infty.$$

The norms in $\mathfrak{S}_\Pi$ *and* $\mathfrak{S}_\pi$ *are defined by*

$$|X|_{\mathfrak{S}_\Pi} = |X|_\Pi, \quad |X|_{\mathfrak{S}_\pi} = |X|_\pi.$$

The s.n. ideal $\mathfrak{S}_\Pi^{(0)}$ $(\subset \mathfrak{S}_\Pi)$ *consists of all operators* $X \in \mathfrak{S}_\infty$ *for which*

$$\lim_{n\to\infty}\left(\sum_{j=1}^{n} s_j(X) \Big/ \sum_{j=1}^{n} \pi_j\right) = 0.$$

A special role will be played in the sequel by the sequence

$$\Omega = \{1/(2j-1)\}_{j=1}^{\infty}.$$

The corresponding s.n. ideal $\mathfrak{S}_\Omega$ consists of all $X \in \mathfrak{S}_\infty$ for which

$$|X|_{\mathfrak{S}_\Omega} = |X|_\Omega = \sup_n \left(\sum_{j=1}^{n} s_j(X) \Big/ \sum_{j=1}^{n} (2j-1)^{-1}\right) < \infty,$$

and the s.n. ideal $\mathfrak{S}_\omega$ consists of all $X \in \mathfrak{S}_\infty$ satisfying the condition

$$|X|_{\mathfrak{S}_\omega} = |X|_\omega = \sum_{j=1}^{\infty} (2j-1)^{-1} s_j(X) < \infty.$$

4. Symmetrically-normed ideals conjugate to separable symmetrically-normed ideals. With every s.n. function $\Phi(\xi)$ ($\xi \in \hat{c}$) there is associated the s.n. function

$$\Phi^*(\eta) = \max_{\xi \in \hat{c}} \frac{1}{\Phi(\xi)} \left|\sum_j \xi_j \eta_j\right| \quad (\eta = \{\eta_j\} \in \hat{c}),$$

called the *conjugate* of $\Phi(\xi)$. Always $(\Phi^*)^* = \Phi$.

THEOREM 1.3. *Let $\Phi(\xi)$ be an s.n. function which is not equivalent to the maximal one.*

Then the general form of a continuous linear functional $F(X)$ on the separable space $\mathfrak{S}_\Phi^{(0)}$ is given by the formula

$$F(X) = \mathrm{sp}(AX),$$

where A is an arbitrary operator from $\mathfrak{S}_{\Phi^*}$, and

$$|F| = \sup_{X \in \mathfrak{S}_\Phi^{(0)}} \frac{|\mathrm{sp}(AX)|}{|X|_\Phi} = |A|_{\Phi^*}.$$

Thus, the space $\mathfrak{S}_{\Phi^*}$ is isometric with the conjugate space of $\mathfrak{S}_\Phi^{(0)}$.

In particular, if both $\Phi(\xi)$ and $\Phi^*(\xi)$ are mononormalizing and neither one is equivalent to the maximal s.n. function, then the space $\mathfrak{S}_\Phi$ is reflexive.

THEOREM 1.4. *The general form of a continuous linear functional $F(X)$ on the Banach space $\mathfrak{S}_1$ is given by the formula*

$$F(X) = \mathrm{sp}(AX),$$

where A is an arbitrary operator from $\mathfrak{R}$, and

$$|F| = \sup_{X \in \mathfrak{S}_1} \frac{|\mathrm{sp}(AX)|}{|X|_1} = |A|.$$

It can be shown that $\Phi_p^*(\xi) = \Phi_q(\xi)$, where $p^{-1} + q^{-1} = 1$, $1 \leqq p \leqq \infty$, and also that $\Phi_{\Pi}^*(\xi) = \Phi_\pi(\xi)$. Consequently $\mathfrak{S}_p^* = \mathfrak{S}_q$ $(1 < p \leqq \infty)$, $\mathfrak{S}_{\Pi}^{(0)*} = \mathfrak{S}_\pi$ and $\mathfrak{S}_\pi^* = \mathfrak{S}_{\Pi}$.

Thus the space $\mathfrak{S}_p$ $(1 < p < \infty)$ is reflexive, and its conjugate space is $\mathfrak{S}_q$ $(p^{-1} + q^{-1} = 1)$.

The conjugate space of $\mathfrak{S}_\omega$ is the nonseparable space $\mathfrak{S}_\Omega$.

5. Conical norms in the real Banach spaces $\hat{\mathfrak{S}}$. Let $\mathfrak{S}$ be an s.n. ideal. We denote by $\hat{\mathfrak{S}}$ the real Banach space consisting of all selfadjoint operators from $\mathfrak{S}$. Let $\mathfrak{S}$ be a separable s.n. ideal; it is easily verified that if $\mathfrak{S}^*$ is the conjugate space of $\mathfrak{S}$, then the space $\hat{\mathfrak{S}}^*$ of all selfadjoint operators from $\mathfrak{S}^*$ is the conjugate space of $\hat{\mathfrak{S}}$.

We can introduce another norm in $\hat{\mathfrak{S}}$:

$$|X|_{\mathfrak{S},\mathscr{K};p} = (|X_+|_{\mathfrak{S}}^p + |X_-|_{\mathfrak{S}}^p)^{1/p} \quad (1 \leqq p \leqq \infty; X \in \hat{\mathfrak{S}}), \tag{1.2}$$

where $X = X_+ - X_-$, and $X_\pm$ are nonnegative and mutually orthogonal. This norm is called a *conical* norm; it is topologically equivalent to the original norm.

The norm in $\hat{\mathfrak{S}}^*$ which is conjugate to the norm (1.2) is the norm

$$|X|_{\mathfrak{S}^*;\mathscr{K};q} \quad (p^{-1} + q^{-1} = 1).$$

For the case of the spaces $\hat{\mathfrak{S}}_{\Pi}^{(0)}$, $\hat{\mathfrak{S}}_\pi$, and $\hat{\mathfrak{S}}_{\Pi}$ a special role is played by the mutually conjugate conical norms

$$|X|_{\pi;\mathscr{K}} = \sum_{j=1}^{\infty} \pi_j(\lambda_j^+(X) + \lambda_j^-(X)) \quad (X \in \hat{\mathfrak{S}}_\pi)$$

and

$$|X|_{\Pi;\mathscr{K}} = \max\left\{ \sup_n \frac{\sum_{j=1}^n \lambda_j^+(X)}{\sum_{j=1}^n \pi_j}, \sup_n \frac{\sum_{j=1}^n \lambda_j^-(X)}{\sum_{j=1}^n \pi_j} \right\} \quad (X \in \hat{\mathfrak{S}}_{\Pi}),$$

where

$$\lambda_j^+(X) = \lambda_j(X_+) \quad \text{and} \quad \lambda_j^-(X) = \lambda_j(X_-) \ (j = 1, 2, \cdots).$$

§2. Volterra operators with nuclear imaginary component

1. This class of operators has already been the object of study in Chapter IV of the book [GK] by analytic methods. Here it will be studied on the basis of the theory of the integral of triangular truncation, and a number of new results which are essential in the sequel will be obtained.

The first investigation of this sort of this class of operators was carried out by the authors in the communication [3], and then in [4].

THEOREM 2.1. *Let A be a Volterra operator with $A_{\mathscr{I}} \in \mathfrak{S}_1$. Then $A \in \mathfrak{S}_\Omega$, and moreover for any positive integer n one has the bounds*

$$\sum_{j=1}^{n} s_j(A) \leqq \frac{4}{\pi} |A_{\mathscr{I}}|_1 \sum_{j=1}^{n} \frac{1}{2j-1} \quad (n = 1, 2, \cdots), \tag{2.1}$$

$$\left.\begin{array}{l} \sum_{j=1}^{n} \lambda_j^{+}(A_{\mathscr{R}}) \\ \sum_{j=1}^{n} \lambda_j^{-}(A_{\mathscr{R}}) \end{array}\right\} \leqq \frac{2}{\pi} |A_{\mathscr{I}}|_1 \sum_{j=1}^{n} \frac{1}{2j-1} \quad (n = 1, 2, \cdots). \tag{2.2}$$

If $A_{\mathscr{I}}$ is one-dimensional, equality holds in these bounds for every n.

By the definition of the norms in the spaces $\mathfrak{S}_\Omega$ and $\hat{\mathfrak{S}}_\Omega$ (see §1), (2.1) and (2.2) can be written in the form

$$|A|_\Omega \leqq \frac{4}{\pi} |A_{\mathscr{I}}|_1 \quad \text{and} \quad |A_{\mathscr{R}}|_{\Omega;\mathscr{K}} \leqq \frac{2}{\pi} |A_{\mathscr{I}}|_1. \tag{2.3}$$

PROOF. By Theorem I.3.2, some inessential extension $\widetilde{A}$ of A has a complete eigenchain $\mathfrak{P}$. For this eigenchain

$$\widetilde{A} = 2i\mathscr{C}(\widetilde{A}_{\mathscr{I}}; \mathfrak{P}) \quad \text{and} \quad \widetilde{A}_{\mathscr{R}} = \mathscr{S}(\widetilde{A}_{\mathscr{I}}; \mathfrak{P}).$$

On the other hand, the numbers $s_j(A)$, $\lambda_j^{\pm}(A_{\mathscr{R}})$ and $\lambda_j(A_{\mathscr{I}})$ do not change when A is replaced by $\widetilde{A}$.

It remains to note that the relations (2.1) are obtained from the general relations (II.3.14) as applied to the transformator $\mathscr{C}$ and the operator $X = \widetilde{A}_{\mathscr{I}}$. Similarly, the relations (2.2) are obtained from the relations (II.4.8), (II.4.9) as applied to the transformator $\mathscr{S}$ and the operator $X = \widetilde{A}_{\mathscr{I}}$. To obtain the last relations we have to take into account that

$$C_{mj}^{\pm}(\mathcal{S}) - C_{mj-1}^{\pm}(\mathcal{S}) \leqq C_{m1}^{\pm}(\mathcal{S}) = \frac{2}{\pi}\sum_{n=1}^{m}\frac{1}{2n-1}.$$

The theorem is proved.

SECOND PROOF OF THEOREM 2.1. We now give a direct proof; in contrast with the preceding proof, this proof proceeds independently of the general theorems on transformators (Chapter II).

We maintain the notation and concepts of the preceding proof.

Let us write down the spectral representation of the operator $\widetilde{A}_{\mathcal{J}}$ $(\in \mathfrak{S}_1)$:

$$\widetilde{A}_{\mathcal{J}} = \sum_{j=1}^{\infty}\lambda_j(A_{\mathcal{J}})K_j,$$

where $K_j = (\cdot, \phi_j)\phi_j$, and the series converges in the norm of $\mathfrak{S}_2$. Then

$$\widetilde{A} = 2i\mathscr{C}(\widetilde{A}_{\mathcal{J}}; \mathfrak{P}) = 2i\sum_{j=1}^{\infty}\lambda_j(A_{\mathcal{J}})\mathscr{C}(K_j; \mathfrak{P}).$$

The operator $L_j = 2i\mathscr{C}(K_j; \mathfrak{P})$ is a Volterra operator with the one-dimensional imaginary component K_j $(|K_j| = 1)$. Hence, by Lemma II.6.1

$$|L_j|_{\Omega} = 4/\pi.$$

It follows that the expansion obtained for $\widetilde{A}$ converges in the norm of $\mathfrak{S}_{\Omega}$ and

$$|\widetilde{A}|_{\Omega} \leqq \sum_{j=1}^{\infty}|\lambda_j(A_{\mathcal{J}})|\,|L_j|_{\Omega} = \frac{4}{\pi}|A_{\mathcal{J}}|_1.$$

As we already mentioned, this relation is equivalent to (2.1).

Similarly, starting from the equality

$$\widetilde{A}_{\mathcal{R}} = \mathcal{S}(\widetilde{A}_{\mathcal{J}}; \mathfrak{P}) = \sum_{j=1}^{\infty}\lambda_j(A_{\mathcal{J}})\mathcal{S}(K_j; \mathfrak{P})$$

and taking into account that the operator $M_j = \mathcal{S}(K_j; \mathfrak{P})$ is the real component of the Volterra operator L_j which has a one-dimensional imaginary component, we obtain by Lemma II.6.1

$$|M_j|_{\Omega;\mathcal{K}} = 2/\pi,$$

from which follows the relation

$$|A_{\mathcal{R}}|_{\Omega;\mathcal{K}} \leqq \frac{2}{\pi}|A_{\mathcal{J}}|_1,$$

which is equivalent to (2.2). This completes the proof.

2. Using result 2 of §1, Chapter II, new relations can be derived from the relations (2.1) and (2.2). For example,

$$\sum_{j=1}^{n} s_j^p(A) \leqq \left(\frac{4}{\pi} |A_{\mathscr{I}}|_1\right)^p \sum_{j=1}^{n} \frac{1}{(2j-1)^p} \quad (p>1;\ n=1,2,\cdots), \tag{2.4}$$

$$\left.\begin{array}{l} \sum_{j=1}^{n} [\lambda_j^+(A_{\mathscr{R}})]^p \\ \sum_{j=1}^{n} [\lambda_j^-(A_{\mathscr{R}})]^p \end{array}\right\} \leqq \left(\frac{2}{\pi} |A_{\mathscr{I}}|_1\right)^p \sum_{j=1}^{n} \frac{1}{(2j-1)^p} \quad (p>1; n=1,2,\cdots). \tag{2.5}$$

The last relations imply the bounds

$$|A|_p \leqq \frac{4}{\pi}\left(\sum_{j=1}^{\infty} \frac{1}{(2j-1)^p}\right)^{1/p} |A_{\mathscr{I}}|_1 \quad (1<p\leqq\infty), \tag{2.6}$$

$$\max(|A_{\mathscr{R}}^-|_p, |A_{\mathscr{R}}^+|_p) \leqq \frac{2}{\pi}\left(\sum_{j=1}^{\infty} \frac{1}{(2j-1)^p}\right)^{1/p} |A_{\mathscr{I}}|_1 \quad (1<p\leqq\infty), \tag{2.7}$$

and therefore

$$|A_{\mathscr{R}}|_p \leqq \frac{2^{1+1/p}}{\pi}\left(\sum_{j=1}^{\infty} \frac{1}{(2j-1)^p}\right)^{1/p} |A_{\mathscr{I}}|_1 \quad (1<p\leqq\infty). \tag{2.8}$$

Theorem 2.1 can be formulated in a more general way, namely: *For any normal operator $X \in \mathfrak{S}_1$ and any complete chain $\mathfrak{P}$ one has the bounds*

$$|\mathscr{C}(X)|_{\Omega} \leqq \frac{2}{\pi}|X|_1, \quad |\mathscr{S}(X)|_{\Omega_1} \leqq \frac{2}{\pi}|X|_1, \tag{2.9}$$

where

$$\Omega_1 = \left\{1, 1, \frac{1}{3}, \frac{1}{3}, \cdots, \frac{1}{2n-1}, \frac{1}{2n-1}, \cdots\right\}.$$

Both of these bounds follow from (II.3.16). They lead, in turn, to the bounds

$$\begin{gathered} |\mathscr{C}(X)|_p \leqq \frac{2\gamma_p}{\pi}|X|_1, \quad |\mathscr{S}(X)|_p \leqq \frac{2^{1+1/p}\gamma_p}{\pi}|X|_1 \\ \gamma_p = \left(\sum_{j=1}^{\infty} \frac{1}{(2j-1)^p}\right)^{1/p} \quad (1<p\leqq\infty), \end{gathered} \tag{2.10}$$

for any normal operator $X \in \mathfrak{S}_1$.

3. If we omit the condition of normality, we can state the following result.

THEOREM 2.2. *Let* $\mathfrak{P}$ *be a maximal chain. Then for any operator* $X \in \mathfrak{S}_1[\mathfrak{P}]$[2)]

$$\left| \int_{\mathfrak{P}} PX\,dP \right|_{\Omega} \leq |X|_1. \tag{2.11}$$

The equality is achieved here; moreover, there always exists an infinite set of one-dimensional operators X *for which*

$$\int_{\mathfrak{P}} PX\,dP = X. \tag{2.12}$$

PROOF. If the chain $\mathfrak{P}$ is continuous, then (2.11) is a simple consequence of relation (II.3.14) and the bound (II.6.8).

We shall reduce the general case to this one by showing that the operators X and $Y = \mathscr{L}(X; \mathfrak{P})$ have inessential extensions $\widetilde{X}$ and $\widetilde{Y}$ (acting in some Hilbert space $\widetilde{\mathfrak{H}} \supset \mathfrak{H}$) for which

$$\widetilde{Y} = \int_{\widetilde{\mathfrak{P}}} P\widetilde{X}\,dP, \tag{2.13}$$

where $\widetilde{\mathfrak{P}}$ is some continuous maximal chain in $\widetilde{\mathfrak{H}}$.

To do this we consider the operators

$$Y_1 = \int_{\mathfrak{P}} PX_{\mathscr{R}}\,dP \quad \text{and} \quad Y_2 = \int_{\mathfrak{P}} PX_{\mathscr{I}}\,dP.$$

By Theorem I.5.1,

$$[2iY_1]_{\mathscr{I}} = 2[Y_1]_{\mathscr{R}} = X_{\mathscr{R}} \quad \text{and} \quad [2iY_2]_{\mathscr{I}} = 2[Y_2]_{\mathscr{R}} = X_{\mathscr{I}}. \tag{2.14}$$

The construction which was carried out in the proof of Theorem I.3.2 enables us to assert that the operators Y_1 and Y_2 have inessential extensions $\widetilde{Y}_1$ and $\widetilde{Y}_2$, acting in the same space $\widetilde{\mathfrak{H}} \supset \mathfrak{H}$, which have a common continuous eigenchain $\widetilde{\mathfrak{P}}$. For these extensions

$$\widetilde{Y}_1 = -i\mathscr{L}([\widetilde{Y}_1]_{\mathscr{R}}; \widetilde{\mathfrak{P}}), \quad \widetilde{Y}_2 = -i\mathscr{L}([\widetilde{Y}_2]_{\mathscr{R}}; \widetilde{\mathfrak{P}}),$$

---

2) We denote by $\mathfrak{S}_1[\mathfrak{P}]$ the set of all operators $X \in \mathfrak{S}_1$ which satisfy the condition (I.10.2) concerning the vanishing of their blocks on the gaps of $\mathfrak{P}$.

i.e.

$$\widetilde{Y}_1 = -i\mathscr{C}(\widetilde{X}_{\mathscr{R}};\widetilde{\mathfrak{P}}) \quad \text{and} \quad \widetilde{Y}_2 = \mathscr{C}(\widetilde{X}_{\mathscr{I}};\widetilde{\mathfrak{P}}).$$

Hence we obtain (2.13) for the operator $\widetilde{Y} = \widetilde{Y}_1 + i\widetilde{Y}_2$.

The construction of a one-dimensional operator $X = (\cdot,\psi)\phi$ for which (2.12) holds was given in the proof of Lemma II.6.1. It goes as follows. Let P_0 $(\neq 0, I)$ be a projector from the chain $\mathfrak{P}$, and let ϕ and ψ be chosen so that

$$P_0\phi = \phi \quad \text{and} \quad P_0\psi = 0.$$

Obviously

$$\int_{\mathfrak{P}} PX\,dP = \int_{\mathfrak{P}} (\cdot, dP\psi)P\phi = (\cdot,\psi)\phi.$$

The theorem is proved.

4. The relations (2.1) and (2.2) are also exact from another point of view; namely, we have the following result.

THEOREM 2.3. *For any selfadjoint operator $H \in \mathfrak{S}_1$ and any $\epsilon > 0$ there exists a Volterra operator A, having H as its imaginary component, such that*

$$|A|_{\Omega} \geqq \frac{4}{\pi}|H|_1 - \epsilon, \quad |A_{\mathscr{R}}|_{\Omega;\mathscr{K}} \geqq \frac{2}{\pi}|H|_1 - \epsilon.$$

PROOF. Starting with the spectral decomposition

$$H = \sum_{j=1}^{\infty} \lambda_j(H)(\cdot, e_j)e_j, \tag{2.15}$$

we form the finite-dimensional operator

$$T = \sum_{j=1}^{n} \lambda_j(H)(\cdot, e_j)e_j,$$

where n is chosen so large that

$$\sum_{j=n+1}^{\infty} |\lambda_j(H)| < \frac{\pi\epsilon}{8}. \tag{2.16}$$

Let

$$\mathfrak{H} = \mathfrak{H}_1 \oplus \mathfrak{H}_2 \oplus \cdots \oplus \mathfrak{H}_n$$

be a decomposition of $\mathfrak{H}$ into the orthogonal direct sum of infinite-dimensional subspaces $\mathfrak{H}_k$ $(k = 1,2,\cdots,n)$ such that $e_j \in \mathfrak{H}_j$ $(j = 1,$

$2, \cdots, n)$. Obviously $T\phi = 0$ for all vectors $\phi \in \mathfrak{H}_j$ which are orthogonal to e_j.

Let $\mathfrak{P}_k$ $(k = 1, 2, \cdots, n)$ be any complete chain in the space $\mathfrak{H}_k$. According to Theorem I.8.1, each of the operators

$$J^{(k)} = 2i \int_{\mathfrak{P}_k} (\cdot, dPe_k) Pe_k \quad (k = 1, 2, \cdots, n)$$

is unitarily equivalent to some inessential extension of the operator of integration J. Let us denote by B the orthogonal direct sum of the operators $\lambda_k J^{(k)}$ $(k = 1, 2, \cdots, n)$. Since the complete system of s-numbers of the operator B consists of the union of the sequences

$$\left\{ |\lambda_k(H)| \frac{1}{(2j-1)\pi} \right\}_{j=1}^{\infty} \quad (k = 1, 2, \cdots, n),$$

it follows that

$$\sum_{j=1}^{nr} s_j(B) \geqq \frac{4}{\pi} \sum_{k=1}^{n} |\lambda_k(H)| \sum_{j=1}^{r} \frac{1}{2j-1} \quad (r = 1, 2, \cdots).$$

This relation implies

$$|B|_\Omega \geqq \frac{4}{\pi} \sum_{k=1}^{n} |\lambda_k(H)| \lim_{r \to \infty} \left\{ \left(\sum_{j=1}^{r} \frac{1}{2j-1} \right) \Big/ \left(\sum_{j=1}^{nr} \frac{1}{2j-1} \right) \right\}. \tag{2.17}$$

Taking into account that

$$\lim_{r \to \infty} \sum_{j=r+1}^{nr} \frac{1}{2j-1} = \frac{1}{2} \ln n,$$

we obtain from (2.17)

$$|B|_\Omega \geqq \frac{4}{\pi} \sum_{k=1}^{n} |\lambda_k(H)| = \frac{4}{\pi} |T|_1. \tag{2.18}$$

Comparing (2.1) and (2.18), we deduce that

$$|B|_\Omega = 4|T|_1/\pi.$$

Let $\mathfrak{P}'_k$ $(k = 1, 2, \cdots, n)$ be the chain which is obtained by adding to every projector from $\mathfrak{P}_k$ the orthoprojector onto the subspace $\mathfrak{H}_1 \oplus \mathfrak{H}_2 \oplus \cdots \oplus \mathfrak{H}_{k-1}$. The union of all the chains $\mathfrak{P}'_k$ forms a complete chain $\mathfrak{P}$ in the space $\mathfrak{H}$. Obviously

$$B = 2i \int_{\mathfrak{P}} PT\,dP.$$

We shall show that the operator

$$A = 2i \int_{\mathfrak{P}} PH\,dP$$

is the desired one. In fact,

$$|A|_\Omega \geqq |B|_\Omega - \left| 2i \int_{\mathfrak{P}} P(H-T)\,dP \right|_\Omega .$$

By the bound (2.9) and the inequality (2.16),

$$\left| 2i \int_{\mathfrak{P}} P(H-T)\,dP \right|_\Omega \leqq \frac{4}{\pi} |H-T|_1 < \frac{\epsilon}{2}.$$

Hence

$$|A|_\Omega \geqq \frac{4}{\pi} |T|_1 - \frac{\epsilon}{2} \geqq \frac{4}{\pi} |H|_1 - \epsilon.$$

We can similarly prove that

$$|A_{\mathscr{R}}|_{\Omega;\mathscr{K}} \geqq \frac{2}{\pi} |H|_1 - \epsilon.$$

The theorem is proved.

This theorem can also be formulated in the following way.

THEOREM 2.3′. *For any selfadjoint operator $H \in \mathfrak{S}_1$ one has*

$$\sup \left| \int_{\mathfrak{P}} PH\,dP \right|_\Omega = \frac{2}{\pi} |H|_1 \tag{2.19}$$

and

$$\sup \left| \left[\int_{\mathfrak{P}} PH\,dP \right]_{\mathscr{R}} \right|_{\Omega;\mathscr{K}} = \frac{1}{\pi} |H|_1, \tag{2.20}$$

where the supremum is taken over all complete chains in the space $\mathfrak{H}$.

We remark that the relation (2.19) remains valid when the operator H is replaced by any normal operator $X \in \mathfrak{S}_1$. For this replacement the relation (2.20) can be replaced by the following:

$$\sup \left| \int_{\mathfrak{P}} (PX\,dP - dPXP) \right|_\Omega = \frac{1}{\pi} |X|_1.$$

5. We present two further results which supplement the results of this section.

1. *If the selfadjoint operator* H $(\in \mathfrak{S}_\infty)$ *does not belong to* $\mathfrak{S}_1$, *then there exists a Volterra operator* A *with a continuous maximal eigenchain* $\mathfrak{P}$ *such that*

$$A_{\mathscr{J}} = H \quad and \quad A_{\mathscr{R}} \notin \mathfrak{S}_\Omega.$$

We may suppose without loss of generality that the point $\lambda = 0$ is an eigenvalue of H of infinite multiplicity. In fact, any selfadjoint $H \in \mathfrak{S}_\infty$ can be represented as the sum $H = H_1 + H_2$ of two selfadjoint operators such that $H_2 \in \mathfrak{S}_1$, and the point $\lambda = 0$ is an eigenvalue of H_1 of infinite multiplicity. By Theorem 2.1, the result which we wish to prove for H is equivalent to the same result for H_1.

Let (2.15) be the spectral decomposition of H, and let $\mathfrak{H} = \mathfrak{H}_1 \oplus \mathfrak{H}_2 \oplus \cdots$ be a decomposition of the space $\mathfrak{H}$ into the orthogonal direct sum of infinite-dimensional subspaces $\mathfrak{H}_j$ $(j = 1, 2, \cdots)$ such that $e_j \in \mathfrak{H}_j$ $(j = 1, 2, \cdots)$. The rest of the construction is carried out in the same way as in the proof of Theorem 2.3; here one has to replace the operator T by H and n by ∞.

The operator B thereby obtained, being the orthogonal direct sum of the operators $J^{(k)}$ $(k = 1, 2, \cdots)$, is a Volterra operator and has a continuous eigenchain $\mathfrak{P}$; and it is also easily verified that the operator $B_{\mathscr{R}}$, and along with it the operator B, does not belong to the space $\mathfrak{S}_\Omega$.

2. *For every complete chain* $\mathfrak{P}$ *there exists a selfadjoint operator* $H \in \mathfrak{S}_\Omega$ *such that*

$$\int_{\mathfrak{P}} PH\,dP \notin \mathfrak{S}_\Omega. \tag{2.21}$$

In fact, let P_j $(j = 1, 2, \cdots)$ be a strictly increasing sequence of projectors from the chain $\mathfrak{P}$, and let $\{e_j\}_1^\infty$ be an orthonormal sequence of vectors such that $(P_{j+1} - P_j)e_j = e_j$ $(j = 1, 2, \cdots)$. We form the operator

$$H = \sum_{j=1}^{\infty} \frac{1}{2j-1} (\cdot, e_j) e_j.$$

Obviously $H \in \mathfrak{S}_\Omega$. Arguing as in the proof of Theorem 2.3, we obtain

$$\left| P_n \int_{\mathfrak{P}} PH\,dPP_n \right|_\Omega = \left| \int_{\mathfrak{P}} PP_n HP_n dP \right|_\Omega = \frac{2}{\pi} \sum_{j=1}^{n-1} \frac{1}{2j-1};$$

thus (2.21) holds.

6. We conclude this section with a theorem on Volterra operators with nuclear imaginary component which was obtained in Chapter IV of the book [GK] (Theorem IV.8.2). In the proof of this theorem we shall use, in addition to considerable analytic apparatus, Theorem 2.1 (and also result 2 of §4).

THEOREM 2.4 (M. G. KREĬN). *Let A be a Volterra operator with nuclear imaginary component. Then the eigenvalues of its real component $A_{\mathscr{R}}$ have the following properties:*

$$\text{(2.22)}\qquad 1)\quad \lim_{\epsilon\to 0} \sum_{|\lambda_j(A_{\mathscr{R}})|\geqq\epsilon} \lambda_j(A_{\mathscr{R}}) = 0;$$

$$\text{(2.23)}\qquad 2)\quad \lim_{r\to\infty} \frac{n_+(r;A_{\mathscr{R}})}{r} = \lim_{r\to\infty} \frac{n_-(r;A_{\mathscr{R}})}{r} = \frac{h}{\pi},$$

where

$$\text{(2.24)}\qquad |\operatorname{sp} A_{\mathscr{I}}| \leqq h \leqq |A_{\mathscr{I}}|_1;$$

3) *there exists a constant γ, not depending upon the operator A, such that*

$$\text{(2.25)}\qquad \frac{n(r;A_{\mathscr{R}})}{r} \leqq \frac{\gamma}{\pi} |A_{\mathscr{I}}|_1.$$

Here $n_+(r;A_{\mathscr{R}})$ and $n_-(r;A_{\mathscr{R}})$ denote, respectively, the number of characteristic numbers of the operator $A_{\mathscr{R}}$ in the closed intervals $[0,r]$ and $[-r,0]$, and

$$n(r;A_{\mathscr{R}}) = n_+(r;A_{\mathscr{R}}) + n_-(r;A_{\mathscr{R}}).$$

We remark that the number h appearing in (2.23) can be calculated from the formula

$$\text{(2.26)}\qquad h = \int_{\mathfrak{P}} |dPA_{\mathscr{I}}dP|_1,$$

where $\mathfrak{P}$ is a maximal eigenchain of A.

This integral is to be understood as the Šatunovskiĭ limit of the sums of the form $\sum_{j=1}^n |\Delta P_j A_{\mathscr{I}} \Delta P_j|_1$ along the chain $\mathfrak{P}$.

The formula (2.26) can also be interpreted as the following equality:

$$h = \inf \sum_{j=1}^{n} |\Delta P_j A_{\mathscr{I}} \Delta P_j|_1,$$

where the infimum is taken over the set of all partitions of the chain $\mathfrak{P}$. The point is that by Theorem III.8.7 of the book [GK], if the partition $\{P_j''\}_0^n$ is a refinement of the partition $\{P_j'\}_0^m$, then

$$\sum_{j=1}^{n} |\Delta P_j'' A_{\mathscr{I}} \Delta P_j''|_1 \leqq \sum_{j=1}^{m} |\Delta P_j' A_{\mathscr{I}} \Delta P_j'|_1.$$

Formula (2.26) is closely related to an asymptotic formula from the theory of selfadjoint boundary-value problems for canonical differential equations. The proof of formula (2.26) requires, besides a number of operator-theoretic considerations, the use of a considerable analytic apparatus. It will eventually be established in Chapter VII.

For a dissipative Volterra operator A ($A_{\mathscr{J}} \geqq 0$) this formula yields

$$h = \int_{\mathfrak{P}} \mathrm{sp}(dPA_{\mathscr{J}}dP) = \mathrm{sp}\, A_{\mathscr{J}}, \tag{2.27}$$

and in this case it is a consequence of the bound (2.24), since in this case we have $|A_{\mathscr{J}}|_1 = \mathrm{sp}\, A_{\mathscr{J}}$.

We remark that for a dissipative Volterra operator A it was shown in the book [GK] (p. 185) that one can put $\gamma = 2e$ in the bound (2.25), i.e.

$$\frac{n(r; A_{\mathscr{R}})}{r} \leqq \frac{2e}{\pi} \mathrm{sp}\, A_{\mathscr{J}} \quad (0 < r < \infty; A_{\mathscr{J}} \geqq 0).$$

In this bound the constant $2e$ is not precise, but we recall (see [GK], pp. 184-185) that this bound was obtained as a consequence of the relation

$$\int_0^r \frac{n(\rho; A_{\mathscr{R}})}{\rho}\, d\rho \leqq \frac{2}{\pi} r\, \mathrm{sp}\, A_{\mathscr{J}} \qquad (0 < r < \infty). \tag{2.28}$$

Since, according to (2.23),

$$\lim_{r \to \infty} \left(\frac{1}{r} \int_0^r \frac{n(\rho; A_{\mathscr{R}})}{\rho}\, d\rho \right) = \frac{2}{\pi} \mathrm{sp}\, A_{\mathscr{J}}, \tag{2.29}$$

the constant $2/\pi$ in (2.28) is precise.

Recently, L. de Branges in his very interesting work [**10, 11**] presented a new method for obtaining bounds for the functions $n_{\pm}(r; A_{\mathscr{R}})/r$. This method is interesting in that it enables one, without using theorems from the theory of entire functions, to obtain for a dissipative Volterra operator the bounds[3)]

[3)] In de Branges' paper [**11**] he asserts that

$$\lambda_n^{\pm}(A_{\mathscr{R}}) \leqq \frac{2}{\pi(2n-1)} \mathrm{sp}\, A_{\mathscr{J}} \quad (n = 1, 2, \cdots). \tag{*}$$

Unfortunately, these bounds are incorrect; it is easily seen that they are already incorrect for the orthogonal sum of two dissipative Volterra operators with one-dimensional imaginary components. However, the "derivation" of the bounds (*) is based on a

$$\frac{n_{\pm}(r; A_{\mathscr{R}})}{r} \leqq \frac{2}{\pi} \operatorname{sp} A_{\mathscr{I}}, \tag{2.30}$$

and hence the bound (2.25) with $\gamma = 4$.

We state a more complete result.

THEOREM 2.5. *Let A be a dissipative Volterra operator with nuclear imaginary component. Then for the eigenvalues of its real component we have the bounds*

$$\lambda_n^{\pm}(A_{\mathscr{R}}) \leqq \frac{2}{\pi n} \operatorname{sp} A_{\mathscr{I}} \quad (n = 1, 2, \cdots). \tag{2.31}$$

If the imaginary component $A_{\mathscr{I}}$ has dimension m, we have the more precise bounds

$$\lambda_{jm+k}^{\pm}(A_{\mathscr{R}}) \leqq \frac{2}{\pi((2j-1)m + 2k)} \operatorname{sp} A_{\mathscr{I}} \tag{2.32}$$
$$(k = 0, 1, \cdots, m-1;\ j = 1, 2, \cdots).$$

The bounds (2.32) are precise. They are attained, for example, when A is the orthogonal sum of m dissipative Volterra operators with one-dimensional imaginary components.

To prove this theorem it is obviously sufficient to prove its second assertion. The theory of models (Chapter V and p. 260) shows that this assertion is equivalent to a theorem concerning the spectrum of a skew-periodic boundary-value problem for a definite canonical system. This theorem, together with its simple proof which uses an idea of de Branges, is given in Chapter VI (see Theorem 2.2).

For nondissipative Volterra operators the bounds (2.31) and (2.32) remain valid if their right sides are doubled. This is easily obtained with the aid of result 2 of §4 (see [GK], p. 195).

§3. General properties of the transformator of triangular truncation in s.n. ideals

Let $\mathfrak{P}$ be a complete (i.e. continuous bordered) chain. It follows from Theorems I.5.1 and I.6.1 that every operator $A \in \mathfrak{S}_\infty$ which has $\mathfrak{P}$ as

---

simple and ingenious idea. Carried out correctly, it enables one to obtain the bounds (2.30). Moreover, as de Branges correctly stated, in deriving these estimates one can omit the requirement that A be a Volterra operator, and replace it by the requirement that the spectrum of A be purely imaginary.

We also note that in connection with the invalid inequalities (*), de Branges' [11] derivation of Theorem 4.2 (of V. I. Macaev) and a generalization of it is invalid, although the generalization itself appears to be true.

an eigenchain is a Volterra operator, and can be reconstructed in unique fashion from its imaginary component by means of the transformator of triangular truncation.

There naturally arises the question whether, given a complete chain $\mathfrak{P}$ and an arbitrary completely continuous selfadjoint operator X, one can find an operator Y which has imaginary component equal to X (i.e. $Y_{\mathscr{J}} = X$) and has $\mathfrak{P}$ as an eigenchain.

By virtue of Theorems I.5.1 and I.6.1, this question is equivalent to the following:

Given a complete chain $\mathfrak{P}$; *will the integral*

$$\int_{\mathfrak{P}} PX\,dP \tag{3.1}$$

converge in the uniform norm for every selfadjoint operator $X \in \mathfrak{S}_\infty$?

M. S. Brodskiĭ [**5**] has shown that, generally speaking, the answer to this question is no. We shall show below that for any complete chain there exists an operator X ($\in \mathfrak{S}_\infty$) for which the integral (3.1) diverges.

This result will be obtained as a corollary of a theorem, concerning the transformator conjugate to the transformator $\mathscr{C}$, which will be important in all that follows. In §5, in connection with a theorem of V. I. Macaev, it will be shown how, given a complete chain, one can effectively construct an entire class of operators X for which the integral (3.1) diverges even in the weak sense.

1. General properties of the transformators $\mathscr{C}$ and $\mathscr{S}$. Henceforth the transformator $\mathscr{C}$ (and accordingly $\mathscr{C}^\perp$ and $\mathscr{S}$) will be regarded as an operator acting in the s.n. ideal $\mathfrak{S}_\infty$.

Let $\mathfrak{P}$ be a fixed maximal continuous chain. The domain $\mathfrak{D}(\mathscr{C})$ of the transformator

$$\mathscr{C}(X) = \int_{\mathfrak{P}} PX\,dP$$

consists of all $X \in \mathfrak{S}_\infty$ for which this integral converges in the uniform norm.

Let us note various properties of the transformator $\mathscr{C}$.[4)]

1. *The range* $\mathfrak{R}(\mathscr{C})$ *of* $\mathscr{C}$ *consists of all operators* $X \in \mathfrak{S}_\infty$ *which have* $\mathfrak{P}$ *as an eigenchain.*

[4)] Some of the properties given here were already noted in §10, Chapter I, for the transformator $\mathscr{C}$ in the s.n. ideal $\mathfrak{S}_2$.

This property follows at once from Theorems I.5.1 and I.6.1.

2. *The transformator* $\mathscr{C}$ *is a projector, i.e.* $\mathfrak{R}(\mathscr{C}) \subseteq \mathfrak{D}(\mathscr{C})$ *and for every* $X \in \mathfrak{D}(\mathscr{C})$

$$\mathscr{C}^2(X) = \mathscr{C}(X). \tag{3.2}$$

Let $Y = \mathscr{C}(X)$ $(X \in \mathfrak{D}(\mathscr{C}))$; taking into account that $\mathfrak{P}$ is an eigenchain of Y, and repeating the arguments of the proof of Theorem I.5.1, it is easy to show that $Y \in \mathfrak{D}(\mathscr{C})$ and $\mathscr{C}(Y) = Y$.

3. *If* $Y \in \mathfrak{R}(\mathscr{C})$, *then* $\mathscr{C}(Y^*) = 0$ *and*

$$Y = 2i\mathscr{C}(Y_{\mathscr{I}}) = 2\mathscr{C}(Y_{\mathscr{R}}). \tag{3.3}$$

Let us denote by $\mathscr{C}^{\perp}$ the projector $I - \mathscr{C}$. Then by Lemma I.6.1,

$$\mathscr{C}^{\perp}(X) = \int_{\mathfrak{P}} dPXP. \tag{3.4}$$

Since $\mathscr{C}(Y) = Y$, we have $\mathscr{C}^{\perp}(Y) = 0$ or

$$[\mathscr{C}^{\perp}(Y)]^* = \int_{\mathfrak{P}} PY^*dP = \mathscr{C}(Y^*) = 0.$$

The equality (3.3) follows from

$$\mathscr{C}(Y) = \mathscr{C}(Y) - \mathscr{C}(Y^*) = 2i\mathscr{C}\left(\frac{Y - Y^*}{2i}\right) = 2i\mathscr{C}(Y_{\mathscr{I}})$$

and

$$\mathscr{C}(Y) = \mathscr{C}(Y) + \mathscr{C}(Y^*) = 2\mathscr{C}\left(\frac{Y + Y^*}{2}\right) = 2\mathscr{C}(Y_{\mathscr{R}}).$$

Let us recall that $\mathfrak{P}^{\perp}$ denotes the chain consisting of all orthoprojectors of the form $P^{\perp} = I - P$ $(P \in \mathfrak{P})$. We shall call the chain $\mathfrak{P}^{\perp}$ the *dual* of $\mathfrak{P}$. Together with $\mathfrak{P}$, the chain $\mathfrak{P}^{\perp}$ is continuous.

It is easily seen that $\mathfrak{P}^{\perp}$ is an eigenchain of A if and only if

$$PAP = PA$$

for all $P \in \mathfrak{P}$. In fact, this equality is equivalent to

$$(I - P)A(I - P) = A(I - P).$$

It follows from this remark that *if* $\mathfrak{P}$ *is an eigenchain of* A, *then the dual chain* $\mathfrak{P}^{\perp}$ *is an eigenchain of* A^*.

4. *The set of zeros of the transformator* $\mathscr{C}$ *consists of all operators* $X \in \mathfrak{S}_{\infty}$ *which have the dual* $\mathfrak{P}^{\perp}$ *of* $\mathfrak{P}$ *as an eigenchain.*

Let X have $\mathfrak{P}^{\perp}$ as an eigenchain; then X^* has $\mathfrak{P}$ as an eigenchain, and so $X^* \in \mathfrak{R}(\mathscr{C})$ and

$$X^* = \mathscr{C}(X^*), \tag{3.5}$$

which is equivalent to

$$X = \int_{\mathfrak{P}} dPXP = \mathscr{C}^{\perp}(X) = X - \mathscr{C}(X);$$

therefore $X \in \mathfrak{D}(\mathscr{C})$ and $\mathscr{C}(X) = 0$.

Conversely, if $\mathscr{C}(X) = 0$, then $\mathscr{C}^{\perp}(X) = X$ or

$$X^* = \int_{\mathfrak{P}} PX^*dP.$$

It follows that X^* has $\mathfrak{P}$ as an eigenchain or, what is the same, X has $\mathfrak{P}^{\perp}$ as an eigenchain.

5. *The domain $\mathfrak{D}(\mathscr{C})$ of the transformator $\mathscr{C}$ consists of all $X \in \mathfrak{S}_\infty$ which can be represented as a sum $X = X_1 + X_2$, where X_1 $(\in \mathfrak{S}_\infty)$ has $\mathfrak{P}$ as an eigenchain and X_2 has the dual $\mathfrak{P}^{\perp}$ as an eigenchain.*

This property of $\mathscr{C}$ follows at once from properties 1, 2 and 4.

Obviously the domain of the transformator $\mathscr{S}$ $(= i(\mathscr{C} - \mathscr{C}^{\perp}))$ coincides with the domain of $\mathscr{C}$ and hence also of $\mathscr{C}^{\perp}$, and for every $X \in \mathfrak{D}(\mathscr{C})$ one has $\mathscr{S}(X) \in \mathfrak{D}(\mathscr{C})$ and

$$\mathscr{S}(\mathscr{S}X)) = \mathscr{S}^2(X) = -X.$$

It is easily verified that for any selfadjoint operator $H \in \mathfrak{D}(\mathscr{C})$ one has

$$\mathscr{S}(H) = [2i\mathscr{C}(H)]_{\mathscr{R}}.$$

It follows that the real subspace $\hat{\mathfrak{S}}_\infty$ consisting of all selfadjoint operators from $\mathfrak{S}_\infty$ is invariant relative to $\mathscr{S}$.

2. An identity for the transformator $\mathscr{S}$. In subsequent considerations (see §§6 and 9) an essential role will be played by the following identity for the transformator $\mathscr{S}$, which is of a purely algebraic nature.

THEOREM 3.1. *For any operator $X \in \mathfrak{D}(\mathscr{S})$ we have $X\mathscr{S}(X) + \mathscr{S}(X)X \in \mathfrak{D}(\mathscr{S})$ and*

$$[\mathscr{S}(X)]^2 = X^2 + \mathscr{S}(X\mathscr{S}(X) + \mathscr{S}(X)X) \quad (X \in \mathfrak{D}(\mathscr{S})). \tag{3.6}$$

PROOF. For the case of selfadjoint X (we can restrict ourselves to this case in subsequent applications) the identity (3.6) can be obtained immediately. In fact, by Theorem I.6.1 the operator

$$A = \mathscr{S}(X) + iX$$

is a Volterra operator which has $\mathfrak{P}$ as an eigenchain. Therefore the operator

$$A^2 = [\mathscr{S}(X)]^2 - X^2 + i(X\mathscr{S}(X) + \mathscr{S}(X)X)$$

is also a Volterra operator which has $\mathfrak{P}$ as an eigenchain. Thus, by Theorem I.5.1, $X\mathscr{S}(X) + \mathscr{S}(X)X \in \mathfrak{D}(\mathscr{C})$ and

$$A^2 = 2i\mathscr{C}(X\mathscr{S}(X) + \mathscr{S}(X)X),$$

or

$$[A^2]_{\mathscr{R}} = [\mathscr{S}(X)]^2 - X^2 = \mathscr{S}[X\mathscr{S}(X) + \mathscr{S}(X)X],$$

which coincides with (3.6).

In the general case of a nonselfadjoint X the proof of (3.6) is not very much more difficult. It is a corollary of a simple algebraic fact which is formulated in terms of rings.

Let R be an arbitrary ring containing subrings R_1 and R_2 whose intersection consists only of zero. We have the following result.

Let σ be the involution which acts on the direct sum $\hat{R} = R_1 + R_2$ according to the rule

$$\sigma(x_1 + x_2) = x_1 - x_2 \quad (x_j \in R_j;\ j = 1, 2).$$

Then for all $x \in \hat{R}$ we have $y = x\sigma(x) + \sigma(x)x \in \hat{R}$, and

$$\sigma[x\sigma(x) + \sigma(x)x] = [\sigma(x)]^2 + x^2. \tag{3.7}$$

In fact, let $x = x_1 + x_2$ $(x_1 \in R_1,\ x_2 \in R_2)$. Then it is easily seen that

$$y = x\sigma(x) + \sigma(x)x = 2(x_1^2 - x_2^2).$$

Since $x_j^2 \in R_j$ $(j = 1, 2)$, we have $y \in \hat{R}$ and $\sigma(y) = 2(x_1^2 + x_2^2)$. On the other hand, it is easily seen that $x^2 + [\sigma(x)]^2 = 2(x_1^2 + x_2^2)$. Therefore (3.7) holds.

The identity (3.6) is obtained from (3.7) if we put $R = \mathfrak{S}_\infty$, and take for R_1 the ring of all completely continuous operators which have $\mathfrak{P}$ as an eigenchain, and for R_2 the ring of all completely continuous operators which have the dual of $\mathfrak{P}$ as an eigenchain. Obviously, then, $\sigma = -i\mathscr{S}$.

The theorem is proved.

3. The action of the transformator $\mathscr{C}$ from one ideal into another. Let $\mathfrak{S}_{\mathrm{I}}$ and $\mathfrak{S}_{\mathrm{II}}$ be two Banach spaces of completely continuous operators. By the transformator $\mathscr{C}_{\mathrm{I}\to\mathrm{II}}$, acting from $\mathfrak{S}_{\mathrm{I}}$ into $\mathfrak{S}_{\mathrm{II}}$ $(\mathscr{C}_{\mathrm{I}\to\mathrm{II}}: \mathfrak{S}_{\mathrm{I}} \to \mathfrak{S}_{\mathrm{II}})$, we shall understand the appropriate restriction of the transformator $\mathscr{C}$, i.e. the transformator which coincides with $\mathscr{C}$ on its domain, which consists of all X in the intersection $\mathfrak{D}(\mathscr{C}) \cap \mathfrak{S}_{\mathrm{I}}$ for which $\mathscr{C}(X) \in \mathfrak{S}_{\mathrm{II}}$.

THEOREM 3.2. *Let $\mathfrak{S}_{\mathrm{I}}$ and $\mathfrak{S}_{\mathrm{II}}$ be two Banach spaces of completely continuous operators, whose norms $|\ |_{\mathrm{I}}$ and $|\ |_{\mathrm{II}}$ satisfy the inequalities*

$$|X| \equiv |X|_{\infty} \leqq C_j|X|_j \quad (X \in \mathfrak{S}_j;\ j = \mathrm{I}, \mathrm{II}), \tag{3.8}$$

where the C_j $(j = \mathrm{I}, \mathrm{II})$ are constants.

Then the restriction $\mathscr{C}_{\mathrm{I}\to\mathrm{II}}$ of the transformator $\mathscr{C}$ is a closed operator.

PROOF. Let X_n $(n = 1, 2, \cdots)$ be a sequence of operators from the domain of $\mathscr{C}_{\mathrm{I}\to\mathrm{II}}$ such that

$$\lim_{n\to\infty} |X_n - X_0|_{\mathrm{I}} = 0 \quad \text{and} \quad \lim_{n\to\infty} |\mathscr{C}(X_n) - Y_0|_{\mathrm{II}} = 0.$$

The elements X_n $(n = 1, 2, \cdots)$ belong to the domain of $\mathscr{C}^{\perp}$ $(= I - \mathscr{C})$. Hence, by (3.8)

$$\lim_{n\to\infty} |\mathscr{C}^{\perp}(X_n) - X_0 + Y_0| = 0.$$

It follows that $\mathfrak{P}^{\perp}$ is an eigenchain of the operator $-Y_0 + X_0$. Since $\mathfrak{P}$ is an eigenchain of Y_0, by result 5 we have

$$X_0 = Y_0 + (X_0 - Y_0) \in \mathfrak{D}(\mathscr{C}) \quad \text{and} \quad \mathscr{C}(X_0) = Y_0.$$

COROLLARY 3.1. *If the hypotheses of Theorem 3.2 are satisfied and the transformator $\mathscr{C}_{\mathrm{I}\to\mathrm{II}}$ is defined on all of $\mathfrak{S}_{\mathrm{I}}$, then it is bounded.*

Henceforth we shall say that $\mathscr{C}_{\mathrm{I}\to\mathrm{II}}$ is bounded if its domain coincides with all of $\mathfrak{S}_{\mathrm{I}}$ and if it is bounded as an operator from $\mathfrak{S}_{\mathrm{I}}$ into $\mathfrak{S}_{\mathrm{II}}$.

LEMMA 3.1. *Let $\mathfrak{S}_{\mathrm{I}}$ and $\mathfrak{S}_{\mathrm{II}}$ be two s.n. ideals. If the transformator $\mathscr{C}_{\mathrm{I}\to\mathrm{II}}(\cdot\,;\mathfrak{P})$ is bounded for every complete chain $\mathfrak{P}$, then*

$$\sup_{\mathfrak{P}} |\mathscr{C}_{\mathrm{I}\to\mathrm{II}}(\cdot\,;\mathfrak{P})| < \infty, \tag{3.9}$$

where the supremum is taken over all complete chains $\mathfrak{P}$.

PROOF. Let us assume that (3.9) does not hold. Then there exists a sequence of complete chains $\mathfrak{P}_n$ $(n = 1, 2, \cdots)$ and a sequence of operators A_n $(\in \mathfrak{S}_{\mathrm{I}})$ $(n = 1, 2, \cdots)$ such that

$$|A_n|_{\mathrm{I}} = 1/n^2 \quad \text{and} \quad |\mathscr{C}(A_n; \mathfrak{P}_n)|_{\mathrm{II}} \geqq n \ \ (n = 1, 2, \cdots).$$

We decompose the space $\mathfrak{H}$ into the orthogonal sum of an infinite number of infinite-dimensional subspaces:

$$\mathfrak{H} = \mathfrak{H}^{(1)} \oplus \mathfrak{H}^{(2)} \oplus \cdots.$$

Let us denote by U_n $(n = 1, 2, \cdots)$ an isometric mapping of $\mathfrak{H}^{(n)}$ onto $\mathfrak{H}$. We put

$$A^{(n)} = U_n A_n U_n^{-1}, \quad \mathfrak{P}^{(n)} = U_n \mathfrak{P}_n U_n^{-1} \quad (n = 1, 2, \cdots)$$

and denote by Q_n $(n = 1, 2, \cdots)$ the orthoprojector onto the subspace $\mathfrak{H}^{(n)}$ (we take $Q_0 = 0$). To each projector of the chain $\mathfrak{P}^{(n)}$ we add the projector $Q_1 + \cdots + Q_{n-1}$ $(n = 1, 2, \cdots)$. The union of all the chains thereby obtained forms a complete chain which we denote by $\mathfrak{P}$. Since

$$\sum_{n=1}^{\infty} |A^{(n)}|_{\mathrm{I}} = \sum_{n=1}^{\infty} |A_n|_{\mathrm{I}} < \infty,$$

the operator

$$A = \sum_{n=1}^{\infty} A^{(n)} Q_n \tag{3.10}$$

belongs to the space $\mathfrak{S}_{\mathrm{I}}$, and so by hypothesis we have $\mathscr{L}(A; \mathfrak{P}) \in \mathfrak{S}_{\mathrm{II}}$.

On the other hand, by virtue of the continuity of the transformator $\mathscr{L}_{\mathrm{I}\to\mathrm{II}}(\cdot\,; \mathfrak{P})$ and (3.10),

$$\mathscr{L}(A; \mathfrak{P}) = \sum_{n=1}^{\infty} \mathscr{L}(A^{(n)} Q_n; \mathfrak{P}). \tag{3.11}$$

Since

$$\mathscr{L}(A^{(n)} Q_n; \mathfrak{P}) = \mathscr{L}(A^{(n)}; \mathfrak{P}^{(n)}) Q_n,$$

it follows from (3.11) that

$$|\mathscr{L}(A; \mathfrak{P})|_{\mathrm{II}} \geqq |\mathscr{L}(A^{(n)}; \mathfrak{P}^{(n)})|_{\mathrm{II}}.$$

This, however, is not possible, since

$$|\mathscr{L}(A^{(n)}; \mathfrak{P})|_{\mathrm{II}} = |\mathscr{L}(A_n; \mathfrak{P}_n)|_{\mathrm{II}} \to \infty \quad \text{as } n \to \infty.$$

The lemma is proved.

4. A theorem on the action of the transformator $\mathscr{L}$ in conjugate ideals. The following theorem plays an important role in the sequel.

THEOREM 3.3 (DUALITY PRINCIPLE). *Let $\mathfrak{S}_{\mathrm{I}}$ be a separable s.n. ideal, and let $\mathfrak{S}_{\mathrm{II}}$ be a separable s.n. ideal whose conjugate $\mathfrak{S}_{\mathrm{II}}^*$ is also separable. If the transformator $\mathscr{L}_{\mathrm{I}\to\mathrm{II}}$ is bounded, then the transformator $\mathscr{L}_{\mathrm{II}^*\to\mathrm{I}^*}$ is bounded. Conversely, if the transformator $\mathscr{L}_{\mathrm{II}^*\to\mathrm{I}^*}$ is bounded and in addition $\mathscr{L}(\mathfrak{R}) \subseteq \mathfrak{S}_{\mathrm{II}}$, then the transformator $\mathscr{L}_{\mathrm{I}\to\mathrm{II}}$ is bounded. In both cases* *the second of the two transformators* $\mathscr{L}_{\mathrm{I}\to\mathrm{II}}$, $\mathscr{L}_{\mathrm{II}^*\to\mathrm{I}^*}$ *is the conjugate of the first.*

In the case where $\mathfrak{S}_{\mathrm{I}}$ *coincides elementwise with* $\mathfrak{S}_1$, *the theorem remains valid if in its statement the space* $\mathfrak{S}_{\mathrm{I}^*}$ *is replaced by* $\mathfrak{S}_\infty$.

Here the symbol $\mathscr{L}_{\mathrm{II}^*\to\mathrm{I}^*}$ denotes the appropriate restriction of the transformator $\mathscr{L}$, acting from $\mathfrak{S}_{\mathrm{II}}^*$ into $\mathfrak{S}_{\mathrm{I}}^*$, and $\mathfrak{R}$ denotes as before the set of all finite-dimensional operators.

PROOF. Let K_1 and K_2 be finite-dimensional operators. Since $\mathscr{L}$ is an orthogonal projector in $\mathfrak{S}_2$, we have

$$\mathrm{sp}(\mathscr{L}(K_1)K_2^*) = \mathrm{sp}(K_1[\mathscr{L}(K_2)]^*). \tag{3.12}$$

If the transformator $\mathscr{L}_{\mathrm{I}\to\mathrm{II}}$ is bounded, then the transformator $\mathscr{L}^*_{\mathrm{I}\to\mathrm{II}} = \mathscr{T}$ is also bounded. Since

$$\mathrm{sp}(\mathscr{L}(K_1)K_2^*) = \mathrm{sp}(K_1[\mathscr{T}(K_2)]^*) \quad (K_1, K_2 \in \mathfrak{R}),$$

it follows that

$$\mathrm{sp}(K_1[\mathscr{T}(K_2) - \mathscr{L}(K_2)]^*) = 0 \quad (K_1, K_2 \in \mathfrak{R}). \tag{3.13}$$

The set $\mathfrak{R}$ of all finite-dimensional operators is dense in $\mathfrak{S}_{\mathrm{I}}$; hence, by virtue of (3.13)

$$\mathscr{T}(K) = \int_{\mathfrak{P}} PKdP = \mathscr{L}_{\mathrm{II}^*\to\mathrm{I}^*}(K) \quad (K \in \mathfrak{R}). \tag{3.14}$$

According to Theorem 3.2, the transformator $\mathscr{L}_{\mathrm{II}^*\to\mathrm{I}^*}$ is closed; moreover, by (3.14) it is a restriction of the transformator $\mathscr{T}$. This, obviously, is possible only when $\mathscr{T} = \mathscr{L}_{\mathrm{II}^*\to\mathrm{I}^*}$.

Let us consider the second case, for which it is given that $\mathscr{L}_{\mathrm{II}^*\to\mathrm{I}^*}$ is bounded. In this case it follows from (3.12) that

$$\mathrm{sp}(\mathscr{L}(K)A^*) = \mathrm{sp}(K[\mathscr{L}(A)]^*) \quad (K \in \mathfrak{R}, A \in \mathfrak{S}_{\mathrm{II}}^*).$$

Hence[5)]

$$|\mathrm{sp}(\mathscr{L}(K)A^*)| \leqq |K|_{\mathrm{I}} |\mathscr{L}(A)|_{\mathrm{I}^*} \leqq |\mathscr{L}_{\mathrm{II}^*\to\mathrm{I}^*}|\,|K|_{\mathrm{I}}\,|A|_{\mathrm{II}^*}.$$

It follows that

$$|\mathscr{L}(K)|_{\mathrm{II}} \leqq |\mathscr{L}_{\mathrm{II}^*\to\mathrm{I}^*}|\,|K|_{\mathrm{I}} \quad (K \in \mathfrak{R}).$$

Thus, the closed transformator $\mathscr{L}_{\mathrm{I}\to\mathrm{II}}$ is bounded on the set $\mathfrak{R}$ of all finite-dimensional operators, which set lies in its domain and is dense

5) Here $|\ |_{\mathrm{I}^*}$ and $|\ |_{\mathrm{II}^*}$ denote the norms in the conjugate spaces $\mathfrak{S}_{\mathrm{I}}^*$ and $\mathfrak{S}_{\mathrm{II}}^*$.

in $\mathfrak{S}_{\mathrm{I}}$. Therefore $\mathscr{L}_{\mathrm{I}\to\mathrm{II}}$ is bounded. It is easily seen that $\mathscr{L}_{\mathrm{II}^*\to\mathrm{I}^*}$ is the conjugate of $\mathscr{L}_{\mathrm{I}\to\mathrm{II}}$.

To conclude the proof we note that for any completely continuous operator A from the domain of $\mathscr{L}$ the operator $\mathscr{L}(A)$ is completely continuous. Therefore, when $\mathfrak{S}_{\mathrm{I}} = \mathfrak{S}_1$ the space $\mathfrak{S}_{\mathrm{I}}^*$ $(= \mathfrak{R})$ can be replaced by $\mathfrak{S}_\infty$ everywhere in the arguments.

The theorem is proved.

REMARK 3.1. Theorems 3.2 and 3.3, and Lemma 3.1, obviously remain valid if in their statements the transformator $\mathscr{L}$ is replaced by the transformator $\mathscr{L}^\perp$ or $i\mathscr{S}$.

As a corollary of Theorem 3.3 we obtain the following result.

6. *For any complete chain* $\mathfrak{P}$ *there exists a completely continuous selfadjoint operator which is not the imaginary component of any Volterra operator having* $\mathfrak{P}$ *as an eigenchain.*

Thus, the transformator $\mathscr{L}(\cdot;\mathfrak{P})$ *is not bounded.*

In fact, let us assume the contrary, i.e. the transformator $\mathscr{L}$ is bounded and thus defined on all of $\mathfrak{S}_\infty$. Then by Theorem 3.3 the transformator $\mathscr{L}_{1\to 1}$ $(=\mathscr{L}\colon \mathfrak{S}_1 \to \mathfrak{S}_1)$ would be bounded and hence defined everywhere on $\mathfrak{S}_1$. We are led to a contradiction, since even for a one-dimensional selfadjoint operator X the integral $\mathscr{L}(X)$ does not belong to $\mathfrak{S}_1$ (see §6, Chapter II).

This proof does not indicate how to construct operators for which the integral $\mathscr{L}(X;\mathfrak{P})$ does not converge. As has already been mentioned, in §4 a constructive description will be given for an entire class of operators for which this integral does not converge even in the weak sense.

We supplement result 6 by the following remark of V. I. Macaev.

Every selfadjoint completely continuous operator is the imaginary component of some Volterra operator.

In fact, let H be a selfadjoint operator from $\mathfrak{S}_\infty$. In view of Theorem I.10.1 we need only consider the case in which $H \notin \mathfrak{S}_2$. We break up the sequence of all eigenvalues of H into an infinite number of sequences $\{\lambda_n^{(j)}\}_{n=1}^\infty$ $(j = 1, 2, \cdots)$ which have the property

$$M_j^2 = \sum_{n=1}^\infty |\lambda_n^{(j)}|^2 < \infty \quad (j = 1, 2, \cdots) \quad \text{and} \quad \lim_{j\to\infty} M_j = 0.$$

Let us denote by $e_n^{(j)}$ $(n, j = 1, 2, \cdots)$ a normalized eigenvector of H corresponding to the eigenvalue $\lambda_n^{(j)}$. We form the operators

$$H_j = \sum_{n=1}^\infty \lambda_n^{(j)}(\cdot, e_n^{(j)})e_n^{(j)} \quad (j = 1, 2, \cdots).$$

Obviously $H_j \in \mathfrak{S}_2$ $(j = 1, 2, \cdots)$, and by Theorem I.10.1 there exists a Volterra operator A_j $(j = 1, 2, \cdots)$, acting in the subspace $\overline{\mathfrak{R}(H_j)}$, with the properties

$$(A_j)_{\mathscr{J}} = H_j \text{ on } \overline{\mathfrak{R}(H_j)}, \quad |A_j| \leqq |A_j|_2 \leqq M_j \quad (j = 1, 2, \cdots).$$

Let $A^{(j)}$ $(j = 1, 2, \cdots)$ be the inessential extension of A_j to all of $\mathfrak{H}$. Obviously

$$A = \sum_{j=1}^{\infty} A^{(j)}$$

is a Volterra operator, and

$$A_{\mathscr{J}} = \sum_{j=1}^{\infty} A^{(j)}_{\mathscr{J}} = \sum_{j=1}^{\infty} H_j = H.$$

5. The transformators $\mathscr{L}$ and $\mathscr{S}$ for a chain with gaps. Let $\mathfrak{P}$ be an arbitrary bordered chain and let (P_j^-, P_j^+) $(j = 1, 2, \cdots, \omega)$ be the sequence of all of its gaps. As has already been noted, the vanishing of the blocks of the operator $X \in \mathfrak{S}_\infty$ on these gaps, i.e. the condition

$$(3.15) \qquad (P_j^+ - P_j^-) X (P_j^+ - P_j^-) = 0 \quad (j = 1, 2, \cdots, \omega),$$

is a necessary condition for the convergence of the integral

$$\mathscr{L}(X; \mathfrak{P}) = \int_{\mathfrak{P}} PX \, dP.$$

Thus, for a chain with gaps the domain of the transformator $\mathscr{L}(\cdot; \mathfrak{P})$ lies in the subspace $\mathfrak{S}_\infty[\mathfrak{P}]$ of all operators $X \in \mathfrak{S}_\infty$ which satisfy the condition (3.15). The range of $\mathscr{L}(\cdot; \mathfrak{P})$ also lies in the subspace $\mathfrak{S}_\infty[\mathfrak{P}]$.

It is easily seen that all the results of subsections 1 and 2 of this section, which have been proved for the case of a continuous chain $\mathfrak{P}$, remain valid for a chain with gaps if in their formulation $\mathfrak{S}_\infty$ is replaced by the subspace $\mathfrak{S}_\infty[\mathfrak{P}]$. The results of subsection 3 also remain valid if in their formulation $\mathfrak{S}_{\mathrm{I}}$ and $\mathfrak{S}_{\mathrm{II}}$ are replaced by their subspaces $\mathfrak{S}_{\mathrm{I}}[\mathfrak{P}] = \mathfrak{S}_\infty[\mathfrak{P}] \cap \mathfrak{S}_{\mathrm{I}}$ and $\mathfrak{S}_{\mathrm{II}}[\mathfrak{P}] = \mathfrak{S}_\infty[\mathfrak{P}] \cap \mathfrak{S}_{\mathrm{II}}$.

These assertions can be supplemented by the following theorem.

THEOREM 3.4. *Let $\mathfrak{S}_{\mathrm{I}}$ and $\mathfrak{S}_{\mathrm{II}}$ be two s.n. ideals. If the transformator $\mathscr{L}_{\mathrm{I}\to\mathrm{II}}(\cdot; \mathfrak{P})$ is bounded for every complete chain $\mathfrak{P}$ and*

$$|\mathscr{L}_{\mathrm{I}\to\mathrm{II}}(\cdot; \mathfrak{P})| \leqq l,$$

where l is a constant which does not depend on $\mathfrak{P}$, then for any bordered chain $\mathfrak{P}$ (with gaps) the transformator $\mathscr{L}_{\mathrm{I}[\mathfrak{P}]\to\mathrm{II}[\mathfrak{P}]}(\cdot; \mathfrak{P}) \equiv \mathscr{L}_{\mathfrak{S}_{\mathrm{I}}[\mathfrak{P}]\to\mathfrak{S}_{\mathrm{II}}[\mathfrak{P}]}$

is likewise bounded, and

$$|\mathscr{L}_{\mathrm{I}[\mathfrak{P}]\to\mathrm{II}[\mathfrak{P}]}(\cdot;\mathfrak{P})| \leqq l.$$

PROOF. Let $\mathfrak{P} = \{P\}$ be *any* bordered chain. Just as in the proof of Theorem I.3.2, we enlarge the chain $\mathfrak{P}$ to a continuous chain $\widetilde{\mathfrak{P}}$ in some extended Hilbert space $\widetilde{\mathfrak{H}} \supset \mathfrak{H}$. Let X belong to $\mathfrak{S}_{\mathrm{I}}[\mathfrak{P}]$, and let $\widetilde{X}$ be the inessential extension of X to the space $\widetilde{\mathfrak{H}}$. It is easy to see that every Riemann sum of the first of the integrals

$$\widetilde{Y} = \int_{\widetilde{\mathfrak{P}}} \widetilde{P}\widetilde{X}\,d\widetilde{P}, \quad Y = \int_{\mathfrak{P}} PX\,dP$$

is the inessential extension from $\mathfrak{H}$ to $\widetilde{\mathfrak{H}}$ of some Riemann sum of the second integral. We conclude that if the first integral converges, then so does the second, and $\widetilde{Y}$ is the inessential extension to $\widetilde{\mathfrak{H}}$ of Y. Since by hypothesis

$$|\widetilde{Y}|_{\mathrm{II}} \leqq l|\widetilde{X}|_{\mathrm{I}}$$

and

$$|\widetilde{X}|_{\mathrm{I}} = |X|_{\mathrm{I}}, \quad |\widetilde{Y}|_{\mathrm{II}} = |Y|_{\mathrm{II}},$$

it follows that

$$|Y|_{\mathrm{II}} \leqq l|X|_{\mathrm{I}}.$$

The theorem is proved.

We note further that the results of subsection 4 concerning the conjugate transformator also can be extended to chains with gaps. For this one has to bear in mind that if $\mathfrak{S}$ is a separable s.n. ideal, then $(\mathfrak{S}[\mathfrak{P}])^* = \mathfrak{S}^*[\mathfrak{P}]$.

Everything that has been said for the transformator $\mathscr{L}$ is, of course, true for the transformators $\mathscr{L}^{\perp}$ and $\mathscr{S}$.

§4. Volterra operators with imaginary component from $\mathfrak{S}_\omega$

1. By result 6 of §3 the integral

$$Y = \int_{\mathfrak{P}} PX\,dP \tag{4.1}$$

with respect to an arbitrary continuous closed chain does not have meaning for every completely continuous linear operator X. In this section we shall show that the class $\mathfrak{S}_\omega$ is the widest class of operators X for which the integral (4.1) converges in the uniform norm for *every* complete chain.

The class $\mathfrak{S}_\omega$, which was introduced by V. I. Macaev, is remarkable for other reasons as well. This class, as its discoverer showed, also plays an unusual role in various questions of the theory of nonselfadjoint perturbations of selfadjoint operators (see, for example, the theorems by V. I. Macaev mentioned in the Appendix).

LEMMA 4.1. *For any orthonormal system* ϕ_j $(j = 0, \pm 1, \pm 2, \cdots)$ *there exists a complete chain* $\mathfrak{P}$ *such that for every completely continuous selfadjoint operator*

$$X = \sum_{j=-\infty}^{\infty} \lambda_j(\cdot, \phi_j)\phi_j \tag{4.2}$$

satisfying the conditions

$$\lambda_j \geqq 0 \ (j = 1, 2, \cdots), \quad \lambda_j \leqq 0 \ (j = 0, -1, \cdots);$$

$$\sum_{j=-\infty}^{\infty} \frac{\lambda_j}{2j-1} = \infty, \tag{4.3}$$

the integral (4.1) *diverges in the weak sense.*

Conversely, for any complete chain $\mathfrak{P}$ *there exists an orthonormal system* ϕ_j $(j = 0, \pm 1, \cdots)$ *such that the above conclusion will hold for the specified class of selfadjoint operators.*

PROOF. We may suppose without loss of generality that the system ϕ_j $(j = 0, \pm 1, \cdots)$ is complete. The space $\mathfrak{H}$ can then be realized as the space $L_2(0,1)$, with

$$\phi_\kappa = e^{(2k-1)\pi i t} \quad (k = 0, \pm 1, \cdots; \ 0 \leqq t \leqq 1).$$

We shall show that the ordinary "truncation" chain $\{P_t\}_{0 \leqq t \leqq 1}$:

$$P_t f(x) = \begin{cases} f(x), & x \leqq t \\ 0, & x > t, \end{cases} \quad (f \in L_2(0,1)), \tag{4.4}$$

satisfies the requirements of the lemma.

In fact, let us consider the sum

$$S_n = \sum_{j=0}^{n-1} P_{\tau_j} X \Delta P_{t_j} \quad (n = 1, 2, \cdots),$$

where $t_j = j/n$ $(j = 0, 1, \cdots, n)$ and $\tau_j = (2j+1)/2n$ $(j = 0, 1, \cdots, n-1)$ and, finally, $\Delta P_{t_j} = P_{t_{j+1}} - P_{t_j}$.

From (4.4) follows

$$S_n = \sum_{j=0}^{n-1} \sum_{k=-\infty}^{\infty} \lambda_k(\cdot, \Delta P_{t_j}\phi_k) P_{\tau_j}\phi_k,$$

and so for $f \equiv 1$ we will have

$$\sigma_n = (S_n f, f) = \sum_{j=0}^{n-1} \sum_{k=-\infty}^{\infty} \lambda_k(f, \Delta P_{t_j}\phi_k)(P_{\tau_j}\phi_k, f). \tag{4.5}$$

Bearing in mind (4.4) and (4.5), we obtain

$$\sigma_n = \sum_{j=1}^{n-1} \sum_{k=-\infty}^{\infty} \lambda_k \int_0^{\tau_j} e^{(2k-1)\pi it} dt \int_{t_j}^{t_{j+1}} e^{-(2k-1)\pi it} dt,$$

or

$$\sigma_n = \sum_{j=0}^{n-1} \sum_{k=-\infty}^{\infty} \frac{\lambda_k}{(2k-1)^2\pi^2} (e^{(2k-1)\pi i\tau_j} - 1)(e^{-(2k-1)\pi it_{j+1}} - e^{-(2k-1)\pi it_j}).$$

After some uncomplicated transformations we arrive at

$$\sigma_n = \sum_{k=-\infty}^{\infty} \frac{2\lambda_k}{(2k-1)^2\pi^2 i} \sin\frac{2k-1}{2n}\pi \sum_{j=0}^{n-1} \left(1 - \exp\left[-(2k-1)\pi i\frac{2j-1}{2n}\right]\right),$$

or

$$\sigma_n = -\sum_{k=-\infty}^{\infty} \frac{2\lambda_k i}{(2k-1)^2\pi^2} \sin\frac{2k-1}{2n}\pi\left(n + \frac{i}{\sin((2k-1)/2n)\pi}\right).$$

Thus

$$\operatorname{Im}\sigma_n = \sum_{k=-\infty}^{\infty} \frac{2\lambda_k n}{\pi^2(2k-1)^2} \sin\frac{2k-1}{2n}\pi$$

and so for $m = 2n$

$$|\operatorname{Im}\sigma_m| \geqq \alpha_n - \beta_n,$$

where

$$\alpha_n = \sum_{k=-n}^{n} \frac{4\lambda_k n}{\pi^2(2k-1)^2} \sin\frac{2k-1}{4n}\pi;$$

$$\beta_n = \left(\sum_{k=-\infty}^{-n-1} + \sum_{k=n+1}^{\infty}\right) \frac{4|\lambda_k|n}{\pi^2(2k-1)^2} \left|\sin\frac{(2k-1)\pi}{4n}\right|.$$

It is easily seen that

$$\beta_n \leqq \frac{8|X|n}{\pi^2} \sum_{k=n+1}^{\infty} \frac{1}{(2k-1)^2} \leqq \frac{4n|X|}{\pi^2(2n-1)} \quad (n = 1, 2, \cdots)$$

and

$$\alpha_n \geqq \frac{2}{\pi} \sum_{k=-n+1}^{n} \frac{\lambda_k}{2k-1} \quad (n = 1, 2, \cdots).$$

Hence

$$\lim_{n\to\infty} |\operatorname{Im} \sigma_{2n}| = \infty.$$

This proves that the integral (4.1) diverges in the weak sense.

Let $\mathfrak{P}$ be any complete chain, and let $\hat{\mathfrak{H}}$ be the closed linear hull of all the vectors Pe ($P \in \mathfrak{P}$), where e is some fixed unit vector from $\mathfrak{H}$. The subspace $\hat{\mathfrak{H}}$ is invariant with respect to all the projectors $P \in \mathfrak{P}$, and therefore the chain $\mathfrak{P}$ generates a complete chain $\hat{\mathfrak{P}}$ in $\hat{\mathfrak{H}}$. We form the simple (by Theorem I.7.1) dissipative Volterra operator

$$J^{(e)} = 2i \int_{\hat{\mathfrak{P}}} (\cdot, dPe)\, Pe,$$

which has the one-dimensional imaginary component $[J^{(e)}]_{\mathscr{I}} = (\cdot, e)e$. By Lemma II.6.1 its real component has a complete orthonormal system of eigenvectors ϕ_j $(j = 0, \pm 1, \cdots)$ such that

$$J^{(e)}_{\mathscr{R}} \phi_j = \frac{2}{\pi(2j-1)} \phi_j.$$

Then the integral (4.1) will diverge for any operator X of the form (4.2) satisfying the conditions (4.3). In fact, by Theorem I.8.1 there exists a unitary operator U, mapping $L_2(0,1)$ onto $\hat{\mathfrak{H}}$, such that

$$J^{(e)} = UJU^{-1},$$

where J is the operator defined in $L_2(0,1)$ by

$$(Jf)(t) = 2i \int_t^1 f(s)\, ds,$$

and $\hat{\mathfrak{P}} = \{UP_tU^{-1}\}_{0 \leqq t \leqq 1}$, while

$$\phi_k = U(e^{-(2k-1)\pi i t}) \quad (k = 0, \pm 1, \cdots).$$

The lemma is proved.

THEOREM 4.1. *Let $X \in \mathfrak{S}_\infty$. In order that the integral* (4.1) *converge in the uniform norm for every continuous chain $\mathfrak{P}$, it is necessary and sufficient that $X \in \mathfrak{S}_\omega$. If $X \in \mathfrak{S}_\omega$ and $\mathfrak{P}$ is a closed chain on all of whose gaps (P^-, P^+) the blocks of X vanish, i.e.*

$$(P^+ - P^-)X(P^+ - P^-) = 0, \tag{4.6}$$

then the integral (4.1) *converges, and one has the precise bound*

$$\left| \int_{\mathfrak{P}} PX\,dP \right| \leqq |X|_\omega. \tag{4.7}$$

PROOF. We shall first prove the second assertion of the theorem for the case in which $\mathfrak{P}$ is a continuous chain. According to Theorem 2.2, the transformator

$$\mathscr{L}_{1\to\Omega}(X) = \int_{\mathfrak{P}} PX\,dP \quad (\mathscr{L}_{1\to\Omega} \equiv \mathscr{L}_{\mathfrak{S}_1\to\mathfrak{S}_\Omega},\ X \in \mathfrak{S}_1)$$

is a bounded linear operator, and

$$|\mathscr{L}_{1\to\Omega}| = 1. \tag{4.8}$$

It follows by virtue of Theorem 3.3 that the integral (4.1) converges for every operator $X \in \mathfrak{S}_\omega$, and

$$\mathscr{L}^*_{\omega\to\infty} = \mathscr{L}_{1\to\Omega} \quad (\mathscr{L}^*_{\omega\to\infty} \equiv \mathscr{L}^*_{\mathfrak{S}_\omega\to\mathfrak{S}_\infty}). \tag{4.9}$$

The relations (4.8) and (4.9) imply the bound (4.7), and also show that (4.7) is precise.

The proof of the second assertion of the theorem for the general case of a chain $\mathfrak{P}$ with gaps can be obtained on the basis of Theorem 3.4 and that which has already been proved.

To complete the proof of the theorem it remains to show that for every completely continuous operator X which does not belong to $\mathfrak{S}_\omega$ there exists a continuous chain $\mathfrak{P}$ such that the integral (4.1) does not converge in the uniform norm. We can obviously restrict ourselves to the case in which X is selfadjoint.

Since adding to X a nuclear operator $T = T^*$ does not affect the convergence or divergence of (4.1), we can assume without loss of generality that X has an infinite number of positive and negative eigenvalues. But then the conditions of Lemma 4.1 will be satisfied by X. Hence there exists a chain $\mathfrak{P}$ for which the integral (4.1) will diverge even in the weak sense.

The theorem is proved.

Thus, for any continuous chain $\mathfrak{P}$ the corresponding transformator $\mathscr{L} = \mathscr{L}(\cdot\,; \mathfrak{P})$ maps all of $\mathfrak{S}_\omega$ into $\mathfrak{S}_\infty$. The question naturally arises: does the set of all values of the transformator $\mathscr{L}_{\omega\to\infty}$ belong to the space $\mathfrak{S}_\omega$? The answer to this question is no; namely, we have the following result.

1. *For any complete chain* $\mathfrak{P}$ *there exists an* $X \in \mathfrak{S}_\omega$ *such that the integral* (4.1) *does not belong to the class* $\mathfrak{S}_\omega$.

This result follows at once from result 2 of §2 and from Theorem 3.3. In fact, if the result we wish to prove did not hold, the transformator $\mathscr{L}_{\omega\to\omega}$ $(\equiv \mathscr{L}_{\mathfrak{S}_\omega\to\mathfrak{S}_\omega})$ would be bounded. By Theorem 3.3, $\mathscr{L}_{\Omega\to\Omega} = (\mathscr{L}_{\omega\to\omega})^*$ $(\mathscr{L}_{\Omega\to\Omega} \equiv \mathscr{L}_{\mathfrak{S}_\Omega\to\mathfrak{S}_\Omega})$; hence the transformator $\mathscr{L}_{\Omega\to\Omega}$ would be bounded. This contradicts result 2 of §2.

2. THEOREM 4.2 (V. I. MACAEV [2]). *Let* A *be a Volterra operator with imaginary component* $A_{\mathscr{I}} = H$ *belonging to the class* $\mathfrak{S}_\omega$. *Then*

$$|A_{\mathscr{R}}| \leqq \frac{2}{\pi}\sum_{j=1}^{\infty}\frac{\lambda_j^+(H)+\lambda_j^-(H)}{2j-1}\ \left(=\frac{2}{\pi}\,|H|_{\omega;\mathscr{K}}\right). \tag{4.10}$$

For any selfadjoint operator $H \in \mathfrak{S}_\omega$ *and any* $\epsilon > 0$ *there exists a Volterra operator* A *with* $A_{\mathscr{I}} = H$ *such that*[6)]

$$|A_{\mathscr{R}}| > \frac{2}{\pi}\,|H|_{\omega;\mathscr{K}} - \epsilon. \tag{4.11}$$

Moreover, we shall show that if both H^+ and H^- are infinite-dimensional, or at least the subspace of zeros of H is infinite-dimensional, then there exists a Volterra operator A with $A_{\mathscr{I}} = H$ for which

$$|A_{\mathscr{R}}| = \frac{2}{\pi}\,|H|_{\omega;\mathscr{K}}. \tag{4.12}$$

PROOF. For the case in which $H \in \mathfrak{S}_2$, the relation (4.10) is a simple corollary of the relation (II.4.8). If we establish (4.10) for operators $H \in \mathfrak{S}_2$, it is easy to extend it to any $H \in \mathfrak{S}_\omega$ by a passage to the limit.

In view of the importance of the bound (4.10), we shall present the original proof of it given by V. I. Macaev. This proof has the virtue that by means of it one can establish the second assertion of the theorem, and also the assertion regarding the equality (4.12).

We first consider the case of finite-dimensional $A_{\mathscr{I}} = H$. Let $\widetilde{A}$ be an inessential extension of A which has a complete eigenchain $\mathfrak{P}$, and let e be any unit vector from $\mathfrak{H}$. We form the operator $J^{(e)}$;

---

6) A finite-dimensional analog of this theorem and of Theorem 2.1, with more precise values for the constants appearing in the bounds, was obtained in a paper of I. C. Gohberg [2].

$$J^{(e)} = 2i \int_{\mathfrak{P}} (\cdot, dPe)\, Pe.$$

Then $J^{(e)}\widetilde{A}$ is a Volterra operator which belongs to $\mathfrak{S}_1$. Hence $\mathrm{sp}(J^{(e)}_{\mathscr{I}}\widetilde{A}_{\mathscr{R}}) = -\,\mathrm{sp}(J^{(e)}_{\mathscr{R}}\widetilde{H})$ or, what is the same,

$$(A_{\mathscr{R}}e, e) = -\frac{2}{\pi}\left(\sum_{j=1}^{\infty}\frac{(\widetilde{H}\phi_j, \phi_j)}{2j-1} + \sum_{j=-\infty}^{-1}\frac{(\widetilde{H}\phi_j, \phi_j}{2j-1}\right), \tag{4.13}$$

where $\{\phi_j\}_{-\infty}^{\infty}$ is a complete orthonormal sequence of eigenvectors of $J^{(e)}_{\mathscr{R}}$, so that

$$J^{(e)}_{\mathscr{R}}\phi_j = \frac{2}{\pi(2j-1)}\phi_j \quad (j = 1, 2, \cdots);$$

$$J^{(e)}_{\mathscr{R}}\phi_j = \frac{2}{(2j+1)\pi}\phi_j \quad (j = -1, -2, \cdots).$$

From Lemma II.2.2 it follows (applying the results of §1 to Φ_ω) that

$$|(A_{\mathscr{R}}e, e)| = \frac{2}{\pi}\left|\sum_{j=1}^{\infty}\frac{(\widetilde{H}^+\phi_j, \phi_j) + (\widetilde{H}^-\phi_{-j}, \phi_{-j})}{2j-1} - \sum_{j=1}^{\infty}\frac{(\widetilde{H}^-\phi_j, \phi_j) + (\widetilde{H}^+\phi_{-j}, \phi_{-j})}{2j-1}\right|;$$

hence

$$|(A_{\mathscr{R}}e, e)| \leq \frac{2}{\pi}\max\left\{\sum_{j=1}^{\infty}\frac{(\widetilde{H}^+\phi_j, \phi_j) + \widetilde{H}^-\phi_{-j}, \phi_{-j})}{2j-1},\ \sum_{j=1}^{\infty}\frac{(\widetilde{H}^-\phi_j, \phi_j) + (\widetilde{H}^+\phi_{-j}, \phi_{-j})}{2j-1}\right\}.$$

From Lemma II.2.2 it follows that

$$\sum_{j=1}^{\infty}\frac{(\widetilde{H}^{\pm}\phi_j, \phi_j)}{2j-1} \leq \sum_{j=1}^{\infty}\frac{\lambda_j^{\pm}(H)}{2j-1}$$

and

$$\sum_{j=1}^{\infty}\frac{(\widetilde{H}^{\pm}\phi_{-j}, \phi_{-j})}{2j-1} \leq \sum_{j=1}^{\infty}\frac{\lambda_j^{\pm}(H)}{2j-1}.$$

Thus

$$|(A_{\mathscr{R}}e, e)| \leq \frac{2}{\pi}\sum_{j=1}^{\infty}\frac{\lambda_j^+(H) + \lambda_j^-(H)}{2j-1}. \tag{4.14}$$

Taking into account the arbitrariness of the unit vector e, we obtain (4.10).

In the general case of an infinite-dimensional H we form the operators

$$A_n = 2i \int_{\mathfrak{P}} PH_n dP \quad (n = 1, 2, \cdots),$$

where H_n $(n = 1, 2, \cdots)$ is the nth partial sum in the spectral decomposition of H. By what has already been proved,

$$|G_n| \leqq \frac{2}{\pi} |H_n|_{\omega; \mathscr{K}} \quad (G_n = (A_n)_{\mathscr{R}}). \tag{4.15}$$

Since the operators $A_n - A_m$ $(n, m = 1, 2, \cdots)$ are Volterra operators and $[A_n - A_m]_{\mathscr{R}} = H_n - H_m$, we have

$$|G_n - G_m| \leqq \frac{2}{\pi} |H_n - H_m|_{\omega; \mathscr{K}}.$$

It follows that the sequence $\{G_n\}_1^\infty$ tends in the uniform norm to the operator $\widetilde{A}_{\mathscr{R}}$; passing to the limit $n \to \infty$ in (4.15), we obtain (4.10).

Let H be any operator from $\mathfrak{S}_\omega$ which satisfies at least one of the following two conditions: 1) both H^+ and H^- are infinite-dimensional, 2) the set of zeros of H is infinite-dimensional. Then H can be represented in the form

$$H = \sum_{j=1}^{\infty} \lambda_j^+(H)(\cdot, \phi_j)\phi_j - \sum_{j=-\infty}^{-1} \lambda_j^-(H)(\cdot, \phi_j)\phi_j,$$

where $(\phi_j, \phi_\kappa) = \delta_{jk}$ $(j, k = \pm 1, \pm 2, \cdots)$.

Taking into account the properties of the operator J (see §8, Chapter I and §6, Chapter II), we can form a simple dissipative Volterra operator $J^{(H)} = J_{\mathscr{R}}^{(H)} + i(\cdot, \psi)\psi$ $(|\psi| = 1)$, with one-dimensional imaginary component, such that

$$J_{\mathscr{R}}^{(H)} = \frac{2}{\pi}\left[\sum_{j=1}^{\infty} \frac{1}{2j-1}(\cdot, \phi_j)\phi_j - \sum_{j=-\infty}^{-1} \frac{1}{2j-1}(\cdot, \phi_j)\phi_j\right].$$

We denote by $\mathfrak{P}$ a maximal continuous eigenchain of $J^{(H)}$ (see result 2 of §3, Chapter I) and form the Volterra operator

$$A = 2i \int_{\mathfrak{P}} PH\, dP.$$

Applying the same considerations to the operators $J^{(H)}$ and A as were applied to the operators $J^{(e)}$ and $\widetilde{A}$ in the proof of the first part of the theorem, we arrive at the relation (4.14) with an equal sign, which in the present case is equivalent to

$$(A_{\mathscr{R}}\psi,\psi) = \frac{2}{\pi}\sum_{j=1}^{\infty}\frac{\lambda_j^+(H)+\lambda_j^-(H)}{2j-1},$$

which together with (4.10) yields (4.12).

Let H be an arbitrary selfadjoint operator from $\mathfrak{S}_\omega$ and let ϵ be any positive number. We denote by T a selfadjoint operator from $\mathfrak{S}_\omega$ with sufficiently small norm ($|T|_{\omega;\mathscr{K}} < \pi\epsilon/4$) such that the subspace of zeros of the operator $H_1 = H - T$ is infinite-dimensional. Then by what has already been proved there exists a complete chain $\mathfrak{P}$ such that for the operator

$$B = 2i\int_{\mathfrak{P}} PH_1dP$$

one has

$$|B_{\mathscr{R}}| = \frac{2}{\pi}|H_1|_{\omega;\mathscr{K}}.$$

Let us form the operator

$$A = 2i\int_{\mathfrak{P}} PH\,dP.$$

Then from the equalities

$$B_{\mathscr{R}} = \mathscr{S}(H_1;\mathfrak{P}) = \mathscr{S}(H;\mathfrak{P}) - \mathscr{S}(T;\mathfrak{P}) = A_{\mathscr{R}} - \mathscr{S}(T)$$

follows the relation

$$|A_{\mathscr{R}}| \geqq |B_{\mathscr{R}}| - |\mathscr{S}(T)| \geqq \frac{2}{\pi}|H-T|_{\omega;\mathscr{K}} - \frac{2}{\pi}|T|_{\omega;\mathscr{K}};$$

thus

$$|A_{\mathscr{R}}| \geqq \frac{2}{\pi}|H|_{\omega;\mathscr{K}} - \frac{4}{\pi}|T|_{\omega;\mathscr{K}} \geqq \frac{2}{\pi}|H|_{\omega;\mathscr{K}} - \epsilon.$$

The theorem is proved.

From this theorem of Macaev it follows at once that the integral (4.1) converges for any $X \in \mathfrak{S}_\omega$ and any continuous chain $\mathfrak{P}$ and, moreover, for any closed chain $\mathfrak{P}$ satisfying the condition (4.6). Macaev also conjectured that the condition $X \in \mathfrak{S}_\omega$ is *necessary* for the convergence of the integral (4.1) for every continuous chain $\mathfrak{P}$. This conjecture has been verified (see Theorem 4.1).

REMARK 4.1. Since in the pairs of spaces $(\hat{\mathfrak{S}}_\omega, \hat{\mathfrak{S}}_\Omega)$ and $(\hat{\mathfrak{S}}_\infty, \hat{\mathfrak{S}}_1)$ the second space is the conjugate of the first, we can assert on the basis of the general Theorem 3.3 that if the transformator

$$\mathscr{S}(X) = i\int_{\mathfrak{P}} (PX\,dP - dP\,XP)$$

generates a bounded operator $\mathscr{S}_{1\to\Omega}$ $(\equiv \mathscr{S}_{\mathfrak{S}_1\to\mathfrak{S}_\Omega})$ acting from $\mathfrak{S}_1$ into $\mathfrak{S}_\Omega$, then the transformator $\mathscr{S}_{\omega\to\infty}$ $(\equiv \mathscr{S}_{\mathfrak{S}_\omega\to\mathfrak{S}_\infty})$, acting from $\mathfrak{S}_\omega$ into $\mathfrak{S}_\infty$, will also be bounded, and

$$|\mathscr{S}_{1\to\Omega}| = |\mathscr{S}_{\omega\to\infty}|.$$

It follows that the relations (4.10) and (2.2) can be obtained as corollaries of one another.

To V. I. Macaev [**3**] is also due the following result, which generalizes the relations (4.10) and (2.2).

THEOREM 4.3. *Let A be a Volterra operator with imaginary component $A_{\mathscr{J}} \in \mathfrak{S}_\omega$. Then*

$$\max\left\{\sum_{j=1}^n \lambda_j^+(A_{\mathscr{R}}), \sum_{j=1}^n \lambda_j^-(A_{\mathscr{R}})\right\}$$

$$\leq \frac{2}{\pi}\left[\left(\sum_{j=1}^n \frac{1}{2j-1}\right)\sum_{j=1}^n (\lambda_j^+(A_{\mathscr{J}}) + \lambda_j^-(A_{\mathscr{J}})) + n\sum_{j=n+1}^\infty \frac{\lambda_j^+(A_{\mathscr{J}}) + \lambda_j^-(A_{\mathscr{J}})}{2j-1}\right] \quad (n = 1, 2, \cdots). \tag{4.16}$$

For the case $A_{\mathscr{J}} \in \mathfrak{S}_2$ this bound follows from the bound (II.4.12) and equalities (II.6.5). From this special case one can easily deduce the bound (4.16) for the general case $A_{\mathscr{J}} \in \mathfrak{S}_\omega$.

3. We give a result which was used in the proof of Theorem 2.4 (see, in this connection, [GK], p. 193).

2. *Let A be a Volterra operator with imaginary component $A_{\mathscr{J}} \in \mathfrak{S}_\omega$. Then A has an inessential extension $\widetilde{A}$ (to some Hilbert space $\widetilde{\mathfrak{H}} \supset \mathfrak{H}$) which can be expressed as the difference of two dissipative Volterra operators B and C ($\widetilde{A} = B - C$) whose imaginary components $B_{\mathscr{J}}$ and $C_{\mathscr{J}}$ are the inessential extensions (from $\mathfrak{H}$ to $\widetilde{\mathfrak{H}}$) of nonnegative mutually orthogonal operators).*

In fact, by Theorem I.3.2 some inessential extension $\widetilde{A}$ of A from $\mathfrak{H}$ to $\widetilde{\mathfrak{H}}$ $(\supset \mathfrak{H})$ has a continuous maximal eigenchain $\mathfrak{P}$. Let $A_{\mathscr{J}} = H_+$

$-H_-$, where $H_\pm$ are nonnegative mutually orthogonal operators. Then the required decomposition of $\widetilde{A}$ is obtained with the operators

$$B = 2i\int_{\mathfrak{P}} P\widetilde{H}_+ dP, \quad C = 2i\int_{\mathfrak{P}} P\widetilde{H}_- dP.$$

§5. Two interpolation theorems in the spaces $\mathfrak{S}_p$

1. The well-known interpolation theorems of M. Riesz [1] (see also A. Zygmund [1]) in the scales of spaces l_p and L_p $(1 \leqq p < \infty)$ can be extended to the scales of spaces $\mathfrak{S}_p$ $(1 \leqq p < \infty)$.

A simple proof of Riesz' theorem together with the logarithmic convexity of the Riesz constant can be obtained by the use of the function-theoretic three lines theorem. An operator analog of this theorem is the theorem on holomorphic operator-functions which wander in the spaces $\mathfrak{S}_p$, which was discussed in [GK], Chapter III, §13. For the convenience of the reader we state this theorem.

THEOREM 5.1. *Let $T(z)$ be a holomorphic operator-function (with values in $\mathfrak{R}$) in the strip $a \leqq \operatorname{Re} z \leqq b$, and suppose that*

$$\ln^+\ln|T(x+iy)| = O(|y|) \quad (|y| \to \infty).$$

If the values of the operator-function T on the line $z = a + iy$ $(-\infty < y < \infty)$ belong to the space $\mathfrak{S}_{r_1}$ $(1 \leqq r_1 < \infty)$, and on the line $z = b + iy$ $(-\infty < y < \infty$; $a < b)$ to the space $\mathfrak{S}_{r_2}$ $(r_1 < r_2 \leqq \infty)$, and if

$$|T(a+iy)|_{r_1} \leqq C_1, \quad |T(b+iy)|_{r_2} \leqq C_2 \quad (-\infty < y < \infty),$$

then on every intermediate line $z = x + iy$ $(a < x < b;\ -\infty < y < \infty)$ the values of T belong to $\mathfrak{S}_r$, where

$$r^{-1} = r_1^{-1} + t_x(r_2^{-1} - r_1^{-1}), \quad t_x = (x-a)/(b-a),$$

and

$$|T(x+iy)|_r \leqq C_1^{1-t_x} C_2^{t_x} \quad (-\infty < y < \infty).$$

This theorem enables us to prove the following analog of the Riesz-Thorin interpolation theorem.

THEOREM 5.2. *Let p_k and r_k $(k = 1, 2)$ be numbers satisfying $1 \leqq p_1 \leqq p_2 \leqq \infty$; $1 \leqq r_1 \leqq r_2 \leqq \infty$. If $\mathscr{T}$ is a bounded linear transformator acting from $\mathfrak{S}_{p_2}$ into $\mathfrak{S}_{r_2}$ and from $\mathfrak{S}_{p_1}$ into $\mathfrak{S}_{r_1}$, then for any p $(p_1 < p < p_2)$ $\mathscr{T}$ maps $\mathfrak{S}_p$ into $\mathfrak{S}_r$, where*

$$(5.1) \qquad r^{-1} = t_p r_2^{-1} + (1 - t_p) r_1^{-1}, \quad t_p = (p - p_1)/(p_2 - p_1),$$

and

$$|\mathscr{T}_{p\to r}| \leqq |\mathscr{T}_{p_1\to r_1}|^{1-t_p} |\mathscr{T}_{p_2\to r_2}|^{t_p}.$$

Here $\mathscr{T}_{p\to r} \equiv \mathscr{T}_{\mathfrak{S}_p \to \mathfrak{S}_r}$.

PROOF. Let X be an operator from $\mathfrak{S}_p$ and let $X = UH$ be its polar decomposition. If

$$H = \sum_j \lambda_j (\cdot, \phi_j) \phi_j$$

is the spectral decomposition of H, then for any z ($\operatorname{Re} z \geqq 0$) we define the holomorphic operator-function H^z by

$$H^z = \sum_j \lambda_j^z (\cdot, \phi_j) \phi_j$$

$$(\lambda^z = \exp(z \ln \lambda), \ \lambda > 0, \ -\infty < \ln \lambda < \infty).$$

Let us consider the holomorphic operator-function

$$T(z) = \mathscr{T}(UH^z) \quad (\operatorname{Re} z \geqq 0).$$

If $\operatorname{Re} z = p/p_2$ $(= a)$, then $UH^z \in \mathfrak{S}_{p_2}$ and so

$$|T(a + iy)|_{r_2} \leqq |\mathscr{T}_{p_2\to r_2}| \, |H^a|_{p_2} = |\mathscr{T}_{p_2\to r_2}| \, |H|_p^a \ (-\infty < y < \infty).$$

If $\operatorname{Re} z = p/p_1$ $(= b)$, then $UH^z \in \mathfrak{S}_{p_1}$ and by hypothesis $T(b + iy) = \mathscr{T}(UH^z) \in \mathfrak{S}_{r_1}$. Hence

$$|T(b + iy)|_{r_1} \leqq |\mathscr{T}_{p_1\to r_1}| \, |H^b|_{p_1} = |\mathscr{T}_{p_1\to r_1}| \, |H|_p^b \quad (-\infty < y < \infty).$$

Thus, by the three lines theorem for the wandering operator-function $T(z)$ we obtain

$$\begin{aligned} |T(1)|_r &\leqq |\mathscr{T}_{p_1\to r_1}|^{1-t_p} |\mathscr{T}_{p_2\to r_2}|^{t_p} |H|_p^{a(1-t_p)+bt_p} \\ &= |\mathscr{T}_{p_1\to r_1}|^{1-t_p} |\mathscr{T}_{p_2\to r_2}|^{t_p} |H|_p. \end{aligned}$$

The theorem is proved.

In the proof just presented we have used the method of A. Calderon and A. Zygmund [1] for proving the M. Riesz theorem. The present theorem together with this proof was obtained jointly with S. G. Kreĭn.

2. The second interpolation theorem, to be given below, is an extension of a well-known theorem by J. Marcinkiewicz (see also A. Zygmund [1] and Dunford and Schwartz [2], Chapter XI) to the case of operators in the scale $\mathfrak{S}_p$. We owe this theorem to M. Š. Birman and M. Z. Solomjak.

Given an operator $T \in \mathfrak{S}_\infty$, we shall denote by $\nu(r; T)$ $(0 \leqq r < \infty)$ the distribution function of the characteristic numbers of the operator $(T^*T)^{1/2}$; in other words, $\nu(r; T)$ denotes the number of numbers $1/s_n(T)$ $(n = 1, 2, \cdots)$ which do not exceed r. Obviously, for any $X \in \mathfrak{S}_p$ $(1 \leqq p < \infty)$

$$|X|_p^p = \int_0^\infty \frac{d\nu(r; X)}{r^p}.$$

Integrating by parts, we obtain

$$|X|_p^p = p \int_0^\infty \frac{\nu(r; X)}{r^{p+1}}\, dr.$$

We need the following simple result.

LEMMA 5.1. *For any operators $A, B \in \mathfrak{S}_\infty$ we have*

$$\nu\left(\frac{r_1 r_2}{r_1 + r_2}; A + B\right) \leqq \nu(r_1; A) + \nu(r_2; B) \quad (0 \leqq r_1, r_2 < \infty), \tag{5.2}$$

in particular,

$$\nu(r; A + B) \leqq \nu(2r; A) + \nu(2r; B). \tag{5.3}$$

PROOF. For any positive numbers t_1 and t_2 there exist nonnegative integers n, m and p such that

$$s_{n+1}(A) < t_1 \leqq s_n(A), \quad s_{m+1}(B) < t_2 \leqq s_m(B)$$

and

$$s_{p+1}(A + B) < t_1 + t_2 \leqq s_p(A + B);$$

for convenience we put

$$s_0(A) = s_0(B) = s_0(A + B) = \infty.$$

Then obviously

$$\nu(t_1^{-1}; A) = n, \quad \nu(t_2^{-1}; B) = m \quad \text{and} \quad \nu\left(\frac{1}{t_1 + t_2}; A + B\right) = p.$$

We use the relations (II.1.2),

$$s_{m+n+1}(A + B) \leqq s_{n+1}(A) + s_{m+1}(B),$$

from which we obtain

$$s_{m+n+1}(A + B) < t_1 + t_2.$$

It follows at once that

$$s_{m+n+1}(A+B) \leqq s_{p+1}(A+B)$$

or $p+1 \leqq m+n+1$; hence

$$\nu\left(\frac{1}{t_1+t_2}; A+B\right) \leqq \nu\left(\frac{1}{t_1}; A\right) + \nu\left(\frac{1}{t_2}; B\right).$$

If we put $r_j = t_j^{-1}$ $(j=1,2)$, we obtain (5.2). The lemma is proved.

THEOREM 5.3. *Let p_1, p_2 $(1 \leqq p_1 < p_2 < \infty)$ be given numbers, and let $\mathscr{T}$ be a bounded linear transformator from $\mathfrak{S}_{p_2}$ into $\mathfrak{S}_\infty$ such that for every finite-dimensional selfadjoint operator X the bounds*

$$s_n(\mathscr{T}X) \leqq C|X|_{p_j} n^{-1/p_j} \quad (j=1,2;\ n=1,2,\cdots), \tag{5.4}$$

with C a constant, are satisfied.

Then for every intermediate p $(p_1 < p < p_2)$ the restriction of $\mathscr{T}$ to the space $\mathfrak{S}_p$ is a bounded linear transformator which maps $\mathfrak{S}_p$ into itself.

PROOF. Let X be any finite-dimensional selfadjoint operator, and let

$$X = \sum_j s_j(X)(\cdot, \phi_j)\psi_j \quad (s_j(X) = |\lambda_j(X)|,\ \psi_j = \pm\phi_j)$$

be its Schmidt series. With every nonnegative number τ we associate two operators

$$Y_\tau = \sum_{s_j(X) \geqq 1/\tau} s_j(X)(\cdot, \phi_j)\psi_j, \quad Z_\tau = \sum_{s_j(X) < 1/\tau} s_j(X)(\cdot, \phi_j)\psi_j.$$

Obviously $X = Y_\tau + Z_\tau$, whence

$$T = \mathscr{T}(X) = V_\tau + W_\tau, \quad \text{where } V_\tau = \mathscr{T}(X_\tau) \text{ and } W_\tau = \mathscr{T}(Z_\tau).$$

As has already been noted, the norm $|T|_p$ can be calculated from

$$(|T|_p)^p = p\int_0^\infty \frac{\nu(t;T)}{t^{p+1}}\,dt.$$

According to the relation (5.3),

$$\nu(t;T) \leqq \nu(2t; V_\tau) + \nu(2t; W_\tau) \quad (0 \leqq t,\ \tau < \infty).$$

In particular, putting $\tau = 2t$, we obtain

$$\nu(t;T) \leqq \nu(2t; V_{2t}) + \nu(2t; W_{2t}).$$

Consequently,

$$(5.5)\qquad (|T|_p)^p \leqq p2^p\left(\int_0^\infty \frac{\nu(t;\,V_t)}{t^{p+1}}\,dt + \int_0^\alpha \frac{\nu(t;\,W_t)}{t^{p+1}}dt\right).$$

It follows from the relations (5.4) that

$$\nu(t;\,V_t) \leqq C^{p_1}(|Y_t|_{p_1})^{p_1}t^{p_1} \quad (t \geqq 0),$$

$$\nu(t;\,W_t) \leqq C^{p_2}(|Z_t|_{p_2})^{p_2}t^{p_2} \quad (t \geqq 0).$$

Taking into account, further, that

$$(|Y_t|_{p_1})^{p_1} = \int_0^t \frac{d\nu(s;X)}{s^{p_1}} \quad \text{and} \quad (|Z_t|_{p_2})^{p_2} = \int_t^\infty \frac{d\nu(s;X)}{s^{p_2}},$$

we obtain from (5.5) the relation

$$(|T|_p)^p \leqq p2^pC^{p_2}\left[\int_0^\infty \frac{1}{t^{p-p_1+1}}\left(\int_t^t \frac{d\nu(s;X)}{s^{p_1}}\right)dt + \int_0^\infty \frac{1}{t^{p-p_2+1}}\left(\int_t^\infty \frac{d\nu(s;X)}{s^{p_2}}\right)dt\right]$$

$$= p2^pC^{p_2}[\mathscr{I}_1 + \mathscr{I}_2].$$

Integrating the two integrals by parts, we find that

$$\mathscr{I}_1 = \frac{1}{p-p_1}\int_0^\infty \frac{d\nu(s;X)}{s^p} = \frac{1}{p-p_1}(|X|_p)^p$$

and

$$\mathscr{I}_2 = \frac{1}{p_2-p}\int_0^\infty \frac{d\nu(s;X)}{s^p} = \frac{1}{p_2-p}(|X|_p)^p.$$

Thus

$$|\mathscr{T}(X)|_p \leqq K|X|_p,$$

where K is a constant which does not depend on X, and

$$K^r \leqq p2^pC^{p_2}\frac{p_2-p_1}{(p_2-p)(p-p_1)}.$$

It follows that for any (nonselfadjoint) finite-dimensional operator X we have

$$|\mathscr{T}(X)|_p \leqq |\mathscr{T}(X_{\mathscr{R}})|_p + |\mathscr{T}(X_{\mathscr{J}})|_p$$

$$\leqq K(|X_{\mathscr{R}}|_p + |X_{\mathscr{J}}|_p) \leqq 2K|X|_p.$$

Finally, it remains to note that the set of finite-dimensional operators is dense in $\mathfrak{S}_p$. The theorem is proved.

§6. Volterra operators with imaginary component from $\mathfrak{S}_p$ $(1 < p < \infty)$

1. By the *order* of an operator A we mean the infimum $p(A)$ of those numbers p for which $A \in \mathfrak{S}_p$. If A $(\in \mathfrak{S}_\infty)$ does not belong to any $\mathfrak{S}_p$ $(p < \infty)$, we will say that A has *infinite* order: $p(A) = \infty$.

The orders of A and A^* are obviously the same: $p(A) = p(A^*)$. As is easily seen, the orders of the Hermitian components of an operator $A \in \mathfrak{S}_\infty$ do not exceed the order of A.

Since the order of any operator from $\mathfrak{S}_\Omega$ does not exceed unity, it follows from Theorem 2.1 that if, for a Volterra operator A, one has $A_{\mathcal{J}} \in \mathfrak{S}_p$ $(p \leqq 1)$, then

$$p(A_{\mathcal{R}}) \leqq 1. \tag{6.1}$$

We emphasize that no matter how small $p < 1$ is, the relation (6.1) cannot be strengthened, as already for the case of one-dimensional $A_{\mathcal{J}}$ equality holds in (6.1).

However for $p \geqq 1$ we have the following theorem.

THEOREM 6.1. *If the order $p(A_{\mathcal{J}})$ of the imaginary component of a Volterra operator A is not less than unity, we have*

$$p(A_{\mathcal{R}}) \leqq p(A) = p(A_{\mathcal{J}}). \tag{6.2}$$

For the case $1 \leqq p(A_{\mathcal{J}}) \leqq 2$ this theorem can be deduced from Theorems I.10.1 and 2.1 by using Theorem 5.1 concerning wandering holomorphic operator-functions. Following this, the validity of the theorem for the case $2 < p(A_{\mathcal{J}}) < \infty$ can be easily established by applying Theorem 3.3.

2. Shortly after the authors established Theorem 6.1, V. I. Macaev obtained a more precise result, namely:

THEOREM 6.2 (V. I. MACAEV [3]). *Let A be a Volterra operator. If its imaginary component $A_{\mathcal{J}}$ belongs to $\mathfrak{S}_p$ $(1 < p < \infty)$, then its real component $A_{\mathcal{R}}$ likewise belongs to $\mathfrak{S}_p$, and*

$$|A_{\mathcal{R}}|_p \leqq \gamma_p |A_{\mathcal{J}}|_p, \tag{6.3}$$

where the constant γ_p depends only on p.

On the basis of the authors' Theorem 3.3 it suffices to prove this theorem of Macaev for values of p lying in one of the two intervals $(1, 2)$,

$(2, \infty)$. V. I. Macaev [**3**] proved the validity of this theorem for $1 < p < 2$ with the help of new theorems on entire functions (see Macaev [**1**]), which are of interest in themselves (see [GK], Chapter IV, §11).

Once one has the interpolation Theorem 5.2, Theorem 6.2 for the interval $1 < p < 2$ turns out to be an immediate consequence of Theorem I.10.1 of L. A. Sahnovič and Theorem 2.4 of M. G. Kreĭn. However, Theorem 2.4 was in turn obtained with the use of a considerable analytic apparatus, also using special theorems of the theory of entire functions.

We shall give below an elementary proof of Theorem 6.2, from the authors' paper [**5**], which does not use refined methods of the theory of entire functions. This proof makes it possible to establish a number of properties of the best constant γ_p as a function of p $(1 < p < \infty)$ and at the same time to obtain simple upper and lower bounds for it.

It is curious that in contrast with the two proofs just mentioned, the proof below first establishes the theorem for the case $2 \leqq p < \infty$ and then extends it on the basis of the general Theorem 3.3 to the case $1 < p \leqq 2$.

PROOF OF THEOREM 6.2. As before, we may suppose that A has a complete eigenchain $\mathfrak{P}$, so that $A_{\mathscr{R}} = \mathscr{S}(A_{\mathscr{I}}; \mathfrak{P})$. Thus the theorem will be proved if for each p $(1 < p < \infty)$ we prove the existence of a constant γ_p such that

$$|\mathscr{S}(H; \mathfrak{P})|_p \leqq \gamma_p |H|_p, \tag{6.4}$$

where H is any selfadjoint operator from $\mathfrak{S}_p$, and $\mathfrak{P}$ is any complete chain.

We shall first prove that if such a constant γ_p exists for $p = r$, then it exists for $p = 2r$. Thus, suppose that for some $p = r$ $(1 < r < \infty)$ and any $H = H^* \in \mathfrak{S}_r$ and any complete chain $\mathfrak{P}$ we have

$$|\mathscr{S}(H; \mathfrak{P})|_r \leqq \gamma_r |H|_r, \tag{6.5}$$

where γ_r does not depend upon H or $\mathfrak{P}$.

First let H be finite-dimensional. By Theorem 3.1, the relation

$$[\mathscr{S}(H)]^2 = H^2 + \mathscr{S}(H\mathscr{S}(H) + \mathscr{S}(H)H) \tag{6.6}$$

holds for any finite-dimensional selfadjoint H. The operator $H\mathscr{S}(H) + \mathscr{S}(H)H$ is finite-dimensional; hence, according to the hypothesis (6.5)

$$|\mathscr{S}(H\mathscr{S}(H) + \mathscr{S}(H)H)|_r \leqq \gamma_r(|H\mathscr{S}(H)|_r + |\mathscr{S}(H)H|_r). \tag{6.7}$$

From (6.6) and (6.7) follows

$$|[\mathscr{S}(H)]^2|_r \leqq |H^2|_r + \gamma_r(|H\mathscr{S}(H)|_r + |\mathscr{S}(H)H|_r).$$

Bearing in mind that

$$|[\mathscr{S}(H)]^2|_r = |\mathscr{S}(H)|_{2r}^2; \ |H^2|_r = |H|_{2r}^2$$

and

$$\max(|H\mathscr{S}(H)|_r, |\mathscr{S}(H)H|_r) \leqq |H|_{2r}|\mathscr{S}(H)|_{2r},$$

we obtain

$$|\mathscr{S}(H)|_{2r}^2 \leqq |H|_{2r}^2 + 2\gamma_r|H|_{2r}|\mathscr{S}(H)|_{2r}. \tag{6.8}$$

Putting $\xi = |\mathscr{S}(H)|_{2r}/|H|_{2r}$, we obtain from (6.8)

$$\xi^2 \leqq 1 + 2\gamma_r\xi,$$

from which

$$\xi \leqq \gamma_r + \sqrt{1+\gamma_r^2}.$$

Thus, for any finite-dimensional selfadjoint operator H we have the relation

$$|\mathscr{S}(H)|_{2r} \leqq \gamma_{2r}|H|_{2r}, \tag{6.9}$$

where for the constant γ_{2r} we can take the number $\gamma_r + \sqrt{1+\gamma_r^2}$.

We now show that (6.9) also holds for any infinite-dimensional selfadjoint $H \in \mathfrak{S}_{2r}$. In fact, let H_n $(n = 1, 2, \cdots)$ be a sequence of finite-dimensional selfadjoint operators such that

$$\lim_{n\to\infty} |H_n - H|_{2r} = 0. \tag{6.10}$$

It follows from (6.9) and (6.10) that the sequence of operators $\mathscr{S}(H_n)$ $(n = 1, 2, \cdots)$ has a limit in the sense of the norm of $\mathfrak{S}_{2r}$. Since by Theorem 3.2 the transformator $\mathscr{S}$ is closed, this limit equals $\mathscr{S}(H)$, i.e.

$$\lim_{n\to\infty} |\mathscr{S}(H_n) - \mathscr{S}(H)|_{2r} = 0.$$

Letting $n \to \infty$ in the relation

$$|\mathscr{S}(H_n)|_{2r} \leqq \gamma_{2r}|H_n|_{2r} \quad (n = 1, 2, \cdots),$$

we see that (6.5) holds for every selfadjoint $H \in \mathfrak{S}_{2r}$.

From Theorem I.10.1 and what we have already proved we conclude that the relation (6.4) holds for all $p = 2^n$ $(n = 1, 2, \cdots)$ with appropriate γ_p. Applying the interpolation Theorem 5.2 for the case $\mathscr{T} = \mathscr{S}$, $p_1 = r_1 = 2^n$, $p_2 = r_2 = 2^{n+1}$ $(n = 1, 2, \cdots)$, we prove (6.4) for all p from the interval $2 \leqq p < \infty$.

Thus, Theorem 6.2 is valid for all $p \geqq 2$. Its validity for the other values of p $(1 < p < 2)$ can now be deduced on the basis of the general Theorem 3.3.

The theorem is proved.

3. Let $\mathfrak{P}$ be any complete chain, and let $\mathscr{S}_{p\to r} = \mathscr{S}_{\mathfrak{S}_p\to\mathfrak{S}_r}(\cdot;\mathfrak{P})$ be the corresponding transformator $\mathscr{S}$, acting from $\mathfrak{S}_p$ *into* $\mathfrak{S}_r$.

THEOREM 6.3. *Each of the norms* $|\mathscr{S}_{p\to r}|$, $|\mathscr{C}_{p\to r}|$ *is a logarithmically convex (and therefore continuous) function of the two-dimensional variable* $u = (p^{-1}, r^{-1})$ *in the region* Δ *obtained from the triangle* $(0 \leqq r^{-1} \leqq p^{-1} \leqq 1)$ *by removing the two vertices* $p = r = 1$, $p = r = \infty$.

Moreover, the functions $|\mathscr{S}_{p\to r}|$ *and* $|\mathscr{C}_{p\to r}|$ *are nondecreasing with respect to the variable* p $(1 \leqq p \leqq r)$ *and nonincreasing with respect to the variable* r $(p \leqq r \leqq \infty)$.

In particular, the quantity

$$\gamma_p = |\mathscr{S}_{p\to p}| \quad (1 < p < \infty)$$

is a logarithmically convex function of the argument $1/p$, *satisfies the equation*

$$\gamma_p = \gamma_{p/(p-1)} \quad (1 < p < \infty), \tag{6.11}$$

is monotonically increasing for $p \geqq 2$, *and admits the bounds*

$$Np \leqq \gamma_p \leqq Mp \quad (2 < p < \infty), \tag{6.12}$$

where N *and* M *are certain constants which do not depend on the chain* $\mathfrak{P}$, *and*

$$N > 1/2\pi, \quad M < 1/e^{2/3}\ln 2 \ (< 1).$$

PROOF. Let us denote by $F(u)$ the norm $|\mathscr{S}_{p\to r}|$, considered in the region Δ, as a function of the two-dimensional vector u with coordinates $u_x = p^{-1}$ and $u_y = r^{-1}$. By Theorem 5.2 we have, for any two vectors u_1, u_2 from the domain in question,

$$F(tu_1 + (1-t)u_2) \leqq [F(u_1)]^t[F(u_2)]^{1-t} \quad (0 \leqq t \leqq 1),$$

or

$$\log F(tu_1 + (1-t)u_2) \leqq t\log F(u_1) + (1-t)\log F(u_2) \quad (0 \leqq t \leqq 1),$$

which shows that the function $\log F(u)$ is convex in the region Δ.

The second assertion of the theorem is obtained from the relations

$$|\mathscr{S}_{p_1\to r}| = \sup_{X\in\mathfrak{S}_{p_1}} \frac{|\mathscr{S}(X)|_r}{|X|_{p_1}} \leqq \sup_{X\in\mathfrak{S}_{p_2}} \frac{|\mathscr{S}(X)|_r}{|X|_{p_2}} = |\mathscr{S}_{p_2\to r}| \quad (p_2 < p_1)$$

and

$$|\mathscr{S}_{p\to r_1}| = \sup_{X\in\mathfrak{S}_p} \frac{|\mathscr{S}(X)|_{r_1}}{|X|_p} \leqq \sup_{X\in\mathfrak{S}_p} \frac{|\mathscr{S}(X)|_{r_2}}{|X|_p} = |\mathscr{S}_{p\to r_2}| \quad (r_1 < r_2),$$

which follow from the fact that for any operator $Y \in \mathfrak{S}_p$

$$|Y|_r \leqq |Y|_p$$

for all $r > p$. The relation (6.11) follows from the equalities

$$|\mathscr{S}_{p\to p}| = |\mathscr{S}^*_{p\to p}| = |\mathscr{S}_{q\to q}| \quad (q = p/(p-1)),$$

which hold on the basis of Theorem 3.3.

That the function γ_p $(2 \leqq p < \infty)$ is monotone increasing is a consequence of its convexity and the relation (6.11).

We turn to the proof of (6.12). Recall that in the proof of Theorem 6.2 we established the relation

$$\gamma_{2r} \leqq \gamma_r + \sqrt{1+\gamma_r^2} \quad (1 < r < \infty),$$

by virtue of which

$$\gamma_{2^{n+1}} \leqq \gamma_{2^n} + \sqrt{1+\gamma_{2^n}^2} \quad (n = 1, 2, \cdots).$$

Let us consider the sequence of numbers ξ_n $(n = 1, 2, \cdots)$, defined recursively by

$$\xi_1 = 1,\ \xi_{n+1} = \xi_n + \sqrt{1+\xi_n^2} \quad (n = 1, 2, \cdots).$$

Obviously

$$\gamma_{2^n} \leqq \xi_n \quad (n = 1, 2, \cdots).$$

Since

$$\frac{\xi_{n+1}}{\xi_n} = 1 + \sqrt{1+\frac{1}{\xi_n^2}} \leqq 2\left(1 + \frac{1}{4}\xi_n^{-2}\right) \quad (n = 1, 2, \cdots),$$

it follows that

$$\xi_{n+1} \leqq 2^n \prod_{j=1}^{n} \left(1 + \frac{1}{4}\xi_j^{-2}\right) \quad (n = 1, 2, \cdots). \tag{6.13}$$

Since $\xi_{n+1} > 2\xi_n$, we have

$$\xi_{n+1} > 2^n \quad (n = 1, 2, \cdots).$$

Replacing in the right side of (6.13) ξ_j by 2^{j-1} and using the inequality $1 + x < e^x$ $(x > 0)$, we obtain

$$\xi_{n+1} \leqq 2^n \prod_{j=1}^{n} e^{1/4^j} < 2^n e^{1/3}.$$

Thus we have established the inequalities

$$\gamma_p < \frac{e^{1/3}}{2} p \quad (p = 2^n;\ n = 1, 2, \cdots). \tag{6.14}$$

Now let p be any number from the interval $(2^n, 2^{n+1})$, i.e.

$$p^{-1} = 2^{-n}t + (1+t)2^{-n-1} \quad (0 < t < 1).$$

Then it follows from the logarithmic convexity of the function γ_p that

$$\gamma_p \leqq \gamma_{2^n}^{t} \gamma_{2^{n+1}}^{1-t},$$

whence, by virtue of (6.14),

$$\gamma_p < \frac{e^{1/3}}{2} 2^{nt + (1-t)(n+1)}.$$

Since $t = (2^{n+1}/p) - 1$,

$$\gamma_p < 2e^{1/3} 2^{n - 2^{n+1}p^{-1}}.$$

It is not hard to show that

$$\max_{2^n \leqq p \leqq 2^{n+1}} (2^{n-2^{n+1}p^{-1}} p^{-1}) = \frac{1}{2e \ln 2};$$

consequently,

$$\gamma_p < \frac{1}{e^{2/3} \ln 2} p \quad (2 \leqq p < \infty).$$

To establish the lower bound in (6.12), we consider the operator

$$H = \mathscr{S}(K),$$

where $K = (\cdot, e)e$ ($e \in \mathfrak{H}$; $|e| = 1$) is a one-dimensional operator. Since $|K|_q = 1$,

$$\gamma_q = |\mathscr{S}_{q \to q}| \geqq |H|_q.$$

By Theorem I.8.1 the operator H is unitarily equivalent to some inessential extension of the real component $J_{\mathscr{R}}$ of the operator J; hence

$$|H|_q = \frac{2}{\pi} \left(2 \sum_{j=0}^{\infty} \frac{1}{(2j+1)^q} \right)^{1/q} = \frac{1}{\pi} \left(2 \sum_{j=0}^{\infty} \frac{1}{(j+1/2)^q} \right)^{1/q}.$$

From the obvious relations

$$\frac{2}{\pi^q}\sum_{j=0}^{\infty}\frac{1}{(j+1/2)^q} > \frac{2}{\pi^q}\int_{1/2}^{\infty}\frac{dt}{t^q} = \frac{2^q}{\pi^q(q-1)}$$

there follows the inequality

$$\gamma_q > \frac{2}{\pi}(q-1)^{-1/q}.$$

Hence, by (6.11)

$$\gamma_p > \frac{2}{\pi}(q-1)^{-1/q} \quad (p = q/(q-1)),$$

or

$$\gamma_p > \frac{2}{\pi}(p-1)^{1-p^{-1}}.$$

Since $p - 1 < p + 1 \leqq 2^p$ for $p \geqq 1$, we have

$$\gamma_p > \frac{p-1}{\pi} \quad (1 < p < \infty);$$

consequently,

$$\gamma_p > p/2\pi \quad (2 \leqq p < \infty).$$

The theorem is proved.

From the inequality

$$\gamma_{2p} \leqq \gamma_p + \sqrt{1+\gamma_p^2}$$

it easily follows that if $\gamma_p \leqq \operatorname{ctg}\phi_p$ $(2 \leqq p < \infty)$, then $\gamma_{2p} \leqq \operatorname{ctg}(\phi_p/2)$. This implies, in particular, that $\gamma_p \leqq \operatorname{ctg}(\pi/2p)$ for $p = 2^n$ $(n = 1, 2, \cdots)$.

Recently I. C. Gohberg and N. Ja. Krupnik [1], using the theory of singular integral equations and also a result of V. I. Macaev (the bound (10.14) on p. 150), showed that for all p $(2 \leqq p < \infty)$ one has $\gamma_p \geqq \operatorname{ctg}(\pi/2p)$. Thus, these authors have established in their paper [1] that

$$\gamma_p = \operatorname{ctg}\frac{\pi}{2p} \qquad (p = 2^n;\ n = 1, 2, \cdots).$$

They also conjecture that this equality holds for all p $(2 \leqq p < \infty)$.

4. In §10 of Chapter I we established that

$$|\mathscr{C}_{2\to 2}| = 1,$$

and in §2 that

$$|\mathscr{C}_{1\to\infty}| = 1.$$

Combining these results with the fact that, by Theorem 6.3, for $1 \leqq p < 2 < r \leqq \infty$

$$|\mathscr{C}_{1\to\infty}| \leqq |\mathscr{C}_{p\to r}| \leqq |\mathscr{C}_{2\to 2}|,$$

we obtain

$$|\mathscr{C}_{p\to r}| = 1 \quad (1 \leqq p \leqq 2 \leqq r \leqq \infty).$$

Without going into details, we note that from the bounds obtained in §10 of Chapter I and §2 for the norms $|\mathscr{C}_{1\to p}|$, $|\mathscr{S}_{1\to p}|$, $|\mathscr{C}_{2\to 2}|$ and $|\mathscr{S}_{2\to 2}|$, using Theorems 6.3 and 3.3 we can derive certain bounds for the norms $|\mathscr{C}_{p\to r}|$ and $|\mathscr{S}_{p\to r}|$.

In particular, from the relation $(\mathscr{C}_{1\to q})^* = \mathscr{C}_{p\to\infty}$ $(p^{-1}+q^{-1}=1)$ and from (2.10) it follows that for any Volterra operator A with $A_{\mathscr{I}} \in \mathfrak{S}_p$ $(1 \leqq p \leqq \infty)$ one has the following precise bound:

$$|A_{\mathscr{R}}|_\infty \leqq \frac{2}{\pi}\left(\sum_{j=-\infty}^{\infty} \frac{1}{|2j-1|^q}\right)^{1/q} |A_{\mathscr{I}}|_p$$

or the still more precise bound

$$|A_{\mathscr{R}}|_\infty \leqq \frac{2}{\pi}\left(\sum_{j=1}^{\infty} \frac{1}{(2j-1)^q}\right)^{1/q} (|A_{\mathscr{I}}^+|_p + |A_{\mathscr{I}}^-|_p).$$

For the case of finite-dimensional $A_{\mathscr{I}}$ the result (4.10) of V. I. Macaev makes it possible to obtain further sharpenings of these bounds. In fact, let $n_\pm$ be the dimensions of the operators $A_{\mathscr{I}}^\pm$. By Hölder's inequality

$$\sum_{j=1}^{n_\pm} \frac{\lambda_j^\pm(A_{\mathscr{I}})}{2j-1} \leqq \sigma_{n_\pm;q}|A_{\mathscr{I}}^\pm|_p,$$

where

$$\sigma_{n;q} = \left(\sum_{j=1}^{n} \frac{1}{(2j-1)^q}\right)^{1/q},$$

and therefore, by virtue of (4.10),

$$|A_{\mathscr{R}}|_\infty \leqq \frac{2}{\pi}\left[\sigma_{n_+;q}|A_{\mathscr{I}}^+|_p + \sigma_{n_-;q}|A_{\mathscr{I}}^-|_p\right].$$

Hence

$$|A_{\mathscr{R}}|_\infty \leqq \frac{2}{\pi}\sigma_{n;q}|A_{\mathscr{I}}|_p \quad (n = \max(n_+; n_-)).$$

It should be noted that all these bounds are independent of the kind of eigenchain possessed by the Volterra operator A.

§7. Convergence of the integral of triangular truncation in the norm of an s.n. ideal

1. Let $\mathfrak{P}$ be some chain. We shall say that the integral

$$Y = \int_{\mathfrak{P}} PX\,dP \tag{7.1}$$

converges weakly if

$$\lim\left(\sum_{j=1}^{n} Q_j X \Delta P_j f, g\right) = (Yf,g) \quad (P_j \leqq Q_j \leqq P_{j+1}) \tag{7.2}$$

for every pair of vectors $f,g \in \mathfrak{H}$, where the limit is understood in the sense of Šatunovskiĭ.

Arguing as in the proof of Lemma I.6.1, it is easy to establish that if $X \in \mathfrak{S}_\infty$ and if the integral (7.1) converges weakly, then the integral

$$\int_{\mathfrak{P}} dP\,XP \tag{7.3}$$

converges weakly, and

$$X = \int_{\mathfrak{P}} PX\,dP + \int_{\mathfrak{P}} dP\,XP. \tag{7.4}$$

From the weak convergence of the integral (7.3) there follows, in turn, that of the integral

$$\int_{\mathfrak{P}} PX^*dP.$$

It follows from (7.4) that if $X^* = X \in \mathfrak{S}_\infty$ and the integral (7.1) converges weakly, then

$$\left[2i\int_{\mathfrak{P}} PX\,dP\right]_{\mathscr{I}} = X.$$

To conclude these introductory remarks we note that in general the weak convergence of the integral (7.1) does not guarantee the complete continuity of the operator Y, even if X is completely continuous.

LEMMA 7.1. *Let $\mathfrak{S}$ be an arbitrary separable s.n. ideal which does not coincide elementwise with $\mathfrak{S}_1$, let $A \in \mathfrak{S}$, and let $\mathfrak{P}$ be a bordered chain. Then the operator-function $\Psi_A(\mathfrak{z})$, defined on the set $\mathfrak{z}_{\mathfrak{P}}$ of all partitions $\mathfrak{z} = \{P_j\}_0^n$ of the chain $\mathfrak{P}$ by*

$$W_A(\mathfrak{z}) = \sum_{j=1}^{n} \Delta P_j A \Delta P_j,$$

tends on $\mathfrak{z}_{\mathfrak{P}}$ in the norm of $\mathfrak{S}$ to the operator

$$\hat{A} = \sum_{j=1}^{\omega} (P_j^+ - P_j^-) A (P_j^+ - P_j^-),$$

where (P_j^-, P_j^+) $(j = 1, 2, \cdots, \omega; \omega \leqq \infty)$ is the sequence of all gaps of the chain $\mathfrak{P}$.

In particular, if $\mathfrak{P}$ is continuous or if

$$(P_j^+ - P_j^-) A (P_j^+ - P_j^-) = 0 \quad (j = 1, 2, \cdots, \omega),$$

then

$$\lim |\Psi_A(\mathfrak{z})|_{\mathfrak{S}} = 0.$$

PROOF. We first consider the simplest case, in which A is a one-dimensional operator $A_\phi = (\cdot, \phi)\phi$ $(\phi \in \mathfrak{H}; |\phi| = 1)$. Let ϵ be a positive number and let N be a positive integer chosen so that

$$\left| \sum_{j=n+1}^{\omega} Q_j \phi \right| < \frac{\epsilon}{6} \quad (Q_j = P_j^+ - P_j^-; j = 1, 2, \cdots, \omega). \tag{7.5}$$

By Theorem 1.2 the s.n. ideal $\mathfrak{S}$ coincides with some s.n. ideal $\mathfrak{S}_\Phi^{(0)}$, where the s.n. function $\Phi(\xi)$ is not equivalent to the maximal one, i.e.

$$\lim_{n \to \infty} \left\{ \frac{1}{n} \Phi(\overbrace{1, 1, \cdots, 1}^{n}, 0, 0, \cdots) \right\} = 0.$$

Let us assume that the positive integer m is so chosen that

$$\frac{1}{m} \Phi(\underbrace{1, 1, \cdots, 1}_{N+m}, 0, 0, \cdots) < \frac{\epsilon}{2}.$$

We put

$$\psi = \phi - \sum_{j=N+1}^{\omega} Q_j \phi.$$

As the projector P ranges over the chain $\mathfrak{P}$, the quantity $|P\psi|^2$ assumes all values from zero to $|\psi|^2$ ($\leqq 1$) with the exception of values from the intervals $(|P_j^-\psi|^2, |P_j^+\psi|^2)$ $(j = 1, 2, \cdots, N)$.

Let $\mathfrak{z}_\epsilon = \{P_j\}_0^n$ be a partition of $\mathfrak{P}$ which contains all the projectors $P_j^\pm$ $(j = 1, 2, \cdots, N)$ and has the property that for any pair of projectors $P_{j-1}, P_j \in \mathfrak{z}_\epsilon$ of which at least one is not contained among the projectors $P_j^\pm$ $(j = 1, 2, \cdots, N)$,

$$|\Delta P_j \psi|^2 < 1/m. \tag{7.6}$$

Obviously, we can require of the partition $\mathfrak{z}_\epsilon$ that

$$n \leqq m + 2N.$$

We introduce the operator

$$B = \sum_{j=1}^{n} (\cdot, \Delta P_j \psi) \Delta P_j \psi - \sum_{j=1}^{N} (\cdot, Q_j \psi) Q_j \psi = \sum (\cdot, \Delta P_j \psi) \Delta P_j \psi,$$

where the last sum is taken over all j for which the pair (P_{j-1}, P_j) does not coincide with any gap (P_k^-, P_k^+) $(k = 1, 2, \cdots, N)$. Obviously B has at most $m + N$ s-numbers different from zero, and by (7.6) none of these exceeds m^{-1}. Hence

$$|B|_\Phi \leqq \frac{1}{m} \Phi(\underbrace{1, 1, \cdots. 1}_{m+N}, 0, 0, \cdots) < \frac{\epsilon}{2}.$$

It follows at once that

$$\left| \sum_{j=1}^{n} (\cdot, \Delta P_j \psi) \Delta P_j \psi - \sum_{j=1}^{\omega} (\cdot, Q_j \psi) Q_j \psi \right| < \frac{\epsilon}{2},$$

and it is easily seen that this relation holds for all partitions $\mathfrak{z} \supset \mathfrak{z}_\epsilon$.

Since for the one-dimensional operators $(\cdot, \phi)\phi$, $(\cdot, \psi)\psi$ $(|\phi| = 1, |\psi| = 1)$ we have

$$|(\cdot, \phi)\phi - (\cdot, \psi)\psi|_\Phi \leqq 2|\phi - \psi| < \epsilon/3,$$

it follows by (7.5) that for any partition $\mathfrak{z} = \{P_j\}_0^n \supseteq \mathfrak{z}_\epsilon$

$$\begin{aligned} \left| \sum_{j=1}^{n} \Delta P_j A_\phi \Delta P_j - \sum_{j=1}^{\omega} Q_j A_\phi Q_j \right|_\Phi &\leqq \left| \sum_{j=1}^{n} \Delta P_j (A_\phi - A_\psi) \Delta P_j \right|_\Phi \\ &+ \left| \sum_{j=1}^{\omega} Q_j (A_\phi - A_\psi) Q_j \right|_\Phi + \left| \sum_{j=1}^{n} \Delta P_j A_\psi \Delta P_j - \sum_{j=1}^{\omega} Q_j A_\psi Q_j \right|_\Phi < \epsilon. \end{aligned}$$

Any finite-dimensional operator is a linear combination of one-dimensional operators of the form A_ϕ; hence for the case of finite-dimensional A the lemma is proved.

Finally, let A be any operator from $\mathfrak{S}$ and let K be a finite-dimensional operator. Then for any partition $\mathfrak{z} = \{P_j\}_0^n$ we have

$$\left| \sum_{j=1}^{n} \Delta P_j A \Delta P_j - \sum_{j=1}^{\omega} Q_j A Q_j \right|_\Phi \leqq 2|A - K|_\Phi + \left| \sum_{j=1}^{n} \Delta P_j K \Delta P_j - \sum_{j=1}^{\omega} Q_j K Q_j \right|_\Phi . \tag{7.7}$$

Since the set of finite-dimensional operators is dense in $\mathfrak{S}_\Phi^{(0)}$, it follows from (7.7) and that which has already been proved that the lemma is valid in the general case. The lemma is proved.

THEOREM 7.1. *Let $\mathfrak{P}$ be a bordered chain and let $\mathfrak{S}$ be any separable s.n. ideal which does not coincide elementwise with $\mathfrak{S}_1$. If for some $X \in \mathfrak{S}$ the integral* (7.1) *converges weakly to an operator Y, and $Y \in \mathfrak{S}$, then the integral converges to Y in the norm of the ideal $\mathfrak{S}$.*

PROOF. By virtue of the opening remarks made in this section, we can assume without loss of generality that X is selfadjoint.

We shall first show that $\mathfrak{P}$ is an eigenchain of Y. In fact, let P be some projector from $\mathfrak{P}$. Considering the limit (7.2) along partitions containing the projector P, we obtain

$$(YPf, g) = \lim \left(\sum_{j=1}^{n} Q_j X \Delta P_j Pf, g \right)$$

$$= \lim \left(P \sum_{j=1}^{n} Q_j X \Delta P_j Pf, g \right) = (YPf, Pg) \quad (f, g \in \mathfrak{H}).$$

Therefore

$$PYP = YP \quad (P \in \mathfrak{P}). \tag{7.8}$$

Moreover, as is easily seen, the weak convergence of the integral (7.1) implies that

$$((P^+ - P^-) X (P^+ - P^-) f, g) \equiv 0 \quad (f, g \in \mathfrak{H})$$

for any gap (P^-, P^+) of $\mathfrak{P}$, so that

$$(P^+ - P^-) X (P^+ - P^-) = 0, \tag{7.9}$$

i.e. the block of the operator X on every gap equals zero.

From the complete continuity of Y and (7.8), (7.9) it follows by Theorem I.5.1 that the integral (7.1) converges in the uniform norm.

The proof of the convergence of the integral (7.1) in the norm of $\mathfrak{S}$ is accomplished by repeating the proof of Theorem I.5.1 with the uniform norm replaced by the norm of $\mathfrak{S}$, and with Lemma I.5.1 replaced by Lemma 7.1.

The theorem is proved.

THEOREM 7.1 combined with Theorem 6.2 leads to the following result.

COROLLARY 7.1. *Let $\mathfrak{P}$ be a bordered chain. For any $X \in \mathfrak{S}_p$ $(1 < p < \infty)$ the integral* (7.1) *converges in the norm of $\mathfrak{S}_p$ as long as the necessary condition for convergence, i.e.*

$$(P^+ - P^-)X(P^+ - P^-) = 0 \tag{7.10}$$

for every gap (P^-, P^+) of $\mathfrak{P}$, is fulfilled.

2. If $X \in \mathfrak{S}_1$, then when the condition (7.10) is satisfied the integral (7.1) will converge in the norm of every $\mathfrak{S}_p$ $(1 < p \leqq \infty)$; however in the norm of $\mathfrak{S}_1$ it will in general diverge. This follows from the fact that there always exists $X \in \mathfrak{S}_1$, satisfying the necessary condition (7.10), for which the value Y of the integral (7.1) certainly does not belong to $\mathfrak{S}_1$. However by Theorem 2.1 the operator Y will always belong to the space $\mathfrak{S}_\Omega$. In accordance with this one can show that for any $X \in \mathfrak{S}_1$ satisfying the condition (7.10), the integral (7.1) will converge in the norm of $\mathfrak{S}_\Omega$.

The fact that the s.n. ideal $\mathfrak{S}$ does not coincide elementwise with $\mathfrak{S}_1$ was used in an essential way in the proof of Lemma 7.1. For this reason the hypothesis of Theorem 7.1 also excluded $\mathfrak{S}$ coinciding elementwise with $\mathfrak{S}_1$. Nevertheless, more refined arguments show that Theorem 7.1 is also true for $\mathfrak{S}_1$.

THEOREM 7.2. *Let $\mathfrak{P}$ be a bordered chain, let $X \in \mathfrak{S}_1$, and suppose that the condition* (7.10) *is satisfied. Then the operator*

$$Y = \int_{\mathfrak{P}} PX\,dP \tag{7.11}$$

belongs to $\mathfrak{S}_1$ if and only if the integral (7.11) *converges in the norm of $\mathfrak{S}_1$.*

PROOF. The sufficiency of the condition is obvious. The proof of its necessity reduces (see the proof of Theorem I.5.1) to proving that when $Y \in \mathfrak{S}_1$,

(7.12) $$\lim \sum_{j=1}^{n} |\Delta P_j X \Delta P_j|_1 = 0,$$

and for this it is sufficient to deal with the case of selfadjoint X. In fact, if $Y \in \mathfrak{S}_1$, then by (7.4)

$$Y^* - X^* = \int_{\mathfrak{P}} PX^* dP.$$

Hence, if we replace X in the integral (7.11) by its Hermitian components, the integral will belong to $\mathfrak{S}_1$. On the other hand, if (7.12) is satisfied for the components $X_{\mathscr{R}}$ and $X_{\mathscr{I}}$, then it will obviously be satisfied for X itself.

That (7.12) is satisfied for the case $X = X^* \in \mathfrak{S}_1$ and $Y \in \mathfrak{S}_1$ follows directly from the basic theorem (Theorem 1.1) of Chapter VII, where more generally the meaning of the functional

$$h(X) = \lim \sum_{j=1}^{n} |\Delta P_j X \Delta P_j|_1$$

will be clarified for all $X \in \mathfrak{S}_1$.

3. The results of this section enable us to supplement Theorem 3.3 (the duality principle) with the following result.

THEOREM 7.3.[7] *Let $\mathfrak{S}_{\mathrm{I}}$ and $\mathfrak{S}_{\mathrm{II}}$ be separable s.n. ideals, with $\mathfrak{S}_{\mathrm{II}}^* \subseteq \mathfrak{S}_\omega$. If the transformator $\mathscr{L}_{\mathrm{I}\to\mathrm{II}}$ $(\equiv \mathscr{L}_{\mathfrak{S}_{\mathrm{I}}\to\mathfrak{S}_{\mathrm{II}}})$ is bounded, then the transformator $\mathscr{L}_{\mathrm{II}^*\to\mathrm{I}^*}$ $(\equiv \mathscr{L}_{\mathfrak{S}_{\mathrm{II}}^*\to\mathfrak{S}_{\mathrm{I}}^*})$ is bounded. Conversely, if $\mathscr{L}_{\mathrm{II}^*\to\mathrm{I}^*}$ is bounded and the additional condition[8] $\mathscr{L}(\mathfrak{R}) \subseteq \mathfrak{S}_{\mathrm{II}}$ is satisfied, then $\mathscr{L}_{\mathrm{I}\to\mathrm{II}}$ is bounded. In both cases the transformator $\mathscr{L}_{\mathrm{II}^*\to\mathrm{I}^*}$ is the conjugate of $\mathscr{L}_{\mathrm{I}\to\mathrm{II}}$.*

PROOF. For any finite-dimensional operator K and any $A \in \mathfrak{S}_{\mathrm{II}}^*$

(7.13) $$\mathrm{sp}\,(\mathscr{L}(K)A^*) = \int_{\mathfrak{P}} \mathrm{sp}\,(K\,dPA^*P).$$

In fact, by Theorem 7.1 the integral

$$\mathscr{L}(K) = \int_{\mathfrak{P}} PK\,dP$$

---

[7] This theorem differs from Theorem 3.3 in that the separability of the s.n. ideal $\mathfrak{S}_{\mathrm{II}}^*$ is not assumed. The separability of $\mathfrak{S}_{\mathrm{II}}^*$ is here replaced by the condition $\mathfrak{S}_{\mathrm{II}}^* \subseteq \mathfrak{S}_\omega$.

[8] The condition $\mathscr{L}(\mathfrak{R}) \subseteq \mathfrak{S}_{\mathrm{II}}$ is obviously satisfied if $\mathfrak{S}_{\mathrm{II}} \supseteq \mathfrak{S}_\Omega$.

converges in the norm of $\mathfrak{S}_{\mathrm{II}}$; hence the integral

$$\mathscr{C}(K)A^* = \int_{\mathfrak{P}} PK\,dPA^*$$

converges in the norm of $\mathfrak{S}_1$. It follows that

$$\mathrm{sp}\left(\int_{\mathfrak{P}} PK\,dPA^*\right) = \int_{\mathfrak{P}} \mathrm{sp}\,(PK\,dPA^*),$$

and so (7.13) holds.

For the same operators K and A one has

$$\mathrm{sp}\,(K[\mathscr{C}(A)]^*) = \int_{\mathfrak{P}} \mathrm{sp}\,(PK\,dPA^*). \tag{7.14}$$

Indeed, since $A \in \mathfrak{S}_\omega$ the integral

$$[\mathscr{C}(A)]^* = \int_{\mathfrak{P}} dPA^*P$$

converges in the uniform norm, and so the integral

$$K[\mathscr{C}(A)]^* = \int_{\mathfrak{P}} K\,dPA^*P$$

converges in the norm of $\mathfrak{S}_1$, and (7.14) holds.

Comparing (7.13) and (7.14), we obtain

$$\mathrm{sp}\,(\mathscr{C}(K)A^*) = \mathrm{sp}\,(K[\mathscr{C}(A)]^*). \tag{7.15}$$

If the transformator $\mathscr{C}_{\mathrm{I}\to\mathrm{II}}$ is bounded, then its conjugate $\mathscr{T} = \mathscr{C}^*_{\mathrm{I}\to\mathrm{II}}$, which acts from $\mathfrak{S}^*_{\mathrm{II}}$ into $\mathfrak{S}^*_{\mathrm{I}}$, is bounded. We have

$$\mathrm{sp}\,(\mathscr{C}(K)A^*) = \mathrm{sp}\,(K[\mathscr{T}(A)]^*) \quad (A \in \mathfrak{S}^*_{\mathrm{II}}, K \in \mathfrak{K}),$$

and therefore by virtue of (7.15)

$$\mathrm{sp}\,(K[\mathscr{C}(A) - \mathscr{T}(A)]^*) = 0 \quad (A \in \mathfrak{S}^*_{\mathrm{II}}, K \in \mathfrak{K}).$$

It follows that

$$\mathscr{T} = \mathscr{C},$$

and thus the transformator $\mathscr{C}_{\mathrm{II}^*\to\mathrm{I}^*}$ is bounded and is the conjugate of $\mathscr{C}_{\mathrm{I}\to\mathrm{II}}$.

Now suppose, conversely, that $\mathscr{C}_{\mathrm{II}^*\to\mathrm{I}^*}$ is bounded. By hypothesis the operator $\mathscr{C}(K)$ $(K \in \mathfrak{K})$ belongs to $\mathfrak{S}_{\mathrm{II}}$; consequently, we can choose an operator $A_0 \in \mathfrak{S}^*_{\mathrm{II}}$ $(|A_0|_{\mathfrak{S}^*_{\mathrm{II}}} = 1)$ such that

$$\mathrm{sp}\,(\mathscr{C}(K)A_0^*) = |\mathscr{C}(K)|_{\mathrm{II}}.$$

Putting $A = A_0$ in (7.15), we obtain

$$\mathrm{sp}\,(K[\mathscr{C}(A_0)]^*) = |\mathscr{C}(K)|_{\mathrm{II}},$$

and so

$$|\mathscr{C}(K)|_{\mathrm{II}} \leqq |K|_{\mathrm{I}} |\mathscr{C}_{\mathrm{II}^* \to \mathrm{I}^*}| \quad (K \in \mathfrak{K}).$$

Since the transformator $\mathscr{C}_{\mathrm{I}\to\mathrm{II}}$ is closed and the lineal $\mathfrak{K}$ is dense in $\mathfrak{S}_{\mathrm{I}}$, this implies that $\mathscr{C}_{\mathrm{I}\to\mathrm{II}}$ is a bounded transformator. But then, by what has already been proved, the transformator $\mathscr{C}_{\mathrm{II}^*\to\mathrm{I}^*}$ will be the conjugate of $\mathscr{C}_{\mathrm{I}\to\mathrm{II}}$.

The theorem is proved.

§8. A general theorem on transformators

In recent times the remarkable ideas of M. Riesz, T. O. Thorin and J. Marcinkiewicz in the theory of the interpolation of operators (see A. Zygmund [**1**]) have experienced a profound development (see the survey papers by S. G. Kreĭn and Ju. I. Petunin [**1**] and E. Magenes [**1**]). This progress has been reflected in the theory of transformators. Recently E. M. Semenov succeeded in establishing a new and very general interpolation theorem for transformators. This theorem has enabled us to free ourselves from the use of the theory of entire functions, whose precise results first made it possible to achieve progress in the study of the boundedness of the transformator of triangular truncation in various s.n. ideals.

Let $\mathscr{T}$ be a linear transformator which maps the set $\mathfrak{K}$ $(\subset \mathfrak{R})$ of all finite-dimensional operators into $\mathfrak{S}_\infty$. We shall say that $\mathscr{T}$ *is a transformator of type* $(\mathfrak{S}_{\mathrm{I}}, \mathfrak{S}_{\mathrm{II}})$ if $\mathfrak{S}_{\mathrm{I}}$ and $\mathfrak{S}_{\mathrm{II}}$ are s.n. ideals and if the following two conditions are satisfied.

a) $\mathscr{T}\mathfrak{K} \subseteq \mathfrak{S}_{\mathrm{II}}$;

b) $\displaystyle\sup_{K\in\mathfrak{K}} \frac{|\mathscr{T}K|_{\mathrm{II}}}{|K|_{\mathrm{I}}} < \infty$.

If $\mathscr{T}$ is a transformator of type $(\mathfrak{S}_{\mathrm{I}}, \mathfrak{S}_{\mathrm{II}})$, we shall write $\mathscr{T} \in (\mathfrak{S}_{\mathrm{I}}, \mathfrak{S}_{\mathrm{II}})$.

If the s.n. ideal $\mathfrak{S}_{\mathrm{I}}$ is separable, then obviously every transformator $\mathscr{T} \in (\mathfrak{S}_{\mathrm{I}}, \mathfrak{S}_{\mathrm{II}})$ can be extended by continuity to a bounded transformator acting from $\mathfrak{S}_{\mathrm{I}}$ into $\mathfrak{S}_{\mathrm{II}}$.

Let $\mathscr{T}$ be a transformator of type $(\mathfrak{S}_{\mathrm{I}}, \mathfrak{S}_{\mathrm{II}})$ and type $(\mathfrak{S}'_{\mathrm{I}}, \mathfrak{S}'_{\mathrm{II}})$, where $\mathfrak{S}_{\mathrm{I}}$ and $\mathfrak{S}'_{\mathrm{I}}$ are separable s.n. ideals. We denote by $\tilde{\mathscr{T}}$ ($\tilde{\mathscr{T}}'$) the extensions by continuity of $\mathscr{T}$ to a bounded transformator acting from $\mathfrak{S}_{\mathrm{I}}$ into $\mathfrak{S}_{\mathrm{II}}$ (from $\mathfrak{S}'_{\mathrm{I}}$ into $\mathfrak{S}'_{\mathrm{II}}$). Then the transformators $\tilde{\mathscr{T}}$ and $\tilde{\mathscr{T}}'$ coincide on the intersection $\mathfrak{S}_{\mathrm{I}} \cap \mathfrak{S}'_{\mathrm{I}}$, i.e.

$$\tilde{\mathscr{T}} X = \tilde{\mathscr{T}}' X \quad (X \in \mathfrak{S}_{\mathrm{I}} \cap \mathfrak{S}'_{\mathrm{I}}).$$

In fact, as is well known (see [GK], Chapter III, § 6), for the partial Schmidt series X_n $(n = 1, 2, \cdots)$ of an operator $X \in \mathfrak{S}_{\mathrm{I}} \cap \mathfrak{S}'_{\mathrm{I}}$ we will have

$$\lim_{n\to\infty} |X_n - X|_{\mathrm{I}} = \lim_{n\to\infty} |X_n - X|_{\mathrm{II}} = 0.$$

Therefore

$$\lim_{n\to\infty} |\mathscr{T} X_n - \tilde{\mathscr{T}} X|_{\mathrm{I}} = \lim_{n\to\infty} |\mathscr{T} X_n - \tilde{\mathscr{T}}' X|_{\mathrm{II}} = 0.$$

These relations imply

$$\lim_{n\to\infty} |\mathscr{T} X_n - \tilde{\mathscr{T}} X|_{\infty} = \lim_{n\to\infty} |\mathscr{T} X_n - \tilde{\mathscr{T}}' X|_{\infty} = 0,$$

and so

$$\tilde{\mathscr{T}} X = \tilde{\mathscr{T}}' X.$$

The following theorem is basic for this section.

THEOREM 8.1 (E. M. SEMENOV). *Let $\mathscr{T}$ be a transformator of type $(\mathfrak{S}_p, \mathfrak{S}_p)$ for every p from the interval $1 < p < \infty$. Then $\mathscr{T}$ is a transformator of type $(\mathfrak{S}_\pi, \mathfrak{S}_\pi)$, where $\Pi = \{\pi_j\}_1^\infty$ is any binormalizing sequence satisfying the conditions*[9)]

[9)] We remark that the conditions (8.1) are in a certain sense precise. E. M. Semenov has shown that one can construct a transformator which is of type $(\mathfrak{S}_p, \mathfrak{S}_p)$ for every p $(1 < p < \infty)$ and is not of type $(\mathfrak{S}_\pi, \mathfrak{S}_\pi)$ if one of the conditions (8.1) is not satisfied.

After this book went to press, E. M. Semenov kindly informed us that he had obtained the following considerably more general theorem.

Let $\mathscr{T}$ be a transformator of type $(\mathfrak{S}_p, \mathfrak{S}_p)$ for every p from the interval $(1 \leqq) r < p < s (\leqq \infty)$. Then $\mathscr{T}$ is a transformator of type $(\mathfrak{S}_\Phi^{(0)}, \mathfrak{S}_\Phi^{(0)})$ for any s.n. function $\Phi(\xi_1, \xi_2, \cdots)$ satisfying the conditions

$$\lim_{n\to\infty} \frac{\chi(2n)}{\chi(n)} > 2^{1/s} \quad \textit{and} \quad \lim_{r\to\infty} \frac{\chi(2n)}{\chi(n)} < 2^{1/r},$$

where

$$\varliminf\left\{\sum_{j=1}^{2n}\pi_j \Big/ \sum_{j=1}^{n}\pi_j\right\} > 1, \quad \varlimsup\left\{\sum_{j=1}^{2n}\pi_j \Big/ \sum_{j=1}^{n}\pi_j\right\} < 2. \tag{8.1}$$

E. M. Semenov established this theorem by carrying over to the case of transformators a certain part of the results from his paper [**1**] concerning the interpolation of operators in scales of functional spaces. We remark that Semenov's theorem was preceded by results of B. S. Mitjagin [**1**], which when carried over to transformators lead to the following result.

If $\mathscr{T} \in (\mathfrak{S}_1, \mathfrak{S}_1)$ *and* $\mathscr{T} \in (\mathfrak{S}_\infty, \mathfrak{S}_\infty)$, *then* $\mathscr{T} \in (\mathfrak{S}, \mathfrak{S})$ *for any separable s.n. ideal* $\mathfrak{S}$.[10)]

We remark that the methods used by Mitjagin and Semenov for proving their theorems were substantially different.

We precede the proof of Theorem 8.1 by a simple result.

1. *Let* $\Pi = \{\pi_j\}_1^\infty$ *be a binormalizing sequence and let* $X \in \mathfrak{S}_\pi$. *Then*

$$|X|_\pi = \int_0^\infty \sigma(\nu(r; X)) \frac{dr}{r^2}, \tag{8.2}$$

where $\sigma(t)$ $(0 \leqq t < \infty)$ *is a step function defined by*

$$\sigma(t) = \sum_{j=0}^{k} \pi_j \quad (k \leqq t < k+1;\ k = 0, 1, \cdots; \pi_0 = 0),$$

and $\nu(r; X)$ *is the number of numbers* $1/s_n(X)$ $(n = 1, 2, \cdots)$ *which do not exceed* r.

In fact,

$$\begin{aligned}\int_0^\infty \sigma(\nu(r; X)) \frac{dr}{r^2} &= \sum_{n=1}^{\infty} \int_{1/s_n(X)}^{1/s_{n+1}(X)} \sigma(\nu(r; X)) \frac{dr}{r^2} \\ &= \sum_{n=1}^{\infty} \sigma(n)(s_n(X) - s_{n+1}(X)).\end{aligned} \tag{8.3}$$

$$\chi(m) = \Phi(\underbrace{1, 1, \cdots, 1}_{m}, 0, 0, \cdots).$$

Obviously, Theorem 8.1 can be obtained from this theorem as the special case corresponding to $r = 1$, $s = \infty$ and

$$\Phi(\xi_1, \xi_2, \cdots) = \sum_{j=1}^{\infty} \pi_j \xi_j.$$

10) The conclusion of the theorem is also true for an s.n. ideal $\mathfrak{S}^*$ which is conjugate to a separable s.n. ideal.

For any positive integer m satisfying the condition $1 \leqq m < n$ we have

$$s_{n+1}(X) \sum_{j=1}^{n} \pi_j \leqq s_{n+1}(X) \sum_{j=1}^{m} \pi_j + \sum_{j=m+1}^{\infty} \pi_j s_j(X),$$

from which it follows that

$$\lim_{n\to\infty} \sigma(n) s_{n+1}(X) = 0 \quad (X \in \mathfrak{S}_\pi). \tag{8.4}$$

Comparing the equality

$$\sum_{n=1}^{\infty} \pi_n s_n(X) = \sum_{n=1}^{\infty} \sigma(n)(s_n(X) - s_{n+1}(X)) - \lim_{n\to\infty} \sigma(n) s_{n+1}(X)$$

with (8.3) and (8.4), we arrive at (8.2).

PROOF OF THEOREM 8.1. It is asserted in the theorem that if the conditions (8.1) are satisfied, then

$$\sup_{K\in\mathfrak{K};|K|_\pi=1} |\mathscr{T}(K)|_\pi < \infty, \tag{8.5}$$

where $\mathfrak{K}$ is the set of finite-dimensional operators.

Let us denote by $\mathfrak{K}_\pi$ the set of all operators $K \in \mathfrak{K}$ whose nonzero s-numbers are all equal and such that $|K|_\pi = 1$, i.e. the set of all operators of the form

$$K = \left\{\sum_{j=1}^{m} \pi_j\right\}^{-1} \sum_{j=1}^{m} (\cdot, \psi_j)\chi_j,$$

where $\{\psi_j\}_1^m$ and $\{\chi_j\}_1^m$ $(m = 1, 2, \cdots)$ are arbitrary orthonormal systems. It turns out that

$$\sup_{K\in\mathfrak{K};|K|_\pi=1} |\mathscr{T}(K)|_\pi = \sup_{K\in\mathfrak{K}_\pi} |\mathscr{T}(K)|_\pi. \tag{8.6}$$

In fact, let us consider the set $\mathfrak{K}'_\pi(\{\psi_j\}, \{\chi_j\})$ of all operators X of the form

$$X = \sum_{j=1}^{n} \alpha_j (\cdot, \psi_j)\chi_j, \tag{8.7}$$

where $\{\psi_j\}_1^n$ and $\{x_j\}_1^n$ are fixed orthonormal sets, and $\alpha_1 \geqq \alpha_2 \geqq \cdots \geqq \alpha_n$ are arbitrary nonnegative numbers for which

$$\sum_{j=1}^{n} \pi_j \alpha_j = 1.$$

It follows from elementary considerations (see [GK], Chapter III, §15.3) that all the extreme points of the set $\mathfrak{R}'_\pi(\{\psi_j\},\{\chi_j\})$ belong to $\mathfrak{R}_\pi$, and so every operator $X \in \mathfrak{R}'_\pi(\{\psi_j\},\{\chi_j\})$ can be represented as a finite sum

$$X = \sum_j \beta_j K_j,$$

where $K_j \in \mathfrak{R}_\pi$, and $\beta_j \geqq 0$, $\sum_j \beta_j = 1$. This at once implies (8.6).

As was already mentioned in §5.2, for any $X \in \mathfrak{S}_p$ $(1 \leqq p < \infty)$

$$(|X|_p)^p = \int_0^\infty \frac{d\nu(r;X)}{r^p};$$

hence

$$(|X|_p)^p \geqq \int_0^t \frac{d\nu(r;X)}{r^p} \geqq \frac{\nu(t;X)}{t^p}$$

or

$$\nu(r;X) \leqq (r|X|_p)^p.$$

Since $\mathscr{T}$ is a bounded linear transformator in each of the spaces $\mathfrak{S}_p$ $(1 < p < \infty)$, for any operator L of the form

$$L = \sum_{j=1}^m (\cdot,\psi_j)\chi_j,$$

where $\{\psi_j\}$ and $\{\chi_j\}$ are orthonormal systems, we will have

$$\nu(r;\mathscr{T}(L)) \leqq (r|\mathscr{T}(L)|_p)^p \leqq |\mathscr{T}|_p^p (r|L|_p)^p \quad (1 < p < \infty),$$

or

$$\nu(r;\mathscr{T}(L)) \leqq |\mathscr{T}|_p^p r^p m. \tag{8.8}$$

By condition (8.1) there exist numbers a and b such that

$$1 < a < \varliminf \frac{\sigma(2n)}{\sigma(n)}, \quad \varlimsup \frac{\sigma(2n)}{\sigma(n)} < b < 2. \tag{8.9}$$

Consequently, for certain numbers p_1, p_2 $(1 < p_j < \infty)$

$$a^{p_2} > 2 \quad \text{and} \quad b^{p_1} < 2.$$

Obviously we may assume without loss of generality that

$$|\mathscr{T}|_{p_j} \leqq 1 \quad (j = 1, 2).$$

We now estimate the norm of the operator $\mathscr{T}(L)$ in the space $\mathfrak{S}_\pi$ with the help of (8.8):

$$|\mathscr{T}(L)|_\pi = \int_0^1 \sigma(\nu(r;\mathscr{T}(L)))\frac{dr}{r^2} + \int_1^\infty \sigma(\nu(r;\mathscr{T}(L)))\frac{dr}{r^2}$$
$$\leqq \int_0^1 \sigma(mr^{p_1})\frac{dr}{r^2} + \int_1^\infty \sigma(mr^{p_2})\frac{dr}{r^2}.$$

After a change of variables we obtain

$$|\mathscr{T}(L)|_\pi \leqq \int_0^1 \sigma\left(\frac{m}{t}\right) d(t^{1/p_1}) + \int_1^\infty \sigma\left(\frac{m}{t}\right) d(t^{1/p_2}). \tag{8.10}$$

For the first term on the right side we have

$$\int_0^1 \sigma\left(\frac{m}{t}\right) d(t^{1/p_1}) = \sum_{k=1}^\infty \int_{2^{-k}}^{2^{-k+1}} \sigma\left(\frac{m}{t}\right) d(t^{1/p_1})$$
$$\leqq \sum_{k=1}^\infty (2^{(-k+1)/p_1} - 2^{-k/p_1})\sigma(2^k m).$$

By virtue of the condition (8.9), for sufficiently large n we will have $\sigma(2n) \leqq b\sigma(n)$, whence

$$\sigma(2^k m) \leqq C_1 b^k \sigma(m) \quad (m = 1, 2, \cdots),$$

where C_1 is a constant. Hence

$$\int_0^1 \sigma\left(\frac{m}{t}\right) d(t^{1/p_1}) \leqq C_1 \sum_{k=0}^\infty b^k \sigma(m)(2^{1/p_1} - 1)2^{-k/p_1}$$
$$= C_1 \frac{2^{1/p_1} - 1}{1 - (b/2^{1/p_1})}\sigma(m).$$

For the second term we similarly obtain

$$\int_1^\infty \sigma\left(\frac{m}{t}\right) d(t^{1/p_2}) = \sum_{k=0}^\infty \int_{2^k}^{2^{k+1}} \sigma\left(\frac{m}{t}\right) d(t^{1/p_2})$$
$$\leqq \sum_{k=0}^\infty \sigma(2^{-k}m)(2^{(k+1)/p_2} - 2^{k/p_2})$$
$$\leqq C_2\left(\sigma(m) + \sum_{k=0}^\infty a^{-k}\sigma(m)(2^{1/p_2} - 1)2^{k/p_2}\right)$$
$$= C_2\left(1 + \frac{2^{1/p_2} - 1}{1 - (2^{1/p_2}/a)}\right)\sigma(m).$$

Thus

$$|\mathcal{T}(L)|_{\pi} \leqq C\sigma(m),$$

where C is a constant which depends only on the sequence $\Pi = \{\pi_j\}$. It follows that

$$|\mathcal{T}(K)|_{\pi} \leqq C$$

for any operator $K \in \mathfrak{R}_{\pi}$. The theorem is proved.

REMARK 8.1. When the conditions of Theorem 8.1 are satisfied, one can also assert that the transformator $\mathcal{T}$ can be extended to a bounded linear transformator acting in the space $\mathfrak{S}_{\pi}$; if, moreover, $\mathcal{T}\mathfrak{R} \subseteq \mathfrak{S}_{\Pi}^{(0)}$, then $\mathcal{T}$ is a transformator of type $(\mathfrak{S}_{\Pi}^{(0)}, \mathfrak{S}_{\Pi}^{(0)})$. This result will not be used in the sequel; without going into details, we shall however clarify the idea of its proof.

Let us consider the transformator $\mathcal{T}^*$, the conjugate of $\mathcal{T}$ in the space $\mathfrak{S}_p$ $(1 < p < \infty)$. Obviously the values of $\mathcal{T}^*$ on the finite-dimensional operators do not depend on p $(1 < p < \infty)$. Since $\mathcal{T}^*$, together with $\mathcal{T}$, is a transformator of type $(\mathfrak{S}_p, \mathfrak{S}_p)$ $(1 < p < \infty)$, by Theorem 8.1 $\mathcal{T}^* \in (\mathfrak{S}_{\pi}, \mathfrak{S}_{\pi})$. It is easily seen that $\mathcal{T}^{**}$, which is a continuous mapping of $\mathfrak{S}_{\Pi}$ into itself, is an extension of $\mathcal{T}$.[11)]

§9. Volterra operators with imaginary component belonging to one of the s.n. ideals $\mathfrak{S}_{\Pi}^{(0)}, \mathfrak{S}_{\pi}, \mathfrak{S}_{\Pi}$

1. The basic theorem of this section is the following.

THEOREM 9.1. *Let $\mathfrak{S}$ be one of the s.n. ideals $\mathfrak{S}_{\Pi}^{(0)}, \mathfrak{S}_{\pi}, \mathfrak{S}_{\Pi}$, where $\Pi = \{\pi_j\}_1^{\infty}$ is a binormalizing sequence with the properties*

$$\text{(9.1)} \qquad \varliminf \left\{ \sum_{j=1}^{2n} \pi_j \Big/ \sum_{j=1}^{n} \pi_j \right\} > 1 \quad \textit{and} \quad \varlimsup \left\{ \sum_{j=1}^{2n} \pi_j \Big/ \sum_{j=1}^{n} \pi_j \right\} < 2.$$

If A is a Volterra operator such that $A_{\mathcal{J}} \in \mathfrak{S}$, then $A_{\mathcal{R}} \in \mathfrak{S}$ and

$$\text{(9.2)} \qquad |A_{\mathcal{R}}|_{\mathfrak{S}} \leqq M_{\mathfrak{S}} |A_{\mathcal{J}}|_{\mathfrak{S}},$$

where $M_{\mathfrak{S}}$ is a constant which depends only on the space $\mathfrak{S}$.[12)]

---

11) Among other things, we note that the elements of $\mathfrak{S}_{\Pi}$ lie in $\mathfrak{S}_{\pi}$ if and only if $\sum_{j=1}^{\infty} \pi_j^2 < \infty$.

12) E. M. Semenov has informed us that he has established the following result (which is in a certain sense definitive).

Let $\mathfrak{S}_{\Phi}^{(0)}$ be a separable s.n. ideal. In order that for any Volterra operator A the condition

We precede the proof of this theorem by two results.

1. *If the binormalizing sequence* $\Pi = \{\pi_j\}$ *satisfies the conditions* (9.1), *then*

$$\lim_{n\to\infty} \left\{ \sum_{j=1}^{n} \frac{1}{2j-1} \Big/ \sum_{j=1}^{n} \pi_j \right\} = 0 \tag{9.3}$$

and

$$\sum_{j=1}^{\infty} \frac{\pi_j}{2j-1} < \infty . \tag{9.4}$$

For sufficiently large n

$$\sum_{j=1}^{2n} \pi_j \geqq a \sum_{j=1}^{n} \pi_j, \quad \text{where} \quad 1 < a < \varliminf \left\{ \sum_{j=1}^{2n} \pi_j \Big/ \sum_{j=1}^{n} \pi_j \right\},$$

and so

$$\sum_{j=1}^{2^m} \pi_j \geqq Ca^m \qquad (m = 1, 2, \cdots),$$

where C is some constant.

Let $2^m \leqq n < 2^{m+1}$, i.e. $m = [\log_2 n]$; then

$$\sum_{j=1}^{n} \frac{1}{2j-1} = O(m) \quad (n \to \infty),$$

and consequently

$$\sum_{j=1}^{n} \frac{1}{2j-1} \Big/ \sum_{j=1}^{n} \pi_j = O(ma^{-m}) (= O(\log_2 n / n^{\log_2 a})) \quad (n \to \infty),$$

from which follows (9.3).

We similarly deduce from (9.1) that

$$\sum_{j=1}^{n} \pi_j = O(b^m) \quad (n \to \infty),$$

where as before $m = [\log_2 n]$, and

$A_{\mathscr{I}} \in \mathfrak{S}_\Phi^{(0)}$ shall imply that $A \in \mathfrak{S}_\Phi^{(0)}$, it is necessary and sufficient that

$$\varliminf \frac{\chi(2n)}{\chi(n)} > 1 \quad \text{and} \quad \varlimsup \frac{\chi(2n)}{\chi(n)} < 2,$$

where $\chi(n) = \Phi(n) = \Phi(\underbrace{1, 1, \cdots, 1}_{n}, 0, 0, \cdots)$.

$$2 > b > \overline{\lim} \left\{ \sum_{j=1}^{2n} \pi_j \Big/ \sum_{j=1}^{n} \pi_j \right\}.$$

Thus

$$\sum_{j=1}^{n} \pi_j = O(n^{\log_2 b}) \quad (n \to \infty). \tag{9.5}$$

Since $\log_2 b < 1$, (9.5) implies the equality

$$\sum_{j=1}^{\infty} \frac{\pi_j}{2j-1} = \sum_{n=1}^{\infty} \left(\sum_{j=1}^{n} \pi_j \right) \frac{2}{4n^2 - 1},$$

and also the convergence of the right side. Thus, the inequality (9.4) is also proved.

2. *If the binormalizing sequence* $\Pi = \{\pi_j\}_1^\infty$ *satisfies the conditions* (9.1), *then* $\mathfrak{S}_{\Pi} \subseteq \mathfrak{S}_{\omega}$ *and* $\mathfrak{S}_{\Omega} \subseteq \mathfrak{S}_{\pi}$.

In fact, if $A \in \mathfrak{S}_{\Pi}$, then

$$\sum_{j=1}^{n} s_n(A) \leqq |A|_{\Pi} \sum_{j=1}^{n} \pi_j \quad (n = 1, 2, \cdots),$$

and so by virtue of the majorant principle (see [GK], Chapter III, §4.2)

$$\sum_{j=1}^{\infty} \frac{1}{2j-1} s_j(A) \leqq |A|_{\Pi} \sum_{j=1}^{\infty} \frac{\pi_j}{2j-1}.$$

By (9.4) the series on the right side converges; hence $A \in \mathfrak{S}_{\omega}$. Thus $\mathfrak{S}_{\Pi} \subseteq \mathfrak{S}_{\omega}$.

If $A \in \mathfrak{S}_{\Omega}$, the relations

$$\sum_{j=1}^{n} s_j(A) \leqq |A|_{\Omega} \sum_{j=1}^{n} \frac{1}{2j-1} \quad (n = 1, 2, \cdots)$$

imply

$$\sum_{j=1}^{\infty} \pi_j s_j(A) \leqq |A|_{\Omega} \sum_{j=1}^{\infty} \frac{\pi_j}{2j-1} (< \infty).$$

Therefore $\mathfrak{S}_{\Omega} \subseteq \mathfrak{S}_{\pi}$.

PROOF OF THEOREM 9.1. Let A be a Volterra operator with $A_{\mathscr{I}} \in \mathfrak{S}$. As has already been noted in similar cases, we may assume without loss of generality that A has a complete eigenchain $\mathfrak{P}$. Since

$$A = A_{\mathscr{R}} + iA_{\mathscr{I}} = 2i\mathscr{C}(A_{\mathscr{I}}; \mathfrak{P}),$$

the theorem will obviously be proved as soon as we show that

$$\mathscr{C}\mathfrak{S} \subseteq \mathfrak{S} \quad \text{and} \quad \sup |\mathscr{C}_{\mathfrak{S}\to\mathfrak{S}}(\cdot;\mathfrak{P})| < \infty,$$

where the supremum is taken over all complete chains. By Lemma 3.1, the above will hold if for every complete chain $\mathfrak{P}$ the transformator $\mathscr{C}(\cdot;\mathfrak{P})$ is a bounded transformator which maps $\mathfrak{S}$ into itself. Since further arguments do not depend upon the chain $\mathfrak{P}$, we will omit it from the symbol $\mathscr{C}(\cdot;\mathfrak{P})$.

By Theorem 6.1, the transformator $\mathscr{C}$ is of type $(\mathfrak{S}_p, \mathfrak{S}_p)$ for any p from the interval $1 < p < \infty$. Hence, by Theorem 8.1 $\mathscr{C} \in (\mathfrak{S}_\pi, \mathfrak{S}_\pi)$ for any binormalizing sequence $\Pi = \{\pi_j\}_1^\infty$ satisfying the conditions (9.1). Thus $\mathscr{C}$ can be extended by continuity to a bounded transformator acting in $\mathfrak{S}_\pi$. Since the transformator $\mathscr{C}$ acting in $\mathfrak{S}_\infty$ is closed, this extension coincides with $\mathscr{C}$, and so the transformator $\mathscr{C}_{\mathfrak{S}_\pi\to\mathfrak{S}_\pi}$ is bounded.

By virtue of result 2, Theorem 7.3 is applicable to the s.n. ideals $\mathfrak{S}_{\mathrm{I}} = \mathfrak{S}_{\Pi} = \mathfrak{S}_\pi$, $\mathfrak{S}_{\mathrm{I}}^* = \mathfrak{S}_{\Pi}^* = \mathfrak{S}_{\Pi}$; by this theorem the transformator $\mathscr{C}_{\mathfrak{S}_\Pi\to\mathfrak{S}_\Pi}$ is bounded. From (9.3) it follows that $\mathscr{C}(\mathfrak{R}) \subseteq \mathfrak{S}_\Pi^{(0)}$. Hence the boundedness of the transformator $\mathscr{C}_{\mathfrak{S}_\Pi\to\mathfrak{S}_\Pi}$ implies that of $\mathscr{C}_{\mathfrak{S}_\Pi^{(0)}\to\mathfrak{S}_\Pi^{(0)}}$. The theorem is proved.

2. We now recall certain concepts and results of the theory of *regular* binormalizing sequences $\{\pi_j\}_1^\infty$ and the associated s.n. ideals (see [GK], Chapter III, §§14—16).

A binormalizing sequence $\Pi = \{\pi_j\}_1^\infty$ is said to be *regular* if the condition

$$\sum_{j=1}^{n} \pi_j = O(n\pi_n) \quad (n \to \infty)$$

is satisfied. If the binormalizing sequence $\Pi = \{\pi_j\}_1^\infty$ is regular, the s.n. ideal $\mathfrak{S}_\Pi$ coincides with the class of all operators $X \in \mathfrak{S}_\infty$ for which

$$s_n(X) = O(\pi_n) \quad (n \to \infty),$$

and the s.n. ideal $\mathfrak{S}_\Pi^{(0)}$ coincides with the class of all operators $X \in \mathfrak{S}_\infty$ for which

$$s_n(X) = o(\pi_n) \quad (n \to \infty).$$

A nonincreasing sequence $\{\pi_j\}_1^\infty$ is regular if it can be represented in the form

$$\pi_n = n^{-1/p}L(n) \quad (n = 1, 2, \cdots;\ 1 < p < \infty), \tag{9.6}$$

where $L(\nu)$ $(1 \leqq \nu < \infty;\ L(1) = 1)$ is a slowly varying function, and the function $\nu^{1/p}L(\nu)$ is nonincreasing.

A function $L(\nu)$ $(1 \leqq \nu < \infty)$ is said to be *slowly varying* if it is positive, continuously differentiable, and

$$\lim_{\nu \to \infty} \{\nu L'(\nu)/L(\nu)\} = 0.$$

An example of a slowly varying function is any function of the form

$$L(\nu) = c(\ln_1 \nu)^{\alpha_1}(\ln_2 \nu)^{\alpha_2} \cdots (\ln_k \nu)^{\alpha_k},$$

where

$$\ln_1 \nu = \ln \nu;\ \ln_r \nu = \ln(\ln_{r-1}\nu) \quad (r = 2, 3, \cdots),$$

and $\alpha_1, \alpha_2, \cdots, \alpha_k$ are arbitrary real numbers.

The sequence $\{\pi_j\}_1^\infty$ defined by (9.6) satisfies the conditions (9.1) of Theorem 9.1. Indeed, it is not hard to establish the relations

$$\lim_{n \to \infty} \left\{ \sum_{j=1}^{n} j^{-1/p} L(j) \Big/ \int_1^n r^{-1/p} L(r)\, dr \right\} = 1$$

and

$$\lim_{n \to \infty} \left\{ \frac{n^{1-1/p} L(n)}{1 - 1/p} \Big/ \int_1^n r^{-1/p} L(r)\, dr \right\} = 1.$$

It follows at once that

$$\lim_{n \to \infty} \left\{ \sum_{j=1}^{2n} j^{-1/p} L(j) \Big/ \sum_{j=1}^{n} j^{-1/p} L(j) \right\} = 2^{1-1/p}.$$

Thus, the following result is a consequence of Theorem 9.1.

THEOREM 9.2. *Let $L(\nu)$ $(1 \leqq \nu < \infty)$ be a slowly varying function. If, for a Volterra operator A,*

$$s_n(A_{\mathscr{I}}) = O(n^{-1/p} L(n)) \quad (n \to \infty;\ 1 < p < \infty),$$

then

$$s_n(A_{\mathscr{R}}) = O(n^{-1/p} L(n)) \quad (n \to \infty),$$

and

$$M'' \sup_n \frac{s_n(A_{\mathscr{I}})}{n^{-1/p} L(n)} \leqq \sup_n \frac{s_n(A_{\mathscr{R}})}{n^{-1/p} L(n)} \leqq M' \sup_n \frac{s_n(A_{\mathscr{I}})}{n^{-1/p} L(n)},$$

where M' and M'' are constants which depend only on the number p and the function $L(\nu)$.

If for a Volterra operator A

$$s_n(A_{\mathscr{I}}) = o(n^{-1/p} L(n)) \quad (n \to \infty;\ 1 < p < \infty),$$

then

$$s_n(A_{\mathscr{R}}) = o(n^{-1/p}L(n)) \quad (n \to \infty).$$

Theorem 9.2, for values of p from the interval $1 < p < 2$, was established by V. I. Macaev [1] using theorems on entire functions of functions which he had proved (see [GK], Chapter IV, §11).

For the remaining values of p $(2 \leq p < \infty)$ the theorem was established by the authors [**5**]. In this same paper Theorem 9.1 was proved for the s.n. ideal $\mathfrak{S}_\pi$ $(\Pi = \{n^{-1/p}L(n)\}_{n=1}^\infty$.

The proof was based upon Macaev's result, an interpolation theorem (see [GK], Chapter III, Theorem 16.1), an identity from §3 and the general theorem on the conjugate of a transformator. Semenov's theorem has made it possible to substantially simplify the proof of the theorem and at the same time generalize it.

§10. Applications to the Riesz-Titchmarsh transform[13)]

1. We begin with some general remarks concerning Volterra operators in $L_2(0,1)$ of the form

$$(Hf)(t) = \int_0^1 \mathscr{A}(t-s)f(s)\,ds, \tag{10.1}$$

where $\mathscr{A}(t)$ $(-1 \leq t \leq 1)$ is a function from $L_1(-1,1)$.

As is well known, for any function f from $L_2(0,1)$ the integral (10.1) is meaningful for almost all t, and if we denote its value by $g(t)$, then by Young's inequality (cf. Zygmund [1], Vol. 1)

$$\int_0^1 |g(t)|^2 dt \leq \int_0^1 |f(t)|^2 dt \left(\int_{-1}^1 |\mathscr{A}(t)|\,dt \right)^2.$$

Thus (10.1) defines a bounded operator H with norm satisfying

$$|H| \leq \int_{-1}^1 |\mathscr{A}(t)|\,dt. \tag{10.2}$$

Moreover, the operator H is completely continuous: it can be approximated arbitrarily closely by finite-dimensional operators.

Indeed, for any $\epsilon > 0$ there exists a polynomial $\mathscr{A}_\epsilon(t)$ such that

---

13) The possibility of these applications was discovered by V. I. Macaev [3].

$$\int_{-1}^{1} |\mathscr{A}(t) - \mathscr{A}_\epsilon(t)|\, dt < \epsilon.$$

Let us denote by H_ϵ the operator corresponding by (10.1) to the kernel $\mathscr{A}_\epsilon(t-s)$. The operator H_ϵ is obviously finite-dimensional. On the other hand, according to (10.2)

$$|H - H_\epsilon| < \epsilon.$$

It is easily seen that if $\mathscr{A}(t)$ is a Hermitian function, i.e. $\overline{\mathscr{A}(t)} = \mathscr{A}(-t)$ $(-1 \leqq t \leqq 1)$, the operator H will be selfadjoint. In this case H is the imaginary component of the Volterra operator

$$(Af)(t) = 2i \int_t^1 \mathscr{A}(t-s) f(s)\, ds \quad (f \in L_2(0,1)). \tag{10.3}$$

The real component G of A is given by

$$(Gf)(t) = \int_0^t \mathscr{G}(t-s) f(s)\, ds \quad (f \in L_2(0,1)),$$

where

$$\mathscr{G}(t) = i(\operatorname{sign} t)\mathscr{A}(t). \tag{10.4}$$

2. Let us now consider the special case in which the Hermitian function $\mathscr{A}(t) \in L_1(-1,1)$ can be extended to a periodic function with period $T = 1$,[14] and hence

$$\mathscr{A}(t) \sim \sum_{k=-\infty}^{\infty} \xi_k e^{2k\pi i t}, \tag{10.5}$$

where $\{\xi_k\}_{-\infty}^{\infty}$ is some sequence of real numbers. In this case the functions

$$\phi_k(t) = e^{2k\pi i t} \quad (k = 0, \pm 1, \pm 2, \cdots)$$

form a complete orthonormal system of eigenvectors of the operator H, and the Fourier coefficients ξ_k $(k = 0, \pm 1, \pm 2, \cdots)$ are the corresponding eigenvalues: $\int_0^1 \mathscr{A}(t-s)\phi_k(s)\, ds = \xi_k \phi_k(t)$.

It is not hard to verify that the system of functions

$$\psi_k(t) = e^{-(2k+1)\pi i t} \quad (k = 0, \pm 1, \pm 2, \cdots)$$

---

[14] This is equivalent to the additional condition

$$\mathscr{A}(t) = \overline{\mathscr{A}(1-t)}, \quad 0 \leqq t \leqq 1.$$

is orthonormal and constitutes a complete in $L_2(0,1)$ [15] system of eigenvectors of the operator G:

$$\int_0^1 \mathcal{G}(t-s)\psi_k(s)\,ds = \eta_k\psi_k(t) \quad (k = 0, \pm 1, \cdots), \tag{10.6}$$

where

$$\eta_k = i\int_0^1 \mathscr{A}(-t)e^{-(2k+1)\pi it}dt \quad (k = 0, \pm 1, \cdots).$$

Inserting here the expansion for $\mathscr{A}(-t)$ obtained from (10.5), we find that

$$\eta_k = \frac{1}{\pi}\sum_{j=-\infty}^{\infty}\frac{\xi_j}{j+k+1/2} \quad (k = 0, \pm 1, \pm 2, \cdots), \tag{10.7}$$

where the sum is understood in the sense of the principal value.[16]

In the cases to be considered in the sequel the series (10.7) will always be absolutely convergent.

Putting $t = 0$ in (10.6) and then taking complex conjugates, we find that to the skewperiodic function $\mathcal{G}(t)$ $(= -\mathcal{G}(t+1))$ there corresponds the following series expansion in the functions $\psi_k(t)$ $(k = 0, \pm 1, \cdots)$:

$$\mathcal{G}(t) \sim \sum_{k=-\infty}^{\infty}\eta_k e^{-(2k+1)\pi it}. \tag{10.8}$$

It is not hard to verify that if we are given an arbitrary locally integrable *skewperiodic Hermitian* function $\mathcal{G}(t) = \overline{\mathcal{G}(-t)} = -\mathcal{G}(t+1)$, then the Fourier coefficients ξ_k $(k = 0, \pm 1, \cdots)$ from the expansion (10.5) of the corresponding *Hermitian* periodic function

---

[15] The system $\{\psi_k\}_{-\infty}^{\infty}$ is a complete system of fundamental functions of the self-adjoint boundary value problem

$$\psi'' + \lambda\psi = 0,\ \psi(0) + \psi(1) = 0,\ \psi'(0) + \psi'(1) = 0.$$

[16] We have used a well-known general theorem to the effect that if $f(t) \in L_1(0,1)$,

$$f(t) \sim \sum_{j=-\infty}^{\infty} c_j e^{2\pi ijt},$$

and $\psi(t)$ $(0 \leqq t \leqq 1)$ is a function of bounded variation, then

$$\int_0^1 f(t)\psi(t)\,dt = \lim_{n\to\infty}\sum_{j=-n}^{n} c_j\int_0^1 e^{2\pi ijt}\psi(t)\,dt.$$

$$\mathscr{A}(t) = -i(\operatorname{sign} t)\mathscr{G}(t) \ (-1 \leqq t \leqq 1), \ \mathscr{A}(t+1) = \mathscr{A}(t)$$

are found from the Fourier coefficients η_k $(k = 0, \pm 1, \cdots)$ in the expansion (10.8) of $\mathscr{G}(t)$ by the same transformation

$$\xi_k = \frac{1}{\pi} \sum_{j=-\infty}^{\infty} \frac{\eta_j}{k+j+1/2}. \tag{10.9}$$

The transformation (10.7) was first studied by M. Riesz [**2**] and E. C. Titchmarsh [**1**], who introduced it as the discrete analog of the familiar singular integral transform of Hilbert:

$$g(x) = \frac{1}{\pi i} \int_{-\infty}^{\infty}{}' \frac{f(s)\,ds}{s-x}, \tag{10.10}$$

where the prime denotes that the integral is taken in the sense of the principal value.

To make clear the analogy between this integral transform and the transformation (10.7), we note that for a different enumeration of the eigenfunctions ψ, where

$$\psi_k(t) = e^{(2k-1)\pi i t} \quad (k = 0, \pm 1, \cdots),$$

the transformation which transforms the spectrum of the operator H into that of G is no longer written in the form (10.7), but rather in the form

$$\eta_k = \frac{1}{\pi} \sum_{j=-\infty}^{\infty} \frac{\xi_j}{j-k+1/2} \quad (k = 0, \pm 1, \cdots).$$

If two sequences $\{\xi_k\}_{-\infty}^{\infty}$ and $\{\eta_k\}_{-\infty}^{\infty}$ are related by (10.7), and all the series on the right side of (10.7) are absolutely convergent, then we shall write $\eta = \mathfrak{T}\xi$.

We denote by $\boldsymbol{l}_p$ $(1 \leqq p < \infty)$ the Banach space of all real sequences $= \{\xi_k\}_{-\infty}^{\infty}$ with finite norm

$$|\xi|_p = \left(\sum_{k=-\infty}^{\infty} |\xi_k|^p \right)^{1/p}.$$

The series in (10.7) are absolutely convergent for any $\xi \in \boldsymbol{l}_p$, and thus the relation $\eta = \mathfrak{T}\xi$ has meaning.

On the other hand, the operator H generated by the Hermitian periodic function $\mathscr{A}(t)$ whose expansion is (10.5) will belong to the class $\mathfrak{S}_p$ $(1 \leqq p \leqq \infty)$ if and only if $\xi = \{\xi_k\} \in \boldsymbol{l}_p$, since

$$|H|_p = \left(\sum_{k=-\infty}^{\infty} |\xi_k|^p \right)^{1/p} = |\xi|_p.$$

For $1 \leqq p \leqq 2$ there corresponds to any vector $\xi = \{\xi_k\} \in \boldsymbol{l}_p$ a periodic Hermitian locally integrable (even square-integrable) function $\mathscr{A}(t)$ with the expansion (10.5). By Theorem 6.2 of V. I. Macaev, for $p > 1$ together with $H \in \mathfrak{S}_p$ we will also have $G \in \mathfrak{S}_p$, and

$$|G|_p \leqq \gamma_p |H|_p. \tag{10.11}$$

Since

$$|G|_p = \left(\sum_{k=-\infty}^{\infty} |\eta_k|^p \right)^{1/p} = |\eta|_p,$$

the inequality (10.11) shows that

$$|\eta|_p \leqq \gamma_p |\xi|_p. \tag{10.12}$$

For $p = 2$ we will have $|\eta|_2 = |\xi|_2$, i.e.

$$\sum_{k=-\infty}^{\infty} |\eta_k|^2 = \sum_{k=-\infty}^{\infty} |\xi_k|^2 \quad (|G|_2 = |H|_2).$$

We need add little to obtain the basic result of M. Riesz [**2**] (see also E. C. Titchmarsh [**1**]) concerning the transformation $\eta = \mathfrak{T}\xi$.

A) *The transformation* $\eta = \mathfrak{T}\xi$ *maps any space* $\boldsymbol{l}_p$ $(1 < p < \infty)$ *continuously and involutively* ($\mathfrak{T}^2 = I$) *onto itself. In* $\boldsymbol{l}_2$, *moreover,* $\mathfrak{T}$ *is a symmetric orthogonal transformation.*

It remains to clarify the validity of the first assertion for $p > 2$. In this case $q = p/(p-1) < 2$, and the spaces $\boldsymbol{l}_p$ and $\boldsymbol{l}_q$ are mutually conjugate. Let us denote by $\mathfrak{T}_q$ the operator defined in $\boldsymbol{l}_q$ by $\mathfrak{T}_q = \mathfrak{T}$. Since, by what we have proved, $\mathfrak{T}_q$ is a bounded operator, its conjugate $\mathfrak{T}_q^*$ will be bounded, and for any $\xi \in \boldsymbol{l}_p$ and $\zeta = \{\zeta_k\} \in \boldsymbol{l}_q$ we will have

$$(\xi, \mathfrak{T}_q \zeta) = (\mathfrak{T}_q^* \xi, \zeta),$$

where we have used the notation

$$(\xi, \eta) = \sum_{k=-\infty}^{\infty} \xi_k \eta_k \quad (\xi \in \boldsymbol{l}_p, \eta \in \boldsymbol{l}_q).$$

On the other hand, if ζ is a finite vector (i.e. only a finite number of its coordinates ζ_k are different from zero), we easily obtain from the symmetry of the matrix of the transformation (10.7) that

$$(\xi, \mathfrak{T}_q \zeta) = (\mathfrak{T}\xi, \zeta),$$

so that for any $\xi \in l_p$ and any finite ζ we have

$$(\mathfrak{T}\xi - \mathfrak{T}_q^* \xi, \zeta) = 0.$$

It now follows that $\mathfrak{T}\xi \in l_p$ for any $\xi \in l_p$, and moreover $\mathfrak{T}\xi = \mathfrak{T}_q \xi$. This proves result A).

At the same time, we have shown that for any $\xi \in l_p$ and $\zeta \in l_q$ $(p^{-1} + q^{-1} = 1)$

$$(\mathfrak{T}\xi, \zeta) = (\xi, \mathfrak{T}\zeta), \tag{10.13}$$

and if we denote by M_p $(1 < p < \infty)$ the norm of the operator $\mathfrak{T}_p$, which is induced in l_p by the transformation (10.7), then

$$M_p = M_{p/(p-1)} \quad (1 < p < \infty).$$

From the interpolation theorems of M. Riesz [**1, 2**] (cf. also Zygmund [**1**]), of which the interpolation theorems proved in this book represent a certain development, it follows that M_p is a logarithmically convex function of the argument $1/p$.

We also present the following result of M. Riesz [**1, 2**], which establishes an analogy and a connection between the integral transformation (10.10) and the discrete transformation (10.7).

It turns out that for any $f \in L_p = L_p(-\infty, \infty)$ $(1 < p < \infty)$ there exists, in the sense of convergence in L_p, the limit

$$g(x) = \underset{\epsilon \to 0}{\text{l.i.m.}} \frac{i}{\pi} \int_{|s-x| > \epsilon} \frac{f(s)}{s - x} ds.$$

The mapping Γ: $f \to g$ thus obtained is a continuous involution which maps L_p onto itself.[17)] The norm of this mapping coincides with the constant $M_p (= |\mathfrak{T}|_p)$. In $L_2(-\infty, \infty)$ the involution Γ, as had been discovered by Hilbert, is a unitary operator.

[17)] Starting from this result and the identity

$$[(\Gamma f)(x)]^2 = [f(x)]^2 + 2\Gamma(f \Gamma f)(x), \tag{!}$$

which holds, for example, for all step functions with compact support, one can prove, using arguments similar to those used in the proof of Theorem 6.2, the boundedness of the integral transform Γ in L_p for $p = 2^n$ $(n = 1, 2, \cdots)$, and then for any $p > 1$. The identity (!) and this method for proving M. Riesz' theorem concerning the transforms Γ in L_p $(1 < p < \infty)$ was pointed out by M. Cotlar [**1**]. The considerations of Cotlar served as a stimulus for the authors to establish the identity (3.6) and use it for the proof of Theorem 6.2.

It follows from (10.12) that

$$M_p \leqq \gamma_p \quad (1 < p < \infty), \tag{10.14}$$

where γ_p is the best constant from Theorem 6.2.

The bound (10.14) is due to V. I. Macaev.

3. In result A) the value $p = 1$ is excluded. Our results for Volterra operators with imaginary component from $\mathfrak{S}_1$ enable us to fill this gap in result A) with the following result.

B) *Let $\xi = \{\xi_k\}_{-\infty}^{\infty}$ be any real sequence from l_1. Then for $\eta = \mathfrak{T}\xi$ we have the relations*

$$\max\left\{\sup_n\left\{\frac{1}{\sigma_n}\sum_{k=1}^{n}\eta_k^+\right\},\ \sup_n\left\{\frac{1}{\sigma_n}\sum_{k=1}^{n}|\eta_k^-|\right\}\right\} \leqq \frac{2}{\pi}\sum_{k=-\infty}^{\infty}|\xi_k| \tag{10.15}$$

and

$$\lim_{n\to\infty} n\eta_n^+ = -\lim_{n\to\infty} n\eta_n^- = \frac{1}{\pi}\left|\sum_{k=-\infty}^{\infty}\xi_k\right|, \tag{10.16}$$

where $\sigma_n = \sum_{k=1}^{n} 1/(2k-1)$, and $\{\eta_k^+\}_1^\infty$ and $\{\eta_k^-\}_1^\infty$ denote respectively the positive and negative terms of the sequence $\{\eta_k\}_{-\infty}^{\infty}$, arranged in decreasing order of their absolute values.[18] *Moreover, we also have the relations*

$$\lim_{r\to 0}\sum_{|\eta_j|\geqq r}\eta_j = 0 \tag{10.17}$$

and

$$\sup_n n\eta_n^* \leqq \gamma\sum_{j=-\infty}^{\infty}|\xi_j|, \tag{10.18}$$

where γ $(\leqq 4/\pi)$ is some absolute constant, and $\{\eta_k^*\}_1^\infty$ denotes the non-increasing sequence consisting of the absolute values $|\eta_k|$ of all the numbers η_k $(k = 0, \pm 1, \cdots)$. Finally, if $\xi_k \geqq 0$ for all $k = 0, \pm 1, \cdots$, then

$$\max(\eta_n^+, |\eta_n^-|) \leqq \frac{2}{\pi}\frac{1}{n}\sum_{j=-\infty}^{\infty}\xi_j \quad (n = 1, 2, \cdots).$$

Indeed, the inequalities (10.15) say the same thing as do the inequalities (2.2) applied to the Volterra operator A, defined in $L_2(0,1)$ by (10.3) with a Hermitian function $\mathscr{A}(t)$ of the form (10.5), where $\sum|\xi_k| < \infty$ (this condition just says that $H = A_{\mathscr{I}} \in \mathfrak{S}_1$).

---

[18] If one or both of the sequences $\{\eta_k^+\}$, $\{\eta_k^-\}$ turn out to be finite, we make them infinite sequences by adding on an infinite sequence of zero terms.

Applying Theorem 2.4 to this operator enables us to assert (10.17) and (10.18), and also the existence of the common finite limit

$$h = \pi \lim_{n\to\infty} n\eta_n^+ = -\pi \lim_{n\to\infty} n\eta_n^- . \tag{10.19}$$

If $\xi_k \geqq 0$ for all $k = 0, \pm 1, \cdots$ (or if, conversely, $\xi_k \leqq 0$ for all k), the operator H will be nonnegative (nonpositive), and so A $(-A)$ will be dissipative. In this case, by Theorem 2.4

$$h = |\operatorname{sp} H| = \left| \sum_{k=-\infty}^{\infty} \xi_k \right|, \tag{10.20}$$

and thus (10.16) holds.

In the general case, where the coefficients ξ_k $(k = 0, \pm 1, \cdots)$ can have arbitrary sign, it is not hard to establish the validity of (10.16) if we show that the quantity h, defined by (10.19), is a continuous function of the vector ξ $(\in l_1)$. This fact is a consequence of the general Lemma VII.1.2 (or Theorem VII.1.1).

By virtue of this fact, it suffices to establish the relation (10.16) for vectors $\xi \in l_1$ which have only a finite number of coordinates ξ_k different from zero, and where $\sum_k \xi_k \neq 0$ and even $\sum_k \xi_k > 0$.

For such a vector ξ it is not hard to deduce from (10.7) that for $n \to \pm\infty$

$$\begin{aligned} \eta_n &= \frac{1}{\pi n} \sum_{k=-\infty}^{\infty} \xi_k + O\left(\frac{1}{n^2}\right), \\ \eta_n - \eta_{n-1} &= \frac{1}{\pi n^2} \sum_{k=-\infty}^{\infty} \xi_k + O\left(\frac{1}{n^3}\right). \end{aligned} \tag{10.21}$$

From these relations it follows that for sufficiently large n (> 0) the numbers η_n are positive and do not increase, and the numbers η_{-n} are negative and do not decrease. We conclude that there exist numbers r and s such that for sufficiently large n (> 0) we have $\eta_n^+ = \eta_{n+r}$ and $\eta_n^- = \eta_{-n-s}$. But then, according to the first of relations (10.21),

$$\lim_{n\to\infty} n\eta_n^+ = \lim_{n\to\infty} n\eta_{n+r} = \frac{1}{\pi} \sum_{k=-\infty}^{\infty} \xi_k = \lim_{n\to-\infty} n\eta_{n-s} = -\lim_{n\to\infty} n\eta_n^- .$$

The last relation of result B) follows from a result of de Branges (see (2.30)). The result B) is proved.

We remark that (10.18) is the discrete analog of a theorem of A. N. Kolmogorov [**1**] on Hilbert transforms (cf. also N. K. Bari [**1**]).

We do not know whether results which might be regarded as analogs of the other results of this theorem have been established for the Hilbert transform.

We should also note the following situation.

While methods of the theory of entire functions are not needed to establish result A), result B) (in particular, as regards the relations (10.16) and (10.17)) required the application of deep theorems of this theory.

Theorem 9.1 enables us to formulate results concerning the boundedness of the Riesz-Titchmarsh transformation for certain spaces generated by binormalizing sequences which are subject to the conditions of Semenov. We shall formulate these results for the most interesting case of a sequence of numbers $\pi_k = k^{-1/p}L(k)$ $(k = 1, 2, \cdots)$, where $L(\nu)$ is a slowly varying function.

In what follows let $L(\nu)$ $(1 \leqq \nu < \infty)$ be an arbitrary slowly varying function, and let p lie in the interval $1 < p < \infty$.

Under these conditions the sequence of numbers

$$\pi_k = k^{1/p}L(k) \quad (k = 1, 2, \cdots) \tag{10.22}$$

will be monotone decreasing from some value of k on.

In the results which follow we may suppose, without loss of generality, that the sequence (10.22) is nonincreasing: $\pi_1 \geqq \pi_2 \geqq \cdots$.

C) *Suppose that the sequence of real numbers* $\xi = \{\xi_k\}_{-\infty}^{\infty}$ *satisfies the condition*

$$\xi_n^* = O(n^{-1/p}L(n)) \quad (n \to \infty). \tag{10.23}$$

Then the sequence $\eta = \mathfrak{T}\xi = \{\eta_k\}_{-\infty}^{\infty}$ *satisfies the same condition*:

$$\eta_n^* = O(n^{-1/p}L(n)) \quad (n \to \infty),$$

and moreover

$$\sup_n n^{1/p}L^{-1}(n)\,\eta_n^* \leqq C \sup_n n^{1/p}L^{-1}(n)\,\xi_n^*, \tag{10.24}$$

where C (> 0) *is a constant which depends only on* p *and the choice of the function* $L(\nu)$.

Finally, if

$$\xi_n^* = o(n^{-1/p}L(n)) \quad (n \to \infty), \tag{10.25}$$

then

$$\eta_n^* = o(n^{-1/p}L(n)) \quad (n \to \infty). \tag{10.26}$$

D) *Let* $\xi = \{\xi_k\}_{k=-\infty}^{\infty}$ *be a sequence of real numbers such that*

$$\sum_{n=1}^{\infty} n^{-1/p} L(n) \xi_n^* < \infty. \tag{10.27}$$

Then for $\eta = \mathfrak{T}\xi$ *we have*

$$\sum_{n=1}^{\infty} n^{-1/p} L(n) \eta_n^* \leqq M \sum_{n=1}^{\infty} n^{-1/p} L(n) \xi_n^*, \tag{10.28}$$

where M (>0) *is a constant which depends only on* p *and the choice of the function* $L(\nu)$.

As in result B), ξ^* denotes the nonincreasing sequence consisting of the absolute values of the numbers ξ_k $(k = 0, \pm 1, \cdots)$.

At first glance, results C) and D) appear to be immediate consequences of Theorems 9.1 and 7.3. It would seem that it is sufficient to apply these theorems to the Volterra operator A, acting in $L_2(0,1)$ according to (10.3), where $\mathscr{A}(t)$ is defined by the vector ξ on the basis of (10.5). However, if $\xi = \{\xi_k\}_{-\infty}^{\infty}$ satisfies the conditions (10.23) or (10.25), then only for $p < 2$ can one guarantee the existence of a periodic Hermitian locally integrable function $\mathscr{A}(t)$ whose expansion is (10.5). Therefore by this route we only prove results C) and D) for $1 < p < 2$.

To establish these results for all p $(1 < p < \infty)$, we shall proceed by a different route, using Theorems 9.1 and 7.3 as applied to the Volterra operator A corresponding to a finite vector $\xi = \{\xi_k\}$. We denote the set of such vectors by $\mathfrak{M}$.

Moreover, we denote by $\mathfrak{B}$ the Banach space of all real sequences $\xi = \{\xi_k\}_{-\infty}^{\infty}$ such that

$$|\xi|_{\Pi} = \sup_n \left(\sum_{k=1}^{n} \xi_k^* \Big/ \sum_{k=1}^{n} \pi_k \right) < \infty \qquad (\pi_k = k^{-1/p} L(k);\ k = 1, 2, \cdots),$$

and by $\mathfrak{B}^0$ the subspace of $\mathfrak{B}$ consisting of those ξ for which

$$\lim_{n\to\infty} \left(\sum_{k=1}^{n} \xi_k^* \Big/ \sum_{k=1}^{n} \pi_k \right) = 0.$$

Since for the present case the sequence $\Pi = \{\pi_k\}_1^{\infty}$ is regular, the last condition is equivalent to the condition (10.25).

The space conjugate to $\mathfrak{B}^0$ will be the space $\mathfrak{B}'$ of those real sequences $\xi = \{\xi_k\}_{-\infty}^{\infty}$ for which

$$|\xi|_{\pi} = \sum_{n=1}^{\infty} n^{-1/p} L(n) \xi_n^* < \infty,$$

and the conjugate of $\mathfrak{B}'$ is the space $\mathfrak{B}$. These assertions are to be understood in the sense that the general linear functional f on $\mathfrak{B}^0$ (on $\mathfrak{B}'$) is given by

$$f(\xi) = (\alpha, \xi) = \sum_{k=-\infty}^{\infty} \alpha_k \xi_k,$$

where $\alpha = \{\alpha_k\}_{-\infty}^{\infty}$ is an arbitrary vector from $\mathfrak{B}'$ (from $\mathfrak{B}$).

We now consider the Volterra operator $A = G + iH$ corresponding to some vector $\xi \in \mathfrak{M}$.

Obviously the imaginary component H of A belongs to $\mathfrak{S}_{\Pi}^{(0)}$, and therefore by Theorem 9.1 also $G \in \mathfrak{S}_{\Pi}^{(0)}$ and, moreover,

$$|G|_{\Pi} \leqq M |H|_{\Pi},$$

where M is a constant which is determined solely by the sequence $\Pi = \{n^{-1/p} L(n)\}$. Hence (10.26) holds for $\eta = \mathfrak{T}\xi$, and

$$|\eta|_{\Pi} \leqq M |\xi|_{\Pi}. \tag{10.29}$$

Of course, in the present case (10.26) follows directly from (10.21).

Let us denote by $\mathfrak{T}_0$ the operator which is induced on the set $\mathfrak{M}$ of all finite vectors by the transformation $\eta = \mathfrak{T}\xi$. By (10.29), $\mathfrak{T}_0$ is bounded (its norm is $\leqq M$) and it admits a continuous extension to the closure $\overline{\mathfrak{M}}$, which as is easily seen coincides with all of $\mathfrak{B}^0$. We denote this extension, which has the same norm ($\leqq M$), by the same symbol $\mathfrak{T}_0$.

The operator $\mathfrak{T}_0'$, the conjugate of $\mathfrak{T}_0$, will act in $\mathfrak{B}'$ and have the same norm as $\mathfrak{T}_0$. For any $\xi \in \mathfrak{B}'$ and $\zeta \in \mathfrak{M}$ we will have

$$(\mathfrak{T}_0'\xi, \zeta) = (\xi, \mathfrak{T}_0\zeta) = (\xi, \mathfrak{T}\zeta).$$

But since $\mathfrak{B}' \subset l_{p_1}$ for $p_1 > p$, by (10.13) we will have $(\xi, \mathfrak{T}\zeta) = (\mathfrak{T}\xi, \zeta)$ and hence $(\mathfrak{T}_0'\xi - \mathfrak{T}\xi, \zeta) = 0$ for any finite ζ. Therefore

$$\mathfrak{T}\xi = \mathfrak{T}_0'\xi \quad (\xi \in \mathfrak{B}');$$

thus the transformation $\eta = \mathfrak{T}\xi$ leaves the space $\mathfrak{B}'$ invariant and induces in $\mathfrak{B}'$ a bounded operator with norm $\leqq M$. The relation (10.28) expresses just this fact, so that result D) is proved.

Similarly, from a consideration of the operator $\mathfrak{T}_0''$ conjugate to $\mathfrak{T}_0'$ we see that $\mathfrak{T}\xi = \mathfrak{T}_0''\xi$ for all $\xi \in \mathfrak{B}$, and consequently the transformation $\eta = \mathfrak{T}\xi$ carries $\mathfrak{B}$ into itself and induces in $\mathfrak{B}$ a bounded operator with norm $\leqq M$. Thus (10.29) holds for any $\xi \in \mathfrak{B}$. Since the sequence $\Pi = \{n^{-1/p} L(n)\}$ is regular, (10.29) implies (10.24).

Since the transformation $\eta = \mathfrak{T}\xi$ is a continuous mapping of $\mathfrak{B}$ into itself, by virtue of its mapping $\mathfrak{M}$ into $\mathfrak{B}^0 \subset \mathfrak{B}$ it will map $\overline{\mathfrak{M}} = \mathfrak{B}^0$ into $\mathfrak{B}^0$, which proves the last assertion of result C).

Both results C) and D) are proved.

In passing we have also proved that for any vector $\xi = \{\xi_k\}_{-\infty}^{\infty}$ satisfying the condition (10.23) and any vector $\eta = \{\eta_k\}_{-\infty}^{\infty}$ satisfying the condition

$$\sum_{n=-\infty}^{\infty} n^{-1/p} L(n) \eta_n^* < \infty$$

we have

$$(\mathfrak{T}\xi, \eta) = (\xi, \mathfrak{T}\eta).$$

We leave it to the reader to observe what results can be formulated for the transformation $\eta = \mathfrak{T}\xi$ on the basis of the other theorems of this chapter.

CHAPTER IV

THE FACTORIZATION OF OPERATORS WHICH ARE CLOSE TO THE UNIT OPERATOR

The method of inverting a matrix which is based upon its "factorization", i.e. its representation as a product of left and right triangular matrices, has long been known in linear algebra. Substantially later the question of the possibility of bringing a matrix to triangular form by means of a unitary transformation was stated and solved. As is known, the second operation requires considerably more complicated operations than the first.

In the abstract theory of operators the situation is the opposite. The idea of the triangular representation of operators was developed first, and until recent times the problem of factorization had not even been stated in its general form. It turned out that its solution, in those cases where it is possible, is based in an essential way on the theory of the abstract triangular representation of an operator.

By means of the latter the authors [7, 8] have been able to work out a theory of factorization for operators which differ from the unit operator by a completely continuous operator. The present chapter is devoted to the presentation of the results of these investigations.

In the following sections of this chapter we shall show that the special problems of factorization, which play a large role in the theories of various classes of integral equations, lie within the scheme of the general operator-theoretic statement of the problem of factorization, and that certain new problems arise in this connection.

We also remark that the theory of the factorization of operators plays an important role in the theory of multiplicative representations of the characteristic operator-functions of contraction operators (cf. Gohberg and Kreĭn [11]). However this topic falls outside the scope of the book.

§1. Statement of the problem of the factorization of operators along a chain

1. As is known (see, for example, Faddeev and Faddeeva [1]), inverting a nonsingular matrix $A = \|a_{jk}\|_1^n$ by Gauss' method shows that under certain conditions the inverse A^{-1} can be represented, and in a unique way, as a product of three factors:

$$A^{-1} = S_+ D S_-,$$

where D is a diagonal matrix, and $S_\pm = \| s_{jk}^\pm \|_1^n$ are right and left triangular matrices whose diagonal elements are equal to unity. Such a representation is possible if and only if all the principal minors $D_r = \det \| a_{jk} \|_1^r$ are different from zero $(r = 0, 1, 2, \cdots, n;\ \ D_0 = 1$ by definition).

In operator form these factors can be obtained from the formulas

$$S_+ = \sum_{j=1}^n \frac{D_j}{D_{j-1}} (P_j A P_j)^{(-1)} \Delta P_j;$$

(1.1)

$$S_- = \sum_{j=1}^n \frac{D_j}{D_{j-1}} \Delta P_j (P_j A P_j)^{(-1)};$$

(1.2)
$$D = \sum_{j=1}^n \frac{D_{j-1}}{D_j} \Delta P_j.$$

Let us clarify the notation. To every nth order matrix there corresponds an operator, acting in the n-dimensional space E_n of complex vectors $\xi = \{\xi_j\}_1^n$ and denoted by the same letter as the matrix. The symbol P_j $(j = 1, 2, \cdots, n)$ denotes the projection operator in E_n which leaves the first j coordinates of a vector ξ unaltered, and annihilates all the others; $\Delta P_j = P_j - P_{j-1}$ $(j = 1, 2, \cdots, n;\ P_0 = 0)$. Finally, $(P_j A P_j)^{(-1)}$ $(j = 1, 2, \cdots, n)$ is the operator which acts in the subspace $P_j E_n$ and is, in this subspace, the inverse of the operator $P_j A P_j$ (the matrix of $(P_j A P_j)^{(-1)}$ is the inverse of the truncated matrix $\|a_{lk}\|_1^j$).

We shall not dwell here on the derivation of formulas (1.1) and (1.2); they follow directly from the more general operator formulas of §6.

In the case where A is a Hermitian matrix, necessarily $S_- = S_+^*$, and the representation $A^{-1} = S_+ D S_+^*$ is easily obtained from the decomposition of the quadratic form $(A\xi, \xi)$ into a sum of squares by the Lagrange-Jacobi method.

2. The continuous analog of the indicated factorization of matrices for a Fredholm integral operator is also well known (cf. §7).

The natural abstract construction which encompasses and generalizes both of these factorizations is the factorization of an operator along a chain, which is defined in the following way.

By the (*left*) *factorization* of an operator $A \in \mathfrak{R}$ along a chain $\mathfrak{P}$ we shall mean the representation of A in the form

$$A = A_+ D A_- \tag{1.3}$$

where A_+, A_- $(\in \mathfrak{R})$ are operators having respectively $\mathfrak{P}$ and $\mathfrak{P}^\perp$ as eigenchains, and D $(\in \mathfrak{R})$ is an operator having both $\mathfrak{P}$ and $\mathfrak{P}^\perp$ as eigenchains, i.e. which commutes with all of the orthoprojectors $P \in \mathfrak{P}$.

If for a given operator $A \in \mathfrak{R}$ a factorization (1.3) is possible, then it is not unique. Indeed, from the given factorization we can obtain an infinite set of others, putting for example $A'_+ = A_+ B$, $A'_- = CA_-$ and $D' = B^{-1}DC^{-1}$, where B and C $(\in \mathfrak{R})$ are arbitrary invertible operators which commute with all the projectors $P \in \mathfrak{P}$.

Let us denote by $\mathfrak{S}_\infty(\mathfrak{P})$ the set of all operators from $\mathfrak{S}_\infty$ which have $\mathfrak{P}$ as an eigenchain. Obviously $\mathfrak{S}_\infty(\mathfrak{P})$ is a closed subring of $\mathfrak{S}_\infty$. If $\mathfrak{P}$ is a maximal chain, then in order that $X \in \mathfrak{S}_\infty(\mathfrak{P})$ be a Volterra operator, it is necessary and sufficient that $X \in \mathfrak{S}_\infty[\mathfrak{P}]$.

By a *special* (*left*) *factorization* of an operator A $(\in \mathfrak{R})$ *along a chain* $\mathfrak{P}$ we shall mean the representation of A in the form

$$A = (I + X_+)D(I + X_-), \tag{1.4}$$

where X_+ and X_- are Volterra operators from $\mathfrak{S}_\infty(\mathfrak{P})$ and $\mathfrak{S}_\infty(\mathfrak{P}^\perp)$ respectively, D commutes with all the projectors $P \in \mathfrak{P}$, and $D - I \in \mathfrak{S}_\infty$.

The factors $I + X_\pm$ are invertible operators. In the case where A is invertible, the diagonal factor $D = I + Z$ is also invertible. In this case, obviously, the operators $P(I + X_\pm)P$ and $P(I + Z)P$, regarded as operators in the subspace $P\mathfrak{H}$, will be invertible. It follows that *in order that an invertible operator A admit a special factorization relative to the chain $\mathfrak{P}$, it is necessary that each of the operators $PA^{-1}P$ $(P \in \mathfrak{P})$ be invertible in the corresponding subspace $P\mathfrak{H}$.*

If an invertible operator $A \in \mathfrak{R}$ admits a special factorization relative to a maximal chain $\mathfrak{P}$, then this factorization will be unique.

Indeed, if

$$(I + X_+)D(I + X_-) = (I + X'_+)D'(I + X_-'),$$

then

$$(I + X'_+)^{-1}(I + X_+)D = D'(I + X'_-)(I + X_-)^{-1}.$$

It follows that the operator $(I+X'_+)^{-1}(I+X_+)D$ has both $\mathfrak{P}$ and $\mathfrak{P}^{\perp}$ as eigenchains; consequently, it commutes with each projector $P\in\mathfrak{P}$. Since D is invertible, the operator $(I+X'_+)^{-1}(I+X_+)$ commutes with each $P\in\mathfrak{P}$. Thus, the Volterra operator[1] $V_+ = Y'_+ + X_+ + Y'_+X_+$ $(Y'_+ = (I+X'_+)^{-1} - I)$, which has $\mathfrak{P}$ as an eigenchain, commutes with all the projectors from $\mathfrak{P}$. This is possible only for $V_+ = 0$, i.e. $X_+ = X'_+$. The proof of $X'_- = X_-$ is similar. As a consequence of these equalities it follows that $D' = D$.

From the uniqueness of a special factorization for an invertible operator relative to a maximal chain it follows that if a selfadjoint invertible operator $A = A^*$ $(\in\mathfrak{R})$ has such a factorization, then $X_+^* = X_-$ and $D^* = D$.

We note that only an operator A which differs from the identity by a completely continuous operator can have a special factorization. Therefore in the future, when we seek formulas for the operators $X_\pm$ and D for an invertible A, we can assume at once without loss of generality that A has the form $A = (I-T)^{-1}$, where $T\in\mathfrak{S}_\infty$.

If $\mathfrak{P}$ is a complete chain, the problem of special factorization simplifies. In this case, as is easily seen,[2] $D = I$ and hence (1.4) assumes the form

$$A = (I+X_+)(I+X_-). \tag{1.5}$$

Thus, if an operator A $(\in\mathfrak{R})$ has a special factorization along a complete (i.e. bordered continuous) chain, obviously it is always invertible. If A is moreover selfadjoint, then $X_+^* = X_-$ and so A is positive.

We further remark that the factorizations used in the theory of scalar and vector Wiener-Hopf equations (cf. M. G. Kreĭn [**2**], I. C. Gohberg and M. G. Kreĭn [**2**]) are also examples of factorizations in the sense of the general definitions of this section. For more details on this matter see §9.

---

[1] That V_+ is a Volterra operator follows from Theorems I.5.1 and I.6.1, once one notes that since Y'_+ and X_+ are Volterra operators having $\mathfrak{P}$ as an eigenchain, and whose blocks on the gaps of $\mathfrak{P}$ therefore equal zero, the blocks of V_+ on the gaps of $\mathfrak{P}$ equal zero.

[2] If an operator $V\in\mathfrak{S}_\infty$ has a complete eigenchain $\mathfrak{P}$, and $\mathfrak{P}^{\perp}$ is also an eigenchain of V, then for any partition $\mathfrak{z} = \{P_j\}_0^n$ of $\mathfrak{P}$ we have

$$V = \sum_j \Delta P_j V \Delta P_j,$$

and so $V = 0$ by virtue of Lemma III.7.1.

§2. General theorems on the factorization of operators

THEOREM 2.1. *In order that an operator* $A = (I - T)^{-1}$ $(A \in \mathfrak{R}; T \in \mathfrak{S}_\infty)$ *have a special factorization along a complete chain* $\mathfrak{P}$, *it is necessary and sufficient that:*

a) *all the operators* $I - PTP$ $(P \in \mathfrak{P})$ *be invertible, and*

b) *at least one of the integrals*

$$X_+ = \int_{\mathfrak{P}} (I - PTP)^{-1} PT\, dP, \tag{2.1}$$

$$X_- = \int_{\mathfrak{P}} dP\, TP (I - PTP)^{-1}, \tag{2.1'}$$

converge in the uniform norm.

If one of these integrals converges, then so will the other, and their values yield the operators appearing in the special factorization (1.5).

PROOF. Let $A = (I - T)^{-1}$ have a special factorization (1.5). Then

$$(I - T) = (I + Y_-)(I + Y_+),$$

where

$$Y_\pm = (I + X_\pm)^{-1} - I \qquad (\in \mathfrak{S}_\infty).$$

The operators Y_+ and Y_- are Volterra operators which have $\mathfrak{P}$ and $\mathfrak{P}^\perp$ as eigenchains, respectively; hence, for any projector $P \in \mathfrak{P}$

$$-PTP = PY_-P + PY_+P + PY_-PY_+P$$

and

$$I - PTP = (I + PY_-P)(I + PY_+P).$$

Since $PY_\pm P$ are Volterra operators, the operator $I - PTP$ is invertible and

$$(I - PTP)^{-1} = (I + PY_+P)^{-1}(I + PY_-P)^{-1} \quad (P \in \mathfrak{P}).$$

It is easily seen that

$$(I + PY_\pm P)^{-1}P = P(I + PY_\pm P)^{-1}P = P(I + Y_\pm)^{-1}P \quad (P \in \mathfrak{P}).$$

Consequently,

$$(I + PY_+P)^{-1}P = (I + Y_+)^{-1}P = (I + X_+)P$$

and

$$(I + PY_-P)^{-1}P = P(I + Y_-)^{-1} = P(I + X_-).$$

Thus

$$(I + PTP)^{-1}P = (I + X_+)P(I + X_-).$$

Multiplying the last relation by the operator $T = -Y_- - (I + Y_-)Y_+$, we obtain

$$(I - PTP)^{-1}PT = (I + X_+)PX_- - (I + X_+)PY_+. \tag{2.2}$$

Since the integrals

$$\int_{\mathfrak{P}} PX_- \, dP \quad \text{and} \quad \int_{\mathfrak{P}} PY_+ \, dP$$

converge in the uniform norm, the first being equal to zero and the second being equal to Y_+, it follows that

$$\int_{\mathfrak{P}} (I - PTP)^{-1}PT \, dP = -(I + X_+)Y_+ = X_+.$$

Similarly, we can prove that the integral in (2.1′) converges in the uniform norm and equals X_-.

We now prove the sufficiency of the conditions of the theorem. Suppose that condition a) is satisfied and that the integral in (2.1) converges in the uniform norm. We denote by $\mathfrak{z} = \{P_j\}_0^n$ an arbitrary partition of the chain $\mathfrak{P}$ and by $X_{\mathfrak{z}}$ the operator

$$X_{\mathfrak{z}} = \sum_{j=1}^{n} (I - P_{j-1}TP_{j-1})^{-1}P_{j-1}T\Delta P_j.$$

Moreover, with any operator $A \in \mathfrak{S}_\infty$ we associate the sum

$$S_{\mathfrak{z}}(A) = \sum_{j=1}^{n} P_{j-1}A\Delta P_j.$$

Since

$$\begin{aligned} S_{\mathfrak{z}}(TX_{\mathfrak{z}}) &= \sum_{k=1}^{n} P_{k-1}T\sum_{j=1}^{n} (I - P_{j-1}TP_{j-1})^{-1}P_{j-1}T\Delta P_j \Delta P_k \\ &= \sum_{j=1}^{n} P_{j-1}TP_{j-1}(I - P_{j-1}TP_{j-1})^{-1}P_{j-1}T\Delta P_j \end{aligned}$$

and

$$P_{j-1}TP_{j-1}(I - P_{j-1}TP_{j-1})^{-1} = (I - P_{j-1}TP_{j-1})^{-1} - I,$$

we have

$$S_{\mathfrak{z}}(TX_{\mathfrak{z}}) = X_{\mathfrak{z}} - S_{\mathfrak{z}}(T),$$

or what is the same,

$$S_{\mathfrak{z}}(T(I+X_{\mathfrak{z}})) = X_{\mathfrak{z}}.$$

With any operator $A \in \mathfrak{S}_\infty$ we also associate the sum

$$S'_{\mathfrak{z}}(A) = \sum_{j=1}^{n} \Delta P_j A P_j.$$

It is easily seen that

$$S_{\mathfrak{z}}(A) + S'_{\mathfrak{z}}(A) = A.$$

Hence

$$-X_{\mathfrak{z}} + T(I+X_{\mathfrak{z}}) = S'_{\mathfrak{z}}(T(I+X_{\mathfrak{z}}))$$

or

$$(I-T)(I+X_{\mathfrak{z}}) = I - Y_{\mathfrak{z}}, \tag{2.3}$$

where

$$Y_{\mathfrak{z}} = S'_{\mathfrak{z}}(T(I+X_{\mathfrak{z}})).$$

The operator $X_{\mathfrak{z}}$ has a uniform limit along the chain $\mathfrak{P}$, and this limit X_+ is a Volterra operator having $\mathfrak{P}$ as an eigenchain. It follows from (2.3) that $Y_{\mathfrak{z}}$ likewise has a uniform limit Y_- along $\mathfrak{P}$. This limit, as is easily seen, is a completely continuous operator having $\mathfrak{P}^\perp$ as an eigenchain. Since $\mathfrak{P}^\perp$ is a continuous chain, Y_- is a Volterra operator.

Thus, the operator $(I-T)^{-1}$ has the special factorization

$$(I-T)^{-1} = (I-Y_-)^{-1}(I+X_+).$$

The proof of the sufficiency of the conditions of the theorem when the integral in (2.1′) converges is similar.

The theorem is proved.

THEOREM 2.2. *Let $\mathfrak{P}$ be a complete chain and let $\mathfrak{S}$ be a separable s.n. ideal.*

If for an operator $T \in \mathfrak{S}$, satisfying condition a) *of Theorem* 2.1, *at least one of the integrals in* (2.1), (2.1′) *converges weakly and its value belongs to $\mathfrak{S}$, then both integrals converge in the norm of $\mathfrak{S}$.*

PROOF. Repeating almost verbatim the proof of the last part of Theorem 2.1, we can show that under the hypotheses of the theorem the operator $(I-T)^{-1}$ has a special factorization (1.5) with operators $X_\pm \in \mathfrak{S}$. Hence, by virtue of (2.2), we obtain

$$\int_{\mathfrak{P}} (I - PTP)^{-1} PT\, dP$$
(2.4)
$$= (I + X_+) \int_{\mathfrak{P}} PX_- dP - (I + X_+) \int_{\mathfrak{P}} PY_+ dP,$$

where $Y_+ = (I + X_+)^{-1} - I\ (\in \mathfrak{S})$.

By Theorem III.7.1, both integrals in the right side of (2.4) converge in the norm of the ideal $\mathfrak{S}$, and so the integral in the left side of (2.4) converges in the norm of $\mathfrak{S}$.

§3. The factorization of operators which differ from the unit operator by a finite-dimensional operator

We start with the following lemma.

LEMMA 3.1. *Let $X = (\cdot, \psi)\chi$ be a one-dimensional operator, where ψ, χ are unit vectors in $\mathfrak{H}$, and let $\mathfrak{P}$ be a complete chain. Then for any continuous complex-valued function $a(P)$ $(P \in \mathfrak{P})$ the integral*

$$Y = \int_{\mathfrak{P}} a(P) PX\, dP \tag{3.1}$$

converges in the uniform norm, and

$$s_j(Y) \leqq \frac{1}{2j-1} \max_{P \in \mathfrak{P}} |a(P)| \quad (j = 1, 2, \cdots). \tag{3.2}$$

Note that it follows from (3.2), in particular, that $Y \in \mathfrak{S}_\Omega$ and $|Y|_\Omega \leqq \max_{P \in \mathfrak{P}} |a(P)|$.

PROOF. We shall show that the integral

$$Y_1 = \int_{\mathfrak{P}} F(P)(\cdot, dP\psi) \tag{3.3}$$

converges in the uniform norm, where $F(P)$ is any continuous[3)] vector-function with values in $\mathfrak{H}$. This will imply the convergence of the integral (3.1), which is obtained from (3.3) for $F(P) = a(P) P\chi$.

For any number $\epsilon > 0$ we denote by $\mathfrak{z}_\epsilon$ a partition of $\mathfrak{P}$ with the property that

[3)] A vector-function $F(P)$ $(P \in \mathfrak{P})$, mapping the chain $\mathfrak{P}$ into $\mathfrak{H}$, is said to be continuous if the strong convergence of a sequence of projectors from $\mathfrak{P}$ implies the convergence of the corresponding sequence of values of F in the norm of $\mathfrak{H}$.

$$|F(P') - F(P'')| < \epsilon$$

for every pair of projectors $P', P'' \in \mathfrak{P}$ lying between two successive projectors in the partition $\mathfrak{z}_\epsilon$.

Let $\mathfrak{z}_1$ and $\mathfrak{z}_2$ be arbitrary refinements of $\mathfrak{z}_\epsilon$ and let S_1 and S_2 be any Riemann sums of the integral (3.3) corresponding to the partitions $\mathfrak{z}_1$ and $\mathfrak{z}_2$. Let us denote by $\mathfrak{z} = \{P_j\}_0^n$ the union of the partitions $\mathfrak{z}_1$ and $\mathfrak{z}_2$. As is easily seen, the difference $\Delta S = S_1 - S_2$ can be represented in the form

$$\Delta S = \sum_{j=1}^{n} (F(P_j') - F(P_j''))(\cdot, \Delta P_j \psi),$$

and

$$|F(P_j') - F(P_j'')| < \epsilon \quad (j = 1, 2, \cdots, n).$$

Thus, for any vector $f \in \mathfrak{H}$

$$|\Delta S f| \leqq \epsilon \sum_{j=1}^{n} |(\Delta P_j f, \Delta P_j \psi)|.$$

Since

$$\sum_{j=1}^{n} |(\Delta P_j f, \Delta P_j \psi)| \leqq \sum_{j=1}^{n} |\Delta P_j f| \, |\Delta P_j \psi|$$
$$\leqq \left(\sum_{j=1}^{n} |\Delta P_j f|^2\right)^{1/2} \left(\sum_{j=1}^{n} |\Delta P_j \psi|^2\right)^{1/2} \leqq |f|,$$

we have

$$|\Delta S| < \epsilon.$$

From this it is easy to show in the usual way the uniform convergence of the integral (3.3), and therefore the uniform convergence of the integral (3.1).

We now establish the bound (3.2). Since the operator Y defined by (3.1) has the form

$$Y = \int_{\mathfrak{P}} a(P)(\cdot, dP\psi) P\chi,$$

we have

$$Y^* = \int_{\mathfrak{P}} \overline{a(P)}(\cdot, P\chi) dP\psi,$$

and so

$$YY^* = \int_{\mathfrak{P}} |a(P)|^2 |dP\psi|^2(\cdot, P\chi)P\chi.$$

It is easily seen that

$$YY^* \leqq \max_{P\in\mathfrak{P}} |a(P)|^2 \int_{\mathfrak{P}} |dP\psi|^2(\cdot, P\chi)P\chi. \tag{3.4}$$

Now

$$\int_{\mathfrak{P}} |dP\psi|^2(\cdot, P\chi)P\chi = Y_1 Y_1^*, \tag{3.5}$$

where

$$Y_1 = \int_{\mathfrak{P}} (\cdot, dP\psi)P\chi = \int_{\mathfrak{P}} PX\,dP.$$

By Lemma II.6.2

$$s_j(Y_1)\ (= s_j(Y_1^*)) \leqq 1/(2j-1) \quad (j = 1, 2, \cdots);$$

hence (3.2) follows from (3.4) and (3.5).

The following theorem is basic for this section.

THEOREM 3.1. *Let $\mathfrak{P}$ be a complete chain and let K be a finite-dimensional operator for which all of the operators $I - PKP$ $(P \in \mathfrak{P})$ are invertible. Then the integral*

$$\int_{\mathfrak{P}} (I - PKP)^{-1} PK\,dP \tag{3.6}$$

converges in the uniform norm, and its value Y belongs to the ideal $\mathfrak{S}_\Omega$.

Hence the operator $(I-K)^{-1}$ has a special factorization (1.5) *relative to the chain $\mathfrak{P}$ with operators $X_\pm \in \mathfrak{S}_\Omega$.*

PROOF. The operator K can be represented in the form

$$K = \sum_{j=1}^{n} (\cdot, \psi_j)\chi_j,$$

where $\{\psi_j\}_1^n$ and $\{\chi_j\}_1^n$ are linearly independent systems of vectors from $\mathfrak{H}$. The equation

$$g - PKPg = f \tag{3.7}$$

can be rewritten in the form of a system:

$$\begin{cases} g = f + \sum_{j=1}^{n} \xi_j P\chi_j, \\ \xi_j = (g, P\psi_j) \quad (j = 1, 2, \cdots, n). \end{cases}$$

It follows that the equation (3.7) is equivalent to the system

$$\xi_j - \sum_{k=1}^{n} (P\chi_k, \psi_j)\xi_k = (f, P\psi_j) \quad (j = 1, 2, \cdots, n).$$

Since the operator $I - PKP$ is invertible, the determinant

$$\Delta_p = \det \| \delta_{jk} - (P\chi_k, \psi_j) \|_1^n \quad (P \in \mathfrak{P})$$

does not vanish anywhere on $\mathfrak{P}$.

It is easily seen that the solution g of (3.7) can be found from the formula

$$g = \frac{1}{\Delta_P} \begin{vmatrix} f & (f, P\psi_1) & \cdots & (f, P\psi_n) \\ P\chi_1 & & & \\ \cdot & & & \\ \cdot & & \Delta_P & \\ \cdot & & & \\ P\chi_n & & & \end{vmatrix},$$

and consequently the operator $(I - PKP)^{-1}PK\Delta P$ will have the form

$$(I - PKP)^{-1}PK\Delta P$$

$$= \frac{1}{\Delta_P} \sum_{j=1}^{n} \begin{vmatrix} P\chi_j & (P\chi_j, \psi_1) & \cdots & (P\chi_j, \psi_n) \\ P\chi_1 & & & \\ \cdot & & & \\ \cdot & & \Delta_P & \\ \cdot & & & \\ P\chi_n & & & \end{vmatrix} (\cdot, \Delta P\psi_j). \tag{3.8}$$

Expanding the determinant on the right side of (3.8) with respect to the elements in the first column, we obtain

$$(I - PKP)^{-1}PK\Delta P = \sum_{j,k=1}^{n} a_{jk}(P)(\cdot, \Delta P\psi_j)P\chi_k, \tag{3.9}$$

where the $a_{jk}(P)$ $(P \in \mathfrak{P})$ are certain continuous complex-valued functions. By virtue of Lemma 3.1 it follows from (3.9) that the integral (3.6) converges in the uniform norm. By the same lemma, each of the operators

$$\int_{\mathfrak{P}} a_{jk}(P)(\cdot, dP\psi_j)P\chi_k \quad (j, k = 1, 2, \cdots, n) \tag{3.10}$$

belongs to the s.n. ideal $\mathfrak{S}_\Omega$; consequently, the operator Y (see formula (3.6)), which is the sum of all the operators (3.10), also belongs to $\mathfrak{S}_\Omega$. The theorem is proved.

§4. An auxiliary lemma concerning factorization in normed rings

In this section we present a lemma concerning the factorization of elements which are "close" to the unit element. This lemma plays an important role in the proof of the basic theorem on the factorization of operators along a chain (§5). Results which are related to this lemma both in content and in method of proof[4)] can be found in papers by P. Masani [1], G. Baxter [1], N. Wiener and P. Masani [1], G. F. Mandžavidze [1], G. F. Mandžavidze and B. V. Hvedelidze [1], I. B. Simonenko [1]. Rather recently there appeared a paper by F. V. Atkinson [1], in which Baxter's results were further developed. We also mention a paper by M. S. Budjanu and I. C. Gohberg [1] (see also M. S. Budjanu [1]), in which a generalization of this lemma on factorization is obtained (for the case $R = \mathscr{I}$).

We precede the basic result of this section by the following auxiliary result.

1. *Let R be a normed ring with unit e, and let $a(\lambda)$ be a continuous function which maps some bounded region (i.e. connected open bounded set) Λ of the complex plane into R.*

If for every point $\lambda \in \Lambda$ the element $a(\lambda)$ has an inverse from one side, and for at least one point $\lambda_0 \in \Lambda$ the element $a(\lambda_0)$ has a two-sided inverse, then $a(\lambda)$ has a two-sided inverse for all points $\lambda \in \Lambda$.

We shall first show that if some element $x \in R$ has an inverse from only one side, then all the elements from some sufficiently small neighborhood of x also are invertible only from that one side.

In fact, suppose for definiteness that x is invertible only from the left. We denote its left inverse by $x_l^{(-1)}$. Let us consider any element $y \in R$ satisfying the condition

$$|x - y| < 1/|x_l^{(-1)}|. \tag{4.1}$$

Then

$$y = x + y - x = (e - (x - y)x_l^{(-1)})x.$$

By virtue of (4.1) the element

4) We particularly mention the work of P. Masani [1], in which was first proved a result analogous to Lemma 4.1, though by methods different from those used here. This proof in a concrete situation had been presented earlier by H. Cartan [1].

$$z = e - (x - y)x_l^{(-1)}$$

has a two-sided inverse, and

$$z^{-1} = \sum_{j=0}^{\infty} ((x - y)x_l^{(-1)})^j.$$

Obviously the element $x_l^{(-1)}z^{-1}$ is the left inverse of y $(= zx)$. The element y does not have a right inverse, as otherwise the relation

$$yy_r^{(-1)} = e = zxy_r^{(-1)},$$

where $y_r^{(-1)}$ is a right inverse of y, would imply

$$xy_r^{(-1)} = z^{-1}, \quad xy_r^{(-1)}z = e,$$

i.e. x has a right inverse.

Let us denote by Λ_b and Λ_u the set of all $\lambda \in \Lambda$ for which the element $a(\lambda)$ has a two-sided inverse or only a one-sided inverse, respectively. The sets Λ_b and Λ_u are disjoint and open and cover all of Λ. This is possible only when one of the sets Λ_b, Λ_u is empty. By hypothesis Λ_b contains at least one point; hence Λ_u is empty.

LEMMA 4.1. *Let R be a normed ring with a unit e ($|e|_R = 1$) which contains two closed subrings $R_\pm$ whose intersection consists only of zero. Suppose, moreover, that there exists a two-sided ideal $\mathscr{I}$ in R with the properties*:

a) *a norm $|c|_{\mathscr{I}}$ is defined in $\mathscr{I}$ which makes $\mathscr{I}$ a Banach space, and*

$$\sup_{c \in \mathscr{I}} \frac{|c|_R}{|c|_{\mathscr{I}}} < \infty; \tag{4.2}$$

b) *there exists a constant $k > 0$ such that for any $x, y \in R$ and $c \in \mathscr{I}$*

$$|xcy|_R \leqq k|x|_R|c|_{\mathscr{I}}|y|_R;$$

c) *any element $c \in \mathscr{I}$ can be represented in a unique way in the form*

$$c = c_+ + c_-,$$

where $c_\pm \in R_\pm$.

Then

$$D = \sup_{c \in \mathscr{I}} \frac{\max(|c_+|_R, |c_-|_R)}{|c|_{\mathscr{I}}} < \infty, \tag{4.3}$$

and for any element $a \in \mathscr{I}$ such that

(4.4) $$|a|_{\mathscr{I}} < \delta \,(= \min(1, 1/kD))^{5)}$$

the following assertions hold:

1) *the element* $(e-a)^{-1}$ *can be represented in the form*

(4.5) $$(e-a)^{-1} = (e+b_+)(e+b_-),$$

where $b_\pm \in R_\pm$.

2) *the formulas*

(4.6) $$b_+ = a_+ + (aa_+)_+ + (a(aa_+)_+)_+ + \cdots$$

(4.7) $$b_- = a_- + (a_-a)_- + ((a_-a)_-a)_- + \cdots,$$

where the series converge in the norm of the ring R, *define the unique elements* $b_\pm \in R_\pm$ *which give a representation of the form* (4.5), *and for which the factors* $(e+b_\pm)$ *are invertible and*

(4.8) $$(e+b_\pm)^{-1} - e \in R_\pm.$$

Before going on to a complete proof of this lemma, let us clarify the meaning of its conditions and the assertion (4.3).

Condition a) says that the imbedding of the Banach space $\mathscr{I}$ into the normed ring R is continuous. Condition b), which consists essentially of the two conditions

$$|xc|_{\mathscr{I}} \leqq \text{const}\,|x|_R\,|c|_{\mathscr{I}}, \quad |cx|_{\mathscr{I}} \leqq \text{const}\,|c|_{\mathscr{I}}\,|x|_R \quad (x \in R;\ c \in \mathscr{I}),$$

says that the operator of multiplication on the right or the left by a fixed element x of the ring R is bounded in the space $\mathscr{I}$, and up to a multiplicative constant its norm does not exceed the norm of x as an element of the ring R.

We remark that in the decomposition of an element $c \in \mathscr{I}$ into the sum of elements $c_\pm \in R_\pm$ (condition c)) the components $c_\pm$ will no longer, generally speaking, belong to the ideal $\mathscr{I}$.

Condition c) enables us to introduce a "projection" operator P_+ from the ideal $\mathscr{I}$ into R_+:

$$P_+x = x_+ \quad (x \in \mathscr{I}).$$

This operator is a closed operator from $\mathscr{I}$ into R. In fact, let

$$\lim_{n\to\infty} |x^{(n)} - x|_{\mathscr{I}} = 0 \quad (x^{(n)} \in \mathscr{I}, x \in \mathscr{I})$$

and

5) The lemma remains valid if we put $\delta = 1/kD$ (cf. Remark 4.1).

$$\lim_{n\to\infty} |x_+^{(n)} - y|_R = 0 \quad (y \in R_+).$$

Then the sequence $x_-^{(n)} = x^{(n)} - x_+^{(n)}$ $(n = 1, 2, \cdots)$ converges in the norm of R and has as its limit the element $z = x - y$, which obviously belongs to the subring R_-. From the relation $x = y + z$ it follows that

$$Px = x_+ = y.$$

Moreover, P_+ is defined on all of $\mathscr{I}$; hence, by a well-known theorem of Banach [1] on operators having closed graphs, P_+ is bounded. It follows that

$$D = \max(|P_+|, |P_-|) < \infty \quad (P_- = I - P_+).$$

By means of the projectors P_+ and P_-, the relation (4.4) which defines δ can be rewritten in the form

$$\delta = \min\left(1, \frac{1}{k|P_+|}, \frac{1}{k|P_-|}\right),$$

and formulas (4.6) and (4.7) can be rewritten in the form

$$b_+ = P_+a + P_+(aP_+a) + P_+(aP_+(aP_+a)) + \cdots$$

and

$$b_- = P_-a + P_-[(P_-a)a] + P_-((P_-((P_-a)a))a) + \cdots.$$

We now proceed to the proof of the lemma. First we shall show that for elements a from the ball $S_\delta = \{a : a \in \mathscr{I}; |a|_{\mathscr{I}} < \delta\}$, where δ is the number defined by (4.4), the elements $b_\pm$ which yield a representation (4.5) and satisfy the conditions (4.8) are unique, and are given by formulas (4.6) and (4.7).

In fact, if we put $y_- = (e + b_-)^{-1} - e$, then from the representation (4.5) we find

$$e + y_- = (e - a)(e + b_+)$$

or

$$a + ab_+ = b_+ - y_-.$$

Since the left side belongs to $\mathscr{I}$, this shows that

$$P_+(a + ab_+) = b_+,$$

whence

$$b_+ - P_+(ab_+) = P_+a.$$

Thus the element b_+ is a solution of the equation

$$x - T_+x = P_+a, \tag{4.9}$$

where T is the operator in R defined by

$$T_+x = P_+(ax) \quad (x \in R).$$

This operator is bounded and maps R into R_+, and

$$|T_+| \leq k|P_+| \, |a|_{\mathscr{I}}.$$

Therefore $|T_+| < 1$ for $a \in S_\delta$, and so the operator $I + T_+$ is invertible, and the solution x of (4.9) is given by the Neumann series

$$x = \sum_{j=0}^{\infty} T^j P_+a. \tag{4.10}$$

It is obvious that the series (4.10) coincides termwise with the series (4.6). The proof of the corresponding assertion for the element b_- is analogous.

It remains to verify that for any $a \in S_\delta$ the elements b_+ and b_- defined by (4.6) and (4.7) do indeed yield the representation (4.5) and satisfy the condition (4.8). As was already clarified, for the indicated choice of the number δ both series converge in the norm of the ring R and define elements $b_\pm$ for which

$$b_+ - P_+(ab_+) = P_+a, \quad b_- - P_-(b_-a) = P_-a.$$

These relations show that there exist $y_\pm \in R_\pm$ such that

$$(e - a)(e + b_+) = e + y_- \tag{4.11}$$

and

$$(e + b_-)(e - a) = e + y_+. \tag{4.12}$$

Multiplying the first of these relations from the left by $e + b_-$ and then using the second relation, we obtain

$$(e + y_+)(e + b_+) = (e + b_-)(e + y_-)$$

or

$$y_+ + b_+ + y_+b_+ = y_- + b_- + b_-y_-. \tag{4.13}$$

Since the intersection $R_+ \cap R_-$ consists solely of the zero element, we conclude that both sides of (4.13) equal zero, and hence

$$(e+y_+)(e+b_+)=e, \tag{4.14}$$

$$(e+b_-)(e+y_-)=e. \tag{4.15}$$

From (4.11) and (4.14), (4.15) we would be able to derive the representation (4.5) and at the same time establish the properties (4.8), if we knew that the elements $e+b_\pm$ were invertible (the relations (4.14) and (4.15) only imply the one-sided invertibility of these elements).

To prove the invertibility of these elements we make use of the fact that all of the foregoing arguments remain valid if the element a ($|a|_{\mathscr{I}}<\delta$) appearing in them is replaced by the element λa, where the complex number λ runs through the disc $C=\{\lambda;\,|\lambda|<\delta/|a|_{\mathscr{I}}\}$. For the replacement $a\to\lambda a$ the elements $b_\pm$ go over into $b_\pm(\lambda)$, certain analytic functions of $\lambda\in C$. According to (4.14) the element $e+b_+(\lambda)$ will have a left inverse at every point $\lambda\in C$. Since the element $e+b_+(0)=e$ has a two-sided inverse, by result 1 the element $e+b_+(\lambda)$ will have a two-sided inverse for all $\lambda\in C$. In particular, the element $e+b_+=e+b_+(1)$ will have a two-sided inverse, as $1\in C$. Similarly, we can prove that $e+b_-$ has a two-sided inverse.

The lemma is proved.

REMARK 4.1. As was already indicated in footnote 5, the lemma remains valid if we put $\delta=1/kD$. We needed the condition $\delta\leqq 1$ only to guarantee that the element $e-\lambda a$ is invertible if $|\lambda a|_{\mathscr{I}}<\delta$. It turns out that the element $e-\lambda a$ is invertible if the condition $|\lambda a|_{\mathscr{I}}<1/kD$ is satisfied.

Indeed, the series

$$b_+(\lambda)=\lambda P_+a+\lambda^2P_+(aP_+a)+\cdots,$$

$$b_-(\lambda)=\lambda P_-a+\lambda^2P_-((P_-a)a)+\cdots$$

certainly converge in the disc $C_1=\{\lambda;\,|\lambda|<1/kD|a|_{\mathscr{I}}\}$ and define analytic functions in this disc. For λ with sufficiently small modulus $(e-\lambda a)^{-1}=(e+b_+(\lambda))(e+b_-(\lambda))$, i.e.

$$(e+b_+(\lambda))(e+b_-(\lambda))(e-\lambda a)=(e-\lambda a)(e+b_+(\lambda))(e+b_-(\lambda))=e.$$

Since the factors are all analytic, these relations will hold everywhere in C_1.

REMARK 4.2. Let us denote by D the set of elements $a\in\mathscr{I}$ for which the element $e-a$ is invertible and $(e-a)^{-1}$ has a factorization (4.5) with invertible factors $e+b_\pm$, where $(e+b_\pm)^{-1}-e\in R_\pm$. Such a factorization for $a\in D$ will be unique.

One can deduce from the lemma that under its conditions a), b) and c) the set D is an open subset of the Banach space $\mathscr{I}$ and that the mappings $a \to b_+$ and $a \to b_-$ are continuous mappings of D into $R_\pm$.

§5. The basic theorem on the factorization of operators in s.n. ideals

1. THEOREM 5.1. *Let $\mathfrak{P}$ be a complete chain, $\mathfrak{S}_{\mathrm{I}}$ an arbitrary separable s.n. ideal, and $\mathfrak{S}_{\mathrm{II}}$ an s.n. ideal which contains $\mathfrak{S}_{\mathrm{I}}$: $\mathfrak{S}_{\mathrm{II}} \supseteq \mathfrak{S}_{\mathrm{I}}$.*

If for every $T \in \mathfrak{S}_{\mathrm{I}}$ the integral

$$S = \int_{\mathfrak{P}} PTdP \tag{5.1}$$

converges in the uniform norm and belongs to $\mathfrak{S}_{\mathrm{II}}$, then for every operator $T \in \mathfrak{S}_{\mathrm{I}}$ for which all of the operators $I - PTP$ $(P \in \mathfrak{P})$ are invertible (the necessary condition a) *from* Theorem 2.1), *the integrals*

$$X_+ = \int_{\mathfrak{P}} (I - PTP)^{-1} PTdP \quad \textit{and} \quad X_- = \int_{\mathfrak{P}} dPTP(I - PTP)^{-1}$$

converge in the uniform norm, and $X_\pm \in \mathfrak{S}_{\mathrm{II}}$. Thus, in this case the operator $A = (I - T)^{-1}$ has a special factorization (1.5) *with operators $X_\pm \in \mathfrak{S}_{\mathrm{II}}$.*

PROOF. Let us denote by R the normed ring obtained by adjoining to $\mathfrak{S}_{\mathrm{II}}$ the unit operator. The norm in R is defined by

$$|\lambda I + A|_R = |\lambda| + |A|_{\mathrm{II}} \quad (|A|_{\mathrm{II}} \equiv |A|_{\mathfrak{S}_{\mathrm{II}}}).$$

We denote by R_+ the normed ring $\mathfrak{S}_{\mathrm{II}}(\mathfrak{P})$ $(= \mathfrak{S}_{\mathrm{II}} \cap \mathfrak{S}_\infty(\mathfrak{P}))$, and by R_- the normed ring $\mathfrak{S}_{\mathrm{II}}(\mathfrak{P}^\perp)$ $(= \mathfrak{S}_{\mathrm{II}} \cap \mathfrak{S}_\infty(\mathfrak{P}^\perp))$. The intersection $R_+ \cap R_-$ consists of the zero operator.

The set $\mathscr{I} = \mathfrak{S}_{\mathrm{I}}$ is a two-sided ideal of the ring R, and every element $C \in \mathscr{I}$ can be represented in a unique way in the form

$$C = C_+ + C_- \quad (C_\pm \in R_\pm),$$

where

$$C_+ = \int_{\mathfrak{P}} PCdP \quad \text{and} \quad C_- = \int_{\mathfrak{P}} dPCP,$$

i.e. the ideal $\mathscr{I}$ has the property c) from Lemma 4.1. Moreover, it is easily seen that $\mathscr{I}$ has the other two properties a) and b) from Lemma 4.1 (see §1, Chapter III). Hence by virtue of this lemma there exists a number $\delta > 0$ such that for all operators $T \in \mathfrak{S}_{\mathrm{I}}$ such that $|T|_{\mathrm{I}} < \delta$ $(|T|_{\mathrm{I}} \equiv |T|_{\mathfrak{S}_{\mathrm{I}}})$ the operator $(I - T)^{-1}$ has a special factorization (1.5) along the chain $\mathfrak{P}$, and $X_\pm \in \mathfrak{S}_{\mathrm{II}}$.

Now let T be any operator from $\mathfrak{S}_{\mathrm{I}}$ satisfying the necessary condition a) from Theorem 2.1, and let K be a finite-dimensional operator such that

$$|T - K|_{\mathrm{I}} < \delta.$$

We put $M = T - K$. As was already noted, by Lemma 4.1 the operator $(I - M)^{-1}$ has a special factorization

$$(I - M)^{-1} = (I + Z_+)(I + Z_-) \quad (Z_\pm \in \mathfrak{S}_{\mathrm{II}}).$$

Hence, for the operator $I - T = I - M - K$ we have the representation

$$I - T = (I + Z_-)^{-1}(I - (I + Z_-)K(I + Z_+))(I + Z_+)^{-1}$$

or

$$(I - T)^{-1} = (I + Z_+)(I - K_1)^{-1}(I + Z_-), \tag{5.2}$$

where K_1 is a finite-dimensional operator, equal to

$$(I + Z_-)K(I + Z_+).$$

It is easy to see from (5.2) that

$$I - PK_1P = (I - PZ_-P)^{-1}(I - PTP)(I - PZ_+P)^{-1} \quad (P \in \mathfrak{P}).$$

Thus the operator $I - PK_1P$ is invertible for any $P \in \mathfrak{P}$.

By Theorem 3.1 the operator $(I - K_1)^{-1}$ has a special factorization

$$(I - K_1)^{-1} = (I + V_+)(I + V_-).$$

Thus

$$(I - T)^{-1} = (I + Z_+)(I + V_+)(I + V_-)(I + Z_-).$$

Obviously the theorem will be proved as soon as we show that $V_\pm \in \mathfrak{S}_{\mathrm{II}}$.

It was established in the proof of Theorem 3.1 that each of the operators $V_\pm$ can be represented as the sum of m^2 ($m = \dim K_1$) terms $L_j^\pm$ such that

$$s_n(L_j^\pm) \leqq k_j/(2n - 1) \quad (n = 1, 2, \cdots; j = 1, 2, \cdots, m^2),$$

where the numbers k_j do not depend on n.

By the hypothesis of the theorem, the s.n. ideal $\mathfrak{S}_{\mathrm{II}}$ contains all Volterra operators having a one-dimensional imaginary component and eigenchain $\mathfrak{P}$. By Lemma I.6.1 the s-numbers of such an operator are $= \mathrm{const}(2n - 1)^{-1}$ ($n = 1, 2, \cdots$). It follows (see [GK], result 2 of §2, Chapter III) that the operators $L_j^\pm$, and with them the operators $V_\pm$, belong to $\mathfrak{S}_{\mathrm{II}}$.

The theorem is proved.

By virtue of the results of Chapter III, it follows from Theorem 5.1 that for $\mathfrak{S}_{\mathrm{I}}$ we can take, for example, the s.n. ideals $\mathfrak{S}_1$, $\mathfrak{S}_p$ $(1 < p < \infty)$, $\mathfrak{S}_\omega$ and then for $\mathfrak{S}_{\mathrm{II}}$ the s.n. ideals $\mathfrak{S}_\Omega$, $\mathfrak{S}_p$, $\mathfrak{S}_\infty$, respectively.

REMARK 5.1. It is not hard to show that if an operator $T \in \mathfrak{S}_{\mathrm{I}}$ satisfies the necessary condition a) of the theorem, then all operators from some neighborhood of T in $\mathfrak{S}_{\mathrm{I}}$ will satisfy this condition and the integrals (2.1) and (2.1′) will be continuous mappings of this neighborhood into $\mathfrak{S}_{\mathrm{II}}$.

2. As is known (see the footnote on page **197**), if the operator $T = H$ is selfadjoint, then condition a) of Theorem 2.1 does not depend on the complete chain $\mathfrak{P}$, and says that the operator $I - H$ is positive.

For this case Theorem 5.1 can be supplemented by the following.

THEOREM 5.2. *Let $\mathfrak{P}$ be a complete chain and $\mathfrak{S}_{\mathrm{I}}$, $\mathfrak{S}_{\mathrm{II}}$ two arbitrary (with $\mathfrak{S}_{\mathrm{I}}$ in general nonseparable) two-sided ideals of the ring $\mathfrak{R}$, and such that the transformator $\mathscr{C}(\cdot\,;\mathfrak{P})$ maps $\mathfrak{S}_{\mathrm{I}}$ into $\mathfrak{S}_{\mathrm{II}}$.*

If some selfadjoint operator $(I - H)^{-1}$ $(H \in \mathfrak{S}_{\mathrm{I}})$ has a special factorization

$$(I - H)^{-1} = (I + X_+)(I + X_+^*), \tag{5.3}$$

then $X_+ \in \mathfrak{S}_{\mathrm{II}}$.

The factorization (5.3) *is always possible if, besides the necessary condition of the positivity of $I - H$, the condition $H \in \mathfrak{S}_\omega$ is satisfied.*

In view of Theorem 5.1, this theorem is of interest only for the case in which $\mathfrak{S}_{\mathrm{I}}$ is not a separable s.n. ideal.

PROOF. Let the operator $(I - H)^{-1}$ $(H \in \mathfrak{S}_{\mathrm{I}})$ have a special factorization (5.3). We form the operator Y,

$$Y = (2I + X_+)^{-1} X_+. \tag{5.4}$$

For this operator we have

$$\begin{aligned} Y_{\mathscr{R}} &= \tfrac{1}{2}(Y + Y^*) \\ &= \tfrac{1}{2}(2I + X_+)^{-1}[X_+(2I + X_+^*) + (2I + X_+)X_+^*](2I + X_+^*)^{-1} \\ &= (2I + X_+)^{-1}(X_+ + X_+^* + X_+X_+^*)(2I + X_+^*)^{-1} \end{aligned}$$

or

$$Y_{\mathscr{R}} = (2I + X_+)^{-1} T_1 (2I + X_+^*)^{-1},$$

where $T_1 = (I - H)^{-1} - I \in \mathfrak{S}_{\mathrm{I}}$.

The operator Y has $\mathfrak{P}$ as an eigenchain, and hence

$$Y = 2\mathscr{C}(Y_{\mathscr{R}}\,;\mathfrak{P}).$$

Thus $Y \in \mathfrak{S}_{\mathrm{II}}$. Taking into account that Y is a Volterra operator, we obtain from (5.4)

$$X_+ = (I - Y)^{-1} 2Y,$$

whence $X_+ \in \mathfrak{S}_{\mathrm{II}}$. The theorem is proved.

Recalling Theorem III.9.2, we can deduce the following from the theorem just proved.

If the eigenvalues $\lambda_j(H)$ of an operator $H = H^* \in \mathfrak{S}_\infty$ conform to the estimate

$$\lambda_n(H) = O(n^{-1/p} L(n)) \quad (n \to \infty), \tag{5.5}$$

where $1 < p < \infty$ and $L(\nu)$ is a slowly varying function, then the operator $(I - H)^{-1}$ has a factorization (5.3), *and*

$$s_n(X_+) = O(n^{-1/p} L(n)) \quad (n \to \infty). \tag{5.6}$$

By virtue of each of Theorems 5.1 and 5.2, this assertion remains valid if in (5.5) and (5.6) we replace O by o.

§6. The special factorization of operators along a chain with gaps

As has already been noted in §1, special factorization along a chain with gaps differs from special factorization along a continuous chain by the presence of a diagonal factor.

We shall give here those changes in the statements and proofs of the theorems of the preceding section which enable us to extend them to discontinuous chains.

1. To clarify the above-mentioned changes we need a new notion of the integral with respect to a chain. Let $\mathfrak{P}$ be an arbitrary chain, $F(P)$ $(P \in \mathfrak{P})$ an operator-function with values in $\mathfrak{R}$, and let $\mathfrak{z} = \{P_j\}_0^n$ be a partition of $\mathfrak{P}$. By the integrals

$$(m) \int_{\mathfrak{P}} F(P)\, dP \quad \text{and} \quad (M) \int_{\mathfrak{P}} F(P)\, dP \tag{6.1}$$

we shall understand, respectively, the Šatunovskiĭ limits of the corresponding Riemann sums

$$S_m = \sum_{j=1}^n F(P_{j-1}) \Delta P_j \quad \text{and} \quad S_M = \sum_{j=1}^n F(P_j) \Delta P_j.$$

If the integral

$$\int_{\mathfrak{P}} F(P)\, dP \tag{6.2}$$

converges in the uniform norm, then obviously the integrals (6.1), in particular, will converge, and

$$\int_{\mathfrak{P}} F(P)\,dP = (m)\int_{\mathfrak{P}} F(P)\,dP = (M)\int_{\mathfrak{P}} F(P)\,dP.$$

It is easily seen that the condition

$$[F(P^+) - F(P^-)](P^+ - P^-) = 0,$$

where (P^-, P^+) is any gap of the chain $\mathfrak{P}$, which is a necessary condition for the convergence of the integral (6.2), is by no means necessary for the convergence of the integrals (6.1).

Let us consider the transformators $\mathscr{C}_m$ and $\mathscr{C}_M$, defined by

$$\mathscr{C}_{m,M}(X) = \mathscr{C}_{m,M}(X;\mathfrak{P}) = (m,M)\int_{\mathfrak{P}} PX\,dP \quad (P\in\mathfrak{P}).$$

It is easily seen that if one of the integrals $\mathscr{C}_{m,M}(X)$ converges in the uniform norm, then so does the other, and they are related by the equality

$$\mathscr{C}_M(X) = \mathscr{C}_m(X) + \sum_j (P_j^+ - P_j^-)X(P_j^+ - P_j^-),$$

where $\{(P_j^-, P_j^+)\}$ is the set of all gaps of the chain $\mathfrak{P}$.

If both of the integrals $\mathscr{C}_{m,M}(X)$ converge in the uniform norm and are equal to each other, then one can see without difficulty that

$$\mathscr{C}_{m,M}(X) = \int_{\mathfrak{P}} PX\,dP.$$

If the integral $\mathscr{C}_m(X)$ converges, then the integral

$$\mathscr{C}(Y) = \int_{\mathfrak{P}} PY\,dP$$

will converge, where

$$Y = X - \sum_j (P_j^+ - P_j^-)X(P_j^+ - P_j^-),$$

and

$$\mathscr{C}(Y) = \mathscr{C}_m(X).$$

Each of the transformators $\mathscr{C}_{m,M}(\cdot;\mathfrak{P})$ is an unbounded projector in $\mathfrak{S}_\infty$. The range of $\mathscr{C}_m$ coincides with the subspace consisting of *all Volterra* operators having $\mathfrak{P}$ as an eigenchain, and the range of $\mathscr{C}_M$

coincides with the subspace of *all* operators from $\mathfrak{S}_\infty$ (*not necessarily Volterra operators*) having $\mathfrak{P}$ as an eigenchain.

One can also easily show that if for every continuous chain the transformator $\mathscr{C}$ maps some separable s.n. ideal $\mathfrak{S}_{\mathrm{I}}$ into an s.n. ideal $\mathfrak{S}_{\mathrm{II}}$, then for every chain $\mathfrak{P}$ each of the transformators $\mathscr{C}_{m,M}$ maps $\mathfrak{S}_{\mathrm{I}}$ into $\mathfrak{S}_{\mathrm{II}}$.

2. The following theorem is the analog of Theorem 2.1 for the case of a chain with gaps.

THEOREM 6.1. *In order that the operator* $A = (I - T)^{-1}$ $(A \in \mathfrak{R};$ $T \in \mathfrak{S}_\infty)$ *have a special factorization along a maximal chain* $\mathfrak{P}$, *it is necessary and sufficient that*

a) *each of the operators* $(I - PTP)^{-1}$ $(P \in \mathfrak{P})$ *be invertible, and*

b) *at least one of the integrals*

$$
X_+ = (m)\int_{\mathfrak{P}} (I - PTP)^{-1} PT\,dP,
$$

$$
X_- = (m)\int_{\mathfrak{P}} dP\,TP(I - PTP)^{-1},
$$

$$
\widetilde{X}_+ = (M)\int_{\mathfrak{P}} (I - PTP)^{-1} PT\,dP, \tag{6.3}
$$

$$
\widetilde{X}_- = (M)\int_{\mathfrak{P}} dP\,TP(I - PTP)^{-1}
$$

converge in the uniform norm.

If at least one of the integrals (6.3) *converges in the uniform norm, then so do all of the others, and one has*

$$
A = (I + X_+)(I + \widetilde{X}_-) = (I + \widetilde{X}_+)(I + X_-) \tag{6.4}
$$

and the special factorization

$$
A = (I + X_+)D(I + X_-), \tag{6.5}
$$

where

$$
D = I + \sum_j (P_j^+ - P_j^-)\,[(I - P_j^+ T P_j^+)^{-1} - I](P_j^+ - P_j^-), \tag{6.6}
$$

and $\{(P_j^-, P_j^+)\}$ *is the set of all gaps of the chain* $\mathfrak{P}$.

PROOF. Suppose that A has the special factorization (6.5). Then putting $D = I + L$, we obtain

$$A = (I + X_+)(I + \widetilde{X}_-) = (I + \widetilde{X}_+)(I + X_-), \tag{6.7}$$

where

$$\widetilde{X}_- = L + X_- + LX_- \quad \text{and} \quad \widetilde{X}_+ = L + X_+ + X_+L. \tag{6.8}$$

Obviously $\widetilde{X}_+ \in \mathfrak{S}_\infty(\mathfrak{P})$ and $\widetilde{X}_- \in \mathfrak{S}_\infty(\mathfrak{P}^\perp)$.

Repeating now for the equalities (6.8) the first part of the proof of Theorem 2.1, but with the integral $\int_{\mathfrak{P}}$ replaced by the integrals $(m)\int_{\mathfrak{P}}$ and $(M)\int_{\mathfrak{P}}$ respectively, we conclude that condition a) is satisfied and that the operators $X_\pm$, $\widetilde{X}_\pm$ are given by the integrals (6.3), which converge in the uniform norm.

It follows from (6.8) that

$$\begin{aligned}(P_j^+ - P_j^-)\widetilde{X}_+(P_j^+ - P_j^-) \\ = (P_j^+ - P_j^-)L(P_j^+ - P_j^-) + (P_j^+ - P_j^-)X_+L(P_j^+ - P_j^-).\end{aligned}$$

Since

$$(P_j^+ - P_j^-)X_+L(P_j^+ - P_j^-) = (P_j^+ - P_j^-)X_+(P_j^+ - P_j^-)L$$

and $(P_j^+ - P_j^-)X_+(P_j^+ - P_j^-) = 0$, we have

$$(P_j^+ - P_j^-)\widetilde{X}_+(P_j^+ - P_j^-) = (P_j^+ - P_j^-)L(P_j^+ - P_j^-).$$

Taking into account that $L \in \mathfrak{S}_\infty(\mathfrak{P}) \cap \mathfrak{S}_\infty(\mathfrak{P}^\perp)$, we obtain

$$L = \sum_j (P_j^+ - P_j^-)L(P_j^+ - P_j^-) = \sum_j (P_j^+ - P_j^-)\widetilde{X}_+(P_j^+ - P_j^-).$$

Inserting into this equality the integral expression for X_+, we easily obtain from (6.3)

$$\begin{aligned}\sum_{j=1}^N (P_j^+ - P_j^-)\widetilde{X}_+(P_j^+ - P_j^-) \\ = \sum_{j=1}^N (P_j^+ - P_j^-)\,[(I - P_j^+TP_j^+)^{-1} - I](P_j^+ - P_j^-).\end{aligned}$$

Passing to the limit $N \to \infty$, we arrive at (6.6).

If at least one of the integrals (6.3) converges, for example the integral X_-, then, repeating the argument of the second part of the proof of Theorem 2.1, we see that A has the factorization

$$A = (I + Y_+)(I + X_-) \quad (Y_+ = \widetilde{X}_+),$$

where $Y_+ \in \mathfrak{S}_\infty(\mathfrak{P})$.

To separate out the diagonal factor from $I + Y_+$, we form the operator

$$L = \sum_j (P_j^+ - P_j^-) Y_+ (P_j^+ - P_j^-) \quad (\in \mathfrak{S}_\infty).$$

It is easily seen that together with $I + Y_+$ its "diagonal" part—the operator $I + L$—will be invertible.

Let us denote by $Z_+ \in \mathfrak{S}_\infty$ the operator defined by

$$I + Z_+ = (I + Y_+)(I + L)^{-1}.$$

Obviously Z_+ has $\mathfrak{P}$ as an eigenchain. Moreover,

$$\begin{aligned}(P_j^+ - P_j^-)(I + Y_+)(I + L)^{-1}(P_j^+ - P_j^-) \\ &= (P_j^+ - P_j^-)(I + Y_+)(P_j^+ - P_j^-)(I + L)^{-1} \\ &= (P_j^+ - P_j^-)(I + L)(I + L)^{-1} = P_j^+ - P_j^-\end{aligned}$$

and so

$$(P_j^+ - P_j^-) Z_+ (P_j^+ - P_j^-) = 0.$$

Thus, Z_+ is a Volterra operator and A has the special factorization

$$A = (I + Z_+)(I + L)(I + X_-).$$

The theorem is proved.

The formulas (1.1) and (1.2) for the factors in the factorization of a matrix are an immediate consequence of Theorem 6.1.

We note also that in the statement of Theorem 6.1 the integrals $\widetilde{X}_\pm$ can be replaced by the following:

$$\widetilde{X}_+ = (M) \int_{\mathfrak{P}} [(I - PTP)^{-1} - I] dP,$$

$$\widetilde{X}_- = (M) \int_{\mathfrak{P}} dP [(I - PTP)^{-1} - I].$$

It is easy to deduce from Theorem 6.1 the following corollary.

COROLLARY 6.1. *Let T $(\in \mathfrak{S}_\infty)$ be an operator for which (with $\mathfrak{P}$ a maximal chain) the operators $I - PTP$ are invertible and the integral*

$$(m) \int_{\mathfrak{P}} (I - PTP)^{-1} PT \, dP$$

converges in the uniform norm. In order that

$$(6.9) \qquad (m)\int_{\mathfrak{P}}(I-PTP)^{-1}PTdP=(M)\int_{\mathfrak{P}}(I-PTP)^{-1}PTdP,$$

it is necessary and sufficient that the condition

$$(6.10) \qquad (P_j^+-P_j^-)(I-P_j^+TP_j^+)^{-1}(P_j^+-P_j^-)=P_j^+-P_j^-$$

be satisfied.

If the condition (6.10) *is satisfied, then both integrals in* (6.9) *are equal to*

$$\int_{\mathfrak{P}}(I-PTP)^{-1}PTdP.$$

For the special factorization of an operator relative to a chain with gaps, a theorem analogous to Theorem 2.2 holds.

3. Theorem 6.1 makes it possible to give a new abstract triangular representation for operators which differ "slightly" from the unit operator. We refer thus to every operator of the form $Y=I+X$, where X is a Volterra operator. In this representation the operator Y can be reconstructed from any one of its maximal eigenchains and from its operator modulus $(YY^*)^{1/2}$.

In fact, let $\mathfrak{P}$ be a maximal eigenchain of the Volterra operator X and let $T=-(YY^*)^{-1}+I$ $(\in\mathfrak{S}_\infty)$. Then the relation

$$(I-T)^{-1}=YY^*=(I+X)(I+X^*)$$

yields a special factorization of the operator $(I-T)^{-1}$, and consequently

$$Y=I+X=I+\int_{\mathfrak{P}}(I-PTP)^{-1}PTdP.$$

This result generalizes to operators which differ "slightly" from unitary operators, i.e. to operators of the form $Y=U+X$ (U a unitary operator, $X\in\mathfrak{S}_\infty$) whose entire spectrum lies on the unit circle. This fact is of fundamental value for the theory of the multiplicative representation of operator-functions (see I. C. Gohberg and M. G. Krein [**11**]).

4. We have the following generalization of Theorem 5.1.

THEOREM 6.2. *Let* $\mathfrak{P}$ *be a maximal chain,* $\mathfrak{S}_{\mathrm{I}}$ *any separable s.n. ideal, and* $\mathfrak{S}_{\mathrm{II}}$ *any s.n. ideal containing* $\mathfrak{S}_{\mathrm{I}}$: $\mathfrak{S}_{\mathrm{I}}\subseteq\mathfrak{S}_{\mathrm{II}}$.

If for every operator $T\in\mathfrak{S}_{\mathrm{I}}$ *the integral*

$$S=(m)\int_{\mathfrak{P}}PTdP$$

converges in the uniform norm and belongs to $\mathfrak{S}_{\mathrm{II}}$, *then for every operator* $T \in \mathfrak{S}_{\mathrm{I}}$ *satisfying the necessary condition that the operators* $I - PTP$ *be invertible* (*condition* a) *from Theorem* 6.1), *the integrals* (6.3) *converge in the uniform norm, the operators* $X_{\pm}$ *and* $D - I$ *belong to* $\mathfrak{S}_{\mathrm{II}}$, *and consequently the operator* $A = (I - T)^{-1}$ *has a special factorization* (6.5) *with operators* $X_{\pm}$ *and* $D - I$ *from* $\mathfrak{S}_{\mathrm{II}}$.

PROOF. This theorem is proved in the same way as Theorem 5.1, once one has the appropriate generalizations to discontinuous chains of the auxiliary Lemma 3.1 and Theorem 3.1.

Lemma 3.1 without the bound (3.2) remains valid for a discontinuous chain $\mathfrak{P}$ if the integral $\int_{\mathfrak{P}}$ is replaced by either of the integrals $(m)\int_{\mathfrak{P}}$, $(M)\int_{\mathfrak{P}}$, and if for the case of a chain $\mathfrak{P}$ with an infinite number of gaps one imposes on the function $a(P)$ $(P \in \mathfrak{P})$ the following additional condition:

$$\lim_{j\to\infty} |a(P_j^+) - a(P_j^-)| = 0, \tag{6.11}$$

where (P_j^-, P_j^+) $(j = 1, 2, \cdots)$ is the set of all gaps of $\mathfrak{P}$.

It is easily seen that with this additional condition the proof of the lemma remains valid, with almost no changes, for a discontinuous chain. The bounds (3.2) can now obviously be replaced by the bounds

$$s_j(Y) \leqq s_j(Z) \max_{P\in\mathfrak{P}} |a(P)| \quad (j = 1, 2, \cdots), \tag{6.12}$$

where

$$Z = (m, M) \int_{\mathfrak{P}} PX\, dP.$$

The proof of Theorem 3.1 proceeded independently of the nature of the chain, up to the point where (3.9) was obtained. From this point we argue as follows.

All the functions $a_{jk}(P)$ from (3.9) satisfy the additional condition (6.11). In fact, replacing the vectors χ_j and ψ_j in the expressions for the functions a_{jk} by the vectors

$$\chi_j' = \chi_j - \sum_{k=N}^{\omega} (P_k^+ - P_k^-)\chi_j \quad \text{and} \quad \psi_j' = \psi_j - \sum_{k=N}^{\omega} (P_k^+ - P_k^-)\psi_j,$$

where N is a sufficiently large number, we obtain functions $b_{jk}(P)$ which differ arbitrarily little from the functions $a_{jk}(P)$ and have the property

$$b_{jk}(P_r^+) = b_{jk}(P_r^-) \quad (r = N, N+1, \cdots).$$

Thus, if the symbol $\int_{\mathfrak{P}}$ in (3.10) is replaced by $(m) \int_{\mathfrak{P}}$, the integral thereby obtained will converge in the uniform norm.

Finally, we note that the relations (3.2) are entirely replaced by the relations (6.12) in the concluding portion of the proof of the theorem.

The theorem is proved.

§7. The factorization of Fredholm operators of the second kind

In this section we shall illustrate the special factorization of operators with the example of Fredholm integral operators of the second kind, defined in $L_2^{(r)}(a,b)$ $(r \leqq \infty)$ by a continuous matrix kernel.

As usual, we denote by $L_2^{(r)}(a,b)$ the Hilbert space consisting of all r-dimensional vector-functions $f(t) = \{f_j(t)\}_1^r$ $(a \leqq t \leqq b)$ with measurable coordinates $f_j(t)$ such that

$$|f|^2 = \int_a^b \sum_{j=1}^r |f_j(t)|^2 dt < \infty .$$

The scalar product in $L_2^{(r)}(a,b)$ is defined in a natural way by

$$(f,g) = \int_a^b g^*(t) f(t)\, dt = \int_a^b \sum_{j=1}^r \overline{g_j(t)} f_j(t)\, dt.$$

For the case at hand it is natural to take for the chain $\mathfrak{P}$ the truncation chain $\hat{\mathfrak{P}} = \{\hat{P}(t)\}_{a \leqq t \leqq b}$, defined by

$$(\hat{P}(s) f)(t) = \begin{cases} f(t) & (a \leqq t \leqq s) \\ 0 & (s < t \leqq b) \end{cases} \quad (f \in L_2^{(r)}(a,b)).$$

1. We begin our considerations with the general case of operators with Hilbert-Schmidt matrix kernel. Such an operator A is defined by

$$(7.1) \qquad (Af)(t) = f(t) - \int_a^b T(t,s) f(s)\, ds \quad (f \in L_2^{(r)}(a,b)),$$

where $T(t,s) = \| T_{jk}(t,s) \|_1^r$ and

$$\int_a^b \int_a^b \sum_{j,k=1}^r |T_{jk}(t,s)|^2 dt\, ds < \infty .$$

An operator A of the form (7.1) will have $\hat{\mathfrak{P}}$ as an eigenchain if and only if

$$(7.2) \qquad T(t,s) = 0 \text{ for } a \leqq s < t \leqq b,$$

and will have $\hat{\mathfrak{P}}^{\perp}$ as an eigenchain if and only if

$$T(t,s) = 0 \text{ for } a \leqq t \leqq s \leqq b.$$

In the first case we shall write $T(t,s) = T_+(t,s)$ and call $T(t,s)$ a *left Volterra* kernel, and in the second case we shall write $T(t,s) = T_-(t,s)$ and call $T(t,s)$ a *right Volterra* kernel.

If the equation

$$g(t) - \int_a^b T(t,s) g(s)\, ds = f(t) \tag{7.3}$$

is solvable in $L_2^{(r)}(a,b)$ for any right side, its solution can be represented in the form

$$g(t) = f(t) + \int_a^b \Gamma(t,s) f(s)\, ds,$$

where $\Gamma(t,s)$ is again a Hilbert-Schmidt kernel, called the *resolvent kernel.*

In the present case the existence of a special factorization of the operator $A^{-1} = (I - T)^{-1} = (I + V_+)(I + V_-)$ implies the existence of square-integrable left and right Volterra kernels $V_+(t,s)$ and $V_-(t,s)$ such that

$$\Gamma(t,s) = V_+(t,s) + V_-(t,s) + V_+(t,s) * V_-(t,s), \tag{7.4}$$

where the symbol $*$ denotes the operation of composition of kernels. Taking into account that $V_\pm(t,s)$ are Volterra kernels, this relation can be written out in greater detail as follows:

$$\Gamma(t,s) = \begin{cases} V_+(t,s) + \displaystyle\int_s^b V_+(t,r)\, V_-(r,s)\, dr & (a \leqq t < s \leqq b), \\ V_-(t,s) + \displaystyle\int_t^b V_+(t,r)\, V_-(r,s)\, dr & (a \leqq s \leqq t \leqq b). \end{cases} \tag{7.5}$$

Theorems I.10.1 and 5.1 enable us to formulate the following result.

1. *In order that the equation* (7.3) *with a Hilbert-Schmidt kernel* $T(t,s)$ *have a resolvent, representable in the form* (7.4), *where* $V_+(t,s)$ *and* $V_-(t,s)$ *are left and right Hilbert-Schmidt Volterra kernels, it is necessary and sufficient that for every* ξ $(a < \xi \leqq b)$ *the equation*

$$g(t) - \int_a^\xi T(t,s) g(s)\, ds = f(t) \quad (a \leqq t \leqq \xi) \tag{7.6}$$

be solvable in the space $L_2^{(r)}(a,\xi)$.

Theorem 5.2 also enables us to assert that if for some p $(1 < p \leqq 2)$

the s-numbers of the operator T satisfy the condition

$$\sum_{j=1}^{\infty} s_j^p(T) < \infty$$

or, for example, the condition

$$s_n(T) = o(n^{-1/p}) \quad (n \to \infty),$$

then the s-numbers of the operators $V_\pm$ will satisfy the same condition.

If T is in fact a nuclear operator, then

$$\sup_n \left\{ \sum_{j=1}^{n} s_j(V_\pm) \Big/ \sum_{j=1}^{n} \frac{1}{2j-1} \right\} < \infty .$$

It is likely that an attempt to establish these facts by the methods of the ordinary theory of integral equations would encounter considerable difficulties.

We also note that if $T(t,s)$ is a Hermitian-positive kernel, then the equation (7.6) will be solvable for any $\xi \leqq b$ as long as it is solvable for $\xi = b$.

2. For the case of a continuous matrix kernel $T(t,s)$ and $r < \infty$ this result can be developed further. In this case, if the equation (7.6) is solvable in the space $L_2^{(r)}(a,\xi)$ (which is equivalent to its solvability in the space $C^{(r)}(a,\xi)$ of continuous vector-functions) the equation (7.6) will have a continuous resolvent kernel $\Gamma_\xi(t,s)$. This kernel will be uniquely determined by each of the integral equations

$$\text{(7.7)} \quad \begin{aligned} \Gamma_\xi(t,s) - \int_a^\xi T(t,r)\,\Gamma_\xi(r,s)\,dr &= T(t,s) \quad (a \leqq t,s \leqq \xi), \\ \Gamma_\xi(t,s) - \int_a^\xi \Gamma_\xi(t,r)\,T(r,s)\,dr &= T(t,s) \quad (a \leqq t,s \leqq \xi). \end{aligned}$$

We remark that if the resolvent $\Gamma_\xi(t,s)$ exists for every ξ $(a \leqq \xi \leqq b)$, then from Fredholm's formula which represents it as the quotient of the Fredholm minor and the Fredholm determinant, it is easy to show that the function $\Gamma_\xi(t,s)$ will be jointly continuous with respect to the variables ξ, t, s and, moreover, will be continuously differentiable with respect to ξ.

In place of result 1 we can now formulate a more complete result.

2. *In order that the equation* (7.3) *with a continuous kernel* $T(t,s)$ *have a resolvent* $\Gamma(t,s)$ *representable in the form* (7.4), *it is necessary and*

sufficient that for every ξ $(a < \xi \leqq b)$ *there exist a resolvent* $\Gamma_\xi(t,s)$. *When this condition is fulfilled the kernels* $V_\pm$ *will be continuous in the corresponding triangles and their values there can be found from the formulas*

$$V_+(t,s) = \Gamma_s(t,s) \quad (a \leqq t \leqq s \leqq b), \tag{7.8}$$

$$V_-(t,s) = \Gamma_t(t,s) \quad (a \leqq s \leqq t \leqq b). \tag{7.9}$$

In fact, going back to operator notation for the relation (7.4),

$$(I - T)^{-1} = (I + V_+)(I + V_-),$$

we rewrite it in the form

$$(I - T)(I + V_+) = (I + V_-)^{-1} = I + Z_-,$$

where Z_- is an operator with a right Volterra kernel.

Hence

$$-T(t,s) + V_+(t,s) - \int_a^b T(t,r)\, V_+(r,s)\, dr = Z_-(t,s).$$

Let us consider this equality in the triangle $(a \leqq t \leqq s \leqq b)$, in which the kernel $Z_-(t,s)$ may be taken equal to zero. Then we obtain

$$V_+(t,s) - \int_a^s T(t,r)\, V_+(r,s)\, dr = T(t,s) \quad (a \leqq t \leqq s \leqq b).$$

This equation **defines** $V_+(t,s)$ uniquely, in its triangle.

On the other hand, if we put $\xi = s$ in the first of equations (7.7), we obtain

$$\Gamma_s(t,s) - \int_a^s T(t,r)\, \Gamma_s(r,s)\, dr = T(t,s) \quad (a \leqq t \leqq s \leqq b),$$

from which follows (7.8). The proof of (7.9) is analogous.

The representation (7.4) for continuous kernels $T(t,s)$ was used by M. G. Krein [**3**] in connection with inverse problems of the spectral theory of differential operators. For the scalar case this representation was used, for other purposes, in papers by N. P. Sergeev [**1**] and S. L. Sobolev [**1**].

The decomposition (7.4) for continuous kernels can be deduced independently of the general theory of the factorization of operators on the basis of the following property of the resolvent $\Gamma_\xi(t,s)$:

$$\frac{\partial \Gamma_\xi(t,s)}{\partial \xi} = \Gamma_\xi(t,\xi)\, \Gamma_\xi(\xi,s). \tag{7.10}$$

To obtain (7.10) we differentiate (7.7) with respect to ξ, obtaining

$$\frac{\partial \Gamma_\xi(t,s)}{\partial \xi} - \int_a^\xi T(t,r)\frac{\partial \Gamma_\xi(r,s)}{\partial \xi}\,dr = T(t,\xi)\,\Gamma_\xi(\xi,s).$$

On the other hand, putting $s = \xi$ in (7.7) and then multiplying from the right by $\Gamma_\xi(\xi,s)$, we discover that the matrix kernel $\Phi(t,s) = \Gamma_\xi(t,\xi) \times \Gamma_\xi(\xi,s)$ is a solution of the equation

$$\Phi(t,s) - \int_a^\xi T(t,r)\,\Phi(r,s)\,dr = T(t,\xi)\,\Gamma_\xi(\xi,s).$$

Since for fixed s $(a \leqq s \leqq b)$ this equation has a unique solution, which is continuous in r, we arrive at (7.10).

To verify the relation (7.4) or the equivalent relations (7.5), where $V_\pm(t,s)$ are defined by formulas (7.8) and (7.9), it remains to make use in (7.5) of the fact that

$$V_+(t,r)\,V_-(r,s) = \Gamma_r(t,r)\,\Gamma_r(r,s) = \frac{\partial \Gamma_r(t,s)}{\partial r}.$$

3. The rules which have been obtained for the factorization of Fredholm operators of the second kind can be used in many cases where a resolvent kernel $\Gamma_\xi(t,s)$ $(a < \xi \leqq b)$ exists and where formulas (7.8) and (7.9) have meaning, although the condition that the kernel $T(t,s)$ be continuous is not satisfied.

We shall clarify this with the example of an operator $I - K$, acting in $L_2^{(r)}(a,b)$ $(r < \infty)$, where K is an arbitrary n-dimensional operator:

$$K = \sum_{j=1}^{n} (\cdot, g_j) f_j,$$

and so $\{g_j\}$ and $\{f_j\}$ are arbitrary systems of linearly independent vector-functions from $L_2^{(r)}(a,b)$.

The kernel $k(t,s)$ of K obviously has the form

$$k(t,s) = F(t)\,G^*(s),$$

where the columns of the matrix-functions $F(t)$ and $G(t)$ coincide respectively with the vector-functions $f_1, f_2, \cdots, f_n$ and $g_1, g_2, \cdots, g_n$.

To determine the resolvent kernel $\Gamma_\xi(t,s)$ of the operator K we consider the equation

$$g(t) - F(t)\int_a^\xi G^*(s)g(s)\,ds = f(t).$$

Since

$$g(t) = F(t)x + f(t), \tag{7.11}$$

where

$$x = \int_a^\xi G^*(s)g(s)\,ds, \tag{7.12}$$

on inserting the expression (7.11) for $g(t)$ into (7.12) we obtain

$$\left(I - \int_a^\xi G^*(s)F(s)\,ds\right)x = \int_a^\xi G^*(s)f(s)\,ds. \tag{7.13}$$

It follows that for the existence of the kernel $\Gamma_\xi(t,s)$ it is necessary that the matrix $I - \int_a^\xi G^*(s)F(s)\,ds$ be invertible.

This condition is also sufficient for the existence of the resolvent kernel $\Gamma_\xi(t,s)$. In fact, it follows from (7.13) that

$$x = \mathfrak{A}_\xi \int_a^\xi G^*(s)f(s)\,ds,$$

where

$$\mathfrak{A}_\xi = \left(1 - \int_a^\xi G^*(s)F(s)\,ds\right)^{-1}.$$

Thus

$$g(t) = F(t)\mathfrak{A}_\xi \int_a^\xi G^*(s)f(s)\,ds + f(t)$$

and

$$\Gamma_\xi(t,s) = F(t)\mathfrak{A}_\xi G^*(s).$$

If the matrix-functions $F(t)$ and $G(t)$ are *continuous,* then by result 2 the kernels $V_\pm(t,s)$ which yield the factorization (7.4) are defined by formulas (7.8) and (7.9), i.e.

$$V_+(t,s) = F(t)\mathfrak{A}_s G^*(s) \quad (s \geqq t) \tag{7.14}$$

and

$$V_-(t,s) = F(t)\mathfrak{A}_t G^*(s) \quad (s \leqq t). \tag{7.15}$$

It turns out that in our case the kernels $V_\pm(t,s)$, defined by (7.14) and (7.15), also always yield a factorization (7.4). This can be seen by direct verification. Indeed, for $a \leqq t \leqq s \leqq b$

$$F(t)\mathfrak{A}_s G^*(s) + \int_s^b F(t)\mathfrak{A}_r G^*(r) F(r)\mathfrak{A}_r G^*(s)\,dr$$

$$= F(t)\left[\mathfrak{A}_s + \int_s^b \mathfrak{A}_r G^*(r) F(r)\mathfrak{A}_r\,dr\right] G^*(s).$$

It is easily seen that

$$\frac{\partial \mathfrak{A}_\xi}{\partial \xi} = \mathfrak{A}_\xi G^*(\xi) F(\xi)\mathfrak{A}_\xi,$$

and so

$$\mathfrak{A}_s + \int_s^b \mathfrak{A}_r G^*(r) F(r)\mathfrak{A}_r\,dr = \mathfrak{A}_b.$$

Since

$$\Gamma(t, s) = \Gamma_b(t, s) = F(t)\mathfrak{A}_b G^*(s),$$

it follows that the first of the relations (7.5) is satisfied. The proof of the validity of the second of these relations is similar.

It is easily seen that the results obtained generalize to the case $r = \infty$. They also certainly generalize to the case $r \leqq \infty$ and $n = \infty$ when K is a nuclear operator. If one constructs the matrices $F(t)$ and $G(t)$ in the appropriate way from the Schmidt expansion of such an operator K, then they will have the property

$$\int_a^b \operatorname{sp} F^*(t) F(t)\,dt = \int_a^b \operatorname{sp} G^*(t) G(t)\,dt = |K|_1 < \infty.$$

Then the matrix $\mathfrak{A}_\xi - I$ will be nuclear and formulas (7.14) and (7.15) will have meaning.

§8. A special method for the solution of linear integral equations[6)]

1. Closely related with the factorization of Fredholm operators is a new method for solving integral equations. We shall first discuss this method for the case of a scalar integral equation of the second kind

$$\text{(8.1)} \qquad \phi(t) - \int_a^b \mathscr{K}(t, s)\phi(s)\,ds = f(t) \quad (a \leqq t \leqq b),$$

and then point out its possible generalizations.

---

6) In this section we give a detailed presentation of a number of results from M. G. Kreĭn's paper [**6**].

Along with the equation (8.1) we shall consider for any ξ $(a \leqq \xi \leqq b)$ the equations

$$g(t,\xi) - \int_a^\xi \mathscr{K}(t,s) g(s,\xi)\, ds = 1 \tag{8.2}$$

$$(a \leqq t \leqq \xi),$$

$$g^*(t,\xi) - \int_a^\xi \mathscr{K}(s,t) g^*(s,\xi)\, ds = 1. \tag{8.2*}$$

Assuming that the truncated equation (8.2) has a resolvent for every ξ $(a < \xi \leqq b)$, we will have

$$g(t,\xi) = 1 + \int_a^\xi \Gamma_\xi(t,s)\, ds, \tag{8.3}$$

$$g^*(t,\xi) = 1 + \int_a^\xi \Gamma_\xi(s,t)\, ds. \tag{8.3*}$$

If the kernel $\mathscr{K}(t,s)$ $(a \leqq t, s \leqq b)$ is continuous, each of these relations may be differentiated with respect to ξ. Using here the property (7.10) of the resolvent Γ_ξ:

$$\frac{\partial \Gamma_\xi(t,s)}{\partial \xi} = \Gamma_\xi(t,\xi)\, \Gamma_\xi(\xi,s), \tag{8.4}$$

upon differentiating every term of (8.3) we obtain

$$\frac{\partial g(t,\xi)}{\partial \xi} = \Gamma_\xi(t,\xi) \left(1 + \int_a^\xi \Gamma_\xi(\xi,s)\, ds \right).$$

Thus

$$\frac{\partial g(t,\xi)}{\partial \xi} = \Gamma_\xi(t,\xi) g(\xi,\xi) \quad (a \leqq t \leqq \xi \leqq b) \tag{8.5}$$

and similarly

$$\frac{\partial g^*(t,\xi)}{\partial \xi} = \Gamma_\xi(\xi,t) g^*(\xi,\xi) \quad (a \leqq t \leqq \xi \leqq b). \tag{8.5*}$$

We introduce the function

$$\mathscr{M}(\xi) = \xi - \int_a^\xi \int_a^\xi \Gamma_\xi(t,s)\, dt\, ds \quad (a \leqq \xi \leqq b; \mathscr{M}(a) = 0).$$

This function can very easily be expressed in terms of each of the functions $g(t,\xi)$ and $g^*(t,\xi)$, namely,

$$\mathscr{M}(\xi) = \int_a^\xi g(t,\xi)\,dt = \int_a^\xi g^*(t,\xi)\,dt.$$

Differentiating the first of these equalities with respect to ξ and using (8.5) yields

$$\mathscr{M}'(\xi) = g(\xi,\xi) + g(\xi,\xi)\int_a^\xi \Gamma_\xi(t,\xi)\,dt,$$

whence, by (8.3*),

$$\mathscr{M}'(\xi) = g(\xi,\xi)g^*(\xi,\xi). \tag{8.6}$$

Thus, we also have

$$\mathscr{M}(\xi) = \int_a^\xi g(s,s)g^*(s,s)\,ds.$$

THEOREM 8.1. *Suppose that for every ξ $(a < \xi \leqq b)$ equation (8.2) and hence equation (8.2*) has a unique solution in $C(0,\xi)$*[7]*, which solutions we denote by $g(t,\xi)$ and $g^*(t,\xi)$, and suppose that $g(\xi,\xi)g^*(\xi,\xi)\neq 0$ $(a\leqq\xi\leqq b)$.*

Then for any function $f(x)\in C(a,b)$ the unique solution $\phi(t)\in C(a,b)$ of equation (8.1) can be obtained from the formula

$$\begin{aligned}\phi(t) = {} & \left[\frac{1}{\mathscr{M}'(\xi)}\frac{d}{d\xi}\int_a^\xi g^*(s,\xi)f(s)\,ds\right]_{\xi=b} g(t,b)\\ & - \int_t^b g(t,\xi)\frac{d}{d\xi}\left[\frac{1}{\mathscr{M}'(\xi)}\frac{d}{d\xi}\int_a^\xi g^*(s,\xi)f(s)\,ds\right]d\xi,\end{aligned} \tag{8.7}$$

where $\mathscr{M}'(\xi) = g(\xi,\xi)g^*(\xi,\xi)$.

PROOF. In assuming the existence for every ξ of unique solutions $g(t,\xi)$ and $g^*(t,\xi)$ of equations (8.2) and (8.2*), we are at the same time assuming the existence of the resolvent $\Gamma_\xi(t,s)$ for every ξ $(a<\xi\leqq b)$. But then all of the relations indicated earlier are valid.

By result 2 of §7 the solution ϕ of equation (8.1) can be obtained from the formulas

---

[7] As usual, we denote by $C(a,b)$ the set of all continuous functions on the interval $[a,b]$.

$$(8.8)\qquad \begin{cases} \phi(t) = g(t) + \displaystyle\int_t^b \Gamma_\xi(t,\xi)g(\xi)\,d\xi \\ \qquad\qquad\qquad\qquad\qquad\qquad (a \leqq t \leqq b), \\ g(t) = f(t) + \displaystyle\int_a^t \Gamma_t(t,s)f(s)\,ds. \end{cases}$$

To prove the equivalence of this pair of formulas with formula (8.7), let us denote the first term in the right side of (8.7) by $\mathscr{I}_1(t)$, and the second term by $\mathscr{I}_2(t)$.

Integration by parts in $\mathscr{I}_2$ yields

$$\mathscr{I}_2 = -\mathscr{I}_1 + \frac{g(t,t)}{\mathscr{M}'(t)}\frac{d}{dt}\int_a^t g^*(s,t)f(s)\,ds$$

$$+ \int_t^b \frac{\partial g(t,\xi)}{\partial \xi}\frac{1}{\mathscr{M}'(\xi)}\frac{d}{d\xi}\int_a^\xi g^*(s,\xi)f(s)\,ds\,d\xi.$$

Carrying out the indicated differentiations and using (8.3) and (8.3*), we obtain

$$\mathscr{I}_1 + \mathscr{I}_2 = f(t) + \int_a^t \Gamma_t(t,s)f(s)\,ds$$

$$+ \int_t^b \Gamma_\xi(t,\xi)\left[f(\xi) + \int_a^\xi \Gamma_\xi(\xi,s)f(s)\,ds\right]d\xi.$$

Comparing with (8.8), we see that $\mathscr{I}_1(t) + \mathscr{I}_2(t) \equiv \phi(t)$.

The theorem is proved.

We remark that if the kernel $\mathscr{K}$ is symmetric: $\mathscr{K}(s,t) = \mathscr{K}(t,s)$, then

$$g^*(t,\xi) = g(t,\xi) \quad (a \leqq \xi \leqq t \leqq b),$$

and if $\mathscr{K}$ is Hermitian: $\mathscr{K}(s,t) = \overline{\mathscr{K}(t,s)}$, then

$$g^*(t,\xi) = \overline{g(t,\xi)} \quad (a \leqq \xi \leqq t \leqq b).$$

2. Formula (8.7) enables us to obtain, from the solutions g and g^* of the set of equations (8.2) and (8.2*) with the *special* right side $f(t) = 1$, the solution of equation (8.1) with "any" right side $f(t)$. However, the application of this formula is restricted by the condition $\mathscr{M}'(\xi) \neq 0$ $(a \leqq \xi \leqq b)$. We shall indicate below some special classes of equations of the form (8.1) for which this condition will be fulfilled automatically (as a consequence of the existence of the solutions g and g^*).

Let us consider the equation

$$\phi(t) + \int_0^{2a} \mathscr{A}(t-s)\phi(s)\,ds = f(t) \quad (0 \leqq t \leqq 2a), \tag{8.9}$$

where $\mathscr{A}(t)$ $(-2a \leqq t \leqq 2a)$ is some function which is summable over the interval $[-2a, 2a]$. As before, we shall suppose that the right side $f(t)$ is continuous.

Since the operator represented by the integral in (8.9) is a completely continuous mapping of the space $C(0,2a)$[8] of continuous functions into itself under the sole condition that $\mathscr{A}(t)$ be summable (cf. Chapter III, §10), we relinquish in the present case the requirement that the kernel be continuous.

Thus, in the present case the interval (a,b) is the interval $(0,2a)$, and

$$\mathscr{K}(t,s) = -\mathscr{A}(t-s) \quad (0 \leqq t, s \leqq 2a).$$

Therefore the kernel $\Gamma_\xi(t,s)$ will now be determined from the equation

$$\Gamma_\xi(t,s) + \int_0^{\xi} \mathscr{A}(t-r)\Gamma_\xi(r,s)\,dr = -\mathscr{A}(t-s) \quad (0 \leqq t,s \leqq \xi) \tag{8.10}$$

or from the transposed equation[9]

$$\Gamma_\xi(s,t) + \int_0^{\xi} \Gamma_\xi(s,r)\mathscr{A}(r-t)\,dr = -\mathscr{A}(t-s) \quad (0 \leqq t,s \leqq \xi). \tag{8.11}$$

Making the substitution $t \to \xi - t$, $s \to \xi - s$, $r \to \xi - r$ in (8.10), we see that the function $\Phi(s,t) = \Gamma_\xi(\xi - t, \xi - s)$ satisfies the second equation (8.11). We conclude that

$$\Gamma_\xi(\xi - t, \xi - s) = \Gamma_\xi(s,t). \tag{8.12}$$

Therefore for $t = \xi$ it follows from (8.3) that

$$g(\xi,\xi) = 1 + \int_0^{\xi} \Gamma_\xi(\xi,s)\,ds = 1 + \int_0^{\xi} \Gamma_\xi(\xi - s, 0)\,ds,$$

so that

---

[8] As well as any space $L_p(0,2a)$ $(1 \leqq p < \infty)$.

[9] Since the right side $-\mathscr{A}(t-s)$ in equations (8.10) and (8.11) can be regarded as a continuous vector-function of $s \in [0,\xi]$ (or of $t \in [0,\xi]$) with values in $L_1(0,\xi)$, it follows that, regarding each of these equations as in $L_1(0,\xi)$, we can assert that $\Gamma_\xi(t,s)$ is a continuous vector-function of $t \in [0,\xi]$ $(s \in [0,\xi])$ *with values in* $L_1(0,\xi)$.

(8.13) $$g(\xi,\xi) = 1 + \int_0^\xi \Gamma_\xi(s,0)\,ds.$$

Similarly we obtain

(8.13*) $$g^*(\xi,\xi) = 1 + \int_0^\xi \Gamma_\xi(0,s)\,ds.$$

Assuming that the function $\mathscr{A}(t)$ is continuous, we differentiate (8.13) with respect to ξ. Using (8.12), we obtain

$$\frac{d}{d\xi}g(\xi,\xi) = \Gamma_\xi(\xi,0)\left[1 + \int_0^\xi \Gamma_\xi(s,\xi)\,ds\right]$$
$$= \Gamma_\xi(\xi,0)\left[1 + \int_0^\xi \Gamma_\xi(0,\xi - s)\,ds\right].$$

Thus,

(8.14) $$\frac{d}{d\xi}g(\xi,\xi) = \Gamma_\xi(\xi,0)g^*(\xi,\xi)$$

and similarly

(8.14*) $$\frac{d}{d\xi}g^*(\xi,\xi) = \Gamma_\xi(0,\xi)g(\xi,\xi).$$

We now consider two cases:

1) $\mathscr{A}(t) = \mathscr{A}(-t)$ $(-2a < t < 2a)$;

2) $\mathscr{A}(t) = \overline{\mathscr{A}(-t)}$ $(-2a < t < 2a)$.

In the first case the kernel $\mathscr{A}(t-s)$ is symmetric, and hence

$$g(\xi,\xi) = g^*(\xi,\xi)$$

so that from (8.14) follows

$$g(\xi,\xi) = \exp\left(\int_0^\xi \Gamma_t(t,0)\,dt\right) \neq 0 \quad (0 \leqq \xi \leqq 2a).$$

In the second case the kernel $\mathscr{A}(t-s)$ is Hermitian, and hence

$$g^*(\xi,\xi) = \overline{g(\xi,\xi)}, \quad \Gamma_\xi(\xi,0) = \overline{\Gamma_\xi(0,\xi)}.$$

Therefore we deduce from (8.14) and (8.14*) that

$$\frac{d}{d\xi}|g(\xi,\xi)|^2 = \frac{d}{d\xi}[g(\xi,\xi)g^*(\xi,\xi)]$$
$$= 2\,\mathrm{Re}[\Gamma_\xi(\xi,0)g^2(\xi,\xi)] \geqq -2\,|\Gamma_\xi(\xi,0)|\,|g(\xi,\xi)|^2,$$

whence

$$|g(\xi,\xi)| \geqq \exp\left(-\int_0^{\xi} |\Gamma_t(t,0)|\,dt\right) \;(>0) \quad (0 \leqq \xi \leqq 2a).$$

Thus, in the two cases under consideration and with the assumption that $\mathscr{A}(t)$ $(0 \leqq t \leqq 2a)$ is *continuous*, we have proved that

$$(8.15) \qquad (\mathscr{M}'(\xi) =)\, g(\xi,\xi)\, g^*(\xi,\xi) \neq 0 \quad (0 \leqq \xi \leqq 2a).$$

We shall show by another method that in these two cases (8.15) holds under the *general* assumption that

$$\mathscr{A}(t) \in L_1(-2a, 2a).$$

This method is based upon consideration of the operator ∂, defined by

$$\partial\phi = \frac{d\phi}{dt} + \phi(2a)\,\Gamma_{2a}(0,t)$$

on the set $\mathfrak{D}$ of all absolutely continuous functions $\phi \in C(0,2a)$ for which $\phi(0) = 0$.

Considering first the case of a Hermitian function $\mathscr{A}(t)$, we introduce the Hermitian form $[\phi,\psi]$ for any two functions $\phi,\psi \in L_1(0,2a)$, of which at least one is continuous:

$$[\phi,\psi] = \int_0^{2a} \phi(t)\overline{\psi(t)}\,dt + \int_0^{2a}\int_0^{2a} \mathscr{A}(t-s)\,\phi(s)\overline{\psi(t)}\,ds\,dt.$$

It turns out that for any $\phi,\psi \in \mathfrak{D}$ one has the relation

$$(8.16) \qquad [\partial\phi,\psi] + [\phi,\partial\psi] = 2\phi(2a)\overline{\psi(2a)}.$$

To see this, we first verify that

$$\left[\frac{d\phi}{dt},\psi\right] + \left[\phi,\frac{d\psi}{dt}\right] = 2\phi(2a)\overline{\psi(2a)} + \phi(2a)\int_0^{2a} \mathscr{A}(t-2a)\overline{\psi(t)}\,dt$$

$$+ \overline{\psi(2a)}\int_0^{2a} \mathscr{A}(2a-t)\,g(t)\,dt.$$

If $\mathscr{A}(t)$ is a smooth (i.e. continuously differentiable) function, the above relation is obtained by integration by parts. But since any function $\mathscr{A}(t) \in L_1(-2a,2a)$ can be approximated arbitrarily closely in the metric of $L_1(-2a,2a)$ by smooth functions, this relation is valid in the general case.

Now with the help of this relation and the functional equations for the resolvent,

$$\Gamma_{2a}(t, 2a) = \overline{\Gamma_{2a}(2a, t)},$$

$$\Gamma_{2a}(t, 2a) + \int_0^{2a} \mathscr{A}(t-s)\,\Gamma_{2a}(s, 2a)\,ds = -\mathscr{A}(t-2a),$$

$$\Gamma_{2a}(2a, t) + \int_0^{2a} \Gamma_{2a}(2a, s)\,\mathscr{A}(s-t)\,ds = -\mathscr{A}(2a-t)$$

it is not hard to see that (8.16) is valid. Obviously, to prove the inequality (8.15) for any ξ $(0 < \xi \leqq 2a)$ it is sufficient to prove it for $\xi = 2a$. Assuming the contrary, i.e. that $\mathscr{M}'(2a) = 0$, we will have

$$g(2a, 2a) = 1 + \int_0^{2a} \Gamma_{2a}(0, t)\,dt = 0.$$

Then for the function

$$\phi_0(t) = \int_0^t \Gamma_{2a}(0, s)\,ds$$

we will have $\phi_0 \in \mathfrak{D}$, $\phi_0(2a) = -1$ and $\partial\phi_0 = 0$.

But this is impossible, by virtue of (8.16), for $\phi = \psi = \phi_0$.

Similarly, one can prove the impossibility of the relation $\mathscr{M}'(2a) = 0$ for the case of an even function $\mathscr{A}(t)$. In this case, instead of the Hermitian form $[\phi, \psi]$ one introduces the symmetric form

$$\{\phi, \psi\} = \int_0^{2a} \phi(t)\psi(t)\,dt + \int_0^{2a}\int_0^{2a} \mathscr{A}(t-s)\phi(s)\psi(t)\,ds\,dt.$$

In place of (8.16) we will now have the relation

$$\{\partial\phi, \psi\} + \{\phi, \partial\psi\} = 2\phi(2a)\psi(2a),$$

from which again, for $\phi = \psi = \phi_0$, will follow the impossibility of the equalities $\partial\phi_0 = 0$, $\phi_0(2a) = -1$, each of which is equivalent to $g(2a, 2a) = 0$.[10)]

Thus, (8.15) is proved for the cases 1) and 2) under the general hypothesis $\mathscr{A}(t) \in L_1(-2a, 2a)$. We have therefore completely established the following result.

[10)] It should be mentioned that for the cases 1) and 2) we have proved $g(2a, 2a) \neq 0$ under the sole assumption of the existence of the resolvent $\Gamma_\xi(t, s)$ for $\xi = 2a$. Thus we have proved that $g(\xi, \xi) \neq 0$ for each ξ $(0 < \xi \leqq 2a)$ for which the resolvent $\Gamma_\xi(t, s)$ exists.

THEOREM 8.2. *Let* $\mathscr{A}(t)$ *be an even or Hermitian function from* $L_1(a-b, b-a)$, *and suppose that for every* ξ $(a \leqq \xi \leqq b)$ *the equation*

$$g(t,\xi) + \int_a^\xi \mathscr{A}(t-s) g(s,\xi)\, ds = 1 \quad (a \leqq t, s \leqq \xi)$$

has a unique continuous solution. Then $g(\xi,\xi) \neq 0$ $(a \leqq \xi \leqq b)$ *and the solution* ϕ *of the equation*

$$\phi(t) + \int_a^b \mathscr{A}(t-s)\phi(s)\, ds = f(t) \tag{8.17}$$

for any $f \in C(a,b)$ *can be obtained from the formula* (8.7), *if in* (8.7) *we put* $g^*(t,\xi) = g(t,\xi)$ *when* $\mathscr{A}(t)$ *is even, and* $g^*(t,\xi) = \overline{g(t,\xi)}$ *when* $\mathscr{A}(t)$ *is Hermitian.*

In the latter case (in particular, for the case of a real even $\mathscr{A}(t)$*), for the existence of a unique solution in* $C(a,b)$ *of the equation* (8.17) *it is necessary and sufficient that for any* $\chi(t) \in C(a,b)$ *not identically zero one have*

$$\int_a^b |\chi(t)|^2 dt + \int_a^b \int_a^b \mathscr{A}(t,s)\chi(s)\overline{\chi(t)}\, ds dt > 0.$$

Let us clarify that the second assertion of the theorem follows from the general assertion noted at the beginning of §5.2.[11)]

The fact that instead of the initial equation (8.9) with special limits we have written equation (8.17) with general limits is not essential, as one equation goes over into the other under the simple translation of variables $s \to s + a$, $t \to t + a$.

3. Let us look more carefully at the case of an even function $\mathscr{A}(t)$. In this case, in place of equation (8.9) it is sometimes more convenient to deal with the equation

$$\phi(t) + \int_{-a}^a \mathscr{A}(t-s)\phi(s)\, ds = f(t). \tag{8.18}$$

Obviously, the equation (8.9) can be transformed to the form (8.18) by the simple change of variables $t \to t + a$, $s \to s + a$. A similar change of variables in the equation

$$g(t, 2\xi) + \int_0^{2\xi} \mathscr{A}(t-s) g(s, 2\xi)\, ds = 1$$

[11)] We are referring to the following assertion: *if* $H = H^* \in \mathfrak{S}_\infty$ *and* $\mathfrak{P}$ *is a complete chain, then for the invertibility of the operator* $I + PHP$ *for every* $P \in \mathfrak{P}$ *it is necessary and sufficient that the operator* $I + H$ *be positive.*

leads to the equation

$$(8.19)\qquad q(t,\xi)+\int_{-\xi}^{\xi}\mathscr{A}(t-s)q(s,\xi)\,ds=1\quad(-\xi\leqq s,t\leqq\xi),$$

where now $0<\xi\leqq a$. The functions $q(t,\xi)$ and $g(t,\xi)$ are related by

$$q(t,\xi)=g(t+\xi,2\xi).$$

In particular, $q(\xi,\xi)=g(2\xi,2\xi)$ and so

$$q(\xi,\xi)\neq 0\quad(0\leqq\xi\leqq a).$$

If we make the change of variables $t\to -t$, $s\to -s$ in (8.18), we see that when the right side $f(t)$ is an even or an odd function, the unique solution ϕ of (8.18) will be even or odd, respectively. In each of these cases it suffices to find the value of $\phi(t)$ on the interval $(0,a)$, and in the first case the function $\phi(t)$ will be, on this interval, the unique solution in $C(0,a)$ of the equation

$$(8.20)\qquad \phi(t)+\int_0^a[\mathscr{A}(t-s)+\mathscr{A}(t+s)]\phi(s)\,ds=f(t)\quad(0\leqq t\leqq a),$$

and in the second case, of the equation

$$(8.21)\qquad \phi(t)+\int_0^a[\mathscr{A}(t-s)-\mathscr{A}(t+s)]\phi(s)\,ds=f(s)\quad(0\leqq t\leqq a).$$

In particular, corresponding to the first case the function $q(t,\xi)$ $(-\xi\leqq t\leqq\xi)$ will be an even function of t for any ξ $(0<\xi\leqq a)$, and on the interval $(0,\xi)$ it will be the unique solution of the equation

$$(8.22)\qquad q(t,\xi)+\int_0^{\xi}[\mathscr{A}(t-s)+\mathscr{A}(t+s)]q(s,\xi)\,ds=1.$$

If we now put[12)]

$$(8.23)\qquad M(\xi)=\int_0^{\xi}q(s,\xi)\,ds\quad(0\leqq\xi\leqq a),$$

[12)] Note that according to (8.6), $M'(\xi)=q^2(\xi,\xi)$, while at the same time, for

$$\mathscr{M}(\xi)=\int_0^{\xi}g(t,\xi)\,dt,$$

we have $\mathscr{M}'(\xi)=g^2(\xi,\xi)=q^2(\xi/2,\xi/2)$ $(0\leqq\xi\leqq 2a)$. We conclude that $2M(\xi)=\mathscr{M}(2\xi)$.

then by Theorem 8.1 the solution $\phi(t)$ of (8.20) can be found from the formula

$$\tag{8.24} \begin{aligned} \phi(t) &= \frac{1}{M'(a)} \left[\frac{d}{d\xi} \int_0^\xi q(s,\xi) f(s)\, ds \right]_{\xi=a} q(t,a) \\ &\quad - \int_t^a q(t,\xi) \frac{d}{d\xi} \left[\frac{1}{M'(\xi)} \frac{d}{d\xi} \int_a^\xi q(s,\xi) f(s)\, ds \right] d\xi. \end{aligned}$$

We shall show that the continuous solution $\phi(t)$ of (8.21) for any continuous $f(t)$ $(0 \leqq t \leqq a)$ with $f(0) = 0$ can be obtained from the formula

$$\tag{8.25} \phi(t) = -\frac{d}{dt} \int_t^a \frac{q(t,\xi)}{M'(\xi)} \left[\int_0^\xi q(s,\xi)\, df(s) \right] d\xi,$$

where the inner integral is a Stieltjes integral. This integral always has meaning, as $q(t,\xi)$ is an absolutely continuous function of t and, moreover,[13)]

$$\tag{8.26} \frac{\partial q(t,\xi)}{\partial t} = -\, q(\xi,\xi) R_\xi(t,\xi) \quad (0 \leqq t \leqq \xi \leqq a),$$

where $R_\xi(t,s)$ $(0 \leqq t, s \leqq \xi)$ is the resolvent of the truncated equation

$$\phi(t) + \int_0^\xi [\mathscr{A}(t-s) - \mathscr{A}(t+s)] \phi(s)\, ds = f(t) \quad (0 \leqq t \leqq \xi),$$

and gives the solution ϕ of this equation in the form

$$\phi(t) = f(t) + \int_0^\xi \boldsymbol{R_\xi(t,s) f(s) ds} \quad (0 \leqq t \leqq \xi).$$

To prove the relation (8.26) we differentiate (8.22) with respect to t, assuming for the moment that $\mathscr{A}(t)$ is continuously differentiable. This yields the relation

$$\frac{\partial q(t,\xi)}{\partial t} + \int_0^\xi [\mathscr{A}'(t-s) + \mathscr{A}'(t+s)] q(s,\xi)\, ds = 0,$$

from which integration by parts gives the equation

13) Let us recall, by way of comparison, that by (8.5)

$$\partial q(t,\xi)/\partial\xi = q(\xi,\xi)\Gamma_\xi(t,\xi),$$

where $\Gamma_\xi(t,s)$ is the resolvent of the truncated equation obtained from (8.19) by replacing a by ξ.

$$\frac{\partial q(t,\xi)}{\partial t} + \int_0^\xi [\mathscr{A}(t-s) - \mathscr{A}(t+s)] \frac{\partial q(s,\xi)}{\partial s}\, ds$$
$$= q(\xi,\xi)\,[\mathscr{A}(t-\xi) - \mathscr{A}(t+\xi)].$$

Recalling the equation which the resolvent $R_\xi(t,\xi)$ satisfies, we see that (8.26) holds.

The relation (8.26) can be rewritten in the form

$$q(\xi,\xi) - q(t,\xi) = q(\xi,\xi) \int_t^\xi R_\xi(s,\xi)\, ds \quad (0 \leqq t \leqq \xi \leqq a). \tag{8.27}$$

On the other hand, if the relation (8.27) has been established for the case of smooth continuously differentiable even functions $\mathscr{A}(t)$, then it will be valid for any even function $\mathscr{A}(t) \in L_1(-a,a)$ (for which, it is understood, the equation (8.19) has a unique solution for every ξ $(0 \leqq \xi \leqq a)$.

The point is (compare with the argument on p. 195) that if the function $\mathscr{A}(t)$ is approximated sufficiently closely in the metric of $L_1(-a,a)$ by a smooth even function $\widetilde{\mathscr{A}}(t)$, then for any fixed ξ $(0 < \xi \leqq a)$ the corresponding functions $\tilde{q}(t,\xi)$ and $\widetilde{R}_\xi(t,\xi)$ will approximate the corresponding functions $q(t,\xi)$ (in the metric of $C(0,\xi)$) and $R_\xi(t,\xi)$ (in the metric of $L_1(0,\xi)$) to any desired degree of accuracy.

Integrating by parts in the inner integral of the right side $\mathscr{I}(t)$ from (8.25) yields

$$\mathscr{I}(t) = -\frac{d}{dt}\int_t^a \frac{q(t,\xi)}{M'(\xi)}\left[q(\xi,\xi)f(\xi) - \int_0^\xi f(s)\frac{\partial q(s,\xi)}{\partial s}\, ds\right] d\xi,$$

and if we now use (8.26) and the fact that $M'(\xi) = q^2(\xi,\xi)$, we obtain

$$\mathscr{I}(t) = f(t) + \int_0^t R_t(s,t) f(s)\, ds$$
$$+ \int_t^a R_\xi(t,\xi)\left[f(\xi) + \int_0^\xi R_\xi(s,\xi) f(s)\, ds\right] d\xi.$$

Recalling formula (8.8) in connection with the present case, we find that $\phi(t) = \mathscr{I}(t)$ is a solution of (8.21).

We combine the two formulas (8.24) and (8.25) in the following result.

THEOREM 8.3. *Let the even function* $\mathscr{A}(t) \in L_1(-2a, 2a)$ *be such that for every* ξ $(0 < \xi \leqq a)$ *the equation* (8.19) *has a unique continuous solution* $q(t,\xi)$ $(-\xi \leqq t \leqq \xi)$. *Then for any* $f \in C(-a,a)$ *the solution*

$\phi \in C(-a,a)$ of (8.18) *can be obtained from the formula*

$$
\begin{aligned}
\phi(t) &= \frac{1}{2M'(a)}\left[\frac{d}{da}\int_{-a}^{a} q(s,a)f(s)\,ds\right] q(t,a) \\
&\quad -\frac{1}{2}\int_{|t|}^{a} q(t,\xi)\frac{d}{d\xi}\left(\frac{1}{M'(\xi)}\frac{d}{d\xi}\int_{-\xi}^{\xi} q(s,\xi)f(s)\,ds\right)d\xi \qquad (8.28)\\
&\quad -\frac{1}{2}\frac{d}{dt}\int_{|t|}^{a}\frac{q(t,\xi)}{M'(\xi)}\left(\int_{-\xi}^{\xi} q(s,\xi)\,df(s)\right)d\xi \quad (-a \leqq t \leqq a).
\end{aligned}
$$

In fact, if $f(t) = f(-t)$, the last integral vanishes identically and formula (8.28) yields an even function $\phi(t)$ which is, for $t > 0$, the same function as that given by (8.24).

If $f(t) = -f(-t)$, then the first two terms of the right side vanish identically, and the formula yields an odd function which is, for $t > 0$, the same function as that given by (8.25).

Since the operator defined by the right side of (8.28) is linear, and any function $f \in C(-a,a)$ can be represented as the sum of an even and an odd function from $C(-a,a)$, the formula is valid for any function $f \in C(-a,a)$.

We remark (cf. Theorem 8.2) that for a *real even function* $\mathscr{A}(t) \in L_1(-2a,2a)$ the hypothesis of Theorem 8.3 (the existence for every ξ $(0 < \xi \leqq a)$ of a unique solution $q(t,\xi)$ in $C(-\xi,\xi)$ of (8.19)) is satisfied if and only if the inequality

$$\int_{-a}^{a}\phi^2(t)\,dt + \int_{-a}^{a}\int_{-a}^{a}\mathscr{A}(t-s)\phi(s)\phi(t)\,ds\,dt > 0$$

holds for every real function $\phi \in C(-a,a)$ $(\phi \not\equiv 0)$.

4. Now let $\mathscr{A}(t)$ be a real even function from $L_1(-2a,2a)$ which is positive definite, i.e.

$$\int_{-a}^{a}\int_{-a}^{a}\mathscr{A}(t-s)\phi(s)\phi(t)\,ds\,dt > 0 \qquad (8.29)$$

for every function $\phi \in C(-a,a)$ not identically zero. In this case, for any $\mu > 0$ and for $\xi \in (0,a)$ the equation

$$\mu q(t,\xi) + \int_{-\xi}^{\xi}\mathscr{A}(t-s)q(s,\xi)\,ds = 1 \qquad (8.30)$$

will have a unique solution $q \in C(-\xi,\xi)$. By means of simple transformations it is not hard to verify that the solution ϕ of the equation

$$\mu\phi(t) + \int_{-a}^{a} \mathscr{A}(t-s)\phi(s)\,ds = f(t) \tag{8.31}$$

is obtained as before from formula (8.28), where $q(t,\xi)$ is determined from the equation (8.30), and $M(\xi)$ as before is defined by

$$M(\xi) = \int_0^{\xi} q(t,\xi)\,dt. \tag{8.32}$$

This suggests that under certain additional conditions concerning the function $\mathscr{A}(t)$ formula (8.28) will remain valid for the integral equation of the first kind

$$\int_{-a}^{a} \mathscr{A}(t-s)\phi(s)\,ds = f(t) \quad (-a \leqq t \leqq a), \tag{8.33}$$

obtained from (8.31) by passing to the limit $\mu \downarrow 0$.

For equation (8.33) the function $q(t,\xi)$ $(0 \leqq \xi \leqq a)$ should of course be defined as the solution in $L_1(-\xi,\xi)$ of the equation

$$\int_{-\xi}^{\xi} \mathscr{A}(t-s)q(s,\xi)\,ds = 1,$$

and $M(\xi)$ should be defined by (8.32).[14]

Indeed, one can show that if the even positive definite function $\mathscr{A}(t)$ has the form

$$\mathscr{A}(t) = \mathscr{F}(t) + \begin{cases} \rho \ln(1/|t|) \\ \rho|t|^{-h}, \end{cases} \quad (-2a \leqq t \leqq 2a),$$

where $\rho > 0$,[15] and $\mathscr{F}(t)$ is a function having an absolutely continuous derivative on the interval $[a,b]$, then for any twice continuously differentiable function $f(t)$ $(-a \leqq t \leqq b)$ the solution of (8.33) can be obtained from the formula (8.28).

Thus, for example, equation (8.33) can be effectively solved, using this formula, for those cases in which $\mathscr{A}(t)$ coincides on the interval $(-2a, 2a)$ with one of the functions listed in the following table.

---

[14] The formula $M'(\xi) = q^2(\xi,\xi)$ will now no longer be valid; for example, in the examples given below always $q(\xi,\xi) = \infty$, while $M'(\xi)$ is a continuous function.

[15] If the condition (8.29) of positive definiteness of $\mathscr{A}(t)$ is satisfied, negative values of ρ are ruled out.

No.	$\mathscr{A}(t)$	$q(t,\xi)$	$M(\xi)$
1)	$\ln \frac{A}{2\|t\|}$	$\frac{1}{\pi} M(\xi)(\xi^2 - t^2)^{-1/2}$	$\left[\ln \frac{A}{\xi}\right]^{-1}$
2)	$\ln \frac{\sin(A/2)}{2\sin(\|t\|/2)}$	$\frac{1}{\pi} M(\xi)(2\cos t - 2\cos\xi)^{-1/2}\cos\frac{\xi}{2}$	$\left[\ln \frac{\sin(A/2)}{\sin(\xi/2)}\right]^{-1}$
3)	$\ln \frac{\mathrm{sh}(A/2)}{2\,\mathrm{sh}(\|t\|/2)}$	$\frac{1}{\pi} M(\xi)(2\,\mathrm{ch}\,\xi - 2\,\mathrm{ch}\,t)^{-1/2}\mathrm{ch}\frac{t}{2}$	$\left[\ln \frac{\mathrm{sh}(A/2)}{\mathrm{sh}(\xi/2)}\right]^{-1}$
4)	$\ln \mathrm{cth}\frac{\|t\|}{4}$	$\frac{1}{\pi} Q^{-1}_{-1/2}(\mathrm{ch}\,\xi)(2\,\mathrm{ch}\,\xi - 2\,\mathrm{ch}\,t)^{-1/2}$	$\frac{P_{-1/2}(\mathrm{ch}\,\xi)}{Q_{-1/2}(\mathrm{ch}\,\xi)}$
5)	$\|t\|^{-h}$	$\frac{1}{\pi}\cos\frac{\pi h}{2}(\xi^2 - t^2)^{(h-1)/2}$	$\frac{2\sqrt{\pi}\,\xi^h}{h\Gamma\left(\frac{h}{2}\right)\Gamma\left(\frac{1-h}{2}\right)}$
	$0 < h < 1$		$0 < \xi \leqq a$

Here $0 < a < A$, and $P_{-1/2}(\mathrm{ch}\,\xi)$ and $Q_{-1/2}(\mathrm{ch}\,\xi)$ are the Legendre functions of the first and second kind.

One is led to cases 1) and 2) of equation (8.33) in dealing with the classical two-dimensional contact problem of elasticity theory (cf., for example, I. Ja. Štaerman [1]). N. H. Arutjunjan [**1**] was recently led to case 5) of equation (8.33) in connection with two-dimensional contact problems of the theory of creep.

For the cases 1) and 2) the solutions were known, but in a different analytic form, containing singular integrals with Cauchy kernels.[16]

For the case 5), i.e. for the equation

$$\int_{-a}^{a} \frac{\phi(s)}{|t-s|^h}\,ds = f(t)$$

the solution was first obtained by T. Carleman [**1**], in the form

$$\phi(t) = \frac{i\cos(\pi h/2)}{2\pi^2}\frac{d}{dt}\oint_{\Gamma_t}\frac{(z^2-a^2)^{(1-h)/2}}{(z-t)^{1-h}}\left[\int_{-a}^{a}\frac{(a^2-s^2)^{(h-1)/2}}{s-z}f(s)\,ds\right]dz$$

$$(-a < t < a).$$

[16] It should be indicated that for the case 1) and an even right side, a solution of the integral equation (8.33), free of singular integrals, was first obtained by N. A. Rostovcev [**1**].

In this formula Γ_t is a smooth contour which intersects the semi-axis $[-a, \infty)$ only at the point $z = t$, and the branches of the functions $(z^2 - a^2)^{(1-h)/2}$ and $(z - t)^{1-h}$, which appear in the integral along Γ_t, are distinguished in the region $\operatorname{Ext}[-a, a]$[17] by the condition that they be positive for $z > a$.

By way of comparison we give the formula for the solution $\phi(t)$, obtained from Theorem 8.3.

For the case of an even function $f(t)$ $(-a \leqq t \leqq a)$ we obtain

$$\phi(t) = C(h) \left\{ \frac{a^{1-h}}{(a^2 - t^2)^{(1-h)/2}} \frac{d}{da} \int_0^a \frac{f(s)\,ds}{(a^2 - s^2)^{(1-h)/2}} - \int_{|t|}^a \frac{1}{(\xi^2 - t^2)^{(1-h)/2}} \frac{d}{d\xi} \left[\xi^{1-h} \frac{d}{d\xi} \int_0^\xi \frac{f(s)\,ds}{(\xi^2 - s^2)^{(1-h)/2}} \right] d\xi \right\} \quad (-a \leqq t \leqq a),$$

where

$$C(h) = \frac{1}{2\pi^{5/2}} \Gamma\left(\frac{h}{2}\right) \Gamma\left(\frac{1-h}{2}\right) \cos^2 \frac{\pi h}{2}.$$

For the case of an odd function $f(t)$ $(-a \leqq t \leqq a)$ we obtain the perfectly simple formula

$$\phi(t) = -C(h) \frac{d}{dt} \int_{|t|}^a \left(1 - \frac{t^2}{\xi^2}\right)^{(1-h)/2} \left(\int_0^\xi \frac{df(s)}{(\xi^2 - s^2)^{(1-h)/2}} \right) d\xi \quad (-a \leqq t \leqq a).$$

5. The considerations which have been applied to the integral equation (8.33) go over to some extent to more general integral equations of the first kind

$$\int_a^b \mathscr{A}(t - s)\phi(s)\,ds = f(t) \quad (a \leqq t \leqq b), \tag{8.34}$$

for which $\mathscr{A}(t)$ $(a - b \leqq t \leqq b - a)$ is not necessarily an even real function, for example, where it is a complex-valued even function or a Hermitian function and so on.

Under certain conditions this equation can be solved by formula (8.7), where the functions $g(t, \xi)$ and $g^*(t, \xi)$ are now determined from the equations

---

[17] We denote by $\operatorname{Ext}[-a, a]$ the entire complex plane with the interval $[-a, a]$ removed.

$$\int_a^\xi \mathscr{A}(t-s)g(s,\xi)\,ds = 1,$$

$$\int_a^\xi \mathscr{A}(s-t)g(s,\xi)\,ds = 1 \quad (a < \xi \leqq b),$$

and the function $\mathscr{M}(\xi)$ is defined by

$$\mathscr{M}(\xi) = \int_a^\xi g(s,\xi)\,ds \quad (a < \xi \leqq b).$$

This method for solving integral equations has found application in various papers on the contact problem of elasticity theory (cf. N. H. Arutjunjan [**1**], N. H. Arutjunjan and M. M. Manukjan [**1**], I. E. Prokopovič [**1**], G. Ja. Popov [**1,2**], V. M. Aleksandrov [**1**]).

In these works one can find expressions for the functions $q(t,\xi)$, $g(t,\xi)$ and $M(\xi)$ for other concrete examples of the function $\mathscr{A}(t)$ (cf. also O. N. Litvinenko [**1**]).

§9. Concluding remarks

Although up to now we have considered the problem of factorization only for operators which differ from the identity operator by a completely continuous operator, nevertheless examples of the solution of this problem are known for certain special classes of bounded operators.

Namely, as we have already mentioned in §1, the factorization of functions and matrix-functions (which is an important tool in the theory of integral equations on a half-line with kernels, depending on the difference of the arguments; cf. M. G. Kreĭn [**10**] and I. C. Gohberg and M. G. Kreĭn [**2**]) can be interpreted as the factorization of certain classes of operators in the sense of the general definition given there.

This applies in equal measure to the theory of the discrete analogs of similar equations.

The indicated situation is easily clarified for these discrete analogs; we shall therefore begin with them.

Let $a(t)$ $(|t| = 1)$ be a function which can be expanded in an absolutely convergent Fourier series

$$a(t) = \sum_{j=-\infty}^{\infty} \alpha_j t^j \ (|t| = 1), \quad \sum_{j=-\infty}^{\infty} |\alpha_j| < \infty. \tag{9.1}$$

If the condition $a(t) \neq 0$ $(|t| = 1)$ is satisfied, then (cf. M. G. Kreĭn [**10**]) the function $a(t)$ can be represented in the form

(9.2) $$a(t) = a_+(t)a_-(t) \quad (|t| = 1),$$

where

$$a_\pm(t) = \sum_{j=0}^{\infty} \alpha_j^{(\pm)} t^{\pm j} \quad (|t| = 1), \quad \sum_{j=0}^{\infty} |\alpha_j^{(\pm)}| < \infty.$$

We shall show that the factorization (9.2) gives rise to a certain operator factorization. With every function $a(t)$ of the form (9.1) we associate the operator A, defined in the space $\boldsymbol{l}_2$ of sequences $\xi = \{\xi_j\}_{-\infty}^{\infty}$ $(\sum_j |\xi_j|^2 < \infty)$ by the matrix $\|\alpha_{j-k}\|_{-\infty}^{\infty}$.

It is easily seen that A is a bounded linear operator. Moreover, if $a_1(t)$ and $a_2(t)$ are functions of the form (9.1), and $a(t) = a_1(t)a_2(t)$, then $A = A_1 A_2$.

Let us denote by P_j $(j = 0, \pm 1, \cdots)$ the orthoprojector in $\boldsymbol{l}_2$ which acts according to the rule

$$P_j\{\xi_n\}_{-\infty}^{\infty} = \{\cdots, \xi_{j-1}, \xi_j, 0, 0, \cdots\}.$$

The set of all the projectors P_j, together with 0 and I, forms a maximal chain, which we denote by $\mathfrak{P}$.

Obviously, the matrices corresponding to the operators A_+ and A_- generated by the functions $a_+(t)$ and $a_-(t)$ are left lower and right upper triangular, respectively. Hence the operators A_+ and A_- have $\mathfrak{P}$ and $\mathfrak{P}^\perp$ respectively as eigenchains. Thus (9.2) implies a factorization of A relative to the chain $\mathfrak{P}$:

(9.3) $$A = A_+ A_-.$$

It follows from the results of the previously cited paper (M. G. Kreĭn [**10**]) that if the condition $a(t) \neq 0$ $(|t| = 1)$ is satisfied and if, moreover,

(9.4) $$\int_{|t|=1} d \arg a(t) = 0,$$

the operators $A_\pm$ will be invertible, and their inverses have the same triangular structure as A_+ and A_-, respectively. Moreover, multiplying A by a suitable constant, we can ensure that only ones will appear on the diagonals of the matrices of $A_\pm$.

We remark that in this example the condition (9.4) is necessary and sufficient for the invertibility of all the operators $P_j A P_j$ in the corresponding spaces $P_j \boldsymbol{l}_2$.

This example has a continuous analog. Let $k(t) \in L_1(-\infty, \infty)$ and let

$$\mathscr{K}(\lambda) = \int_{-\infty}^{\infty} k(t) e^{i\lambda t} dt.$$

According to the previously cited paper, if the condition

$$1 - \mathscr{K}(\lambda) \neq 0 \quad (-\infty < \lambda < \infty)$$

is satisfied, the function $1 - \mathscr{K}(\lambda)$ admits the following factorization:

$$1 - \mathscr{K}(\lambda) = (1 - \mathscr{K}_+(\lambda))(1 - \mathscr{K}_-(\lambda)), \tag{9.5}$$

where

$$\mathscr{K}_+(\lambda) = \int_0^{\infty} k_+(t) e^{i\lambda t} dt, \ \mathscr{K}_-(\lambda) = \int_{-\infty}^{0} k_-(t) e^{i\lambda t} dt \quad (k_\pm \in L_1).$$

With every function

$$1 - k(t) \qquad (k(t) \in L_1(-\infty, \infty))$$

we associate the bounded linear operator A which acts in $L_2(-\infty, \infty)$ according to

$$(Af)(t) = f(t) - \int_{-\infty}^{\infty} k(t-s) f(s)\, ds.$$

We consider the orthoprojector P_s $(-\infty < s < \infty)$, defined in $L_2(-\infty, \infty)$ by

$$(P_s f)(t) = \begin{cases} f(t) & (t \leqq s) \\ 0 & (t > s). \end{cases}$$

We denote by $\mathfrak{P}$ the continuous chain consisting of all the orthoprojectors P_s $(-\infty < s < \infty)$, 0 and I. Obviously the operators $I + K_+$ and $I + K_-$ generated by the functions $1 + k_+(t)$ and $1 + k_-(t)$ have $\mathfrak{P}$ and $\mathfrak{P}^\perp$ respectively as eigenchains.

It is easily seen that in operator notation (9.5) gives the following factorization of the operator A relative to the chain $\mathfrak{P}$:

$$A = A_+ A_-.$$

If the condition

$$\int_{-\infty}^{\infty} d \arg(1 - \mathscr{K}(\lambda)) = 0 \tag{9.6}$$

is satisfied, the operators $A_\pm^{-1}$ will be operators of the same type as the operators $A_\pm$.

The condition (9.6) is necessary and sufficient for the invertibility of all the operators PAP ($P \in \mathfrak{P}$) in the respective spaces PL_2.

In conclusion we mention two circumstances which in our view merit attention.

Although the operators A considered in this section do not differ from the identity operator by a completely continuous operator, they do have the following property relative to the corresponding chain $\mathfrak{P}$: $PA(I-P) \in \mathfrak{S}_\infty$ for every $P \in \mathfrak{P}$. This property proved to be very essential for the construction of factorizations of matrix-functions (cf. I. C. Gohberg and M. G. Kreĭn [2]).

The second circumstance is that all the results of this section remain valid if we replace the space $\boldsymbol{l}_2$ by any one of the Banach spaces $\boldsymbol{l}_p$ ($1 \leqq p \leqq \infty$), $\boldsymbol{c}_0$ and $\boldsymbol{c}$, and the space L_2 by one of the spaces L_p ($1 \leqq p \leqq \infty$), C, C_0 and others. Thus, examples of the factorization of operators of certain classes in Banach spaces are comparatively well known.

In recent papers by M. A. Barkar′ and I. C. Gohberg [**1, 2**] a number of results of this chapter have been generalized to the case of operators in Banach spaces.

§10. Appendix. A general theorem on the convergence of the integral along a chain

In previous chapters as well as the present chapter we have encountered, in various concrete forms, the following integrals along a chain:

$$\int_{\mathfrak{P}} F(P)\,X\,dP, \qquad \int_{\mathfrak{P}} dP\,X\,F(P),$$

where $F(P)$ ($P \in \mathfrak{P}$) is some operator-function and $X \in \mathfrak{S}_\infty$. In Chapters I—III we considered the case $F(P) = P$, and in this chapter the case $F(P) = (I - PXP)^{-1}P$. As was already mentioned in §4, Chapter I, for the convergence of the integral

$$\int_{\mathfrak{P}} F(P)\,X\,dP \tag{10.1}$$

it is necessary that the condition

$$(F(P^+) - F(P^-))\,X(P^+ - P^-) = 0 \tag{10.2}$$

be satisfied for every gap (P^-, P^+) of the chain $\mathfrak{P}$. It turns out that for operators $X \in \mathfrak{S}_2$ this condition is also sufficient.

We have the following general theorem.

THEOREM 10.1 (V. M. BRODSKIĬ [1]). *Let* $\mathfrak{P}$ *be a maximal chain, and* $F(P)$ $(P \in \mathfrak{P})$ *a strongly continuous operator-function with values from* $\mathfrak{R}$.

If X *belongs to the ideal* $\mathfrak{S}_2$ *and the condition* (10.2) *is satisfied, then the integral* (10.1) *converges in the norm of* $\mathfrak{S}_2$, *and*

$$\left| \int_{\mathfrak{P}} F(P)XdP \right|_2 \leqq \sup_{P\in\mathfrak{P}} |F(P)|\, |X|_2. \tag{10.3}$$

Let us clarify that an operator-function $F(P)$ $(P \in \mathfrak{P})$ is said to be *strongly continuous* if it maps every strongly convergent sequence from $\mathfrak{P}$ into a strongly convergent sequence of operators, i.e. if

$$\lim_{P\to P_0} |F(P)f - F(P_0)f| = 0 \quad (P, P_0 \in \mathfrak{P})$$

for every $f \in \mathfrak{H}$.

PROOF. For simplicity we restrict ourselves to the proof of the convergence of the integral (10.1) in $\mathfrak{S}_2$ for the case of a continuous chain $\mathfrak{P}$. The general case of a chain with gaps can be reduced to this special case (cf. V. M. Brodskiĭ [1]).

We first consider the case of a one-dimensional operator X:

$$Xh = (h,f)g \qquad (|f| = |g| = 1).$$

Let $\epsilon > 0$; then there exists a partition $\mathfrak{z}_\epsilon = \{P_j\}_0^m$ of $\mathfrak{P}$ such that for any refinement $\mathfrak{z}' = \{P_j'\}_0^n$ of $\mathfrak{z}_\epsilon$ one has the inequality

$$|(F(P_j') - F(P_{j-1}'))g| < \epsilon \qquad (j = 1, 2, \cdots, n). \tag{10.4}$$

Let us denote by S_ϵ and S' any two Riemann sums of the integral (10.1) corresponding to the partitions $\mathfrak{z}_\epsilon$ and $\mathfrak{z}'$:

$$S_\epsilon = \sum_{j=1}^{m} F(Q_j)X\Delta P_j \qquad (P_{j-1} \leqq Q_j \leqq P_j;\ Q_j \in \mathfrak{P}),$$

$$S' = \sum_{j=1}^{n} F(Q_j')X\Delta P_j' \quad (P_{j-1}' \leqq Q_j' \leqq P_j';\ Q_j' \in \mathfrak{P}).$$

Obviously

$$S' - S_\epsilon = \sum_{j=1}^{n} (F(Q_j') - F(Q_j''))X\Delta P_j' = \sum_{j=1}^{n} \Delta F_j X \Delta P_j,$$

where each of the projectors Q_j'' coincides with one of the projectors

Q_j. Let $\{e_k\}_1^\infty$ be an orthonormal basis of $\mathfrak{H}$. Then

$$|S' - S_\epsilon|_2^2 = \sum_{j=1}^{n} \mathrm{sp}(\Delta P_j' X^* \Delta F_j^* \Delta F_j X \Delta P_j') = \sum_{j=1}^{n} \sum_{k=1}^{\infty} |\Delta F_j X \Delta P_j' e_k|_2^2.$$

Inserting the value of the operator X, we obtain

$$|S' - S_\epsilon|_2^2 = \sum_{j=1}^{n} \sum_{k=1}^{\infty} |(\Delta P_j' e_k, f) \Delta F_j g|^2 = \sum_{j=1}^{n} \sum_{k=1}^{\infty} |(\Delta P_j' e_k, f)|^2 |\Delta F_j g|^2.$$

Thus, by virtue of (10.4) we have

$$|S' - S_\epsilon|_2^2 \leqq \epsilon^2 \sum_{j=1}^{n} \sum_{k=1}^{\infty} |(\Delta P_j' f, e_k)|^2 = \epsilon^2 \sum_{j=1}^{n} |\Delta P_j' f| = \epsilon^2.$$

The convergence of the integral (10.1) for a one-dimensional operator X follows at once. Since every finite-dimensional operator can be represented as a sum of one-dimensional operators, the convergence of the integral (10.1) is also proved for such operators.

Let us now consider a Riemann sum of the integral (10.1) for an arbitrary X. From the relation

$$|S|_2^2 = \sum_{j=1}^{m} \mathrm{sp}(\Delta P_j X^* F^*(Q_j) F(Q_j) X \Delta P_j) \leqq \sum_{j=1}^{m} |F(Q_j)|^2 \mathrm{sp}(\Delta P_j X^* X \Delta P_j)$$

it follows that

$$|S|_2^2 \leqq \sup_{P \in \mathfrak{P}} |F(P)|^2 |X|_2^2. \tag{10.5}$$

We now choose a finite-dimensional operator X_0 which approximates X sufficiently closely in the norm of $\mathfrak{S}_2$, and denote by S_0' and S_0 two Riemann sums of the integral

$$\int_{\mathfrak{P}} F(P) X_0 dP$$

corresponding to two partitions. Then for these partitions we have

$$\left| \sum_{j=1}^{n} F(Q_j') X \Delta P_j' - \sum_{j=1}^{m} F(Q_j) X \Delta P_j \right|_2$$

$$\leqq \left| \sum_{j=1}^{n} F(Q_j')(X - X_0) \Delta P_j' \right|_2 + \left| \sum_{j=1}^{m} F(Q_j)(X - X_0) \Delta P_j \right|_2 + |S_0' - S|_2,$$

and so

$$\left| \sum_{j=1}^{n} F(Q_j') X \Delta P_j - \sum_{j=1}^{m} F(Q_j) X \Delta P_j \right|_2$$

$$\leqq \sup_{P \in \mathfrak{P}} |F(P)| \, |X - X_0|_2 + |S_0' - S_0|_2.$$

This implies the convergence of the integral (10.1). The bound (10.3) obviously follows from the bound (10.5).

The theorem is proved.

The conditions of the theorem are in a certain sense precise. That is, for any operator X which does not belong to $\mathfrak{S}_2$ one can choose a maximal chain $\mathfrak{P}$ and a strongly continuous function $F(P)$ $(P \in \mathfrak{P})$ (with one-dimensional values) which satisfies the condition (10.2), such that the integral (10.1) does not converge even in the weak sense.

CHAPTER V

TRIANGULAR MODELS OF VOLTERRA OPERATORS

Volterra operators first appeared in analysis as integral operators, defined by

$$(*)\qquad (Af)(t) = \int_t^b \mathscr{A}(t,s) f(s)\, ds;$$

here certain restrictions are imposed upon the kernel $\mathscr{A}(t,s)$ $(a \leqq t \leqq s \leqq b)$. We arrive at a more general class of Volterra operators by considering operators of the type (*) in which the kernel $\mathscr{A}(t,s)$ is a matrix kernel of finite or infinite order r, and the integral operator, to be precise, acts in the Hilbert space $L_2^{(r)}(a,b)$.

It is natural to call an operator A of the form (*) a *model* of a given (generally speaking, abstract) Volterra operator $\boldsymbol{A}$, if the operator A is unitarily equivalent to some inessential extension of $\boldsymbol{A}$ (a more general and precise definition of *triangular model* is given in §2.1).

The question arises: to what extent is the model (*) universal?

If we admit integral operators (*) with matrix kernels consisting of generalized functions, the model (*) turns out to be universal (cf. L. A. Sahnovič [**2, 3**]).

We shall be interested in classes of Volterra operators which admit models with kernels $\mathscr{A}(t,s)$ having sufficiently simple analytic characteristics.

We have already indicated a model for a simple dissipative Volterra operator with one-dimensional imaginary component (cf. §8, Chapter I). This model is given by Theorem I.8.1 of M. S. Livšic, which is a special case of more general results by this author regarding models. These initial and important results in the theory of models (M. S. Livšic [**2**]) have been further developed in papers by L. A. Sahnovič [**2, 3**] and others.[1)]

At one time the construction of models for abstract operators played a considerable heuristic, and sometimes a decisive role in various kinds

[1)] In the papers cited, models of bounded operators were also considered. These results will not be elucidated here.

of investigations. Now, however, theorems concerning nonselfadjoint operators which were previously obtained with the help of models can in many cases be deduced, and indeed more elegantly, by general methods.

The theory of triangular models of Volterra operators none the less plays a large role at present. It is used in an essential way in Chapter VII in establishing the basic theorem concerning the functional $h(\cdot;\mathfrak{P})$. By specialization of M. S. Livšic's model to Volterra operators, an important connection between the theory of nonselfadjoint operators and the theory of *canonical differential equations* was realized. This connection is studied and used in Chapters VI and VII and in the Appendix.

We remark that the method discussed here for obtaining triangular models of Volterra operators is based on the theory of the integral of triangular truncation and has a number of advantages when compared with the original method of M. S. Livšic.

§1. Eigen-projector-functions of a completely continuous operator

1. An operator-function $P(t)$, defined on a closed bounded set Ω of real numbers, will be called a *standard projector-function* on Ω if the following conditions are satisfied:

a) the values of $P(t)$ $(t \in \Omega)$ are orthoprojectors;

b) the projector-function $P(t)$ is strictly increasing on Ω, i.e. $t_1 < t_2$ $(t_1, t_2 \in \Omega)$ implies

$$P(t_1) < P(t_2);$$

c) the projector-function $P(t)$ is strongly continuous on Ω, i.e. for any vector $f \in \mathfrak{H}$ the vector-function $P(t)f$ is continuous in the norm of $\mathfrak{H}$ on Ω.

Obviously the range of a standard projector-function is a closed chain.

Let $\mathfrak{P}$ be a chain, and $P(t)$ $(t \in \Omega)$ a standard projector-function. We shall say that $P(t)$ is obtained by a *parametrization* of $\mathfrak{P}$ by means of Ω, if $\mathfrak{P}$ coincides with the range of $P(t)$.

A standard projector-function $P(t)$ $(t \in \Omega)$ is said to be *continuous* if its range forms a continuous chain.

A standard projector-function $P(t)$ $(t \in \Omega)$ is continuous if and only if the set Ω is an interval. A projector-function $P(t)$ $(a \leqq t \leqq b)$ is said to be *absolutely continuous* if for every $f \in \mathfrak{H}$ the (monotone) numerical function $(P(t)f, f)$ $(a \leqq t \leqq b)$ is absolutely continuous.

This means that for any vector $f \in \mathfrak{H}$ and any $\epsilon > 0$ there exists $\delta > 0$ such that for any points t_j, s_j $(j = 1, 2, \cdots, n;\ a \leqq s_1 < t_1 < \cdots < s_n < t_n \leqq b)$

$$\sum_{j=1}^{n} (\Delta P_j f, f) = \sum_{j=1}^{n} ([P(t_j) - P(s_j)]f, f) < \epsilon$$

whenever

$$\sum_{j=1}^{n} (t_j - s_j) < \delta.$$

The following result holds.

THEOREM 1.1. *Every chain* $\mathfrak{P}$ [2] *can be parametrized by an appropriate set* Ω $(\subseteq [0,1])$. *Under this parametrization the chain* $\mathfrak{P}$ *becomes a standard projector-function* $P(t)$ *on the closed bounded set* Ω.

If $\mathfrak{P}$ *is continuous, then its parametrization can be so chosen that* $\Omega = [0,1]$ *and* $P(t)$ $(0 \leqq t \leqq 1)$ *is an absolutely continuous projector-function.*

PROOF. Let us denote by P_0 and Q_0 respectively the minimal and maximal projectors of the chain $\mathfrak{P}$, and by $\{\phi_j\}$ an orthonormal basis of the subspace

$$\mathfrak{H}_0 = \mathfrak{R}(Q_0) \ominus \mathfrak{R}(P_0).$$

With every projector $P \in \mathfrak{P}$ we associate a real number t $(0 \leqq t \leqq 1)$ according to the rule

$$t = t(P) = \sum_j 2^{-j}(P\phi_j, \phi_j) \qquad (P \in \mathfrak{P}). \tag{1.1}$$

We denote the set of all numbers $t = t(P)$ of the form (1.1) by Ω. To distinct projectors $P_1, P_2 \in \mathfrak{P}$ there correspond distinct numbers $t(P_1)$, $t(P_2)$; moreover, if $P_1 < P_2$, then $t(P_1) < t(P_2)$. Indeed, if $P_1 < P_2$, then

$$(P_1\phi_j, \phi_j) \leqq (P_2\phi_j, \phi_j) \quad (j = 1, 2, \cdots), \tag{1.2}$$

and for at least one value of j the inequality is strict [as otherwise we would have

$$|(P_2 - P_1)\phi_j|^2 = ((P_2 - P_1)\phi_j, \phi_j) = 0 \quad (j = 1, 2, \cdots),$$

which would imply $P_2\phi_j = P_1\phi_j$, and thus $P_2 = P_1$].

Thus the correspondence $P \to t(P)$ is one-to-one, and the inverse function $P(t)$ $(t \in \Omega)$, which has meaning, is a strictly increasing projector-function.

Let t_j $(j = 1, 2, \cdots)$ be a monotone sequence of points of Ω with

---

[2] Recall that we are considering only closed chains.

limit t'. The monotone sequence of orthoprojectors $P(t_j)$ $(\in \mathfrak{P};$ $j = 1, 2, \cdots)$ has as limit some orthoprojector P' $(\in \mathfrak{P})$. Thus, there exists a point $t'' \in \Omega$ such that $P(t'') = P'$. From (1.1) it follows that $t'' = t' = \lim t_j$. Hence $t' \in \Omega$ and $P(t') = P'$.

We have thus proved that the set Ω is closed, and that the projector-function $P(t)$ is continuous on Ω.

If the chain $\mathfrak{P}$ is continuous, then obviously the parameter t takes on all values from the interval $[0,1]$; consequently, in this case $\Omega = [0,1]$ and $P(t)$ is continuous.

We shall prove, moreover, that in this case $P(t)$ is absolutely continuous. In fact, it follows from (1.1) that

$$(\Delta P\phi_n, \phi_n) = ([P(t) - P(s)]\phi_n, \phi_n) \leqq 2^n \Delta t \quad (0 \leqq s < t \leqq 1;\ \Delta t = t - s),$$

and so for any vector $g = \sum_{j=1}^{k} c_j \phi_j$

$$(\Delta P g, g) = |\Delta P g|^2 \leqq \sum_{j=1}^{k} |c_j|^2 \sum_{j=1}^{k} |\Delta P \phi_j|^2$$

$$\leqq 2^k |g|^2 \sum_{j=1}^{k} 2^{-j} (\Delta P \phi_j, \phi_j) = 2^k |g|^2 \Delta t \quad (k = 1, 2, \cdots),$$

from which follows the absolute continuity of the function $(P(t)g,g)$. Now let f be any vector from $\mathfrak{H}_0$, let $g = \sum_{k=1}^{N} (f, \phi_k)\phi_k$ and $h = f - g$; then

$$\sum_{j=1}^{n} (\Delta P_j f, f) = \sum_{j=1}^{n} |\Delta P_j f|^2 \leqq 2 \sum_{j=1}^{n} |\Delta P_j g|^2 + 2 \sum_{j=1}^{n} |\Delta P_j h|^2,$$

and consequently

$$\sum_{j=1}^{n} (\Delta P_j f, f) \leqq 2 \sum_{j=1}^{n} (\Delta P_j g, g) + 2 |h|^2. \tag{1.3}$$

Since for N sufficiently large the norm of h is arbitrarily small, and for any N the function $(P(t)g,g)$ is absolutely continuous, it follows from (1.3) that the function $(P(t)f,f)$ $(0 \leqq t \leqq 1)$ is absolutely continuous for $f \in \mathfrak{H}_0$. Since $(\Delta P f^{\perp}, f^{\perp}) = 0$ for any $f^{\perp} \in \mathfrak{H} \ominus \mathfrak{H}_0$, the theorem is proved.

A standard projector-function $P(t)$ $(t \in \Omega)$ is said to be *maximal* if the set of its values forms a maximal chain. If the standard projector-function $P(t)$ is maximal, then for every interval (α, β) contiguous to Ω the projector $P(\beta) - P(\alpha)$ is one-dimensional.

2. A standard projector-function $P(t)$ $(t \in \Omega)$ is said to be an *eigen-projector-function* of an operator A if every subspace $P(t)\mathfrak{H}$ $(t \in \Omega)$ is an invariant subspace of A, i.e. if

$$P(t)AP(t) = AP(t) \qquad (t \in \Omega).$$

THEOREM 1.2. *Let A be a Volterra operator. If A has a maximal eigenchain of rank r, then some inessential extension of A has an absolutely continuous eigen-projector-function of rank r.*

By virtue of Theorem I.3.1, this theorem is a consequence of the following.

THEOREM 1.3. *Suppose that the bordered chain $\mathfrak{P}$ is an **eigenchain** of a bounded linear operator A. If the block of A on each gap (P^-, P^+) of $\mathfrak{P}$ equals zero:*

$$(P^+ - P^-)A(P^+ - P^-) = 0,$$

then A has an inessential extension $\widetilde{A}$ which has an absolutely continuous eigen-projector-function $\widetilde{P}(t)$ $(a \leqq t \leqq b)$ having the same rank as the chain $\mathfrak{P}$.

PROOF. Let the standard projector-function $P(t)$ $(t \in \Omega)$ be obtained by a parametrization of the chain $\mathfrak{P}$. For simplicity we restrict ourselves to the case in which there is only one interval (α, β) contiguous to Ω. Proceeding as in the proof of Theorem I.3.2, we construct a continuous eigen-projector-function $\widetilde{P}(t)$ $(a \leqq t \leqq b)$ of some inessential extension $\widetilde{A}$ of A. Recall that by construction

$$\widetilde{P}(t) = P(t) \quad (t \leqq \alpha); \qquad \widetilde{P}(t) = P(\alpha) + Q(t) \quad (\alpha \leqq t \leqq \beta)$$

and

$$\widetilde{P}(t) = P(t) + P_{\mathfrak{M}} \quad (t \geqq \beta),$$

where $Q(t)$ $(\alpha \leqq t \leqq \beta)$ is any bordered continuous projector-function in the subspace $\mathfrak{M} \oplus \mathfrak{R}(P(\beta) - P(\alpha))$, and $P_{\mathfrak{M}}$ is the orthoprojector onto the subspace $\mathfrak{M} = \widetilde{\mathfrak{H}} \ominus \mathfrak{H}$.

Let r be the rank of the projector-function $P(t)$, and $\{\chi_j\}_1^r$ a generating system of $P(t)$. The projector $P(\beta) - P(\alpha)$ does not annihilate all of the vectors χ_j $(j = 1, 2, \cdots, r)$. In fact, assuming the contrary, we would have for any vector $g \in \mathfrak{R}(P(\beta) - P(\alpha))$

$$\begin{aligned}(P(t)\chi_j, g) &= ([P(\beta) - P(\alpha)]P(t)\chi_j, g) \\ &= (P(t)[P(\beta) - P(\alpha)]\chi_j, g) = 0 \quad (t \in \Omega;\ j = 1, 2, \cdots, r).\end{aligned}$$

This says that the vector g is orthogonal to the linear hull of the vectors $P(t)\chi_j$, which contradicts the assumption that $\{\chi_j\}_1^r$ is a generating system.

Suppose for definiteness that $(P(\beta) - P(\alpha))\chi_1 = \psi \neq 0$. We choose the complete operator-function $Q(t)$ $(\alpha \leqq t \leqq \beta)$, whose choice was in no way restricted, so as to have ψ as a generating vector. Then

$$\widetilde{P}(t)\chi_1 = P(\alpha)\chi_1 + Q(t)\chi_1 \quad (\alpha \leqq t \leqq \beta).$$

Since

$$Q(t)\chi_1 = Q(t)\,(P(\beta) - P(\alpha))\chi_1 = Q(t)\psi,$$

the linear hull of all the vectors

$$\widetilde{P}(t)\chi_1 - P(\alpha)\chi_1 \quad (\alpha \leqq t \leqq \beta)$$

is dense in $\mathfrak{M}$.

It follows that the system $\{\chi_j\}_1^r$ is a generating system for the projector-function $\widetilde{P}(t)$. Taking into account that the rank of $\widetilde{P}(t)$ is not less than the rank r of $P(t)$, we find that it equals precisely r.

Finally, by Theorem 1.1, we can if necessary change the parametrization of the chain $\{\widetilde{P}(t)\}$ $(a \leqq t \leqq b)$ so as to obtain an absolutely continuous projector-function.

The theorem is proved.

3. To conclude this section we shall consider another definition of the triangular integral, a definition using parametrization.

Let Ω be any closed bounded set of real numbers. A set of points

$$a = t_0 < t_1 < \cdots < t_m = b \quad (a = \min_{t \in \Omega} t;\ b = \max_{t \in \Omega} t)$$

belonging to Ω is called a *δ-partition* of Ω if for each $k = 1, 2, \cdots, m$ either the numbers t_k and t_{k-1} are the end points of an interval contiguous to Ω, or $t_k - t_{k-1} < \delta$.

Suppose that $\Phi(t)$ and $\Psi(t)$ are operator-functions defined on the closed set Ω, and the values of $\Phi(t)$ belong to some s.n. ideal $\mathfrak{S}$.

If the sums

$$S_\tau(\delta) = \sum_{j=1}^{m} \Phi(\tau_k)\left[\Psi(t_k) - \Psi(t_{k-1})\right] \quad (t_{k-1} \leqq \tau_k \leqq t_k;\ \tau_k \in \Omega)$$

have a limit A in the norm of $\mathfrak{S}$ as $\delta \to 0$, where $\{t_j\}_0^m$ is an arbitrary δ-partition of the set Ω, we write this limit in the form

$$A = \int_\Omega \Phi(t)\,d\Psi(t) \tag{1.4}$$

and say that the integral on the right side converges in the norm of $\mathfrak{S}$.[3)]

Let $P(t)$ $(t \in \Omega)$ be a standard projector-function, and $\mathfrak{P}$ the chain consisting of its values. We denote by $F(P)$ $(P \in \mathfrak{P})$ an operator-function with values in $\mathfrak{S}$.

1. *If the integral*

$$\int_\Omega F(P(t))\, dP(t) \tag{1.5}$$

converges in the norm of $\mathfrak{S}$, *then the integral*

$$\int_\mathfrak{P} F(P)\, dP \tag{1.5'}$$

converges in the same norm, and

$$\int_\mathfrak{P} F(P)\, dP = \int_\Omega F(P(t))\, dP(t). \tag{1.6}$$

If the operator-function $F(P(t))$ is continuous on Ω in the norm of $\mathfrak{S}$, then, conversely, the convergence of the integral (1.5′) *implies the convergence of the integral* (1.5), *and so by the first part of the theorem* (1.6) *holds.*

The first assertion is obvious.

Let us prove the second assertion. Suppose that the integral (1.5′) converges in the norm of $\mathfrak{S}$. For any $\epsilon > 0$ there exists a partition $\mathfrak{z}_\epsilon = \{P_j = P(\xi_j)\}_0^r$ of the chain $\mathfrak{P}$ such that for all refinements $\mathfrak{z}$ $(\mathfrak{P} \supset \mathfrak{z} \supset \mathfrak{z}_\epsilon)$ one has

$$| S_Q(\mathfrak{z}) - A |_\mathfrak{S} < \epsilon/2,$$

where $S_Q(\mathfrak{z})$ is a Riemann-Stieltjes sum for the integral (1.5′) corresponding to the partition $\mathfrak{z}$.

The function $F(P(t))$ is uniformly continuous; hence for any $\epsilon > 0$ there exists $\delta > 0$ such that

$$| F(P(t')) - F(P(t''))|_\mathfrak{S} < \epsilon/4r$$

whenever $|t' - t''| < \delta$ $(t', t'' \in \Omega)$.

Let $\delta_0 = \min\{\delta, \xi_j - \xi_{j-1}\}$. Suppose we are given two partitions of Ω, the first one $a = t_0 < t_1 < \cdots < t_m = b$ an arbitrary δ-partition with $\delta < \delta_0$, and the second of which is obtained from the first by adding to it the points ξ_j $(j = 1, 2, \cdots, r)$.

Let us denote by $S_\tau(\delta)$ any Riemann-Stieltjes sum of (1.5) corre-

---

3) Precisely this concept of the integral was first introduced by M. S. Brodskiĭ [**3—5**].

sponding to the first partition, and by S any Riemann-Stieltjes sum corresponding to the second partition. We choose S so that its terms corresponding to intervals (t_{j-1}, t_j) which do not contain a point ξ_j $(j = 1, 2, \cdots, r)$ are the same as those from the sum $S_\tau(\delta)$.

Obviously, each interval (t_{j-1}, t_j) $(j = 1, 2, \cdots, m)$ contains at most one point ξ_j. Consequently, the difference $S - S_\tau(\delta)$ is the sum of at most $2r$ terms of the form

$$(F(P(\tau'')) - F(P(\tau')))(P(\tau'') - P(\tau')),$$

where $|\tau'' - \tau'| < \delta$. Thus

$$|S - S_\tau(\delta)|_{\mathfrak{S}} < \epsilon/2. \tag{1.7}$$

Since the sum S can be regarded as a Riemann-Stieltjes sum $S_Q(\mathfrak{z})$ of (1.5′) corresponding to a certain partition $\mathfrak{z} \supset \mathfrak{z}_\epsilon$ of the chain $\mathfrak{P}$, it follows from (1.7) that

$$|S_\tau(\delta) - A|_{\mathfrak{S}} < \epsilon/2 + |S_Q(\mathfrak{z}) - A|_{\mathfrak{S}} < \epsilon.$$

The result is proved.

§2. Triangular model of a Volterra operator

1. For brevity we denote by $L_2^{(r)}$ $(1 \leqq r \leqq \infty)$ the Hilbert space $L_2^{(r)}(Q)$, where $Q = [0, 1]$ (see §7, Chapter IV). Thus, an element $f \in L_2^{(r)}$ is an r-dimensional vector-function $f = \{f_\nu(t)\}_1^r$ with measurable components $f_\nu(t)$ $(0 \leqq t \leqq 1)$ such that

$$|f|^2 = \int_0^1 \sum_{\nu=1}^r |f_\nu(t)|^2 dt < \infty.$$

For the scalar product of the elements $f, g \in L_2^{(r)}$ we will have

$$(f, g) = \int_0^1 g^*(t) f(t)\, dt = \int_0^1 \sum_{\nu=1}^r f_\nu(t) \overline{g_\nu(t)}\, dt.$$

Let $\hat{P}(t)$ $(0 \leqq t \leqq 1)$ be the *truncation projector-function*, defined by the conditions $\hat{P}(0) = 0$, $\hat{P}(1) = I$ and

$$(\hat{P}(s) f)(t) = \begin{cases} f(t), & 0 \leqq t < s, \\ 0, & s < t \leqq 1 \end{cases} \qquad (0 < t < 1).$$

Obviously this projector-function is continuous and maximal.

Let $\boldsymbol{A}$ be some (abstract) Volterra operator acting in a Hilbert space $\mathfrak{H}$. A (concrete) Volterra operator A, acting in $L_2^{(r)}$ and having $\hat{P}(t)$ as an eigen-projector-function, is called a *triangular model* of the operator $\boldsymbol{A}$, if A is unitarily equivalent to $\boldsymbol{A}$ or to an inessential extension of $\boldsymbol{A}$.

The preceding results enable us to prove the existence of a triangular model for every Volterra operator. To do this it remains to make use of the following result, which is well known from the theory of spectral types of selfadjoint operators (see A. I. Plesner [**1**] or A. I. Plesner and V. A. Rohlin [**1**] and L. A. Sahnovič [**2, 3**]).

Let $\mathfrak{H}$ *be a separable Hilbert space, and* $P(t)$ $(0 \leqq t \leqq 1)$ *an absolutely continuous projector-function in* $\mathfrak{H}$ *of rank* r $(\leqq \infty)$. *Then there exists a unitary mapping* U *of* $\mathfrak{H}$ *onto some subspace* $\mathfrak{L}$ *of* $L_2^{(r)}$ *by which* $P(t)$ $(0 \leqq t \leqq 1)$ *goes over into the truncation projector-function* $\hat{P}(t)$ *on* $\mathfrak{L}$, *so that*

$$UP(t)f = \hat{P}(t)\,Uf \qquad (0 \leqq t \leqq 1) \tag{2.1}$$

for every $f \in \mathfrak{H}$.

We have the following theorem.

THEOREM 2.1. *Every Volterra operator of rank* r *has a triangular model acting in the space* $L_2^{(r)}$.

PROOF. Let $\boldsymbol{A}$ be a Volterra operator in an abstract separable Hilbert space $\mathfrak{H}$, and let $\mathfrak{P} = \{P\}$ be a maximal eigenchain of $\boldsymbol{A}$ of rank r.

By Theorem 1.3 there always exists an inessential extension $\widetilde{\boldsymbol{A}}$ of $\boldsymbol{A}$ in some $\widetilde{\mathfrak{H}} \supset \mathfrak{H}$ which has an absolutely continuous eigen-projector-function $\widetilde{P}(t)$ $(0 \leqq t \leqq 1)$ of the same rank r as $\mathfrak{P}$. If U is the unitary mapping of $\widetilde{\mathfrak{H}}$ into $L_2^{(r)}$ under which $\widetilde{P}(t)$ $(0 \leqq t \leqq 1)$ goes over into the truncation operator $\hat{P}(t)$ $(0 \leqq t \leqq 1)$, we conclude that $\widetilde{\boldsymbol{A}}$ is unitarily equivalent to an operator A_1 which acts in the subspace $\mathfrak{L} = U\mathfrak{H}$. By (2.1) the subspace $\mathfrak{L}$ (and hence $\mathfrak{L}^{\perp}$ also) is invariant with respect to all the projectors $\hat{P}(t)$ $(0 \leqq t \leqq 1)$.

The relation

$$\hat{P}(t)A_1\hat{P}(t)f = A_1\hat{P}(t)f \qquad (f \in \mathfrak{L};\ 0 \leqq t \leqq 1)$$

remains valid for the inessential extension $A = \widetilde{A}_1$ of the operator A_1 to all of $L_2^{(r)}$. Thus, A is a Volterra operator which has $\hat{P}(t)$ $(0 \leqq t \leqq 1)$ as an eigen-projector-function.

The operator A is unitarily equivalent to some inessential extension of $\widetilde{\boldsymbol{A}}$, which is in turn an inessential extension of the original operator $\boldsymbol{A}$.

The theorem is proved.

2. Let us recall that if A is a Hilbert-Schmidt operator acting in $L_2^{(r)}$, then (see [GK], §9, Chapter III) there is a corresponding measurable matrix kernel $\mathscr{A}(t,s) = \|a_{\mu\nu}(t,s)\|_1^r$ such that

$$(2.2)\qquad (Af)(t) = \int_0^1 \mathscr{A}(t,s) f(s)\, ds \qquad (f \in L_2^{(r)}),$$

and

$$\int_0^1 \int_0^1 \operatorname{sp}(\mathscr{A}^* \mathscr{A})\, dt ds = \int_0^1 \int_0^1 \sum_{\mu,\nu=1}^{r} |a_{\mu\nu}(t,s)|^2 dt ds = |A|_2^2 \ (< \infty).$$

One establishes directly that the truncation projector-function $\hat{P}(t)$ $(0 \leqq t \leqq 1)$ will be an eigen-projector-function of the operator A defined by (2.2) if and only if

$$(2.3)\qquad \mathscr{A}(t,s) = 0 \text{ almost everywhere in the triangle } 0 \leqq s \leqq t \leqq 1.$$

Thus, the general form of a Hilbert-Schmidt operator in $L_2^{(r)}$ which has $\hat{P}(t)$ as an eigen-projector-function is given by (2.2) with the supplementary condition (2.3).

Taking account of the general form of a Hilbert-Schmidt Volterra operator, we arrive at the following theorem of L. A. Sahnovič [**5**] in a sharper form.[4)]

THEOREM 2.2. *Every Hilbert-Schmidt Volterra operator* $\boldsymbol{A}$ *of rank* r *has as a triangular model an integral operator* A, *which acts in the space* $L_2^{(r)}$ *according to the formula*

$$(2.4)\qquad (Af)(t) = \int_t^1 \mathscr{A}(t,s) f(s)\, ds,$$

where

$$\mathscr{A}(t,s) = \|a_{\mu\nu}(t,s)\|_1^r \qquad (0 \leqq t \leqq s \leqq 1)$$

is a Hilbert-Schmidt matrix kernel, i.e.

$$\int_0^1 \int_t^1 \sum_{\mu,\nu=1}^{r} |a_{\mu\nu}(t,s)|^2 ds dt < \infty.$$

3. Let $\boldsymbol{A}$ be a Volterra operator with imaginary component $\boldsymbol{A}_{\mathscr{J}} \in \mathfrak{S}_2$. It is remarkable that this condition alone enables us to establish that

4) The notion of the rank of a Volterra operator and related concepts did not appear in Sahnovič's work.

$\boldsymbol{A}$ has a model of the form (2.4). Thus, we will obtain a second proof of Theorem I.10.2 of L. A. Sahnovič. Indeed, the theorem was first obtained by this route.

Let A be a triangular model, acting in $L_2^{(r)}$, of the operator $\boldsymbol{A}$. Then $A_{\mathscr{I}}$ is a selfadjoint Hilbert-Schmidt operator, and hence acts according to the formula

$$(2.5)\qquad (A_{\mathscr{I}}\, f)(t) = \int_0^1 \mathscr{A}_{\mathscr{I}}\,(t,s) f(s)\, ds \qquad (f \in L_2^{(r)}),$$

where $\mathscr{A}_{\mathscr{I}}\,(t,s) = \| h_{\mu\nu}(t,s) \|_1^r$ is some Hermitian Hilbert-Schmidt matrix kernel.

The operator A is expressed in terms of the kernel $\mathscr{A}_{\mathscr{I}}\,(t,s)$ by

$$(2.6)\qquad (Af)(t) = 2i \int_t^1 \mathscr{A}_{\mathscr{I}}\,(t,s) f(s)\, ds \qquad (f \in L_2^{(r)}).$$

In fact, the right side obviously defines a Hilbert-Schmidt operator which has $\hat{P}(t)$ $(0 \leqq t \leqq 1)$ as a (continuous) eigen-projector-function and has imaginary component defined by (2.5). Consequently, it is a Volterra operator, and (2.6) holds by virtue of Theorem I.6.1.

It follows at once from (2.6) that A, and along with it $\boldsymbol{A}$, is a Hilbert-Schmidt operator.

Finally, the operator $A_{\mathscr{R}}$ obviously has the form

$$(A_{\mathscr{R}}\, f)(t) = \int_0^1 \mathscr{A}_{\mathscr{R}}\,(t,s) f(s)\, ds \qquad (f \in L_2^{(r)}),$$

where

$$\mathscr{A}_{\mathscr{R}}\,(t,s) = i\, \mathrm{sign}(s-t) \mathscr{A}_{\mathscr{I}}\,(t,s).$$

It follows that

$$\int_0^1 \int_0^1 \mathrm{sp}(\mathscr{A}_{\mathscr{R}}^* \mathscr{A}_{\mathscr{R}}\,)\, dt ds = \int_0^1 \int_0^1 \mathrm{sp}(\mathscr{A}_{\mathscr{I}}^* \mathscr{A}_{\mathscr{I}}\,)\, dt ds.$$

This relation is equivalent to

$$| \boldsymbol{A}_{\mathscr{I}} \,|_2^2 = | \boldsymbol{A}_{\mathscr{R}} \,|_2^2.$$

We remark that (2.6) says the same thing as does the relation

$$A = 2i \int_0^1 \hat{P}(t) A_{\mathscr{I}}\, d\hat{P}(t).$$

This follows from Theorem I.6.1; however, we can see it directly.

4. A comparison of Theorem I.7.3 (of G. È. Kisilevskiĭ) and Theorem 2.2 leads to the following conclusion.

THEOREM 2.3. *Every dissipative Volterra operator* $\boldsymbol{A}$ *with nuclear imaginary component has a triangular model* A, *defined in the space* $L_2(0,1)$ *of scalar functions by*

$$(Af)(t) = 2i\int_t^1 \mathscr{K}(t,s)f(s)\,ds \qquad (f\in L_2(0,1)),$$

where $\mathscr{K}(t,s)$ $(0 \leqq s,t \leqq 1)$ *is a Hermitian-positive Hilbert-Schmidt kernel for which*

$$\overline{\lim_{h\to 0}}\frac{1}{4h^2}\int_a^b\int_a^b [2h-|t-s|\,]_+\mathscr{K}(t,s)\,dtds < \infty .$$

Let us clarify that the last condition expresses the nuclearity of the imaginary component of A (see [GK], Theorem 10.1, Chapter III); we also recall the notation: $y_+ = \max(y,0)$.

Of course we have the converse:

If an operator $\boldsymbol{A}$ *admits a model* A *of the form indicated, then it is a dissipative Volterra operator with nuclear imaginary component.*

5. Let $\boldsymbol{A}$ be a Volterra operator, and A a triangular model of it in some space $L_2^{(r)}(0,1)$.

Theorem III.4.1 enables us to state that if the condition $\boldsymbol{A}_{\mathscr{J}} \in \mathfrak{S}_\omega$ is satisfied, then the model A is in a certain sense a generalized integral operator.

In fact, let K_n $(n = 1,2,\cdots)$ be a sequence of finite-dimensional selfadjoint operators in $L_2^{(r)}$ such that

$$\lim_{n\to\infty} |\, K_n - A_{\mathscr{J}}\,|_\omega = 0. \tag{2.7}$$

There correspond to the operators K_n degenerate Hermitian matrix kernels $\mathscr{K}_n(t,s)$ such that

$$(K_nf)(t) = \int_0^1 \mathscr{K}_n(t,s)f(s)\,ds \qquad (f\in L_2^{(r)}).$$

We form the operators A_n, setting

$$(A_nf)(t) = 2i\int_t^1 \mathscr{K}_n(t,s)f(s)\,ds \qquad (f\in L_2^{(r)}).$$

By Theorem III.4.1 the sequence of operators A_n tends to the operator A in the uniform norm:

$$Af = 2i \lim_{n\to\infty} \int_t^1 \mathscr{K}_n(t,s) f(s)\,ds \qquad (f \in L_2^{(r)}).$$

Let us put

$$f_\tau(t) = \begin{cases} 1, & 0 \leqq t \leqq \tau, \\ 0, & \tau < t \leqq 1. \end{cases}$$

On the basis of (2.6) we can assert that the sequence of vector-functions

$$F_n(t,s) = \int_0^s \mathscr{K}_n(t,r)\,dr \left(= \int_0^1 \mathscr{K}_n(t,r) f_s(r)\,dr \right)$$

in $L_2^{(r)}(0,1)$ converges strongly (i.e., relative to the norm) and indeed uniformly in $s \in [0,1]$ to some vector-function $F(t,s)$. Obviously we will have

$$|F(t,s_2) - F(t,s_1)| < C|s_2 - s_1|^{1/2} \qquad (0 \leqq s_1, s_2 \leqq 1).$$

In a certain sense we may regard A as an integral operator with the kernel

$$2i\mathscr{K}(t,s) = 2i \frac{d}{ds} F(t,s) \qquad (0 \leqq t \leqq s \leqq 1).$$

Other triangular generalized integral models can be found in the original papers of L. A. Sahnovič [2, 3].

§3. Canonical triangular model of a Volterra operator

1. The results discussed in §2 were preceded by the investigations of M. S. Livšic [2] in the theory of bounded nonselfadjoint operators with nuclear imaginary components, in which, in particular, a model of a specialized type was obtained.

In order to formulate Livšic's results for the case of Volterra operators, we need some new definitions.

We shall call a matrix-function $\mathscr{H}(t) = \| h_{jk}(t) \|_1^r$ $(0 \leqq t \leqq 1)$ of finite or infinite order r with elements from $L_1(0,1)$ an *H-matrix*, if for almost all t the matrix $\mathscr{H}(t)$ is a Hermitian matrix which defines a bounded operator in the space $l_2^{(r)}$.[5)]

We shall call an H-matrix $\mathscr{H}(t)$ $(0 \leqq t \leqq 1)$ an H-matrix of *positive type* if the following two conditions are satisfied:

1) For any $t \in [0,1]$ the Hermitian form $\xi^* \mathscr{H}(t)\xi$ is nonnegative, i.e.

---

5) We denote by $l_2^{(r)}$ $(r \leqq \infty)$ the r-dimensional Hilbert space of sequences $\xi = \{\xi_j\}_1^r$ of complex numbers with the norm $|\xi|^2 = \sum_j |\xi_j|^2$.

(3.1) $$\xi^*\mathscr{A}(t)\xi=\sum_{j,k=1}^{r}h_{jk}(t)\xi_k\overline{\xi_j}\geqq 0.$$

2) For each $\xi\in l_2^{(r)}$ this Hermitian form has an average over the entire interval, and the averaged form is strictly positive, i.e.[6)]

(3.2) $$\int_0^1\xi^*\mathscr{A}(s)\xi\,ds>0\quad\text{for }\xi\neq 0.$$

We shall call a skew-Hermitian matrix $\mathscr{Y}$ $(\mathscr{Y}^*=-\mathscr{Y})$ which generates a bounded operator in $l_2^{(r)}$ a *signature* matrix, if $\mathscr{Y}^2=-I$. If $\mathscr{Y}$ is a signature matrix, then $i\mathscr{Y}$ is Hermitian and $(i\mathscr{Y})^2=I$. Hence the spectrum of the matrix $i\mathscr{Y}$ consists of at most two points 1 and -1. It follows that every signature matrix is unitarily equivalent to a matrix of the form $i\|\epsilon_j\delta_{jk}\|_1^r$, where $\epsilon_j=\pm 1$.

We now restrict ourselves to consideration of the case $r=n<\infty$.

Let $\mathscr{A}(t)$ be an H-matrix of positive type of order n. For any t from the interval $[0,1]$ there exists a unique nonnegative matrix whose square equals $\mathscr{A}(t)$. We shall denote this matrix by $\mathscr{A}^{1/2}(t)$. The columns of $\mathscr{A}^{1/2}(t)$ form vector-functions from $L_2^{(n)}(0,1)$.[7)]

---

6) Basically, only H-matrices of finite order will be considered in the sequel. In this case the average always exists. We shall not dwell here on a deeper analysis of the meaning of condition 2) for $r=\infty$.

7) Since the jth diagonal element of the matrix $\mathscr{A}(t)$ is equal to the sum of the squared moduli of the elements of the jth column of the matrix $\mathscr{A}^{1/2}(t)$ and belongs to $L_1(0,1)$, to justify the assertion made above it remains to clarify why the elements of $\mathscr{A}^{1/2}(t)$ are measurable. By the well-known Lusin theorem there exists, for any $\epsilon>0$, a closed set $\Omega\subseteq[0,1]$ with measure $>1-\epsilon$ on which the H-matrix $\mathscr{A}(t)$ will be continuous. We choose the number R so that $|\mathscr{A}(t)|<R/2$ $(t\in\Omega)$.

Let us consider the contour γ in the complex plane consisting of the interval $[-iR,iR]$ and the semicircle $z=Re^{i\theta}$ $(-\pi/2\leqq\theta\leqq\pi/2)$. It is easy to show that

(*) $$\mathscr{A}^{1/2}(t)=\frac{1}{2\pi i}\oint_\gamma\sqrt{\lambda}(\mathscr{A}(t)-\lambda I)^{-1}d\lambda\quad(t\in\Omega).$$

The uniform convergence of the contour integral is assured by the fact that

$$|(\mathscr{A}(t)-irI)^{-1}|\leqq 1/|r|\qquad(-\infty<r<\infty).$$

Let us denote by γ_δ the contour which is obtained from γ by deleting the interval $[-i\delta,i\delta]$. Since, with γ replaced by γ_δ, the integral in (*) yields a continuous function of $t\in\Omega$, letting $\delta\to 0$ we conclude that the matrix-function $\mathscr{A}^{1/2}(t)$ is continuous on Ω. This proves the measurability of the matrix-function $\mathscr{A}^{1/2}(t)$.

We note that we have at the same time proved the following result: if the H-matrix $\mathscr{A}(t)$ $(0\leqq t\leqq 1)$ is continuous, then so is the H-matrix $\mathscr{A}^{1/2}(t)$.

Let $\mathscr{A}(t)$ be an H-matrix of positive type, and $\mathscr{Y}$ a signature matrix. We form the operator A_M,[8] defined in the space $L_2^{(n)}$ by

$$(A_M f)(t) = \mathscr{A}^{1/2}(t)\,\mathscr{Y}\int_t^1 \mathscr{A}^{1/2}(s)f(s)\,ds \quad (f \in L_2^{(n)}). \tag{3.3}$$

It is easily seen that A_M is a Volterra operator which has the eigen-projector-function $\hat{P}(t)$ $(0 \leqq t \leqq 1)$. The dimension of its imaginary component does not exceed n; this component is defined by

$$(A_{M\mathscr{I}} f)(t) = \frac{1}{2i}\mathscr{A}^{1/2}(t)\mathscr{Y}\int_0^1 \mathscr{A}^{1/2}(s)f(s)\,ds, \tag{3.4}$$

and the real component, by

$$(A_{M\mathscr{R}} f)(t) = -\frac{1}{2}\mathscr{A}^{1/2}(t)\mathscr{Y}\int_0^1 \operatorname{sign}(t-s)\,\mathscr{A}^{1/2}(s)f(s)\,ds. \tag{3.5}$$

We shall call the operator A_M *the canonical model of an operator.*

Let A be a Volterra operator in $\mathfrak{H}$. If a canonical model A_M is a triangular model for A, we shall call A_M a *canonical triangular model* of A.

2. We have the following theorem.

THEOREM 3.1 (M. S. LIVŠIC [**2**]). *Every simple Volterra operator A with an n-dimensional imaginary component $(n < \infty)$ has a canonical triangular model in the space $L_2^{(n)}$.*

PROOF.[9] By virtue of Theorem 2.2, it is obviously sufficient to prove this theorem for the case of an operator A_0 having the same simple part as the integral operator A, defined in $L_2^{(r)}$ by

$$(Af)(t) = 2i\int_t^1 \mathscr{K}(t,s)f(s)\,ds, \tag{3.6}$$

where

$$\mathscr{K}(t,s) = \sum_{j=1}^n \epsilon_j \psi_j(t)\psi_j^*(s);$$

$\epsilon_j = \pm 1$ $(j = 1, 2, \cdots, n)$, and $\{\psi_j(t)\}_1^n$ is some linearly independent system of vector-functions.

---

[8] The index M comes from the word "model".

[9] The arguments which follow have many points of contact with arguments of M. S. Livšic which appear at the end of his paper [**2**], and with arguments presented by L. A. Sahnovič in his doctoral dissertation, where the question of the connection between two models is also investigated.

In this case the operator $A_{\mathscr{J}}$ has the form

$$(A_{\mathscr{J}} f)(t) = \int_0^1 \mathscr{K}(t,s) f(s)\, ds.$$

We associate with A the signature matrix $\mathscr{Y} = i \| \epsilon_j \delta_{jk} \|_1^n$ and the H-matrix

$$\mathscr{A}(t) = 2 \| \psi_j^*(t) \psi_k(t) \|_1^n \qquad (0 \leqq t \leqq 1).$$

Since

$$\frac{1}{2} \int_0^1 \mathscr{A}(t)\, dt = \left\| \int_0^1 \psi_j^*(t) \psi_j(t)\, dt \right\|_1^n \tag{3.6$'$}$$

is the Gram matrix of the system of vector-functions $\{\psi_j(t)\}_1^n$, the H-matrix $\mathscr{A}(t)$ is of positive type.

We denote by $\mathscr{L}$ the set of all vector-functions $f(t) \in L_2^{(r)}$ of the form

$$f(t) = \sum_{j=1}^n a_j(t) \psi_j(t) \qquad (0 \leqq t \leqq 1), \tag{3.7}$$

where the $a_j(t)$ $(j = 1, 2, \cdots, n)$ are arbitrary continuous functions.

The closure $\overline{\mathscr{L}}$ of $\mathscr{L}$ in $L_2^{(r)}$ obviously contains the range $\mathfrak{R}(A)$ of A and is therefore invariant with respect to A. It is also obvious that $\overline{\mathscr{L}}$ contains the linear hull of the set of vector-functions of the form

$$\hat{P}(s) \psi_j(t) \qquad (0 \leqq s \leqq 1;\ j = 1, 2, \cdots, n).$$

Therefore (see result 2, Chapter I, §7) the operator A_0 induced in $\overline{\mathscr{L}}$ by A has the same simple part as A.

Let us denote by U the (a priori multivalued) operator which associates with each vector-function $f \in \mathscr{L}$ of the form (3.7) the vector-function $\xi(t) = \{\xi_j(t)\}_1^n \in L_2^{(n)}$ defined by

$$\xi(t) = \mathscr{A}^{1/2}(t) a(t) \qquad (\boldsymbol{a}(t) = \{a_j(t)\}_1^n;\ 0 \leqq t \leqq 1). \tag{3.8}$$

Taking (3.7) and (3.8) into account, we easily obtain

$$\int_0^1 \xi^*(t) \xi(t)\, dt = \int_0^1 a^*(t) \mathscr{A}(t) a(t)\, dt = \int_0^1 f^*(t) f(t)\, dt.$$

It follows that U maps $\mathscr{L}$ (and, following **continuous** extension, $\overline{\mathscr{L}}$) isometrically onto some lineal (onto some subspace) of the space $L_2^{(n)}$. We put $\mathfrak{M} = U\overline{\mathscr{L}}$; it is easily seen that the subspace $\mathfrak{M}$ is invariant with respect to all of the projectors $\hat{\boldsymbol{P}}^{(n)}(t)$ $(0 \leqq t \leqq 1)$.[10)]

[10)] $\hat{P}^{(n)}(t)$ denotes the projector-function of "truncation" in the space $L_2^{(n)}$.

If $g(t) = (A_0 f)(t)$, where $f \in \mathscr{L}$, then inserting the expression (3.7) for f into (3.6), we find that

$$g(t) = 2i \int_t^1 \sum_{j,k=1}^{n} \epsilon_j \psi_j(t) \psi_j^*(s) a_k(s) \psi_k(s)\, ds,$$

which according to (3.8) can be rewritten in the form

$$g(t) = \sum_{j=1}^{n} b_j(t) \psi_j(t), \tag{3.9}$$

where $b(t) = \{b_j(t)\}_1^n$ is the vector-function defined by

$$b(t) = \mathscr{Y} \int_t^1 \mathscr{A}(s) a(s)\, ds = \mathscr{Y} \int_t^1 \mathscr{A}^{1/2}(s) \xi(s)\, ds. \tag{3.10}$$

If, on the other hand, we put $\eta(t) = Ug(t)$, then by (3.9) and the definition of the operator U we will have

$$\eta(t) = \mathscr{A}^{1/2}(t) b(t).$$

Taking account of (3.10), we finally obtain

$$\eta(t) = \mathscr{A}^{1/2}(t) \mathscr{Y} \int_t^1 \mathscr{A}^{1/2}(s) \xi(s)\, ds. \tag{3.11}$$

The foregoing transformations prove that the operator A_0 is unitarily equivalent to the operator B_0 which is defined on $\mathfrak{M}$ by (3.11): $B_0 = U^{-1} A_0 U$.

We shall show that the operator B, defined on all of $L_2^{(n)}$ by

$$(Bf)(t) = \mathscr{A}^{1/2}(t) \mathscr{Y} \int_t^1 \mathscr{A}^{1/2}(s) f(s)\, ds, \tag{3.12}$$

is an inessential extension of the operator B_0. To do so, we note that the unitary equivalence of A_0 and B_0 implies that the inessential extension $\widetilde{B}_0$ of B_0 to all of $L_2^{(n)}$ is a Volterra operator which has $\hat{P}^{(n)}(t)$ $(0 \leqq t \leqq 1)$ as an eigen-projector-function and has imaginary component

$$(\widetilde{B}_0)_{\mathscr{J}} = A_{\mathscr{J}} = \sum_{j=1}^{n} \epsilon_j (\cdot, \chi_j) \chi_j,$$

where $\chi_j = U\psi_j$ $(j = 1, 2, \cdots, n)$. On the other hand, B is also a Volterra operator which has $\hat{P}^{(n)}(t)$ $(0 \leqq t \leqq 1)$ as an eigen-projector-function, and $B_{\mathscr{J}} = A_{\mathscr{J}}$. Thus, by Theorem I.6.1 we have $B = \widetilde{B}_0$. The theorem is proved.

3. This theorem is easily generalized to the case of a Volterra operator with an infinite-dimensional imaginary component $A_{\mathscr{J}}$.

In this case the model (3.11) is to be considered in $L_2^{(\infty)}$ and it is convenient to assume that the vector-functions $\{\psi_j(t)\}_1^\infty$ constitute an orthogonal system.

As before, the H-matrix $\mathscr{A}(t)$ is defined by

$$\mathscr{A}(t) = 2\,\| \psi_j^*(t)\psi_k(t)\|_1^\infty,$$

and the signature matrix $\mathscr{Y}$ by $\mathscr{Y} = i\,\|\epsilon_j\delta_{jk}\|_1^\infty$. Since

$$\mathrm{sp}\int_0^1 \mathscr{A}(t)\,dt = \sum_{j=1}^\infty \int_0^1 \psi_j^*(t)\psi_j(t)\,dt = \sum_{j=1}^\infty |\psi_j|^2,$$

and the last sum coincides with the number $|A_{\mathscr{J}}|_1$, it follows that $\mathrm{sp}\,\mathscr{A}(t)$ exists for almost all t and that

$$\int_0^1 \mathrm{sp}\,\mathscr{A}(t)\,dt = |A_{\mathscr{J}}|_1.$$

Thus, the H-matrix $\mathscr{A}(t)$ has a trace for almost all t, and so for almost all t the matrix $\mathscr{A}^{1/2}(t)$ will define a Hilbert-Schmidt operator in the space l_2 of all vectors $\xi = \{\xi_j\}_1^\infty$ $(\sum_j |\xi_j|^2 < \infty)$.

Following these supplementary remarks concerning the case $n = r(A_{\mathscr{J}}) = \infty$, it is easy to extend to this case the arguments which led to Theorem 3.1.

We mention that M. S. Livšic obtained Theorem 3.1 for Volterra operators from a completely different circle of ideas. In his investigations he started out with the concept of the characteristic matrix-function of a nonselfadjoint operator (see §4) which he had introduced, and used the deep results of V. P. Potapov [1] in the theory of analytic matrix-functions (see the Appendix, §3).

REMARK 3.1. The method which has been used to obtain the model (3.11) enables us to make the following important addition to Theorem 3.1 of M. S. Livšic.

One can always arrange that for all t the rank $r_{\mathscr{A}}(t)$ of the H-matrix $\mathscr{A}(t)$ in the model (3.11) *does not exceed the rank $r(A)$ of the operator A.*

In particular, if A is a dissipative operator, one can arrange that $r_{\mathscr{A}}(t) = 1$.

In fact, this assertion requires proof only for the case $r < n$. In this case it follows from the formula for $\mathscr{A}(t)$ and from the fact that the vectors $\psi_k(t)$ $(k = 1, 2, \cdots, n)$ are r-dimensional that any minor of $\mathscr{A}(t)$ of order $r + 1$ equals zero. The second assertion now follows from Theorem 2.3.

REMARK 3.2. We note further that once one has obtained a model of the form (3.11), then by the change of variables

$$\tau = \int_0^t \operatorname{sp}\mathscr{A}(s)\,ds$$

and subsequent simple transformations this model can be transformed to the model

$$(Af)(t) = \mathscr{A}^{1/2}(t)\mathscr{Y}\int_t^l \mathscr{A}^{1/2}(s)f(s)\,ds,$$

where $\operatorname{sp}\mathscr{A}(t) = 1$ $(0 \leqq t \leqq l)$.

If in this construction we take for $\{\psi_j(t)\}_1^\infty$ an orthogonal system of functions, then by (3.6) we will have

$$l = \int_0^l \operatorname{sp}\mathscr{A}(t)\,dt = |A_{\mathscr{I}}|_1. \tag{3.13}$$

In the other cases[11)]

$$l = \operatorname{sp}\int_0^l \mathscr{A}(t)\,dt = \left|\int_0^l \mathscr{A}(t)\,dt\right|_1 = \left|\mathscr{Y}\int_0^l \mathscr{A}(t)\,dt\right|_1 \geqq |A_{\mathscr{I}}|_1.$$

To derive this we have used the following lemma.

LEMMA 3.1. *The spectrum of the imaginary component $A_{M\mathscr{I}}$ of a canonical model A_M of an operator coincides with the spectrum of the matrix*

$$\frac{1}{2i}\mathscr{Y}\int_0^l \mathscr{A}(t)\,dt.$$

PROOF. In fact, the equation

$$\frac{1}{2i}\mathscr{A}^{1/2}(t)\mathscr{Y}\int_0^1 \mathscr{A}^{1/2}(s)\phi(s)\,ds = \lambda\phi(t)$$

is equivalent to the system

$$\begin{cases} \dfrac{1}{2i}\mathscr{A}^{1/2}(t)\,\xi = \lambda\phi(t), \\[2ex] \xi = \mathscr{Y}\displaystyle\int_0^1 \mathscr{A}^{1/2}(s)\phi(s)\,ds. \end{cases}$$

---

11) See, in this connection, the transformation (4.4).

Inserting the value of ϕ from the first equation into the second, we find that

$$\lambda\xi = \frac{1}{2i}\mathscr{Y}\int_0^1 \mathscr{A}(s)\,ds\xi.$$

The lemma is proved.

We also note that for the case of a dissipative operator (when $\mathscr{Y} = iI$) formula (3.13) assumes a simpler form:

$$l = \int_0^l \operatorname{sp}\mathscr{A}(t)\,dt = \operatorname{sp} A_{\mathscr{J}}.$$

This equality holds for any choice of the vector-functions $\{\psi_j(t)\}_1^\infty$.

4. Let us compare the general model (2.6) with a canonical model.

The general model (2.6) of an operator can be written in the form (cf. Chapter IV, §7.3)

$$(Af)(t) = G(t)\int_t^1 \mathscr{Y}\,G^*(s)f(s)\,ds, \tag{3.14}$$

where

$$G(t) = \|\psi_1(t), \psi_2(t), \cdots, \psi_n(t)\|$$

is a rectangular matrix-function with n ($\leqq \infty$) columns and r ($\leqq \infty$) rows, and $\mathscr{Y}$ is the previous signature matrix.

Thus, the canonical model is the special case of the general model obtained when $G(t)$ is a square matrix and an H-matrix of positive type.

This specialization of the kernel, which is convenient in certain cases, is achieved at the cost of not having the rank r of the space $L_2^{(r)}$, in which the model of the given Volterra operator is constructed, as small as possible.[12)]

[12)] Thus, for example, for a general Volterra operator in $L_2(0,1)$

$$(Vf)(t) = \int_t^1 \mathscr{V}(t,s)f(s)\,ds \tag{3.15}$$

with a continuous nondegenerate Hermitian kernel $\mathscr{V}(t,s)$ $(0 \leqq t, s \leqq 1)$ the canonical model is in the form of an integral operator in the space $L_2^{(\infty)}$ of infinite-dimensional matrix-functions.

§4. The monodromy matrix-function as an invariant of a model representation. The characteristic matrix-function

1. Let $A = G + iK$ be a Volterra operator with an n-dimensional imaginary component K which, consequently, can be represented in the form

$$K = \sum_{j=1}^{n} \epsilon_j(\cdot, \psi_j)\psi_j, \tag{4.1}$$

where $\epsilon_j = \pm 1$, and the ψ_j $(j = 1, 2, \cdots, n)$ are linearly independent vectors which form a basis of $\Re(K)$. In this representation the numbers ϵ_j are determined up to their order of enumeration. Once they are fixed, and hence the signature matrix $\mathscr{Y} = i \| \epsilon_j \delta_{jk} \|_1^n$ is fixed, then, as is easily seen, the vectors ψ_j $(j = 1, 2, \cdots, n)$ in (4.1) are determined up to a linear transformation

$$\psi_j \rightarrow \psi_j' = \sum_{k=1}^{n} l_{jk}\psi_k \qquad (j = 1, 2, \cdots, n) \tag{4.2}$$

with a $\mathscr{Y}$-unitary matrix $L = \| l_{jk} \|_1^n$:

$$L^*\mathscr{Y} L = \mathscr{Y}.$$

If now the operator A is realized as an operator in some space $L_2^{(r)}(a, b)$ of vector-functions,[13] then according to §3 we associate with it an H-matrix of positive type

$$\mathscr{A}(t) = 2 \| \psi_j^*(t)\psi_k(t) \|_1^n \qquad (a \leqq t \leqq b)$$

and by Theorem 3.1 the operator A turns out to be unitarily equivalent, up to inessential extension, to the model A_M defined in $L_2^{(n)}(a, b)$ by

$$(A_M f)(t) = \mathscr{A}^{1/2}(t)\mathscr{Y} \int_t^b \mathscr{A}^{1/2}(s) f(s)\, ds. \tag{4.3}$$

If, in the representation (4.1), we go over from the vector-functions ψ_j to the vector-functions ψ_j', then the H-matrix $\mathscr{A}(t)$ goes over into the H-matrix

$$\mathscr{A}'(t) = 2 \| \psi_j'^*(t)\psi_k'(t) \|_1^n \qquad (a \leqq t \leqq b),$$

which is related to $\mathscr{A}(t)$ by the transformation

---

[13] For generality, in this section we shall consider models in a space $L_2^{(r)}(a, b)$ instead of in $L_2^{(r)}(0, 1)$ or $L_2^{(r)}(0, l)$.

$$\mathscr{A}'(t) = L^*\mathscr{A}(t)L \qquad (a \leqq t \leqq b). \tag{4.4}$$

On the other hand, to the pair $\mathscr{A}(t)$ and $\mathscr{Y}$ there corresponds a matrix-function $\mathscr{U}(t;\mu)$ of order n, called a matrizant and defined as the solution of the Cauchy problem[14]

$$\mathscr{Y}\frac{d\mathscr{U}}{dt} = \mu\mathscr{A}(t)\mathscr{U}, \qquad \mathscr{U}(a;\mu) = I_n, \tag{4.5}$$

where μ is a complex parameter.

If the H-matrix $\mathscr{A}(t)$ in (4.5) is replaced by $\mathscr{A}'(t)$, the matrizant $\mathscr{U}(t;\mu)$ goes over into the matrizant $\mathscr{U}'(t;\mu) = L\mathscr{U}(t;\mu)L^{-1}$; in particular, we will have

$$\mathscr{U}'(b;\mu) = L\mathscr{U}(b;\mu)L^{-1}.$$

We shall call the matrix-function $\mathscr{U}(\mu) = \mathscr{U}(b;\mu)$ the *monodromy matrix-function*[15] of the operator model A_M.

In our construction, the nonuniqueness of the model A_M of a given abstract operator A results not only from the nonuniqueness of the representation (4.1). Its nonuniqueness (and indeed immeasurably more complicated nature) results also from the nonuniqueness of a maximal eigenchain $\mathfrak{P}$ of the operator A and from the nonuniqueness of the operation of extending it to an absolutely continuous maximal chain and the subsequent operation of imbedding the abstract Hilbert space $\mathfrak{H}$ in some space $L_2^{(r)}(a,b)$.

Nevertheless, we have the following remarkable theorem, which follows from general results of M. S. Livšic [2].

THEOREM 4.1. *For all operators A_M with a fixed matrix $\mathscr{Y}$ which are models of a given abstract simple Volterra operator, the monodromy matrix-function $\mathscr{U}(\mu)$ is defined up to a transformation $\mathscr{U}(\mu) \to L\mathscr{U}(\mu)L^{-1}$, where L is some $\mathscr{Y}$-unitary matrix.*

PROOF. The adjoint A_M^* of A_M is given by

$$(A_M^* f)(t) = -\mathscr{A}^{1/2}(t)\mathscr{Y}\int_a^t \mathscr{A}^{1/2}(s)f(s)\,ds. \tag{4.6}$$

---

[14] For details on this problem see Chapter VI.

[15] The term monodromy matrix-function is taken from the theory of differential equations with periodic coefficients (see Chapter VI, §3).

We shall show that

$$\mathscr{U}(b;\mu) = I_n - \mu\mathscr{Y}\,\|((I-\mu A_M^*)^{-1}\chi_j,\chi_k)\|_1^n, \tag{4.7}$$

where the vector-functions $\chi_1, \chi_2, \cdots, \chi_n$ are the columns of the matrix-function $\mathscr{A}^{1/2}(t)$:

$$\mathscr{A}^{1/2}(t) = \|\chi_1, \chi_2, \cdots, \chi_n\|.$$

To this end we note first that (4.5) implies

$$\mathscr{U}(t;\mu) = I_n - \mu\mathscr{Y}\int_a^t \mathscr{A}(s)\mathscr{U}(s;\mu)\,ds. \tag{4.8}$$

With $\xi \in E_n$ an arbitrary fixed vector, we form the vector-function

$$\phi_\xi(t) = \mathscr{A}^{1/2}(t)\mathscr{U}(t;\mu)\xi.$$

Multiplying both sides of (4.8) from the left by $\mathscr{A}^{1/2}(t)$ and from the right by ξ, we obtain

$$\phi_\xi(t) = \mathscr{A}^{1/2}(t)\xi - \mu\mathscr{A}^{1/2}(t)\mathscr{Y}\int_a^t \mathscr{A}^{1/2}(s)\phi_\xi(s)\,ds\xi$$

or, what is by (4.6) the same,

$$\phi_\xi = \chi_\xi + \mu A_M^*\phi_\xi \quad (\chi_\xi = \mathscr{A}^{1/2}(t)\xi).$$

Hence

$$\phi_\xi = (I-\mu A_M^*)^{-1}\chi_\xi,$$

and so for any $\xi, \eta \in E_n$

$$((I-\mu A_M^*)^{-1}\chi_\xi,\chi_\eta) = (\phi_\xi,\chi_\eta) = \int_a^b \chi_\eta^*(t)\phi_\xi(t)\,dt$$

$$= \eta^*\int_a^b \mathscr{A}(t)\mathscr{U}(t;\mu)\,dt\xi.$$

Multiplying the first and last terms by μ and using (4.8) for $t=b$, we find that

$$\mu((I-\mu A_M^*)^{-1}\chi_\xi,\chi_\eta) = \eta^*\mathscr{Y}(\mathscr{U}(b,\mu) - I_n)\xi. \tag{4.9}$$

Putting $\xi = e_k$, $\eta = e_j$ $(j,k = 1,2,\cdots,n)$, where e_k denotes the unit vector whose coordinates all vanish, with the exception of the kth which is equal to unity, we see that (4.9) is equivalent to (4.7).

We now note that the imaginary component K_M of the operator $A_M = G_M + iK_M$ has the representation

$$K_M = \frac{1}{2} \sum_{j=1}^{n} \epsilon_j(\cdot, \chi_j)\chi_j. \tag{4.10}$$

Now let $A = G + iK$ be an abstract Volterra operator in the Hilbert space $\mathfrak{H}$ which has A_M as a model. Let us first suppose that A and A_M are simply unitarily equivalent, and let V be a unitary mapping of $L_2^{(n)}(a,b)$ onto $\mathfrak{H}$ which takes A into A_M, so that $A_M = V^{-1}AV$.

If we put

$$\psi_j = (\sqrt{2})^{-1}V\chi_j \qquad (j = 1, 2, \cdots, n),$$

then by virtue of the fact that the isomorphism V not only carries A into A_M, but also A^*, G, K into A_M^*, G_M, K_M, respectively, by (4.10) we will have

$$K = \sum_{j=1}^{n} \xi_j(\cdot, \psi_j)\psi_j, \tag{4.11}$$

and the relation (4.7) for $\mathscr{U}(\mu) = \mathscr{U}(b;\mu)$ can be rewritten in the form

$$\mathscr{U}(\mu) = I_n - \mu\mathscr{Y} \,\|((I - \mu A^*)^{-1}\psi_j, \psi_k)\|_1^n. \tag{4.12}$$

It is easily seen that both (4.11) and (4.12) remain valid if A is unitarily equivalent to A_M only after some inessential extension $A \to \widetilde{A}$.

As we know, for given ϵ_j $(j = 1, 2, \cdots, n)$ the vectors ψ_j $(j = 1, 2, \cdots, n)$ in the representation (4.11) are determined up to a linear transformation (4.2) with a $\mathscr{Y}$-unitary matrix L. It is easily seen that under such a transformation the matrix $\mathscr{U}(\mu)$ transforms according to

$$\mathscr{U}(\mu) \to L^{-1}\mathscr{U}(\mu)L.$$

The theorem is proved.

2. An analysis of the arguments which led us to (4.12) shows that no properties of the matrix $\mathscr{Y}$ were used in them other than the following:

$$\mathscr{Y}^* = -\mathscr{Y}, \quad \mathscr{Y}^2 = -I_n.$$

Moreover, if A_M is an operator in $L_2^{(n)}(a,b)$ defined by (4.3), in which $\mathscr{H}(t)$ is an H-matrix of positive type, and $\mathscr{Y}$ is an arbitrary nonsingular skew-Hermitian matrix ($\mathscr{Y}^* = -\mathscr{Y}$; $\det\mathscr{Y} \neq 0$), then (4.12) continues to have meaning so long only as the matrizant $\mathscr{U}(t;\mu)$ which defines the monodromy matrix-function $\mathscr{U}(\mu) = \mathscr{U}(b;\mu)$ of the operator A_M is defined as the solution of the system

$$\frac{d\mathscr{U}}{dt} = -\mu\mathscr{J}\mathscr{A}(t)\mathscr{U}, \quad \mathscr{U}(a;\mu) = I_n.$$

For $\mathscr{J}^2 = -I_n$ this system is equivalent to the system (4.5). If we put

$$\mathscr{J} = \| j_{pq} \|_1^n,$$

then in the more general case under consideration ($\mathscr{J}^* = -\mathscr{J}$, $\det\mathscr{J} \neq 0$) we will have for the imaginary component K_M, in place of (4.10),

$$K_M = \frac{1}{2i} \sum_{p,q=1}^{n} (\cdot, \chi_p) j_{pq} \chi_q. \tag{4.13}$$

If the simple parts of A_M and the abstract Volterra operator $A = G + iK$ are unitarily equivalent, then arguing as before we deduce from (4.13) that there exists a system $\{\psi_j\}_1^n$ of vectors in $\mathfrak{H}$ such that

$$K = \frac{1}{i} \sum_{p,q=1}^{n} (\cdot, \psi_p) j_{pq} \psi_q \tag{4.14}$$

and for which, as before, (4.12) holds.

It is easily seen that once the skew-Hermitian matrix $\mathscr{J} = \| j_{pq} \|_1^n$ is fixed, the vectors $\psi_1, \psi_2, \cdots, \psi_n$ in the representation (4.14) will be determined up to a transformation (4.2) with some $\mathscr{J}$-unitary matrix $L = \| l_{jk} \|_1^n$. Under such a transformation the matrix-function $\mathscr{U}(\mu)$, defined by (4.12), will transform according to $\mathscr{U}(\mu) \to L^{-1}\mathscr{U}(\mu)L$.

We conclude that *Theorem* 4.1 *remains valid for operator models* A_M *with an arbitrary fixed nonsingular skew-Hermitian matrix* $\mathscr{J}$.

Let us remark that if we are given an abstract Volterra operator $A = G + iK$ with an n-dimensional imaginary component K, then in order that K have a representation (4.14) with a given nonsingular skew-Hermitian matrix $\mathscr{J} = \| j_{pq} \|_1^n$ it is necessary and sufficient that K and the matrix $\mathscr{J}/i$ have the same number of positive and negative eigenvalues.

If the matrix $\mathscr{J}$ satisfies this condition and a representation (4.14) has been obtained, then when the operator A is realized in the form of an integral operator in some space $L_2^{(n)}(a,b)$ by the method indicated in §3, the vectors ψ_j $(j = 1, 2, \cdots, n)$ go over into vector-functions $\psi_j(t) \in L_2^{(r)}(a,b)$ $(j = 1, 2, \cdots, n)$. Setting, as before in §3,

$$\mathscr{A}(t) = 2\| \psi_j^*(t)\psi_k(t) \|_1^n,$$

we obtain an H-matrix of positive type which together with the given matrix $\mathscr{Y}$ enables us to construct the operator model A_M by means of the formula (4.3). The simple part of the operator A_M will be unitarily equivalent to the simple part of the given abstract operator A.

3. Let $A = G + iK$ be an arbitrary bounded operator in $\mathfrak{H}$ with a finite-dimensional imaginary component K. We associate, with the representation (4.14) of K with a skew-Hermitian matrix $\mathscr{Y}$, the matrix-function

$$W_A(\lambda) = I - \mathscr{Y} \| ((A^* - \lambda I)^{-1}\psi_j, \psi_k) \|_1^n, \tag{4.15}$$

called the *characteristic matrix-function* of the operator A. This important concept was introduced by M. S. Livšic [2]; in his original definition it was assumed that the ψ_j $(j = 1, 2, \cdots, n)$ are eigenvectors of K: $K\psi_j = \epsilon_j |\psi_j| \psi_j$ $(\epsilon_j = \pm 1; \; j = 1, 2, \cdots, n)$, and that $\mathscr{Y}$ is the corresponding signature matrix: $\mathscr{Y} = i \| \epsilon_j \delta_{jk} \|_1^n$. This definition of $W_A(\lambda)$ generalizes at once to the case of a bounded operator A with a completely continuous imaginary component $A_{\mathscr{J}}$. M. S. Livšic [2] showed that a nonselfadjoint simple operator A of this type *is completely characterized* by its function $W_A(\lambda)$ up to unitary equivalence.

Study of the function $W_A(\lambda)$ for an operator A with nuclear imaginary component $A_{\mathscr{J}}$ led M. S. Livšic [2] to the construction of a certain triangular model of such an operator, which in the special case of a Volterra operator A reduces to the canonical model. In this case a comparison of (4.15) and (4.12) shows that

$$\mathscr{U}(\mu) = W_A(1/\mu), \tag{4.16}$$

where $\mathscr{U}(\mu)$ is the monodromy matrix-function of the operator A_M.

If we use the concept of a multiplicative integral, then (4.16) can be written in the form

$$W_A\left(\frac{1}{\mu}\right) = \overset{\curvearrowleft}{\int_a^b} e^{-\mu \mathscr{Y} \mathscr{A}(s)\, ds};$$

this was first obtained by Livšic on the basis of a theorem of V. P. Potapov [1] (see Theorem 3.1 of the Appendix).

Subsequently in a paper by M. S. Brodskiĭ [1] and a joint paper by M. S. Brodskiĭ and M. S. Livšic [1] the concept of the characteristic matrix-function was generalized and modified in various directions (see also A. V. Štraus [1]). In particular, in the above-mentioned joint paper the characteristic matrix-function $W_A(\lambda)$ was defined by (4.15) with A^* replaced by A, with $\mathscr{Y}$ satisfying the requirement $\mathscr{Y}^* = -\mathscr{Y}$, $\mathscr{Y}^2 = -I_n$ and, together with the vectors $\psi_1, \psi_2, \cdots, \psi_n$ (which were not assumed linearly independent), being taken from the representation (4.14). The vectors $\psi_1, \psi_2, \cdots, \psi_n$ in the representation (4.14) were called channel vectors[16)] by Brodskiĭ and Livšic.

§5. Triangular models of real Volterra operators

1. Suppose that we are given, in a Hilbert space $\mathfrak{H}$, an involution S of the type complex conjugation, i.e., an operator S which maps $\mathfrak{H}$ onto itself and has the following properties:

1) $S^2 = I$ (the property of being an involution);
2) $S(f_1 + f_2) = Sf_1 + Sf_2$ $(f_1, f_2 \in \mathfrak{H})$;
3) $S(\lambda f) = \bar{\lambda} Sf$ $(f \in \mathfrak{H};\ \lambda$ a complex scalar);
4) $(Sf_1, Sf_2) = \overline{(f_1, f_2)}$ $(f_1, f_2 \in \mathfrak{H})$.

An element $f \in \mathfrak{H}$ is said to be *S-real* if $Sf = f$. Every vector $f \in \mathfrak{H}$ can be represented in a unique way in the form

$$f = g + ih,$$

where the vectors g and h are S-real. They can be found from the formulas

$$g = (f + Sf)/2, \qquad h = (f - Sf)/2i.$$

The elements $f = g + ih$ and $Sf = g - ih$ are said to be *S-conjugate*. **The set of all S-real vectors forms a real Hilbert space which we** shall denote by $\mathfrak{H}_S$.

An operator $A \in \mathfrak{R}$ is said to be *S-real* if it commutes with the operator S: $AS = SA$. Obviously this condition is equivalent to the condition that A map the space $\mathfrak{H}_S$ into itself.

An operator $A \in \mathfrak{R}$ is said to be *S-symmetric* if $A^* = SAS$, and *S-skewsymmetric* if $A^* = -SAS$. Every operator $A \in \mathfrak{R}$ can be represented in a unique way as the sum of an S-symmetric and an S-skewsymmetric operator:

$$A = A_\sigma + A_\alpha.$$

In this representation

$$A_\sigma = \tfrac{1}{2}(A + SA^*S) \quad \text{and} \quad A_\alpha = \tfrac{1}{2}(A - SA^*S).$$

If A is an S-real operator, then

$$A_\sigma = A_{\mathscr{R}} = (A + A^*)/2 \quad \text{and} \quad A_\alpha = iA_{\mathscr{I}} = (A - A^*)/2,$$

---

16) The general definition of the abstract characteristic operator-function $W_A(\lambda)$, given by Livšic for any bounded operator $A \in \mathfrak{R}$, is presented in §2 of the Appendix.

and both of these operators are S-real.

We shall call a subspace $\mathfrak{L} \subset \mathfrak{H}$ a *subspace with S-real support* if it is the complex hull of the set $\mathfrak{L}_S$ of all its S-real vectors (obviously $\mathfrak{L}_S$ is always a subspace of $\mathfrak{H}_S$). It is easily seen that for a subspace $\mathfrak{L}$ to have S-real support, it is necessary and sufficient that the orthoprojector $P_{\mathfrak{L}}$ which projects $\mathfrak{H}$ onto $\mathfrak{L}$ be S-real.

We shall say that a chain of orthoprojectors is S-real, if it consists of S-real projectors. We shall say that such a chain is *real-maximal*, if it is not a proper subset of any other S-real chain.

It is easily seen that for an S-real chain the notion of real maximality coincides with the notion of ordinary maximality.

2. If A is a completely continuous S-real operator, then restricting it to $\mathfrak{H}_S$ and applying to this restriction the arguments of Aronszajn and Smith (see the proof of Theorem 2.1, Chapter I), it is not hard to see that it will have a nontrivial invariant subspace in $\mathfrak{H}_s$. Taking the complex hull of this subspace in $\mathfrak{H}$, we see that there exists a real orthoprojector P $(P \neq 0, I)$ such that

$$AP = PAP.$$

Following this it is not hard to show (see Theorem 3.1, Chapter I) that every S-real Volterra operator has a maximal S-real eigenchain.

If A is an S-real Volterra operator with an S-real maximal eigenchain $\mathfrak{P} = \{P\}$, then

$$A = 2\int_{\mathfrak{P}} PA_\alpha \, dP.$$

Let $A \in \mathfrak{R}$ be an S-real operator, and $\widetilde{A}$ its inessential extension to some space $\widetilde{\mathfrak{H}} \supset \mathfrak{H}$ having an involution S_1. We shall call the extension $\widetilde{A}$ S-real, if the involutions S_1 and S coincide on $\mathfrak{H}$.

For S-real Volterra operators Theorem 1.2 admits the following strengthening.

Let A be an S-real Volterra operator with an S-real maximal eigenchain of rank r $(\leqq \infty)$. Then A has an S-real inessential extension $\widetilde{A}$ which has an absolutely continuous S-real eigen-projector-function of the same rank r.

To obtain models for real Volterra operators we need a slight sharpening (connected with the notion of S-real) of the result from the theory of spectral types formulated in §2.1.

Let $P(t)$ $(0 \leqq t \leqq 1)$ be an absolutely continuous S-real projector-function of rank r. Then there exists a unitary mapping U of $\mathfrak{H}$ onto some subspace $\mathfrak{L} \subseteq L_2^{(r)}$ which takes $P(t)$ $(0 \leqq t \leqq 1)$ into the projector-function of truncation

$\hat{P}(t)$, *and takes* S *into the operator of complex conjugation, so that*

$$UP(t)f = \hat{P}(t)\,Uf \quad \textit{and} \quad USf = \overline{Uf} \qquad (f \in \mathfrak{H}).$$

We shall call an operator A in $L_2^{(r)}$ *real* if it is S-real relative to the involution operator S of ordinary complex conjugation.

Let $\boldsymbol{A}$ be an S-real Volterra. A real operator A in $L_2^{(r)}$ which has $\hat{P}(t)$ as an eigen-projector-function will be called a *real triangular* model of the operator $\boldsymbol{A}$, if A is unitarily equivalent to $\boldsymbol{A}$ or to an S-real extension $\widetilde{\boldsymbol{A}}$ of $\boldsymbol{A}$: $U\widetilde{\boldsymbol{A}} = AU$, where the operator U which establishes this equivalence transforms S into the operator of complex conjugation, i.e.

$$USf = \overline{Uf} \qquad (f \in \mathfrak{H}).$$

It is now easy to obtain a strengthening of Theorem 2.1 for S-real Volterra operators.

THEOREM 5.1. *Every S-real Volterra operator of S-rank r has a real triangular model in* $L_2^{(r)}$.

Let us clarify that by the S-*rank* of an S-real operator we mean the smallest of the ranks of all its S-real maximal eigenchains.

Of course, the S-rank of an S-real operator is not less than its ordinary rank.

Let us also mention a strengthening of Theorem 2.2 for S-real Hilbert-Schmidt operators.

THEOREM 5.2. *Every S-real Hilbert-Schmidt Volterra operator* $\boldsymbol{A}$ *of S-rank r has as a real triangular model the integral operator A in* $L_2^{(r)}$, *defined by*

$$(Af)(t) = 2\int_t^1 \mathscr{G}(t,s) f(s)\,ds, \tag{5.1}$$

where

$$\mathscr{G}(t,s) = \| g_{\mu\nu}(t,s) \|_1^r \qquad (0 \leqq t, s \leqq 1)$$

is a real symmetric matrix-function which satisfies the condition

$$\int_0^1 \int_0^1 \mathrm{sp}(\mathscr{G}\mathscr{G}^*)\,dt\,ds < \infty.$$

3. For an S-real Volterra operator with a nuclear skewsymmetric component one can obtain an interesting detailing of its canonical triangular model.

For the operator (5.1) the skewsymmetric component A_α is given by

$$(A_\alpha f)(t) = \int_0^1 \mathscr{G}(t,s) f(s)\, ds \quad (f \in L_2^{(r)}),$$

where $\mathscr{G}(t,s)$ $(0 \leqq t, s \leqq 1)$ is obtained by skewsymmetrically extending the kernel appearing in (5.1), i.e.

$$g_{\mu\nu}(t,s) = -g_{\nu\mu}(s,t) \quad (\mu, \nu = 1, 2, \cdots, r;\ 0 \leqq t \leqq s \leqq 1).$$

Since A_α is S-real, and selfadjoint upon multiplication by i, its spectrum lies on the imaginary axis and is symmetric with respect to the origin. The nonzero spectrum of A_α consists of a sequence of pairs $\{i\omega_j, -i\omega_j\}$ $(\omega_j > 0)$, to which corresponds a sequence of complex conjugate eigenvector-functions $\{\phi_j(x), \overline{\phi_j(x)}\}$ which form an orthonormal system, so that

$$\mathscr{G}(t,s) = \sum_j i\omega_j(\phi_j(t)\phi_j^*(s) - \phi_j(s)\phi_j^*(t)), \tag{5.2}$$

where the series converges in quadratic mean on the square $0 \leqq t, s \leqq 1$.

Putting

$$\psi_j(t) = \frac{\sqrt{\omega_j}}{\sqrt{2}}(\phi_j(t) + \overline{\phi_j(t)}),$$

$$\chi_j(t) = \frac{\sqrt{\omega_j}}{i\sqrt{2}}(\phi_j(t) - \overline{\phi_j(t)}),$$

we obtain a real orthonormal system $\{\psi_j(t), \chi_j(t)\}$, and (5.2) goes over into

$$\mathscr{G}(t,s) = \sum_j (\psi_j(t)\chi_j^*(s) - \chi_j(t)\psi_j^*(s)). \tag{5.3}$$

For simplicity we continue our discussion for the case where A_α is finite-dimensional, and so j runs through a finite number of values $1, 2, \cdots, m$, where $n = 2m$ is the dimension of A_α.

Introducing for $\chi_j(t)$ new notation

$$\psi_{j+m}(t) = \chi_j(t) \quad (j = 1, 2, \cdots, m), \tag{5.4}$$

we can write (5.3) in the form

$$\mathscr{G}(t,s) = \sum_{p,q=1}^{2m} \psi_p(t)\, j_{pq} \psi_q^*(s), \tag{5.5}$$

where the signature matrix $\mathscr{Y}_m = \| j_{pq} \|_1^{2m}$ has the form

$$\mathscr{Y}_m = \left\| \begin{matrix} 0 & I_m \\ -I_m & 0 \end{matrix} \right\|. \tag{5.6}$$

We shall further use only the fact that the kernel $\mathscr{G}(t,s)$ can be represented in the form (5.5) with linearly independent (but not necessarily orthogonal) vector-functions $\psi_j(t)$ $(j = 1, 2, \cdots, 2m)$, and the matrix $\| j_{pq} \|_1^{2m}$ has the form (5.6).

Arguing as in the proof of Theorem 3.1, we obtain the following result.

THEOREM 5.3. *Every simple S-real Volterra operator of S-rank r with a 2m-dimensional skewsymmetric component* A_α *has as a real triangular model the operator A, defined in* $L_2^{(2m)}$ *by*

$$(5.7) \qquad (A_0 f)(t) = \mathscr{A}^{1/2}(t)\, \mathscr{Y}_m \int_t^1 \mathscr{A}^{1/2}(s) f(s)\, ds,$$

where $\mathscr{A}(t)$ *is a real*[17] *H-matrix of positive type, whose rank does not exceed r for any t, and* $\mathscr{Y}_m$ *is a signature matrix of the form* (5.6).

We shall call the model (5.7) the *canonical real model*.

For a kernel $\mathscr{G}(t,s)$ defined by (5.5), the H-matrix $\mathscr{A}(t)$ is given by

$$\mathscr{A}(t) = 2 \| \psi_j^*(t) \psi_k(t) \|_1^{2m}.$$

If we represent the kernel $\mathscr{G}(t,s)$ in the form

$$\mathscr{A}(t,s) = \sum_{p,q=1}^{2m} \widetilde{\psi}_p(t)\, j_{pq} \widetilde{\psi}_q^*(s),$$

where

$$\widetilde{\psi}_{2j-1}(t) = \psi_j(t) \quad \text{and} \quad \widetilde{\psi}_{2j}(t) = \chi_j(t) \qquad (j = 1, 2, \cdots, m),$$

and define the H-matrix $\mathscr{A}(t)$ accordingly, then we arrive at a model (5.7) where instead of $\mathscr{Y}_m$ there stands a matrix all of whose diagonals, with the exception of those alongside the principal diagonal, consist entirely of zeros, while those alongside the principal diagonal consist of $+1$ and -1, respectively. In this form this model can be generalized to the case of S-real Volterra operators with nuclear infinite-dimensional skewsymmetric component. In this case $\mathscr{A}(t)$ will be an infinite-dimensional H-matrix for which, however, $\operatorname{sp} \mathscr{A}(t)$ will have meaning for almost all t and, moreover,

$$\int_0^1 \operatorname{sp} \mathscr{A}(t)\, dt = | A_\alpha |_1 \quad (< \infty).$$

---

[17] An H-matrix is said to be *real* if all of its elements are real.

Just as in the general case, by appropriate transformations the model (5.7) can be transformed to the model

$$(A_1 f)(t) = \mathscr{A}^{1/2}(t)\,\mathscr{Y}_m \int_t^l \mathscr{A}^{1/2}(s) f(s)\, ds,$$

where now $l = |A_\alpha|_1$, and

$$\operatorname{sp} \mathscr{A}(t) = 1 \quad (0 \leqq t \leqq l).$$

CHAPTER VI

SELFADJOINT BOUNDARY VALUE PROBLEMS FOR A CANONICAL EQUATION. TESTS FOR THE STABLE BOUNDEDNESS OF THE SOLUTIONS OF A CANONICAL EQUATION WITH A PERIODIC H-MATRIX

In this chapter we introduce the concept of a *canonical differential equation* and study boundary value problems for such equations.

One can show that every selfadjoint boundary value problem for one ordinary differential equation of any order or for a system of such equations with parameters which appear linearly in the equation can be transformed into a selfadjoint boundary value problem for a canonical differential equation.

The connection between the theory of Volterra operators and the theory of canonical differential equations was already noted in the theory of models (Chapter V).

It is very striking that the power of the methods of the theory of *nonselfadjoint* operators discussed here is clearly apparent in the theory of *selfadjoint* boundary value problems for a canonical equation. In turn, the theory of these boundary value problems turns out to be necessary for establishing one of the deepest and most difficult results of the general theory of Volterra operators: the fundamental theorem on the density of the spectrum of a Volterra operator with nuclear imaginary component (Chapter VII).

In §1 of this chapter we introduce the notions of a canonical differential equation and a homogeneous selfadjoint boundary value problem for such an equation, and we establish some simple properties of the spectrum of such a problem.

In §2 methods of the abstract theory of operators are used to establish a fundamental property of the spectrum of a *skewperiodic* boundary value problem, namely, that it coincides with the spectrum of the real component of a model operator. This enables us at once to make important deductions concerning the spectrum of such a problem.

The three following sections are devoted to certain applications of theorems concerning transformators (Chapter II) in the theory of canonical differential equations with periodic coefficients. As a consequence of lack of space, we shall present certain results of the theory of canonical differential equations with periodic coefficients without proofs, referring the reader to the original sources.

In §6 we establish various results concerning the asymptotic behavior of the spectrum of a canonical selfadjoint boundary value problem. Theorem 6.1 on the asymptotics of the spectrum is stated without proof; its proof is obtained in Chapter VII simultaneously with the establishment of the fundamental theorem for abstract Volterra operators with nuclear imaginary component. Here the "mutual efficacy" of the theory of abstract Volterra operators and the theory of canonical differential equations becomes brilliantly clear. The remaining results of §6 are obtained as direct corollaries of general results from the theory of abstract Volterra operators. Some of these results are, possibly, discussed here for the first time.

In §7 we establish asymptotic characteristics of the spectrum of the traditional selfadjoint boundary value problems for second order equations under very general assumptions regarding the coefficients. However, these results are now obtained by a simple reduction of the second order boundary value problem to a selfadjoint boundary value problem for a canonical equation. This reduction is carried out in two ways. The second method, which involves the substitution $\lambda = \mu^2$, is possible only for problems with a positive spectrum.

It is interesting that the formula for the density of the frequency spectrum of an inhomogeneous string for an arbitrary distribution of its mass (the distribution function contains an absolutely continuous, a discrete and a singular part) can be obtained rather simply in this way. Its generalization to the case of the corresponding boundary value problem for vector-functions (§8) is then obtained "free of charge".

The reader who is interested in the characteristics of the spectra of boundary value problems can, after reading the first two sections, turn to the last three (§§6—8).

In studying the spectra of various boundary value problems for a canonical differential equation the knowledgeable reader may discern behind the scenes an imposing figure—some selfadjoint unbounded (!) operator acting in some Hilbert space. The space arises as the completion of the set of continuous vector-functions $x(t) = \{x_j(t)\}_1^n$, $a \leqq t \leqq b$, in the $\mathscr{A}$-metric: $|x|^2 = \int_a^b x^*(t)\mathscr{A}(t)x(t)\,dt$, while the operator in this space is generated by the differential expression $\mathscr{A}^{-1}(t)\mathscr{Y}\,d/dt$ (and, of course, the boundary conditions of the problem).

However, an attempt to fit out this operator in rigorous mathematical clothing (the matrix-function $\mathscr{H}(t)$ may, alas, be degenerate for all t) would require an effort not justified under the present conditions.

All the same, if we enlarge the circle of problems being considered, for example, if we include in it the question of the completeness of the system of eigen-vector-functions, then the introduction of this operator would become very desirable. It would become simply necessary for the study of boundary value problems on an infinite or semi-infinite interval, where a continuous spectrum comes on the scene.

We mention, finally, that the content of this chapter, with the exception of the initial results of the first section, does not overlap the content of the recent book by F. V. Atkinson [2], where for the first time the spectral theory and oscillation properties of canonical selfadjoint boundary value problems are discussed systematically.

§1. Selfadjoint boundary value problem for a canonical equation

1. In the sequel we shall frequently have to deal with matrix-functions $\mathscr{H}(t)$, defined on some interval $\Delta = [a, b]$ of the real line. We shall for brevity call such a matrix-function an *H-matrix* (of order n), if its values are Hermitian matrices (of order n).

We shall always assume that all the elements $h_{jk}(t)$ $(j, k = 1, 2, \cdots, n)$ of an H-matrix $\mathscr{H}(t) = \| h_{jk}(t) \|_1^n$ are integrable over the interval Δ.

By a *canonical differential equation* of phase dimension n (with parameters) we shall henceforth mean any differential equation of the form

$$\mathscr{J} \frac{dx}{dt} - \mu \mathscr{H}(t) x = 0, \qquad a \leq t \leq b, \tag{1.1}$$

where $x(t)$ is an n-dimensional vector-function, $\mathscr{H}(t)$ is an H-matrix, $\mathscr{J}$ is a constant signature matrix, i.e., a matrix with the properties

$$\mathscr{J} = -\mathscr{J}^*, \quad \mathscr{J}^2 = -I_n, \tag{1.2}$$

and μ is a complex parameter.

By virtue of the second property of the matrix $\mathscr{J}$, the equation (1.1) can also be written as

$$dx/dt = -\mu \mathscr{J} \mathscr{H}(t) x.$$

We shall call (1.1) a *Hamiltonian* equation if $n = 2m$ is even, $\mathscr{H}(t)$ is a real symmetric matrix for all $t \in \Delta$, and $\mathscr{J} = \mathscr{J}_m$, where

$$\mathscr{J}_m = \begin{pmatrix} 0 & I_m \\ -I_m & 0 \end{pmatrix}.$$

Under a linear transformation

$$\tilde{x} = \mathscr{S} x, \tag{1.3}$$

where $\mathscr{S}$ is a constant unitary matrix ($\mathscr{S}^*\mathscr{S} = I_n$), equation (1.1) goes over into

$$\begin{gathered} \tilde{\mathscr{Y}} \frac{d\tilde{x}}{dt} - \mu \tilde{\mathscr{A}}(t)\, x = 0, \\ \tilde{\mathscr{Y}} = \mathscr{S}^* \mathscr{Y} \mathscr{S}, \quad \tilde{\mathscr{A}}(t) = \mathscr{S}^* \mathscr{A}(t) \mathscr{S}, \end{gathered} \tag{1.4}$$

and thus remains a canonical equation.

By virtue of (1.2) the unitary matrix $\mathscr{S}$ can always be chosen so that $\tilde{\mathscr{Y}} = i \| \epsilon_j \delta_{jk} \|_1^n$, where $\epsilon_j = \pm 1$ $(j = 1, 2, \cdots, n)$.

If the matrix $\mathscr{Y}$ and the matrix-function $\mathscr{A}(t)$ in (1.1) are real, then (1.1) does not differ essentially from a Hamiltonian equation. We mean by this that it can be transformed to a Hamiltonian equation by means of an appropriate substitution (1.3) with a real orthogonal matrix $\mathscr{S}$.

If (1.1) is a Hamiltonian equation, then it remains Hamiltonian under a transformation (1.3) if $\mathscr{S}$ is a real orthogonal matrix, and moreover symplectic, i.e. has the property

$$\mathscr{S}^{\tau} \mathscr{Y}_m \mathscr{S} = \mathscr{Y}_m,$$

where $\mathscr{S}^{\tau}$ is the transpose of $\mathscr{S}$.

2. Let us consider the ***matrizant*** $\mathscr{U}(t; \mu)$ of equation (1.1), i.e., the absolutely continuous solution of the matrix equation

$$\mathscr{Y} \frac{d\mathscr{U}}{dt} - \mu \mathscr{A}(t) \mathscr{U} = 0 \qquad (a \leqq t \leqq b) \tag{1.5$'$}$$

with the initial condition

$$\mathscr{U}(a; \mu) = I_n. \tag{1.5$''$}$$

It can be shown by well-known methods that this solution is unique and can be obtained in the form of a series

$$\mathscr{U}(t; \mu) = I_n + \mu \mathscr{A}_1(t) + \mu^2 \mathscr{A}_2(t) + \cdots, \tag{1.6}$$

where

$$\mathscr{A}_k(t) = -\int_a^t \mathscr{Y} \mathscr{A}(s) \mathscr{A}_{k-1}(s)\, ds \quad (k = 1, 2, \cdots; \mathscr{A}_0(t) = I_n),$$

and the series (1.6) converges absolutely and uniformly on the product of the interval $[a,b]$ and any disc $|\mu| \leqq R$.

To every n-dimensional vector ξ there corresponds a unique absolutely continuous solution $x(t)$ of (1.1) satisfying the initial condition $x(0) = \xi$. This solution is given by the formula

$$x(t) = \mathscr{U}(t;\mu)\,x(0). \tag{1.7}$$

We shall show that

1. *For real μ the matrizant $\mathscr{U}(t;\mu)$ is a $\mathscr{Y}$-unitary matrix, i.e.*

$$\mathscr{U}^*(t;\mu)\mathscr{Y}\mathscr{U}(t;\mu) = \mathscr{Y} \qquad (a \leqq t \leqq b), \tag{1.8'}$$

and hence

$$\mathscr{U}(t;\mu)\mathscr{Y}\mathscr{U}^*(t;\mu) = \mathscr{Y} \qquad (a \leqq t \leqq b). \tag{1.8''}$$

Indeed, using equation (1.5′) and the equation obtained from it by taking Hermitian conjugates, we find that

$$\begin{aligned}\frac{d}{dt}\mathscr{U}^*\mathscr{Y}\mathscr{U} &= \frac{d\mathscr{U}^*}{dt}\mathscr{Y}\mathscr{U} + \mathscr{U}^*\mathscr{Y}\frac{d\mathscr{U}}{dt}\\ &= -\mu\mathscr{U}^*\mathscr{A}\mathscr{U} + \mu\mathscr{U}^*\mathscr{A}\mathscr{U} = 0.\end{aligned}$$

Thus $\mathscr{U}^*\mathscr{Y}\mathscr{U} = \mathscr{C}$, where $\mathscr{C}$ is a constant matrix. Using (1.5″), we obtain $\mathscr{C} = \mathscr{Y}$.

Multiplying (1.8′) from the left by the matrix $-\mathscr{Y}$, we see that $\mathscr{Y}\mathscr{U}(t;\mu)$ and $-\mathscr{Y}\mathscr{U}^*(t;\mu)$ are mutually inverse, which is just what (1.8″) says.

Using (1.8′), we now show that the equation

$$\mathscr{Y}\frac{dx}{dt} - (\mathscr{F}(t) + \mu\mathscr{A}(t))\,x = 0, \tag{1.9}$$

where $\mathscr{A}(t)$ and $\mathscr{F}(t)$ are Hermitian matrix-functions, can be transformed to the canonical form (1.1) with the same $\mathscr{Y}$.

To do this we introduce the matrizant $\mathscr{V}(t)$, defined by the system

$$\mathscr{Y}\frac{d\mathscr{V}}{dt} - \mathscr{F}(t)\mathscr{V} = 0, \quad \mathscr{V}(0) = I_n.$$

According to result 1 we will have

$$\mathscr{V}(t)\mathscr{Y}\mathscr{V}^*(t) = \mathscr{Y}, \quad \mathscr{V}^{-1}(t)\mathscr{Y}(\mathscr{V}^*(t))^{-1} = \mathscr{Y}. \tag{1.10}$$

We then put

$$x(t) = \mathscr{V}(t)\,\tilde{x}(t)$$

in (1.9). Using (1.10), we easily deduce that this substitution transforms (1.9) into

$$\mathscr{Y}\frac{d\tilde{x}}{dt} - \mu\tilde{\mathscr{H}}(t)\tilde{x} = 0, \tag{1.11}$$

where

$$\tilde{\mathscr{H}}(t) = \mathscr{V}^*(t)\mathscr{H}(t)\mathscr{V}(t). \tag{1.12}$$

In particular, if we make a parameter shift $\mu \to \mu + \alpha$ (α a real number) in equation (1.1), then the role of the matrix $\mathscr{F}$ will be played by the matrix $\alpha\mathscr{H}(t)$, and that of the matrix $\mathscr{V}$ by $\mathscr{U}(t;\mu)$, so that in this case we are led to equation (1.11) with $\tilde{\mathscr{H}} = \tilde{\mathscr{H}}_\alpha$, where

$$\tilde{\mathscr{H}}_\alpha(t) = \mathscr{U}^*(t;\alpha)\mathscr{H}(t)\mathscr{U}(t;\alpha) \qquad (a \leqq t \leqq b). \tag{1.13}$$

3. All further applications will be connected with equations of the form (1.1) with H-matrices of positive type.

This is related, in particular, to the fact that in reducing the corresponding boundary value problems for canonical equations to integral equations the positivity of the H-matrix turns out to be necessary for the Hermitivity of the integral operator which arises.

We recall that an H-matrix $\mathscr{H}(t)$ ($a \leqq t \leqq b$) is said to be of *positive type* if the following two conditions are satisfied:

1) *for each* $t \in [a, b]$ *the corresponding Hermitian form is nonnegative, i.e.*

$$\xi^*\mathscr{H}(t)\xi \geqq 0 \qquad (\xi \in E_n); \tag{1.14}$$

2) *the average of this form over the entire interval is a positive form, i.e.*

$$\int_a^b \xi^*\mathscr{H}(t)\xi\, dt > 0 \quad \textit{for } \xi \neq 0. \tag{1.15}$$

We note, first of all, that for all the transformations which we have considered (see (1.4) and (1.12)), the positivity of type of the H-matrices is not violated.

It is not hard to see that

for every solution $x(t;\mu)$ ($\not\equiv 0$) *of equation* (1.1) *with an H-matrix of positive type*

$$\int_a^b x^*(t;\mu)\mathscr{H}(t)x(t;\mu)\, dt > 0. \tag{1.16}$$

Indeed, if we assume the contrary, then by virtue of (1.14) we will have almost everywhere

$$x^*(t;\mu)\mathscr{H}(t)x(t;\mu) = 0, \tag{1.17}$$

and hence $\mathscr{H}(t)x(t;\mu) = 0$. Together with (1.1), this yields $x(t;\mu) = x(0) = \xi$ $(t \in [a,b])$. But then (1.17) contradicts (1.15). Just here we have used the strict positivity of the averaged Hermitian form.

We shall say that the boundary value problem

$$\mathscr{Y}\frac{dx}{dt} - \mu\mathscr{H}(t)x = 0, \tag{1.18}$$

$$\mathscr{A}x(a) + \mathscr{B}x(b) = 0 \tag{1.18'}$$

is *selfadjoint,* if 1) the H-matrix $\mathscr{H}(t)$ is of positive type, 2) the $n \times n$ matrices $\mathscr{A}$ and $\mathscr{B}$ are such that

$$\mathscr{A}\mathscr{Y}\mathscr{A}^* = \mathscr{B}\mathscr{Y}\mathscr{B}^*, \tag{1.19}$$

and 3) the rectangular matrix $[\mathscr{A},\mathscr{B}]$ has rank n.

A complete justification of the application to this system of the term "selfadjoint" will be given later. In particular, the condition (1.19) is necessary in order that the operator $\mathscr{Y}d/dt$ be *formally* selfadjoint on the set $\mathfrak{A}(\mathscr{A},\mathscr{B})$ of all absolutely continuous functions $x(t)$ $(a \leqq t \leqq b)$ satisfying the boundary condition (1.18′), i.e., that

$$\int_a^b y^*\mathscr{Y}\frac{dx}{dt}\,dt = \int_a^b \left(\mathscr{Y}\frac{dy}{dt}\right)^* x\,dt \left(= -\int_a^b \frac{dy^*}{dt}\mathscr{Y}x\,dt\right) \tag{1.20}$$

$$(x, y \in \mathfrak{A}(\mathscr{A},\mathscr{B})).$$

Integration by parts shows that this equality is equivalent to

$$y^*(a)\mathscr{Y}x(a) = y^*(b)\mathscr{Y}x(b). \tag{1.21}$$

This relation can be obtained very simply as a consequence of the conditions

$$\mathscr{A}x(a) + \mathscr{B}x(b) = 0, \quad \mathscr{A}y(a) + \mathscr{B}y(b) = 0, \tag{1.22}$$

if we assume that one of the matrices $\mathscr{A}$ and $\mathscr{B}$ is nonsingular. Suppose, for example, that $\mathscr{B}$ is nonsingular. Then the matrix $\mathscr{E} = -\mathscr{B}^{-1}\mathscr{A}$ is defined, and the condition (1.19) says that $\mathscr{E}^*$ is a $\mathscr{Y}$-unitary matrix $(\mathscr{E}\mathscr{Y}\mathscr{E}^* = \mathscr{Y})$, and hence $\mathscr{E}$ itself will be $\mathscr{Y}$-unitary:

$$\mathscr{E}^*\mathscr{Y}\mathscr{E} = \mathscr{Y}. \tag{1.23}$$

In this case the conditions (1.22) are equivalent to

$$x(b) = \mathscr{E}\, x(a), \quad y(b) = \mathscr{E}\, y(a)$$

which, together with (1.23), obviously guarantee (1.21). In the general case both $\mathscr{A}$ and $\mathscr{B}$ are degenerate,[1] and we therefore have to proceed somewhat differently.

We associate with the vector-functions $x(t)$ and $y(t)$ the vectors

$$X = \begin{pmatrix} x(a) \\ x(b) \end{pmatrix}, \qquad Y = \begin{pmatrix} y(a) \\ y(b) \end{pmatrix}$$

in the $2n$-dimensional space E_{2n}, and moreover we introduce the n-dimensional subspace $M \subset E_{2n}$, defined as the linear hull of the columns of the rectangular matrix

$$\begin{bmatrix} \mathscr{A}^* \\ \mathscr{B}^* \end{bmatrix}.$$

It is easily seen that the boundary condition (1.18′) says that the vectors X and Y are orthogonal to M.

Let us further denote by $\mathscr{I}$ the involution in E_{2n} ($\mathscr{I}^2 = I_{2n}$) defined by

$$\mathscr{I}\begin{pmatrix} \xi \\ \eta \end{pmatrix} = \begin{pmatrix} i\mathscr{Y}\xi \\ -i\mathscr{Y}\eta \end{pmatrix} \qquad (\xi, \eta \in E_n).$$

Then the condition (1.19) can be interpreted as the orthogonality of $\mathscr{I}M$ and M:

$$\mathscr{I}M \perp M.$$

Together with the conditions $X, Y \perp M$, the last condition shows that the vectors X and Y belong to $\mathscr{I}M$, and since $\mathscr{I}$ is an involution ($\mathscr{I}^2 = I_{2n}$), it follows that the vectors $\mathscr{I}X$ and $\mathscr{I}Y$ belong to M, and so $\mathscr{I}X \perp Y$, which is just (1.21).

4. In the theory of selfadjoint boundary value problems for a canonical equation an important role is played by the $\mathscr{H}$-metric associated with the scalar product

$$(x, y)_{\mathscr{H}} = \int_a^b y^*(t)\,\mathscr{H}(t)\,x(t)\,dt.$$

We shall say that two continuous vector-functions $x(t)$ and $y(t)$ ($a \leqq t \leqq b$) are *$\mathscr{H}$-orthogonal,* if

[1] This is the case, in particular, for the so-called Stürmian conditions, where $\mathscr{A}\mathscr{Y}\mathscr{A}^* = 0$ and $\mathscr{B}\mathscr{Y}\mathscr{B}^* = 0$.

$$\int_a^b y^*(t)\,\mathscr{A}(t)\,x(t)\,dt = 0.$$

We have the following theorem.

THEOREM 1.1. *Every eigenvalue of the selfadjoint boundary value problem* (1.1) *is real and has multiplicity* $\leqq n$. *Any two eigen-vector-functions of this problem corresponding to distinct eigenvalues* μ, *are* $\mathscr{A}$*-orthogonal.*

PROOF. If $x_{\mathrm{I}}(t)$ ($\not\equiv 0$) and $x_{\mathrm{II}}(t)$ ($\not\equiv 0$) are two eigenfunctions of the boundary value problem (1.18), (1.18′), then

$$(1.24)\qquad \begin{aligned} &\mathscr{Y}\frac{dx_{\mathrm{I}}}{dt} - \mu_{\mathrm{I}}\mathscr{A}(t)\,x_{\mathrm{I}} = 0, \quad \mathscr{A}x_{\mathrm{I}}(a) + \mathscr{B}x_{\mathrm{I}}(b) = 0,\\ &\mathscr{Y}\frac{dx_{\mathrm{II}}}{dt} - \mu_{\mathrm{II}}\mathscr{A}(t)\,x_{\mathrm{II}} = 0, \quad \mathscr{A}x_{\mathrm{II}}(a) + \mathscr{B}x_{\mathrm{II}}(b) = 0. \end{aligned}$$

Taking the Hermitian conjugate of every term in the second differential equation yields

$$-\frac{dx_{\mathrm{II}}^*}{dt}\mathscr{Y} - \overline{\mu}_{\mathrm{II}}x_{\mathrm{II}}^*\mathscr{A}(t) = 0.$$

We take the scalar product of this relation with x_{I} from the right, and the scalar product of the first relation in (1.24) with x_{II} from the left, and subtract the resulting equalities, obtaining

$$(\mu_{\mathrm{I}} - \overline{\mu}_{\mathrm{II}})\,x_{\mathrm{II}}^*\mathscr{A}(t)\,x_{\mathrm{I}} + x_{\mathrm{II}}^*\mathscr{Y}\frac{dx_{\mathrm{I}}}{dt} - \left(\mathscr{Y}\frac{dx_{\mathrm{II}}}{dt}\right)^* x_{\mathrm{I}} = 0.$$

Integrating this relation over the interval $[a,b]$ and using (1.20) yields

$$(1.25)\qquad (\mu_{\mathrm{I}} - \overline{\mu}_{\mathrm{II}})\int_a^b x_{\mathrm{II}}^*\mathscr{A}(t)\,x_{\mathrm{I}}\,dt = 0.$$

If the eigenfunctions $x_{\mathrm{I}}(t)$ and $x_{\mathrm{II}}(t)$ coincide, then $\mu_{\mathrm{I}} = \mu_{\mathrm{II}}$, and by virtue of (1.16) it follows from (1.25) that $\mu_{\mathrm{I}} = \overline{\mu}_{\mathrm{I}}$. Thus, all eigenvalues of (1.18) are real.

For $\mu_{\mathrm{I}} \neq \mu_{\mathrm{II}} = \overline{\mu_{\mathrm{II}}}$, (1.25) implies the $\mathscr{A}$-orthogonality of $x_{\mathrm{I}}(t)$ and $x_{\mathrm{II}}(t)$.

It now remains to note that to a given eigenvalue $\mu = \mu_{\mathrm{I}}$ there can correspond not more than n linearly independent eigenfunctions, for the simple reason that for any μ the number of linearly independent solutions of (1.1) is precisely n.

The theorem is proved.

5. It is not difficult to obtain an equation from which the eigenvalues μ_j of (1.18), (1.18′) can be determined.

Representing the solution $x(t)$ of (1.1) in the form $x(t) = \mathscr{U}(t;\mu)\xi$ ($\xi = x(a)$) and inserting this expression in the boundary condition (1.18′), we obtain

$$(\mathscr{A} + \mathscr{B}\mathscr{U}(b;\mu))\xi = 0.$$

It is thus evident that a value $\mu = \mu_0$ will be an eigenvalue of the problem (1.18), (1.18′) if and only if it is a zero of the entire function

$$\mathscr{D}(\mu) = \det(\mathscr{A} + \mathscr{B}\mathscr{U}(b;\mu)).$$

To each zero μ_0 of the function $\mathscr{D}(\mu)$ there correspond exactly d_0 ($\leqq n$) linearly independent eigenfunctions of the boundary value problem (1.18), (1.18′), where d_0 is the defect (the difference of the order and the rank) of the matrix

$$\mathscr{V}(\mu) = \mathscr{A} + \mathscr{B}\mathscr{U}(b;\mu)$$

for $\mu = \mu_0$. Thus,

the multiplicity of any eigenvalue μ_0 of the boundary value problem (1.18), (1.18′) coincides with the defect of the matrix $\mathscr{A} + \mathscr{B}\mathscr{U}(b;\mu_0)$.

A more precise result is the following.

THEOREM 1.2. *The multiplicity of any eigenvalue μ_0 of (1.18), (1.18′) coincides with its multiplicity as a root of the equation $\mathscr{D}(\mu) = 0$.*

In other words, we are asserting that the defect of the matrix $\mathscr{A} + \mathscr{B}\mathscr{U}(b;\mu_0)$ coincides with the multiplicity of μ_0 as a zero of $\mathscr{D}(\mu)$.

This result can be proved, using exclusively the $\mathscr{Y}$-unitarity and $\mathscr{Y}$-contractiveness of the entire matrix-function $\mathscr{U}(b;\mu)$ (see M. G. Kreĭn [8], where it is deduced in this way for the case $\det\mathscr{B} \neq 0$).

We shall give a more elementary proof for the same case $\det\mathscr{B} \neq 0$,[2] in which (1.18′) can be rewritten in the form

$$\mathscr{E}x(a) + x(b) = 0,$$

where $\mathscr{E} = \mathscr{B}^{-1}\mathscr{A}$ is a $\mathscr{Y}$-unitary matrix.

PROOF. Assuming the contrary, i.e. that the defect of the matrix

[2] In the singular case, when $\det\mathscr{B} = 0$, Theorem 1.2 can be obtained by a passage to the limit from the case $\det\mathscr{B} \neq 0$. In the sequel Theorem 1.2 will be used only for the nonsingular case.

$\mathscr{W}(\mu) = \mathscr{E} + \mathscr{U}(b;\mu)$ for $\mu = \mu_0$ does not coincide with the multiplicity of μ_0 as a zero of det $\mathscr{W}(\mu)$, we can assert[3] that for $\mu = \mu_0$ there exists an eigenvector ξ_0 ($\neq 0$) and a corresponding *associated* vector η_0 such that

$$\mathscr{W}(\mu_0)\xi_0 = 0, \ \mathscr{W}(\mu_0)\eta_0 + \mathscr{W}'(\mu_0)\xi_0 = 0.$$

Written out in detail for the case being considered, this says that

$$(1.26) \quad (\mathscr{E} + \mathscr{U}(b;\mu_0))\xi_0 = 0, \ (\mathscr{E} + \mathscr{U}(b;\mu_0))\eta_0 + \dot{\mathscr{U}}(b;\mu_0)\xi_0 = 0,$$

where $\dot{\mathscr{U}}(b;\mu)$ is the derivative $\partial\mathscr{U}(b;\mu)/\partial\mu$. Since the matrices $\mathscr{E}$ and $\mathscr{U}$ are $\mathscr{Y}$-unitary, we have

$$\mathscr{Y}\mathscr{E} = \mathscr{E}^{*-1}\mathscr{Y}, \quad \mathscr{U}^*\mathscr{Y}\mathscr{U} = \mathscr{Y}.$$

Therefore if we multiply the terms of the second relation in (1.26) from the left by $\xi_0^*\mathscr{U}^*(b;\mu_0)\mathscr{Y}$, we obtain

$$(1.27) \qquad \xi_0^*(\mathscr{U}_0^*\mathscr{E}^{*-1} + I)\mathscr{Y}\eta_0 + \xi_0^*\mathscr{U}_0^*\mathscr{Y}\dot{\mathscr{U}}_0\xi_0 = 0 \quad (\mathscr{U}_0 = \mathscr{U}(b;\mu_0)).$$

On the other hand, since by the first relation in (1.26)

$$(I + \mathscr{E}^{-1}\mathscr{U}_0)\xi_0 = 0, \quad \xi_0^*(I + \mathscr{U}_0^*\mathscr{E}^{*-1}) = 0,$$

(1.27) yields

$$(1.28) \qquad \xi_0^*\mathscr{U}^*(b;\mu_0)\mathscr{Y}\dot{\mathscr{U}}(b;\mu_0)\xi_0 = 0.$$

We put

$$x(t;\mu) = \mathscr{U}(t;\mu)\xi_0, \quad y(t;\mu) = \frac{\partial x(t;\mu)}{\partial\mu} = \dot{\mathscr{U}}(t;\mu)\xi_0,$$

$$x_0(t) = x(t;\mu_0), \quad y_0(t) = y(t;\mu_0).$$

Then we have

$$(1.29) \qquad \mathscr{Y}\frac{dx_0}{dt} = \mu_0\mathscr{A}(t)x_0,$$

$$(1.30) \qquad \mathscr{Y}\frac{dy_0}{dt} = \mu_0\mathscr{A}(t)y_0 + \mathscr{A}(t)x_0.$$

To obtain (1.30), one differentiates (1.1) with respect to μ and then puts $\mu = \mu_0$.

---

[3] This follows from the theory of the elementary divisors of polynomial μ-matrices, which can be carried over to the case of any analytic matrix-functions. A more complete statement regarding the connection between the number and the lengths of chains of eigenvectors and associated vectors of a polynomial (analytic) μ-matrix $\mathscr{W}(\mu)$ and the multiplicity of the zeros of det$\mathscr{W}(\mu)$ is given in a paper by V. M. Eni [1].

Now the relation (1.28) implies that

$$x_0^*(b)\,\mathscr{Y}\,y_0(b) = 0. \tag{1.31}$$

On the other hand, it is easily deduced from (1.29) and (1.30) that

$$\frac{dx_0^*}{dt}\,\mathscr{Y}\,y_0 + x_0^*\,\mathscr{Y}\,\frac{dy_0}{dt} = x_0^*\,\mathscr{A}(t)\,x_0$$

and therefore

$$\int_a^b \left(\frac{dx_0^*}{dt}\,\mathscr{Y}\,y_0 + x_0^*\,\mathscr{Y}\,\frac{dy_0}{dt}\right) dt = \int_a^b x_0^*(t)\,\mathscr{A}(t)\,x_0(t)\,dt\;(>0).$$

The left side equals $x_0^*(b)\,\mathscr{Y}\,y_0(b) - x_0^*(a)\,\mathscr{Y}\,y_0(a)$, and hence, by virtue of (1.31) and the relation $y_0(a) = 0$,[4)] it equals zero.

We have arrived at a contradiction. The theorem is proved.

6. The foregoing implies that $\mu = 0$ fails to be an eigenvalue of the boundary value problem (1.18), (1.18′) if and only if

$$\mathscr{D}(0) = \det(\mathscr{A} + \mathscr{B}) \neq 0.$$

In this case the boundary value problem can easily be transformed to an equivalent weighted integral equation

$$x(t) = \mu \int_a^b \mathscr{G}(t,s)\,\mathscr{A}(s)\,x(s)\,ds, \tag{1.32}$$

where

$$\mathscr{G}(t,s) = -\tfrac{1}{2}\,\mathscr{Y}\,\operatorname{sign}(t-s) - \tfrac{1}{2}\,\mathscr{Y} + (\mathscr{A}+\mathscr{B})^{-1}\mathscr{B}\mathscr{Y} \tag{1.33}$$

is a Hermitian kernel, i.e.

$$\mathscr{G}(t,s) = \mathscr{G}^*(s,t). \tag{1.34}$$

Indeed, let us consider the system

$$\mathscr{Y}\,\frac{dx}{dt} = f(t), \quad \mathscr{A}x(a) + \mathscr{B}x(b) = 0, \tag{1.35}$$

where $f(t)$ is some vector-function with components from $L_1(a,b)$.

It follows from (1.35) that

$$x(t) = -\,\mathscr{Y}\int_a^t f(s)\,ds + \xi, \tag{1.36}$$

---

4) Since $y(a;\mu) = \dot{\mathscr{U}}(a;\mu)\,\xi_0 = 0$.

where the constant vector ξ is determined from the condition

$$\mathscr{A}\xi + \mathscr{B}\left(\xi - \mathscr{Y}\int_a^b f(s)\,ds\right) = 0.$$

Determining ξ from this condition and inserting its value in (1.36), we find that

$$x(t) = \int_a^b \mathscr{G}(t,s) f(s)\,ds, \tag{1.37}$$

where the matrix Green's function $\mathscr{G}(t,s)$ has the form (1.33).

In order to obtain the equation (1.32), it suffices to note that for $f(t) = \mu\mathscr{H}(t)x(t)$ the system (1.35) goes over into the system (1.18), (1.18′).

The property (1.34) of the Green's function $\mathscr{G}(t,s)$ can be verified directly by means of algebraic computations. But it can be obtained more simply by an argument of a general character. Let $f(t)$ and $g(t)$ be two arbitrary n-dimensional continuous vector-functions. We define $x(t)$ by (1.37), and $y(t)$ by the corresponding relation

$$y(t) = \int_a^b \mathscr{G}(t,s) g(s)\,ds. \tag{1.38}$$

Then $x(t)$ will satisfy the system (1.35), and $y(t)$ will satisfy a similar system:

$$\mathscr{Y}\frac{dy}{dt} = g(t), \quad \mathscr{A}y(a) + \mathscr{B}y(b) = 0. \tag{1.39}$$

The functions $x(t)$ and $y(t)$ will satisfy the relation (1.20). Using the relations (1.35), (1.37), (1.38) and (1.39) which hold among the functions x and f, y and g, the relation (1.20) can be rewritten in the following form:

$$\int_a^b \left(\int_a^b g^*(s)\mathscr{G}^*(t,s)\,ds\right) f(t)\,dt = \int_a^b g^*(t)\left(\int_a^b \mathscr{G}(t,s) f(s)\,ds\right) dt.$$

Since $g(t)$ and $f(t)$ are arbitrary continuous vector-functions, (1.34) follows.

The weighted integral equation (1.32) can be reduced to an ordinary integral equation with a Hermitian matrix-kernel.

Since for each $t \in [a,b]$ the matrix $\mathscr{H}(t)$ is nonnegative definite, there exists for each $t \in [a,b]$ a unique matrix of the same type whose square is $\mathscr{A}(t)$. We denote this matrix by $\mathscr{A}^{1/2}(t)$. The columns of this matrix obviously form vector-functions from $L_2^{(n)}(a,b)$ (see the second footnote on p. 225).

If in (1.32) we make the substitution

$$\phi(t) = \mathscr{A}^{1/2}(t)\, x(t) \tag{1.40}$$

and multiply the resulting equation from the left by $\mathscr{A}^{1/2}(t)$, we obtain an integral equation in $L_2^{(n)}(a,b)$,

$$\phi(t) = \mu \int_a^b \mathscr{A}(t,s)\phi(s)\, ds, \tag{1.41}$$

with a Hermitian Hilbert-Schmidt matrix kernel

$$\mathscr{A}(t,s) = \mathscr{A}^{1/2}(t)\mathscr{G}(t,s)\mathscr{A}^{1/2}(s). \tag{1.42}$$

Conversely, if $\phi(t)$ ($\in L_2^{(n)}(a,b)$) is a solution of (1.41), then it follows from (1.41) and (1.42) that $\phi(t)$ must have the form (1.40), where

$$x(t) = \mu \int_a^b \mathscr{G}(t,s)\mathscr{A}^{1/2}(s)\phi(s)\, ds \tag{1.43}$$

is a continuous function (the continuity of $x(t)$ is easy to show if one breaks up the integral (1.43) into the sum of two integrals: from a to t, and from t to b). If we insert the value of ϕ from (1.40) into (1.43), we come back to the equation (1.32).

Thus, equations (1.32) and (1.41) are equivalent (despite the fact that the matrix $\mathscr{A}^{1/2}(t)$ in the substitution (1.40) can happen to be degenerate for all t).

Among other things, this at once implies Theorem 1.1[5] for the case under consideration ($\det(\mathscr{A}+\mathscr{B}) \neq 0$). For the case $\det(\mathscr{A}+\mathscr{B}) = 0$ one can make the substitution $\mu \to \mu + \alpha$ in equation (1.18), where $\mathscr{D}(\alpha) \neq 0$, and then transform it to canonical form. Thus, Theorem 1.1 will be proved in the exceptional case as well.

Let us note that in reducing the boundary value problem (1.18), (1.18′) to the integral equation (1.41) with a Hermitian matrix kernel $\mathscr{A}(t,s)$, we used precisely those three conditions under which the boundary value problem (1.18), (1.18′) was said to be selfadjoint.

7. The possibility of reducing the boundary value problem (1.18), (1.18′) to an integral equation (1.41) enables us to prove the following result very simply.

[5] With the exception of the assertion regarding the multiplicities of the eigenvalues.

THEOREM 1.3 (ON THE INTERLACING OF THE SPECTRA). *Let*

$$\text{(I)} \qquad \mathscr{Y}\frac{dx}{dt} - \mu\mathscr{H}(t)x = 0; \quad \mathscr{A}_1 x(a) + \mathscr{B}_1 x(b) = 0,$$

$$\text{(II)} \qquad \mathscr{Y}\frac{dx}{dt} - \mu\mathscr{H}(t)x = 0; \quad \mathscr{A}_2 x(a) + \mathscr{B}_2 x(b) = 0$$

be two selfadjoint boundary value problems which differ only in their boundary conditions, and let $\mu_1^{(1)} \leqq \mu_2^{(1)} \leqq \cdots$ *and* $\mu_1^{(2)} \leqq \mu_2^{(2)} \leqq \cdots$ *be, respectively, the set of all nonnegative eigenvalues of the first and second problems. Then*

$$\mu_j^{(1)} \leqq \mu_{j+n}^{(2)} \text{ and } \mu_j^{(2)} \leqq \mu_{j+n}^{(1)} \qquad (j = 1, 2, \cdots). \tag{1.44}$$

The analogous inequalities hold for the nonpositive eigenvalues of (I) *and* (II)

PROOF. Let us first assume that $\mu = 0$ is not an eigenvalue for either of (I) and (II). We denote by $\mathscr{H}_1(t,s)$ and $\mathscr{H}_2(t,s)$ the kernels of the integral equations corresponding to these problems; then

$$\mathscr{H}_j(t,s) = \mathscr{H}^{1/2}(t)\mathscr{G}_j(t,s)\mathscr{H}^{1/2}(s),$$

where, according to (1.33),

$$\mathscr{G}_j(t,s) = -\tfrac{1}{2}\mathscr{Y}\operatorname{sign}(t-s) - \tfrac{1}{2}\mathscr{Y} - (\mathscr{A}_j + \mathscr{B}_j)^{-1}\mathscr{B}_j \qquad (j = 1,2).$$

The difference

$$\mathscr{C} = \mathscr{G}_1(t,s) - \mathscr{G}_2(t,s)$$

is obviously a constant matrix, and therefore to the difference

$$\mathscr{R}(t,s) = \mathscr{H}_1(t,s) - \mathscr{H}_2(t,s) = \mathscr{H}^{1/2}(t)\mathscr{C}\mathscr{H}^{1/2}(s)$$

there corresponds the operator

$$(Rf)(t) = \mathscr{H}^{1/2}(t)\mathscr{C}\int_a^b \mathscr{H}^{1/2}(s)f(s)\,ds,$$

which has dimension $\leqq n$. Indeed, this operator maps $L_2^{(n)}(a,b)$ into the linear hull of the vector-functions which constitute the columns of the matrix $\mathscr{H}^{1/2}(t)$.

By virtue of the minimax properties of eigenvalues, we obtain the inequalities (1.44) for the case under consideration.[6)]

The excluded case, in which $\mu = 0$ is an eigenvalue of at least one of the boundary value problems (I) and (II), can be reduced to the case just considered by means of a parameter shift $\mu \to \mu - \epsilon$, where ϵ (> 0) is less than the smaller of the absolute values of the first negative eigenvalues of both problems (if at least one such eigenvalue exists; otherwise we can take any $\epsilon > 0$). The theorem is proved.

§2. Skewperiodic boundary value problems and model operators

1. Among selfadjoint boundary value problems of the form (1.18), (1.18′), a special place is occupied by the skewperiodic problem:

$$\mathscr{Y} \frac{dx}{dt} - \mu \mathscr{A}(t) x = 0, \quad x(a) + x(b) = 0. \tag{2.1}$$

This problem is singled out from the others because one can state a whole series of additional characteristics of its spectrum.

Obviously $\mu = 0$ is not an eigenvalue of this boundary value problem. Since the Green's function $\mathscr{G}(t,s)$ of the problem

$$\mathscr{Y} \frac{dx}{dt} = f(t), \quad x(a) + x(b) = 0$$

has the form

$$\mathscr{G}(t,s) = -\tfrac{1}{2} \mathscr{Y} \operatorname{sign}(t-s) \qquad (a \leqq t, s \leqq b),$$

the boundary value problem (2.1) is equivalent (in the sense indicated in the previous section) to an integral equation in $L_2^{(n)}(a,b)$:

[6)] We have in mind the following result, which follows from the minimax properties of the eigenvalues of a selfadjoint completely continuous operator.

Let $H_k = H_k^* \in \mathfrak{S}_\infty$ $(k = 1, 2)$, and let $H_2 - H_1$ be a finite-dimensional operator of dimension $\leqq n$. Then for the complete (nondecreasing) sequences $\{\mu_j^+(H_1)\}$ and $\{\mu_j^+(H_2)\}$ of positive characteristic values of these operators we have the following "generalized interlacing relations":

$$\mu_j^+(H_2) \leqq \mu_{j+n}^+(H_1), \quad \mu_j^+(H_1) \leqq \mu_{j+n}^+(H_2) \quad (j = 1, 2, \cdots).$$

The analogous relations hold for the negative characteristic values of H_1 and H_2.

For the case in which H_1 and H_2 are nonnegative operators, this result is contained in Corollary 2.1, Chapter II of [GK]. For the general case it can be proved by the same methods.

$$(2.2) \qquad \phi(t) = -\frac{\mu}{2}\int_a^b \mathscr{A}^{1/2}(t)\,\mathscr{Y}\,\operatorname{sign}(t-s)\,\mathscr{A}^{1/2}(s)\phi(s)\,ds.$$

We associate with the problem (2.1) the Volterra operator

$$(2.3) \qquad (A_M f)(t) = \int_t^b \mathscr{A}^{1/2}(t)\,\mathscr{Y}\,\mathscr{A}^{1/2}(s) f(s)\,ds,$$

which we called, for reasons which were clarified earlier (see Chapter V, §3), a *model* operator.

The following simple relation holds between the boundary value problem (2.1) and model operators.

LEMMA 2.1. *The spectrum of eigenvalues of the skewperiodic boundary value problem* (2.1) *coincides with the spectrum of characteristic values of the real component of a model operator.*

In fact, since $\mathscr{Y}^* = -\mathscr{Y}$ it follows, as was already mentioned (see Chapter V, §3), that the real component $A_{M\mathscr{R}} = (A_M + A_M^*)/2$ of the operator A_M is given by

$$(A_{M\mathscr{R}} f)(t) = -\frac{1}{2}\int_a^b \mathscr{A}^{1/2}(t)\mathscr{Y}\operatorname{sign}(t-s)\mathscr{A}^{1/2}(s) f(s)\,ds,$$

and hence equation (2.2) can be written as $\phi = \mu A_{M\mathscr{R}}\phi$. The lemma is proved.

For the imaginary component of A_M we obviously have

$$\frac{1}{2i}(A_M - A_M^*) f(t) = \frac{1}{2i}\int_a^b \mathscr{A}^{1/2}(t)\mathscr{Y}\mathscr{A}^{1/2}(s) f(s)\,ds,$$

and therefore it is not more than n-dimensional.

On the other hand, as we know (see Theorem 2.4, Chapter III), the principal value of the trace of the real component of any Volterra operator $A = A_{\mathscr{R}} + iA_{\mathscr{I}}$ with nuclear imaginary component equals zero, i.e.

$$\lim_{r\to\infty} \sum_{|\mu_j| \leqq r} \frac{1}{\mu_j(A_{\mathscr{R}})} = 0.$$

Applying this result to a model operator leads to the following statement.

THEOREM 2.1. *The spectrum of a selfadjoint skewperiodic boundary value problem* (2.1) *contains at least one positive and one negative eigenvalue.*

This theorem was first established by M. G. Kreĭn [8] (see also M. G. Kreĭn [7]) using methods of the theory of entire functions. Subsequently V. B. Lidskiĭ [1] drew attention to the fact that this theorem of the theory of differential equations can be interpreted as a statement concerning the theory of Volterra operators (following this, the theorem leads, in conjunction with Lemma 2.1 and results of M. S. Livšic in the theory of models of nonselfadjoint operators, to certain conclusions regarding the completeness of the systems of root vectors of nonselfadjoint operators).

Here we have followed the opposite route: we have obtained Theorem 2.1 as a simple corollary of a general statement regarding Volterra operators with nuclear imaginary component.

We remark that we could have associated with every (not only skewperiodic) selfadjoint boundary value problem (1.18), (1.18′) a Volterra operator, namely the operator

$$(Af)(t) = \int_t^b \mathscr{A}^{1/2}(t)\mathscr{G}(t,s)\mathscr{A}^{1/2}(s)f(s)\,ds \quad (f \in L_2^{(n)}(a,b));$$

then the spectrum of characteristic numbers of the real component of A coincides with the spectrum of eigenvalues of the problem (1.18), (1.18′). Here the kernel $\mathscr{G}(t,s)$ is the Green's function of the problem (1.18), (1.18′), by means of which one obtains the integral equation (1.41) which is equivalent to the given boundary value problem.

However, if the boundary value problem is other than skewperiodic, the imaginary component of this Volterra operator will not be finite-dimensional (nor, generally speaking, nuclear).

It turns out that the interval (μ_{-1}, μ_1), where μ_1 and μ_{-1} are respectively the first positive and first negative eigenvalues of the skewperiodic boundary value problem (2.1), plays an important role in the theory of canonical equations having an H-matrix which can be periodically extended: $\mathscr{A}(t+l) = \mathscr{A}(t)$ $(l = b - a)$. This will be clarified in the next section.

2. The eigenvalues of a skewperiodic boundary value problem in the *definite case* $\mathscr{Y} = -iI_n$ admit a simple lower bound.

THEOREM 2.2. *Let* $\mu_1^+ \leqq \mu_2^+ \leqq \cdots$ *and* $-\mu_1^- \geqq -\mu_2^- \geqq \cdots$ *be respectively the positive and negative parts of the spectrum of eigenvalues of the definite boundary value problem*

$$dx/dt - i\mu\mathscr{A}(t)x = 0, \quad x(a) + x(b) = 0. \tag{2.4}$$

Then

$$\mu_{jn+k}^{\pm} \geqq \left(\int_a^b \operatorname{sp} \mathscr{A}(t)\, dt \right)^{-1} ((2j-1)n + 2k)\pi \tag{2.5}$$

$$(j = 1, 2, \cdots;\ k = 0, 1, \cdots, n-1).$$

PROOF. The spectrum of the problem (2.4) is determined from the equation

$$\det(I_n + \mathscr{U}(\mu)) = 0 \quad (\mathscr{U}(b;\mu) = \mathscr{U}(\mu)). \tag{2.6}$$

In the case being considered ($\mathscr{Y} = -iI_n$) the matrix $\mathscr{U}(b;\mu)$ will be unitary for real μ. Let us denote its eigenvalues (*multiplicators*) by $\rho_1(\mu), \rho_2(\mu), \cdots, \rho_n(\mu)$; we can regard them as continuous functions of $\mu \in (-\infty, \infty)$, numbered so that

$$\arg \rho_1(\mu) \geqq \arg \rho_2(\mu) \geqq \cdots \geqq \arg \rho_n(\mu).$$

We assume here that $\arg \rho_j(\mu)$ $(j = 1, 2, \cdots, n)$ is also defined as a continuous function of the real variable μ, and that

$$\arg \rho_1(0) = \arg \rho_2(0) = \cdots = \arg \rho_n(0) = 0.$$

This is possible since $\mathscr{U}(0) = I_n$ and $\rho_1(0) = \rho_2(0) = \cdots = \rho_n(0) = 1$. Since, on the one hand,

$$\det \mathscr{U}(\mu) = \rho_1(\mu)\rho_2(\mu) \cdots \rho_n(\mu),$$

and on the other hand, by the theorem of Liouville-Ostrogradskiĭ,

$$\det \mathscr{U}(\mu) = \exp \left(i\mu \int_a^b \operatorname{sp} \mathscr{A}(t)\, dt \right),$$

we can assert that

$$\arg \rho_1(\mu) + \cdots + \arg \rho_n(\mu) = \mu h \quad \left(h = \int_a^b \operatorname{sp} \mathscr{A}(t)\, dt \right). \tag{2.7}$$

We state, for the moment without proof, the following property of multiplicators.

1. *As μ (real) increases, the multiplicators $\rho_1(\mu), \cdots, \rho_n(\mu)$ can only move in a counter-clockwise direction on the unit circle.*

Equation (2.6) is equivalent to the equation

$$\prod_{k=1}^{n} (1 + \rho_k(\mu)) = 0.$$

Thus, a number $\mu = \mu_0$ will be an eigenvalue of the boundary value problem (2.4) if and only if for $\mu = \mu_0$ some of the multiplicators ρ_k are found at the point -1. The number of these is the multiplicity of the eigenvalue $\mu = \mu_0$, since it coincides with the defect of the matrix $I_n + \mathscr{U}(b; \mu_0)$.

For $\mu = 0$ the multiplicators ρ_k lie at the point 1; as μ increases they start to move on the unit circle in a counter-clockwise direction. If $\mu_0 = \mu_k^+$ $(k = 1, 2, \cdots, n)$, we can assert that $\rho_k(\mu_0) = -1$ and

$$\arg \rho_k(\mu_0) = \pi,$$

and that

$$\arg \rho_j(\mu_0) \geqq \pi \quad (j = 1, 2, \cdots, k-1),$$
$$0 \leqq \arg \rho_j(\mu_0) \leqq \pi \quad (j = k+1, \cdots, n).$$

Thus for $\mu = \mu_k^+$ it follows from (2.7) that

$$\mu_k^+ h \geqq \pi k \quad (k = 1, 2, \cdots, n).$$

As μ increases from 0 to μ_{n+k}^+ $(k = 1, 2, \cdots, n)$ all n multiplicators traverse the upper semicircle and at least k "avant garde" multiplicators, having traversed the entire circle once and the upper semicircle a second time, have passed or at least reached the point -1. Therefore for $\mu = \mu_{n+k}^+$ the relation (2.7) yields the bound

$$\mu_{n+k}^+ h \geqq n\pi + 2k\pi \quad (k = 1, 2, \cdots, n).$$

Arguing further in the same way, we find that

$$\mu_{jn+k}^+ h \geqq (2j-1)n\pi + 2k\pi \quad (j = 1, 2, \cdots; \ k = 1, 2, \cdots, n).$$

Analogous arguments yield the same bounds for μ_{jn+k}^-. Thus, to complete the proof of Theorem 2.2 it remains to prove result 1.

To this end we compute the derivative $\dot{\mathscr{U}}(t; \mu) = \partial \mathscr{U}(t; \mu) / \partial \mu$. Differentiating each term of the system (1.5′), with $\mathscr{Y} = -iI_n$, with respect to μ, we obtain

$$\frac{d\dot{\mathscr{U}}(t; \mu)}{dt} = i\mu \mathscr{A}(t) \dot{\mathscr{U}}(t; \mu) + i\mathscr{A}(t) \mathscr{U}(t; \mu),$$
$$\dot{\mathscr{U}}(a; \mu) = 0.$$

Hence we find that

$$\dot{\mathscr{U}}(t; \mu) = i\mathscr{U}(t; \mu) \int_a^t \mathscr{U}^{-1}(s; \mu) \mathscr{A}(s) \mathscr{U}(s; \mu) \, ds.$$

Setting $t = b$, we obtain

(2.8)
$$\frac{d\mathscr{U}(\mu)}{d\mu} = i\mathscr{U}(\mu)\mathscr{R}(\mu),$$

where

$$\mathscr{R}(\mu) = \int_a^b \mathscr{U}^*(s;\mu)\mathscr{A}(s)\mathscr{U}(s;\mu)\,ds \quad (-\infty < \mu < \infty).$$

Since the H-matrix $\mathscr{A}$ is of positive type, the matrix $\mathscr{R}(\mu)$ is Hermitian-nonnegative (one can show that it is indeed Hermitian-positive) for all $\mu \in (-\infty, \infty)$. Let ϵ be an arbitrary number of modulus unity. We consider those values of μ for which $\det(\mathscr{U}(b;\mu) - \epsilon I_n) \neq 0$, and so the Cayley transform

(2.9)
$$\mathscr{Q}(\mu) = i(\epsilon I_n + \mathscr{U}(b;\mu))(\epsilon I_n - \mathscr{U}(b;\mu))^{-1} \quad (-\infty < \mu < \infty)$$

is defined. The matrix $\mathscr{Q}(\mu)$ will be Hermitian with eigenvalues

$$q_j(\mu) = i(\epsilon + \rho_j(\mu))(\epsilon - \rho_j(\mu))^{-1} \quad (j = 1, 2, \cdots, m).$$

On the other hand, it follows from (2.8) and (2.9) that

$$\frac{d\mathscr{Q}}{d\mu} = +2i\epsilon(\mathscr{U}(\mu) - \epsilon I_n)^{-1}\frac{d\mathscr{U}}{d\mu}(\mathscr{U}(\mu) - \epsilon I_n)^{-1}$$
$$= 2[(\mathscr{U}(\mu) - \epsilon I_n)^*]^{-1}\mathscr{R}(\mu)(\mathscr{U}(\mu) - \epsilon I_n)^{-1}.$$

Since the last expression is a Hermitian-nonnegative function, $\mathscr{Q}(\mu)$ is a Hermitian-nondecreasing function of the real variable μ. This means that the eigenvalues $q_j(\mu)$ $(j = 1, 2, \cdots, n)$ are nondecreasing as μ increases. From this it now follows that as μ increases the multiplicators $\rho_j(\mu)$ $(j = 1, 2, \cdots, n)$ can only move in a counter-clockwise direction on the unit circle.

Since by an appropriate choice of ϵ $(|\epsilon| = 1)$ the transformation (2.9) can be made meaningful in some neighborhood of any prescribed point $\mu \in (-\infty, \infty)$, result 1 (and hence Theorem 2.2) is proved.

The bounds (2.5) are exact. If $\mathscr{A}(t)$ is a diagonal H-matrix of positive type, equality holds in (2.5) for $k = 0$ and $j = 1, 2, \cdots$.

Since for any $k = 0, 1, \cdots, n-1$; $n = 1, 2, \cdots$ one has $(2j-1)n + 2k > jn + k$ $(j = 1, 2, \cdots)$, we have

COROLLARY 2.1. *For any boundary value problem* (2.4) *with phase dimension n, the nonnegative quantities $\mu_j^{\pm}$ satisfy the bounds*

$$\frac{1}{\mu_j^{\pm}} \leq \frac{1}{\pi j}\int_a^b \operatorname{sp}\mathscr{A}(t)\,dt \quad (j = 1, 2, \cdots).$$

Theorem 2.2, together with the theorem concerning models for dissipative Volterra operators with an n-dimensional imaginary component, immediately yields the second assertion of Theorem 2.5, Chapter III.

The remark made at the end of §2, Chapter III, enables us to state that the bounds (2.5) hold for the spectrum of *any* (not necessarily definite) skewperiodic boundary value problem (2.1), if we divide their right sides by two.

3. Let us introduce the following notation. We shall denote by $\Lambda_p(X)$ the sum of the pth powers of the moduli of all the eigenvalues of a square matrix X.[7]

The following result is a simple corollary of Theorem 2.1, Chapter III.

THEOREM 2.3. *The relations*

$$\left\{\sum_{j=1}^{k} \frac{1}{\mu_j}\right\} \Big/ \left\{\sum_{j=1}^{k} \frac{1}{2j-1}\right\} \leqq \frac{1}{\pi} \Lambda_1\left(\mathscr{Y}\int_a^b \mathscr{H}(t)\,dt\right), \quad (k = 1, 2, \cdots)$$

as well as the analogous relations obtained by replacing μ_j *by* $|\mu_{-j}|$,[8] *hold for the spectrum* $\{\mu_j\}$ *of a skewperiodic boundary value problem* (2.1).

These relations are exact in the sense that equality holds in them when, for example, $\mathscr{H}(t)$ and $\mathscr{Y}$ are diagonal matrices and k is a multiple of n.

By virtue of Theorem 1.3 on the interlacing of the spectrum, for the spectrum $\{\mu_j\}$ of the general problem (1.18), (1.18′) one can only assert that

$$\left\{\sum_{j=1}^{k} \frac{1}{\mu_{j+n}}\right\} \Big/ \left\{\sum_{j=1}^{k} \frac{1}{2j-1}\right\} \leqq \frac{1}{\pi} \Lambda_1\left(\mathscr{Y}\int_a^b \mathscr{H}(t)\,dt\right) \quad (k = 1, 2, \cdots)$$

and that the analogous relations hold for the negative μ_j $(j = -1, -2, \cdots)$.

§3. Central zone of stability of a canonical differential equation with a periodic H-matrix

1. We consider the canonical equation (with $\mu = 1$)

$$\mathscr{Y}\frac{dx}{dt} - \mathscr{H}(t)\,x = 0 \tag{3.1}$$

with a periodic H-matrix

$$\mathscr{H}(t + l) = \mathscr{H}(t). \tag{3.2}$$

As before, we assume that all the elements $h_{jk}(t)$ $(j, k = 1, 2, \cdots, n)$ of $\mathscr{H}(t) = \| h_{jk}(t) \|_1^n$ are locally integrable.

[7] In what follows we shall use only the functionals Λ_1 and $\Lambda_{1/2}$.

[8] We have used here the simplified notation $\mu_j = \mu_j^+$, $\mu_{-j} = -\mu_j^-$ $(j = 1, 2, \cdots)$.

Any solution $x(t)$ of (3.1) can be represented in the form

$$x(t) = \mathscr{U}(t)\,x(0), \tag{3.3}$$

where $\mathscr{U}(t)$ is the matrizant:

$$\mathscr{Y}\frac{d\mathscr{U}}{dt} - \mathscr{A}(t)\mathscr{U} = 0, \quad \mathscr{U}(0) = I_n.$$

Making the change of variables $t \to t + l$ in (3.1), we see that $\mathscr{U}(t+l)$ is the solution of the Cauchy problem

$$\mathscr{Y}\frac{d\mathscr{V}}{dt} - \mathscr{A}(t)\mathscr{V} = 0, \quad \mathscr{V}(0) = \mathscr{U}(l).$$

On the other hand, the solution of this problem is the matrix $\mathscr{V}(t) = \mathscr{U}(t)\mathscr{U}(l)$, whence

$$\mathscr{U}(t+l) = \mathscr{U}(t)\mathscr{U}(l), \tag{3.4}$$

and therefore

$$\mathscr{U}(t+nl) = \mathscr{U}(t)\mathscr{U}^n(l) \quad (n = 0, \pm 1, \cdots). \tag{3.5}$$

The matrix $\mathscr{U}(l)$ is called the *monodromy matrix* of equation (3.1), and its eigenvalues ρ_j $(j = 1, 2, \cdots, n)$ are called the *multiplicators* of this equation.

With every ρ_0 we associate the eigenspace $\mathfrak{L}_{\rho_0}$ $(\subseteq E_n)$ of the matrix $\mathscr{U}(l)$. To each vector $\xi \in \mathfrak{L}_{\rho_0}$ $(\xi \neq 0)$ there corresponds a solution $x(t) = \mathscr{U}(t)\xi$ of (3.1) with the property

$$x(t+l) = \rho_0 x(t) \quad (-\infty < t < \infty), \tag{3.6}$$

since by (3.4) we will have

$$x(t+l) = \mathscr{U}(t+l)\xi = \mathscr{U}(t)\mathscr{U}(l)\xi = \rho_0\mathscr{U}(t)\xi = \rho_0 x(t).$$

Conversely, if some solution $x(t)$ $(\not\equiv 0)$ of (3.1) has the property (3.6), then in particular $x(l) = \rho_0 x(0)$, and by (3.3) ρ_0 will be a multiplicator (an eigenvalue of the matrix $\mathscr{U}(l)$), and $\xi = x(0)$ will be a corresponding eigenvector.

It is hence clear that for a given ρ_0 the number of linearly independent solutions $x(t)$ of (3.1) having the property (3.6) coincides with the geometric multiplicity of the number ρ_0, i.e., with the dimension of $\mathfrak{L}_{\rho_0}$, which in turn coincides with the defect of the matrix $\mathscr{U}(l) - \rho_0 I$.

If the algebraic multiplicity of ρ_0 (i.e. its multiplicity as a root of the equation $\det(\mathscr{U}(l) - \rho I) = 0$) coincides with its geometric multiplicity, then ρ_0 is called a *multiplicator of simple type.*[9]

It is not hard to conclude from (3.3) and (3.5) that all solutions of (3.1) will be bounded on the entire real axis if and only if for every vector $\xi \neq 0$ the sequence of vectors $\{\mathscr{U}^k(l)\xi\}$ $(k = 0, \pm 1, \cdots)$ is bounded. This condition, as is easily seen, is equivalent to the condition that *all multiplicators of* (3.1) *are of simple type and have modulus unity.*

Thus, this last condition is a necessary and sufficient condition for the boundedness of all solutions of (3.1) on the entire real axis.

2. All of the preceding considerations are valid for the equation (3.1) with any periodic locally integrable matrix-function $\mathscr{A}(t)$ (not necessarily Hermitian).

If $\mathscr{A}(t)$ is a Hermitian matrix-function, then (cf. result 1, §1) the monodromy matrix will be a $\mathscr{Y}$-unitary matrix,

$$\mathscr{U}^*(l)\mathscr{Y}\mathscr{U}(l) = \mathscr{Y},$$

and consequently

$$\mathscr{Y}\mathscr{U}(l)\mathscr{Y}^{-1} = (\mathscr{U}^*(l))^{-1}.$$

This relation implies the following theorem.

THEOREM OF POINCARÉ-LJAPUNOV. *For a canonical equation* (3.1) *with a periodic H-matrix the multiplicators are situated symmetrically with respect to the unit circle $|\rho| = 1$. To any two multiplicators which are situated symmetrically with respect to the unit circle there correspond identical elementary divisors.*[10]

Strictly speaking, the classical formulation of the Poincaré-Ljapunov theorem is concerned with the case of a Hamiltonian equation (3.1), where $n = 2m$, $\mathscr{Y} = \mathscr{Y}_n$ and $\mathscr{A}(t)$ is a real symmetric matrix-function. In this case the monodromy matrix will also be real, and therefore the multiplicators will also be situated symmetrically with respect to the real axis, and to any two multiplicators which are situated symmetrically with respect to the real axis there correspond the same elementary divisors.

The theorem just stated enables us, inter alia, to assert that all solutions of a canonical equation (3.1) with a periodic H-matrix are bounded on the real axis so long as they are all bounded on the half-axis $(0, \infty)$ (or $(-\infty, 0)$).

---

[9] If $\mathscr{Y} = \pm iI_n$, then $\mathscr{U}(l)$ will be a unitary matrix and all of its eigenvalues will be multiplicators of simple type.

[10] That is, elementary divisors with the same system of exponents.

3. We shall present below various concepts and results (in general, without proof) taken from a paper by M. G. Kreĭn [8].

A multiplicator ρ of the canonical equation (3.1) is said to be *definite* if the Hermitian form

$$i\xi^*\mathcal{Y}\xi \tag{3.7}$$

is nondegenerate and is of one sign on the eigenspace $\mathfrak{L}_\rho$ $(\subseteq E_n)$ of the monodromy matrix $\mathcal{U}(l)$ corresponding to the eigenvalue ρ.

Since for $x(t) = \mathcal{U}(t)\xi$ we have $x^*(t)\mathcal{Y}x(t) = \xi^*\mathcal{Y}\xi$, the *definiteness* of the multiplicator ρ means, in other words, that for all solutions $x(t)$ $(\not\equiv 0)$ $(x(t+l) = \rho x(t))$ corresponding to it the form (3.7) has a fixed sign.

A definite multiplicator ρ is called a *multiplicator of the first kind* or the *second kind,* according as the form (3.7) is positive or negative on $\mathfrak{L}_\rho$.

It turns out that *if a multiplicator is definite, then it is of simple type and lies on the unit circle.*

Thus, if all multiplicators of the equation (3.1) are definite, then all solutions of this equation are bounded. But it turns out that even more can be said.

A) *If all the multiplicators of equation* (3.1) *are definite, then all the solutions of this equation are stably bounded.*[11)]

Let us make clear that the solutions of equation (3.1) are said to be *stably bounded,* if all solutions of any canonical equation

$$\mathcal{Y}\frac{dx}{dt} - \mathcal{A}_1(t)x = 0 \quad (\mathcal{A}_1(t+l) = \mathcal{A}(t))$$

with an H-matrix $\mathcal{A}_1(t)$ sufficiently close to the H-matrix $\mathcal{A}(t)$, are bounded (on the entire real axis).

Here, closeness is understood in the sense of the metric $d(\mathcal{A}_1, \mathcal{A}) = \|\mathcal{A}_1 - \mathcal{A}\|$ of the real Banach space $\mathbf{H}_n(l)$ of all l-periodic H-matrices $\mathcal{A}(t)$ with the norm

$$\|\mathcal{A}\| = \int_0^l |\mathcal{A}(t)|\,dt, \tag{3.8}$$

where for definiteness[12)] we put

---

11) It is shown in a paper of I. M. **Gel′fand** and V. B. Lidskiĭ [1] that this assertion admits a converse (cf. also V. A. Jakubovič [1] and M. G. Kreĭn and G. Ja. Ljubarskiĭ [1]).

12) Since in the finite-dimensional (n^2-dimensional) space of all Hermitian (numerical) matrices of order n all norms are topologically equivalent, the topology in the space $\mathbf{H}_n(l)$ clearly is not changed if in (3.8) we choose any other norm in place of $|\cdot| = |\cdot|_2$.

$$|\mathscr{H}(t)| = |\mathscr{H}(t)|_2 = \left(\sum_{j,k=1}^{n} |h_{jk}(t)|^2\right)^{1/2}.$$

It is easily shown that if all the multiplicators of (3.1) are definite, then there will be (counting multiplicity) exactly n_+ of the first kind and n_- of the second kind, where n_+ and n_- are the number of eigenvalues of the matrix $-i\mathscr{Y}$ equal to $+1$ and -1, respectively.

4. We shall say that a canonical equation (with a parameter)

$$\mathscr{Y}\frac{dx}{dt} - \mu\mathscr{H}(t)x = 0 \quad (\mathscr{H}(t+l) = \mathscr{H}(t)) \tag{3.9}$$

with a periodic H-matrix $\mathscr{H}(t)$ is of *positive type*, if $\mathscr{H}(t)$ is of positive type in the interval $(0, l)$ (cf. the definition in §1.3).

The definite multiplicators of an equation (3.9) of positive type have, as functions of the parameter μ, a number of important properties.

If for some $\mu = \mu_0$ $(-\infty < \mu_0 < \infty)$ all the multiplicators of (3.9) are definite, then this will be the case for all values of μ from some neighborhood $\mu' < \mu < \mu''$ of the point μ_0. For μ increasing in this neighborhood all multiplicators of the first kind move on the unit circle (in a group or separately) in a counter-clockwise direction, and those of the second kind in a clockwise direction. The open interval (μ', μ'') can be extended on both sides as long as two multiplicators of different kinds do not happen to come together for $\mu = \mu'$ and $\mu = \mu''$.

We remark that only the second assertion of the above paragraph is related to the positivity of type of the equation.[13)]

An equation (3.9) of positive type has the further essential property that for real μ of sufficiently small absolute value all of its multiplicators are definite, and for $\mu > 0$ $(\mu < 0)$ exactly n_+ multiplicators of the first kind will lie on the open upper (lower) semicircle of the unit circle, and exactly n_- multiplicators of the second kind will lie on the open lower (upper) semicircle.

Assuming for definiteness that μ is positive (small), we denote by $\rho_j^+(\mu) = \exp(i\theta_j^+(\mu))$ $(\pi > \theta_1^+(\mu) \geqq \cdots \geqq \theta_{n_+}^+(\mu) > 0)$ the multiplicators of the first kind of an equation (3.9) of positive type, and by $\rho_j^-(\mu) = \exp(i\theta_j^-(\mu))$ $(-\pi < \theta_1^-(\mu) \leqq \cdots \leqq \theta_{n_-}^-(\mu) < 0)$ its multiplicators of the second kind. By what was said above, as μ increases the multiplicators of the first kind $\rho_j^+(\mu)$ $(j = 1, 2, \cdots, n_+)$ will move on the upper semicircle in a counter-clockwise direction, while the multiplicators $\rho_j^-(\mu)$ $(j = 1, 2, \cdots, n_-)$ will move on the lower semicircle in a clockwise direction. As long as the

---

13) In a paper by M. G. Kreĭn and G. Ja. Ljubarskiĭ [1] one can find additional rules for the motion of multiplicators (relating to the motion of indefinite multiplicators of simple type and others) for equations of positive type.

avant-garde multiplicators $\rho_1^{\pm}(\mu)$ do not reach the point -1, multiplicators of different kinds cannot come together. Let $\mu = \mu_1$ (> 0) be the first value of μ for which at least one of the avant-garde multiplicators $\rho_1^{\pm}(\mu)$ happens to equal -1. Obviously the number μ_1 will be the first (smallest) positive eigenvalue of the boundary value problem

$$\mathscr{Y}\frac{dx}{dt} - \mu\mathscr{A}(t)\,x = 0, \quad x(0) + x(l) = 0. \tag{3.10}$$

It is clear from the foregoing that for $0 < \mu < \mu_1$ all the solutions of equation (3.9) are stably bounded. Carrying out the arguments for negative μ, we are led to the following conclusion.

B) THEOREM ON THE CENTRAL ZONE OF STABILITY. *All solutions of a canonical equation* (3.9) *of positive type are stably bounded for the values of μ from the interval*

$$\mu_{-1}(\mathscr{A}) < \mu < \mu_1(\mathscr{A}),$$

where $\mu_1(\mathscr{A})$ and $\mu_{-1}(\mathscr{A})$ are, respectively, the first positive and first negative eigenvalues of the skewperiodic boundary value problem (3.10).

We remark that for the case in which (3.9) is a Hamiltonian equation, this theorem is exact, i.e. the interval (μ_{-1}, μ_1) cannot be replaced by any larger interval. In fact, in this case $n = 2m$, $n_+ = n_- = m$, the multiplicators are situated symmetrically with respect to the real axis: $\rho_j^-(\mu) = \overline{\rho_j^+(\mu)}$ $(j = 1, 2, \cdots, m)$, and for example for $\mu = \mu_1$ the two avant-garde multiplicators of unlike kinds come together at the point -1: $\rho_1^+(\mu_1) = \rho_1^-(\mu_1) = -1$. Hence for $\mu = \mu_1$ either (3.9) already has an unbounded solution (if $\rho = -1$ is a multiplicator of nonsimple type for this equation), or this can be achieved by an arbitrarily small perturbation of the H-matrix $\mathscr{A}(t)$ (arbitrarily small in the metric of the Banach space $\mathbf{H}_n(l)$.[14])

The arguments which led to Theorem B) enable us to supplement it with the following assertions.

C) *If $0 < \mu < \mu_1(\mathscr{A})$, then exactly n_+ multiplicators of the first kind lie on the upper semicircle $\rho = \exp(i\theta)$ $(0 < \theta < \pi)$, and exactly n_- multiplicators of the second kind lie on the lower semicircle $\rho = \exp(i\theta)$ $(-\pi < \theta < 0)$, for a canonical equation* (3.9) *of positive type.*

---

[14] See footnote 11, above.

A similar assertion can be formulated for the values of μ from the interval $\mu_{-1}(\mathscr{A}) < \mu < 0$.

§4. Tests for the stable boundedness of the solutions of a periodic canonical equation[15)]

1. According to the theorem on the central zone of stability (Theorem B) of §3), the solutions of a canonical equation of positive type

$$\mathscr{Y}\frac{dx}{dt} - \mathscr{A}(t)\,x = 0 \quad (\mathscr{A}(t+l) = \mathscr{A}(t)) \tag{4.1}$$

will be stably bounded whenever $\mu_1(\mathscr{A}) > 1$, where $\mu_1(\mathscr{A})$ is the first (smallest) positive eigenvalue of the skewperiodic boundary value problem

$$\mathscr{Y}\frac{dx}{dt} - \mu\mathscr{A}(t)\,x = 0, \quad x(0) + x(l) = 0. \tag{4.2}$$

Therefore, if we are able to find in explicit form some positive functional $\Phi(\mathscr{A})$ $(\mathscr{A}(t) \in \mathbf{H}_n(l))$ which is a lower bound for $\mu_1(\mathscr{A})$: $\mu_1(\mathscr{A}) \geqq \Phi(\mathscr{A})$, then the inequality

$$\Phi(\mathscr{A}) > 1$$

furnishes a test for the stable boundedness of the solutions of the equation (4.1).[16)]

It is not hard to construct such a functional on the basis of the results of §§3 and 4.

In fact, as was shown in §1.6, the boundary value problem (4.2) is equivalent to the integral equation

$$y(t) = \mu\int_0^l g_{1/2}(t-s)\,\mathscr{K}(t,s)\,y(s)\,ds,$$

where

---

15) The basic content of this section and the following one is taken from the authors' paper [**6**].

16) Such tests are called Z-tests in M. G. Kreĭn's paper [**7**]. On the basis of result A) of §3, those H-matrices $\mathscr{A} \in \mathbf{H}_n(l)$ for which all solutions of (4.1) with a given $\mathscr{Y}$ are stably bounded form a countable number of regions in $\mathbf{H}_n(l)$.

For the case of Hamiltonian equations the corresponding regions were studied by I. M. Gel′fand and V. B. Lidskiĭ [**1**], and in the complex case by V. A. Jakubovič [**1, 2**] (see also the review by M. G. Kreĭn and V. A. Jakubovič [**1**]). A Z-test distinguishes those H-matrices which belong to the simplest, in a certain sense, of these regions.

(4.3) $$\mathscr{K}(t,s) = i\mathscr{A}^{1/2}(t)\mathscr{Y}\mathscr{A}^{1/2}(s), \qquad g_{1/2}(t) = \frac{i}{2}\operatorname{sign} t.$$

Let us consider the integral equation with Hermitian kernel

$$\phi(t) = \frac{1}{\omega}\int_0^l \mathscr{K}(t,s)\phi(s)\,ds$$

or, what is the same, the equation

$$\phi(t) = \frac{i}{\omega}\int_0^l \mathscr{A}^{1/2}(t)\mathscr{Y}\mathscr{A}^{1/2}(s)\phi(s)\,ds.$$

The latter is equivalent to the system

$$\phi(t) = \mathscr{A}^{1/2}(t)\xi, \quad \xi = \frac{i}{\omega}\int_0^l \mathscr{Y}\mathscr{A}^{1/2}(s)\phi(s)\,ds,$$

where ξ is a constant vector. Eliminating the unknown function ϕ from this system, we obtain for ξ the equation

$$\xi = \frac{i}{\omega}\, l\mathscr{Y}\mathscr{A}_{\mathrm{av}}\xi \quad \left(\mathscr{A}_{\mathrm{av}} = \frac{1}{l}\int_0^l \mathscr{A}(t)\,dt\right).$$

Hence we conclude that the nonzero eigenvalues ω_j, $j = 1,2,\cdots,n$, of the finite-dimensional operator generated in $L_2^{(n)}(0,l)$ by the kernel $\mathscr{K}(t,s)$ are the roots of the secular equation

$$\det(il\mathscr{Y}\mathscr{A}_{\mathrm{av}} - \omega I_n) = 0,$$

or

(4.4) $$\det(l\mathscr{A}_{\mathrm{av}} - \omega i\mathscr{Y}) = 0.$$

By hypothesis the H-matrix $\mathscr{A}(t)$ is of positive type, and therefore the matrix $\mathscr{A}_{\mathrm{av}}$ is Hermitian-positive. Since the matrix $i\mathscr{Y}$ is likewise Hermitian, all the roots of (4.4) are real (which was to be expected in advance), and there are among them exactly n_+ positive ones $\omega_1^+ \geqq \omega_2^+ \geqq \cdots \geqq \omega_{n_+}^+$ (>0) and exactly n_- negative ones $-\omega_1^- \leqq -\omega_2^- \leqq \cdots \leqq -\omega_{n_-}^-$ (<0).

Let us recall that according to the notation of §3.3, n_+ and n_- represent the multiplicities of the eigenvalues $+1$ and -1, respectively, of the matrix $i\mathscr{Y}$.

Since the operator B, generated by the kernel $\mathscr{B}(t,s) = g_{1/2}(t-s)\mathscr{K}(t,s)$, is related to the operator K by $B = \mathscr{T}_{1/2}(K)$, where $\mathscr{T}_{1/2}$ is the transformator, considered in Chapter II, §5.4, corresponding to the factor

$g_{1/2}(t-s)$, Theorem II.4.2 can be applied to the operators B and K.

Taking into account that the number $\mu_1^{-1}(\mathscr{A})$ is the first positive eigenvalue of the operator B, on the basis of Theorem II.4.1 and result 4 of §5, Chapter II we obtain the bound

$$\frac{1}{\mu_1(\mathscr{A})} \leqq \sum_{j=1}^{n_+} (C_{1j}^+ - C_{1j-1}^+)\omega_j^+ + \sum_{j=1}^{n_-} (C_{1j}^- - C_{1j-1}^-)\omega_j^-,$$

where

$$C_{1j}^+ = C_{1j}^- = \frac{1}{\pi} \sum_{k=1}^{j} \frac{1}{2k-1} \qquad (j = 1, 2, \cdots).$$

Thus, we always have

$$\frac{1}{\mu_1(\mathscr{A})} \leqq \frac{1}{\pi}\left(\sum_{j=1}^{n_+} \frac{\omega_j^+}{2j-1} + \sum_{j=1}^{n_-} \frac{\omega_j^-}{2j-1}\right) \tag{4.5}$$

and the reciprocal of the right side of this bound can be taken as the functional $\Phi(\mathscr{A})$.

We have arrived at the following result.

TEST I. *All the solutions of a canonical equation* (4.1) *of positive type are stably bounded whenever*

$$\sum_{j=1}^{n_+} \frac{\omega_j^+}{2j-1} + \sum_{j=1}^{n_-} \frac{\omega_j^-}{2j-1} < \pi, \tag{4.6}$$

where $\omega_1^+ \geqq \omega_2^+ \geqq \cdots \geqq \omega_{n_+}^+$ *and* $-\omega_1^- \leqq -\omega_2^- \leqq \cdots \leqq -\omega_{n_-}^-$ *are respectively all the positive and all the negative eigenvalues of the matrix*

$$i\mathscr{Y} \int_0^l \mathscr{A}(t)\,dt.$$

We remark that for a Hamiltonian equation (4.1) of positive type with phase dimension $n = 2m$ the condition (4.6) assumes a simpler form:

$$\sum_{j=1}^{m} \frac{\omega_j^+}{2j-1} < \frac{\pi}{2}. \tag{4.7}$$

In fact, in this case the numbers $\pm\,\omega_j^{\mp}$ are the roots of the equation $\det(l\mathscr{A}_{\mathrm{av}} - \omega i \mathscr{Y}_m) = 0$, which has real coefficients, and therefore $\omega_j^+ = \omega_j^-$ $(j = 1, 2, \cdots, m)$.

From the results of §4, Chapter III it follows that Test I is "exact" (i.e., the constant π in (4.6) cannot be replaced by any larger constant for any fixed positive integers n_+ and n_- and positive numbers ω_j^+ $(j = 1, 2, \cdots, n_+)$ and ω_j^- $(j = 1, 2, \cdots, n_-)$).

For the case in which one of n_+ and n_- is zero (i.e., $\mathscr{Y} = \pm\, iI_n$), Test I is superfluous, as in this case all solutions of (4.1) are always stably bounded.

It is possible that for Hamiltonian systems the test (4.7) is "exact";[17] at any rate this is so for $m = 1$ (see §5.5). Let us make clear that in verifying the exactness of a test for Hamiltonian systems the class of competing canonical equations (4.1) is narrowed as a result of the requirement that the equation be real, and also that the equalities $n_+ = n_-$, $\omega_j^+ = \omega_j^-$ $(j = 1, 2, \cdots, m)$ are indeed possible for complex canonical equations (4.1).

2. By using a rough estimate in the test (4.6), we can obtain a test which does not require calculation of the quantities $\omega_j^\pm$. Indeed, the square of the left side of (4.6) does not exceed

$$\left(\sum_{j=1}^{n_+} (\omega_j^+)^2 + \sum_{j=1}^{n_-} (\omega_j^-)^2\right) \left(\sum_{j=1}^{n_+} \frac{1}{(2j-1)^2} + \sum_{j=1}^{n_-} \frac{1}{(2j-1)^2}\right),$$

and since

$$\sum_{j=1}^{n_+} (\omega_j^+)^2 + \sum_{j=1}^{n_-} (\omega_j^-)^2 = -\, l^2 \mathrm{sp}(\mathscr{Y}\mathscr{A}_{\mathrm{av}})^2,$$

it follows that *all solutions of a canonical equation* (4.1) *of positive type will be stably bounded whenever*

$$-\, l^2 \mathrm{sp}(\mathscr{Y}\mathscr{A}_{\mathrm{av}})^2 \leqq \frac{\pi^2}{\displaystyle\sum_{j=-n_-+1}^{n_+} \frac{1}{(2j-1)^2}}. \tag{4.8}$$

Since

$$(2 \leqq) \sum_{j=-q+1}^{p} \frac{1}{(2j-1)^2} < \sum_{j=-\infty}^{\infty} \frac{1}{(2j-1)^2} = \frac{\pi^2}{4},$$

for n_+ and n_- sufficiently large it is convenient to use the somewhat cruder test

$$-\, l^2 \mathrm{sp}(\mathscr{Y}\mathscr{A}_{\mathrm{av}})^2 < 4. \tag{4.9}$$

In the Hamiltonian case ($\mathscr{Y} = \mathscr{Y}_m$, $\mathscr{A}(t)$ a real symmetric matrix of order $n = 2m$) the inequality (4.9) can be written in a simpler form.

---

[17] A "proof" of this is given in the authors' paper [**6**], but it contains an error.

We break up the Hamiltonian $\mathscr{H}(t)$ into square blocks

$$\mathscr{H}(t) = \begin{pmatrix} \mathscr{H}_{11}(t) & \mathscr{H}_{12}(t) \\ \mathscr{H}_{21}(t) & \mathscr{H}_{22}(t) \end{pmatrix}$$

and put

$$\mathscr{A}_{jk} = \int_0^l \mathscr{H}_{jk}(t)\,dt \quad (j,k = 1,2).$$

Simple computations show that

$$-\,l^2 \mathrm{sp}(\mathscr{Y}\mathscr{H}_{\mathrm{av}})^2 = 2\,\mathrm{sp}(\mathscr{A}_{11}\mathscr{A}_{22} - \mathscr{A}_{12}^2).$$

Thus, the test (4.9) can be rewritten in the form

$$\mathrm{sp}(\mathscr{A}_{11}\mathscr{A}_{22} - \mathscr{A}_{12}^2) < 2. \tag{4.10}$$

This test was given in a paper by M. G. Kreĭn [8]. In this paper it was also shown that for the case of a Hamiltonian equation with an even Hamiltonian $\mathscr{H}(t) = \mathscr{H}(-t)$ the test (4.10) can be replaced by the stronger test

$$\mathrm{sp}(\mathscr{A}_{11}\mathscr{A}_{22} - \mathscr{A}_{12}^2) < 4. \tag{4.11}$$

For the special case in question, this test is "exact". In agreement with this, the inequality (4.8) is more restrictive than (4.11) for any $n_- = n_+ \geqq 1$.

3. For the general case of a nonreal canonical equation (4.1) Test I is one test of a one-parameter series; generally speaking, it can be strengthened.

This series of tests is obtained from the following considerations (compare with the considerations in §3.4).

For $\mu = 0$ all the multiplicators of the canonical equation of positive type

$$\mathscr{Y}\frac{dx}{dt} - \mu\mathscr{H}(t)\,x = 0 \tag{4.12}$$

are equal to unity. For positive and sufficiently small μ ($<\mu_1(\mathscr{H})$) this equation will have exactly n_+ multiplicators of the first kind on the semicircle $\rho = \exp(i\theta)$ $(0 < \theta < \pi)$ and exactly n_- multiplicators of the second kind on the semicircle $\rho = \exp(i\theta)$ $(-\pi < \theta < 0)$. As μ increases the first group of multiplicators ρ_j^+ $(j = 1, 2, \cdots, n_+)$ will move on the unit circle in a counter-clockwise direction, and the second group ρ_j^-

$(j = 1, 2, \cdots, n_-)$—in a clockwise direction. This motion will continue as long as the avant-garde multiplicators of different kinds do not come together.

If (4.12) is a Hamiltonian equation, then for an appropriate enumeration we have $\rho_j^- = \overline{\rho_j^+}$ $(j = 1, 2, \cdots, m;\ n = 2m)$, and such an encounter will take place for $\mu = \mu' = \mu_1(\mathscr{A})$.

In the general case this encounter will take place for $\mu = \mu' > \mu_1(\mathscr{A})$. In either event, for $\mu < \mu'$ $(\mu > 0)$ the solutions of (4.12) are stably bounded.

Suppose that $0 < \kappa < 1$ and for $\mu = \mu_\kappa^+$ (> 0) either the argument of the avant-garde multiplicator ρ_1^+ first becomes equal to $2\pi\kappa$ or the argument of the avant-garde multiplicator ρ_1^- first becomes equal to $-2\pi(1-\kappa)$, or both occur. In other words, let $\mu = \mu_\kappa^+$ be the first positive value of μ for which at least one of the avant-garde multiplicators of the first or second kind is found at the point $\rho = e^{2\pi\kappa i}$.

Obviously the number μ_κ^+ can be characterized as the first positive eigenvalue of the selfadjoint boundary value problem

$$\mathscr{Y}\frac{dx}{dt} - \mu\mathscr{A}(t)\,x = 0, \quad x(l) = e^{2\pi i\kappa}x(0).$$

This boundary value problem is equivalent to the integral equation

$$x(t) = \mu\int_0^l g_\kappa(t-s)\,\mathscr{Y}\,\mathscr{A}(s)\,x(s)\,ds, \tag{4.13}$$

where

$$g_\kappa(t) = \frac{1}{2\pi}\sum_{j=-\infty}^{\infty}\frac{e^{2\pi i(j+\kappa)t/l}}{j+\kappa},$$

i.e. $g_\kappa(t-s)$ is the Green's function of the boundary value problem

$$-\frac{dy}{dt} = f(t), \quad y(l) = e^{2\pi i\kappa}y(0) \qquad (0 \leqq t \leqq l).$$

Arguing further as in §1.6, we transform (4.13) into the equation

$$\phi(t) = \mu\int_0^l g_\kappa(t-s)\,\mathscr{K}(t,s)\,\phi(s)\,ds, \tag{4.14}$$

where as before the matrix kernel $\mathscr{K}(t,s)$ is defined by (4.3).

Now applying to equation (4.14) Theorem 4.2 of Chapter II and result 4 from §5, Chapter II concerning the transformator $\mathscr{T}_\kappa$, we conclude, analogously to what was done in §4.1 for the number $\mu_1(\mathscr{A})$, that

$$\frac{1}{\mu_\kappa^+} \leqq \sum_{j=1}^{n_+} \frac{\omega_j^+}{j+\kappa-1} + \sum_{j=1}^{n_-} \frac{\omega_j^-}{j-\kappa}.$$

It is now easy (compare with the derivation of Test I) to obtain the following result.

TEST I_κ. *Suppose that for some* κ $(0<\kappa<1)$ *the condition*

$$\sum_{j=1}^{n_+} \frac{\omega_j^+}{j+\kappa-1} + \sum_{j=1}^{n_-} \frac{\omega_j^-}{j-\kappa} < 2\pi \tag{4.15}$$

holds for the canonical equation of positive type

$$dx/dt = -\mathscr{Y}\mathscr{A}(t)x \quad (\mathscr{A}(t+l) = \mathscr{A}(t)). \tag{4.1}$$

Then all the solutions of (4.1) *are stably bounded, and moreover all the multiplicators of the first kind of* (4.1) *lie on the open arc* $\rho = e^{i\theta}$ $(0<\theta<2\pi\kappa)$, *and those of the second kind lie on the open arc* $\rho = e^{i\theta}$ $(-2\pi(1-\kappa)<\theta<0)$.

Test I is obviously obtained from test I_κ for $\kappa = \frac{1}{2}$.

We remark that while the second part of the assertion is of interest in all cases (even if $\mathscr{Y} = \pm iI_n$), the first part of the assertion is interesting only when both numbers n_+ and n_- are positive ($\mathscr{Y} \neq \pm iI_n$).

Assuming that this last condition is satisfied, let us study the function

$$\Phi_\omega(\kappa) = \sum_{j=1}^{n_+} \frac{\omega_j^+}{j+\kappa-1} + \sum_{j=1}^{n_-} \frac{\omega_j^-}{j-\kappa}, \tag{4.16}$$

appearing in the left side of (4.15). Since

$$\frac{d^2\Phi_\omega(\kappa)}{d\kappa^2} = 2\left[\sum_{j=1}^{n_+} \frac{\omega_j^+}{(j+\kappa-1)^3} + \sum_{j=1}^{n_-} \frac{\omega_j^-}{(j-\kappa)^3}\right] > 0$$

and $\Phi_\omega(+0) = \Phi_\omega(1-0) = \infty$, the function $\Phi_\omega(\kappa)$ is convex and has a unique minimum in the interval $(0,1)$, which is its smallest value. This value will be attained for $\kappa = \kappa_\omega$, the unique root in $(0,1)$ of the equation

$$-\sum_{j=1}^{n_+} \frac{\omega_j^+}{(j+\kappa-1)^2} + \sum_{j=1}^{n_-} \frac{\omega_j^-}{(j-\kappa)^2} = 0.$$

Obviously, for given ω_j^+ $(j=1,2,\cdots,n_+)$ and ω_j^- $(j=1,2,\cdots,n_-)$ the test (4.15) will be strongest if for κ we take the value κ_ω.

On the other hand, if the inequality (4.15) is fulfilled for some κ (and, consequently, for $\kappa = \kappa_\omega$), then the equation

$$\sum_{j=1}^{n_+} \frac{\omega_j^+}{j+\kappa-1} + \sum_{j=1}^{n_-} \frac{\omega_j^-}{j-\kappa} = 2\pi$$

will have precisely two roots $\kappa_1 < \kappa_2$, with $\kappa_1 < \kappa_\omega$ and $\kappa_2 > \kappa_\omega$. The interval $\kappa_1 < \kappa < \kappa_2$ will contain all values of κ for which the inequality (4.15) is fulfilled.

Hence, from the second assertion of the Test I_κ we can conclude the following.

Each multiplicator of the first kind of equation (4.1) *lies on the arc* $\rho = \exp(i\theta)$ $(0 < \theta \leqq 2\pi\kappa_1)$, *and of the second kind—on the arc* $\rho = \exp(i\theta)$ $(-2\pi(1-\kappa_2) \leqq t < 0)$.

We remark that these two arcs of the unit circle, which have a common initial point $\rho = 1$, do not overlap anywhere; they are separated on the unit circle by the arc $\rho = \exp(i\theta)$ $(2\pi\kappa_1 \leqq \theta \leqq 2\pi\kappa_2)$.

4. Everything simplifies significantly for the case where (4.1) is a Hamiltonian equation. In this case $n_+ = n_- = m$, $\omega_j^+ = \omega_j^-$ $(j = 1, 2, \cdots, m)$,

$$\Phi_\omega(\kappa) = \sum_{j=1}^{m} \omega_j^+ \left(\frac{1}{j+\kappa-1} + \frac{1}{j-\kappa}\right),$$

and consequently

$$\Phi_\omega(\tfrac{1}{2} + \xi) = \Phi_\omega(\tfrac{1}{2} - \xi).$$

Therefore in this case $\kappa_\omega = \frac{1}{2}$ and $\kappa_2 - \frac{1}{2} = \frac{1}{2} - \kappa_1$ (> 0). We have arrived at the following result.

TEST II. *If the condition*

$$\sum_{j=1}^{m} \frac{\omega_j^+}{2j-1} < \frac{\pi}{2} \tag{4.7}$$

is satisfied for a Hamiltonian equation (4.1) *of positive type, then all solutions of this equation are stably bounded and all m multiplicators of the first kind of this equation lie on the arc* $\rho = \exp(i\theta)$ $(0 < \theta \leqq 2\pi\kappa_1)$, *where* κ_1 *is the unique root in the interval* $(0 < \kappa < \frac{1}{2})$ *of the equation*

$$\sum_{j=1}^{m} \omega_j^+ \left(\frac{1}{j+\kappa-1} + \frac{1}{j-\kappa}\right) = 2\pi.$$

This rule could have been formulated under more general hypotheses.

All that was used in its derivation was that $n_+ = n_-$ and $\omega_j^+ = \omega_j^-$ $(j = 1, 2, \cdots, m)$; the fact that (4.1) was a Hamiltonian equation was needed only because it guaranteed that these conditions are satisfied. This remark will be used in §5.

Formulated in this more general way, Test II should be supplemented by the assertion that the multiplicators of the second kind lie on the arc $\rho = \exp(i\theta)$ $(-2\pi\kappa_1 \leqq \theta < 0)$, since for a non-Hamiltonian equation the multiplicators are not necessarily situated symmetrically with respect to the real axis.

5. Let us consider the special case of a Hamiltonian equation (4.1) for a two-dimensional vector-function $(n = 2m = 2)$, when

$$\mathscr{A}(t)\ (\equiv \mathscr{A}(t+l)) = \begin{pmatrix} a(t) & b(t) \\ b(t) & c(t) \end{pmatrix}, \quad \mathscr{Y} = \begin{pmatrix} 0 & 1 \\ -1 & 0 \end{pmatrix}$$

and the vector equation (4.1) can be written as a system of two scalar equations with real coefficients:

$$\begin{aligned} \frac{dx_1}{dt} &= -b(t)x_1 - c(t)x_2, \\ \frac{dx_2}{dt} &= a(t)x_1 + b(t)x_2. \end{aligned} \tag{4.17}$$

This system will be of positive type if

$$a(t) \geqq 0, \quad c(t) \geqq 0, \quad a(t)c(t) - b^2(t) \geqq 0$$

and, moreover,

$$\mathscr{I}_a\mathscr{I}_c - \mathscr{I}_b^{\,2} > 0,$$

where

$$\mathscr{I}_a = \int_0^l a(t)\,dt, \quad \mathscr{I}_b = \int_0^l b(t)\,dt, \quad \mathscr{I}_c = \int_0^l c(t)\,dt.$$

It is now elementary to calculate the value of $\omega = \omega_1^+ = \omega_1^-$:

$$\omega = \sqrt{\mathscr{I}_a\mathscr{I}_c - \mathscr{I}_b^{\,2}}. \tag{4.18}$$

For the case being considered, Test II says the following:

All solutions of the system (4.17) *are stably bounded whenever* $\omega < \pi/2$, *i.e.*

$$\mathscr{I}_a\,\mathscr{I}_c - \mathscr{I}_b^{\,2} < \pi^2/4. \tag{4.19}$$

When this condition is satisfied, the multiplicator of the first kind of this system lies on the arc $\rho = \exp(i\theta)$ $(0 < \theta \leqq 2\pi\kappa_1)$, *where*

$$\kappa_1 = \frac{1}{2}\left(1 - \sqrt{1 - \frac{2\omega}{\pi}}\right).$$

We shall show that we can obtain a stronger test if we introduce the functional

$$\Delta = \int_0^l \sqrt{ac - b^2}\,dt. \tag{4.20}$$

In fact, let us form the new canonical system

$$\begin{aligned} \frac{d\tilde{x}_1}{dt} &= -\overline{b_q(t)}\,\tilde{x}_1 - c(t)\,\tilde{x}_2, \\ \frac{d\tilde{x}_2}{dt} &= a(t)\,\tilde{x}_1 + b_q(t)\,\tilde{x}_2, \end{aligned} \tag{4.21}$$

where

$$b_q(t) = b(t) + iq\sqrt{a(t)c(t) - b^2(t)} \quad (0 < q < 1).$$

Since

$$a(t)c(t) - |b_q(t)|^2 = (1 - q^2)(a(t)c(t) - b^2(t)) \geqq 0,$$

the H-matrix $\mathscr{A}_q(t)$ of the system (4.21) will be a Hermitian-nonnegative matrix-function of t, and hence its integral

$$\int_0^l \mathscr{A}_q(t)\,dt = \begin{pmatrix} \mathscr{I}_a & \mathscr{I}_{b_q} \\ \overline{\mathscr{I}_{b_q}} & \mathscr{I}_c \end{pmatrix} \quad (\mathscr{I}_{b_q} = \mathscr{I}_b + iq\Delta)$$

will be a Hermitian-nonnegative matrix. Therefore

$$\mathscr{I}_a\mathscr{I}_c - \mathscr{I}_b{}^2 - q^2\Delta^2 > 0 \quad (0 \leqq q < 1). \tag{4.22}$$

We have discarded the equal sign in this relation in view of the fact that it is by hypothesis valid for $q = 0$ or $\Delta = 0$, and for $q > 0$ and $\Delta > 0$ it holds in the weaker form (with $\geqq$) as q (< 1) increases. For $q = 1$ we will have $\mathscr{I}_a\mathscr{I}_c - \mathscr{I}_b{}^2 \geqq \Delta^2$, i.e.

$$\int_0^l a(t)\,dt \int_0^l c(t)\,dt - \left(\int_0^l b(t)\,dt\right)^2 \geqq \left(\int_0^l \sqrt{a(t)c(t) - b^2(t)}\,dt\right)^2.$$

This inequality can be regarded as a generalization of the Cauchy-Bunjakovskiĭ inequality, since the latter is obtained from it for $b(t) \equiv 0$ (or $b(t) = \sqrt{a(t)c(t)}$).

In our concise notation this inequality can be written as $\omega \geqq \Delta$.

It follows from (4.22) that the system (4.21) is of positive type for any q $(0 \leqq q < 1)$. The numbers $\omega_1^{\pm}$ for the system (4.21) will have the values

$$\omega_1^{\pm} = \omega \pm q\Delta.$$

The function $\Phi_\omega(\kappa)$ for the system (4.21) will have, by (4.16), the form

$$\Phi_\omega(\kappa) = \frac{\omega + q\Delta}{\kappa} + \frac{\omega - q\Delta}{1 - \kappa},$$

and it is not hard to see that

$$\min_{0<\kappa<1} \Phi_\omega(\kappa) = (\sqrt{\omega + q\Delta} + \sqrt{\omega - q\Delta})^2 = 2[\omega + \sqrt{\omega^2 - q^2\Delta^2}].$$

Therefore on the basis of Test I_κ we can assert that all solutions of the system (4.17) will be stably bounded whenever

$$\omega + \sqrt{\omega^2 - q^2\Delta^2} < \pi. \tag{4.23}$$

We can thus conclude the following.

All solutions of the Hamiltonian system (4.17) *are stably bounded whenever*

$$\omega + \sqrt{\omega^2 - \Delta^2} < \pi \tag{4.24}$$

or, what is equivalent,

$$\mathscr{I}_a\mathscr{I}_c - \mathscr{I}_b^2 < \frac{\pi^2}{4}\left(1 + \frac{\Delta^2}{\pi^2}\right)^2.$$

In fact, if the inequality (4.24) is satisfied, one can find a positive q so close to unity that inequality (4.23) will be satisfied, and consequently all solutions of the system (4.21) are stably bounded.

On the other hand, as is not hard to see, the substitutions

$$x_j(t) = \exp\left(-iq\int_0^t \sqrt{ac - b^2ds}\right)\tilde{x}_j(t) \quad (j = 1, 2) \tag{4.25}$$

transform the system (4.21) into the Hamiltonian system (4.17).

Therefore the stable boundedness of the solutions of (4.21) implies the stable boundedness of the solutions of (4.17), which thus proves the validity of the test (4.24).

Obviously the test (4.24) is stronger than the test (4.18) whenever $\Delta > 0$; for $\Delta = 0$ they coincide.

One can pose the question:

Suppose that for an l-periodic system (4.17) of positive type only the quantities ω (>0) and Δ ($0 \leqq \Delta \leqq \omega$) are known. Will the inequality (4.24) be an exact test for the stable boundedness of the solutions of this system?[18)]

It follows from a result of T. M. Karaseva [1] (see §5.5) that for any $\omega > 0$ and $\Delta = 0$ the test (4.24) is exact.

It is also easy to see that (4.24) is exact for any $\omega = \Delta > 0$.[19)] This condition will be satisfied, for example, when the real l-periodic functions $a(t)$, $b(t)$ and $c(t)$ are constants a, b, and c $(a, c > 0,\ ac - b^2 > 0)$. In this case $\omega = \Delta = l(ac - b^2)$, and the test (4.24) reduces to the condition

$$l(ac - b^2) < \pi. \tag{4.26}$$

On the other hand, elementary calculations show that for constant a, b, c the constant μ_1 for the corresponding skewperiodic boundary value problem

$$\frac{dx_1}{dt} = -\mu(bx_1 + cx_2) \quad x_1(0) + x_1(l) = 0,$$

$$\frac{dx_2}{dt} = \mu(ax_1 + bx_2), \quad x_2(0) + x_2(l) = 0$$

will have the value

$$\mu_1 = \pi / l(ac - b^2),$$

and thus the condition (4.26) is equivalent to the condition $\mu_1 > 1$, which guarantees that the test is exact (see §3).

We shall use the transformation (4.25) to make another deduction. It follows from this transformation that if $\rho^{\pm}$ $(\rho^+\rho^- = 1)$ are the multiplicators of the Hamiltonian system (4.17), then the numbers

$$\tilde{\rho}^{\pm} = \exp(iq\Delta)\rho^{\pm} \tag{4.27}$$

will be the multiplicators of the system (4.21).

If, by the method of §4.3, we find κ_q such that the multiplicator $\tilde{\rho}^+$ lies on the arc $\rho = \exp(i\theta)$ $(0 \leqq \theta \leqq 2\pi\kappa_q)$, then by (4.27) the multi-

---

18) That is, can the constant in (4.24) be replaced by a larger constant, which possibly depends upon ω and Δ?

19) One can show that for a periodic system (4.17) of positive type the condition $\omega = \Delta$ will be satisfied if and only if the functions $a(t)$, $b(t)$ and $c(t)$ are proportional to one nonnegative periodic function. By means of an appropriate change of the independent variable t, this more general case can be reduced to the case of constant a, b, c.

plicator ρ^+ will lie on the arc obtained by shortening the latter by $q\Delta$. Letting q tend to 1, we arrive at the following rule.

If the condition (4.24) *is satisfied, then the multiplicator of the first kind of the Hamiltonian system* (4.17) *will lie on the arc* $\rho = \exp(i\theta)$ $(0 < \theta < \vartheta)$, *where* $\vartheta = \kappa_1 - \Delta$, *and* κ_1 *is the smallest root in the interval* $(0,1)$ *of the equation*

$$\frac{\omega + \Delta}{\kappa} + \frac{\omega - \Delta}{1 - \kappa} = 2\pi.$$

We remark that we will have to deal with the Hamiltonian system (4.17) in connection with a difficult question in the theory of unicellular Volterra operators (see §5 of the Appendix).

6. To conclude this section, we again consider the case $\mathscr{Y} = -iI$ (the case $\mathscr{Y} = iI$ is entirely analogous), for which all solutions of (4.1) are bounded (the matrizant $\mathscr{U}(t)$ is a unitary matrix-function). In this case the numbers ω_j^- are absent $(i\mathscr{Y}\mathscr{A}_{\text{av}} = \mathscr{A}_{\text{av}})$, and therefore the Test I_κ implies, for this case, the following.

Let $\mathscr{A}(t)$ *be a Hermitian-nonnegative matrix-function with elements from* $L_1(0,1)$, *and let* $\mathscr{U}(t)$ *be the solution of the matrix differential equation*

$$d\mathscr{U}/dt = i\mathscr{A}(t)\mathscr{U} \quad (\mathscr{U}(0) = I).$$

Then if the condition

$$\sum_{j=1}^{n} \frac{\omega_j}{2j-1} < 2\pi$$

is satisfied, where $\omega_1 \geqq \omega_2 \geqq \cdots \geqq \omega_n \geqq 0$ *are the eigenvalues of the matrix*

$$\int_0^l \mathscr{A}(t)\,dt \quad (= l\mathscr{A}_{\text{av}}),$$

it follows that all the eigenvalues of the unitary matrix $\mathscr{U}(l)$ *lie on the arc* $\rho = e^{i\theta}$ $(0 \leqq \theta \leqq 2\pi\kappa_1)$, *where* κ_1 *is the unique root in the interval* $(0,1)$ *of the equation*[20]

$$\sum_{j=1}^{n} \frac{\omega_j}{j + \kappa - 1} = 2\pi.$$

[20] It follows from (2.7) that

$$\kappa_1 \geqq \frac{1}{2\pi n} \sum_{j=1}^{n} \omega_j \left(= \frac{1}{2\pi n} \int_0^l \operatorname{sp}\mathscr{A}(t)\,dt\right).$$

In this assertion the requirement that the H-matrix $\mathscr{A}(t)$ be of positive type has been weakened. The point is that every Hermitian-nonnegative matrix-function $\mathscr{A}(t)$ can be approximated by the H-matrix $\mathscr{A}_\epsilon(t) = \mathscr{A}(t) + \epsilon I_n$ $(\epsilon > 0)$ of strictly positive type, and if the assertion is true for every H-matrix $\mathscr{A}_\epsilon(t)$ $(\epsilon > 0)$, then it will be true for $\mathscr{A}_0(t) = \mathscr{A}(t)$.

§5. A test for the stable boundedness of the solutions of a differential system of second order with periodic coefficients

Let us now consider the second order differential equation

$$d^2y/dt^2 + \mathscr{P}(t)y = 0 \tag{5.1}$$

for an m-dimensional vector-function $y(t) = \{y_j(t)\}_1^m$, where $\mathscr{P}(t) = \| p_{jk}(t) \|_1^m$ is a Hermitian matrix-function (an H-matrix) of period l $(\mathscr{P}(t+l) = \mathscr{P}(t))$ with locally integrable elements.

The concept of the stable boundedness of the solutions of the system (5.1) is introduced in complete analogy with its definition for the solutions of equation (1.1).

1. The theorem on the central zone of stability (§4, result A)) enables us to formulate the following assertion.

D) *If the periodic H-matrix $\mathscr{P}(t)$ $(\equiv \mathscr{P}(t+l))$ is of positive type,*[21] *then all solutions of the equation*

$$d^2y/dt^2 + \lambda\mathscr{P}(t)y = 0 \tag{5.2}$$

are stably bounded as long as

$$0 < \lambda < \lambda_1(\mathscr{P}),$$

where $\lambda_1(\mathscr{P})$ is the smallest[22] *eigenvalue of the skewperiodic boundary value problem*

$$d^2y/dt^2 + \lambda\mathscr{P}(t)y = 0, \quad y(0) + y(l) = y'(0) + y'(l) = 0. \tag{5.3}$$

In fact, putting $\lambda = \mu^2$ $(\mu > 0)$, we can rewrite (5.3) in the form of a system:

$$dy/dt = -\mu z, \qquad dz/dt = \mu\mathscr{P}(t)y.$$

---

[21] A periodic matrix-function $\mathscr{P}(t)$ $(\equiv \mathscr{P}(t+l))$ is called an H-matrix of *positive type* if it is an H-matrix of positive type in the interval $(0, l)$ (see the definition in §1.3).

[22] It will be clear from what follows that all the eigenvalues of the boundary value problem (5.3) are positive.

This system can be written as a canonical equation

$$dx/dt = -\mu \mathscr{Y} \mathscr{A}(t)\, x, \tag{5.4}$$

where $x = y \dotplus z$ (the direct sum of the vector-functions y and z), and

$$\mathscr{A}(t) = \begin{pmatrix} \mathscr{P}(t) & 0 \\ 0 & I_m \end{pmatrix}, \quad \mathscr{Y} = \mathscr{Y}_m = \begin{pmatrix} 0 & I_m \\ -I_m & 0 \end{pmatrix}. \tag{5.5}$$

Obviously $\mathscr{A}(t)$ is, along with $\mathscr{P}(t)$, an H-matrix of positive type. Under the indicated transformations the boundary value problem (5.3) goes over into a skewperiodic boundary value problem for the equation (5.4), so that $\lambda_1(\mathscr{P}) = \mu_1^2(\mathscr{A})$.

It is also clear that the stable boundedness of the solutions of (5.4) for $\mu = \mu_0$ ($\neq 0$) implies the stable boundedness of the solutions of (5.2) for $\lambda = \mu_0^2$.

This completes the proof of the result D). We remark that for the scalar case this result was established by A. M. Ljapunov [2] as far back as 1899.

2. The methods of §§4 and 5 of Chapter II enable us to obtain a simple lower bound for the number λ_1, and more generally for the first eigenvalue $\lambda_1(\kappa)$ of the boundary value problem

$$\frac{d^2y}{dt^2} + \lambda \mathscr{P}(t) y = 0, \; y(l) - e^{2\pi i \kappa} y(0) = y'(l) - e^{2\pi i \kappa} y'(0) = 0 \quad (0 < \kappa < 1),$$

whose spectrum consists of positive numbers if $0 < \kappa < 1$.

In fact, this problem is equivalent to the integral equation

$$y(t) = \lambda \int_0^l g_\kappa^{(2)}(t-s) \mathscr{P}(s) y(s)\, ds, \tag{5.6}$$

where

$$g_\kappa^{(2)}(t) = \frac{l}{4\pi^2} \sum_{j=-\infty}^{\infty} \frac{\exp[(2\pi i/l)(j+\kappa)t]}{(j+\kappa)^2}$$

is the Green's function for the scalar boundary value problem

$$\phi'' + \lambda \phi = f(t), \; \phi(l) - e^{2\pi\kappa i}\phi(0) = \phi'(l) - e^{2\pi\kappa i}\phi'(0) = 0.$$

We note that the kernel $g_\kappa^{(2)}(t-s)$ is Hermitian-positive, and

$$g_\kappa^{(2)}(0) = \frac{l}{4\pi^2} \sum_{j=-\infty}^{\infty} \frac{1}{(j+\kappa)^2} = \frac{l}{4\sin^2 \pi\kappa}. \tag{5.7}$$

By means of the substitution $z(t) = \mathscr{P}^{1/2}(t) y$ the integral equation (5.6) can be transformed into the equation

$$z(t) = \lambda \int_0^l g_\kappa^{(2)}(t-s)\mathscr{L}(t,s)z(s)\,ds,$$

where

$$\mathscr{L}(t,s) = \mathscr{P}^{1/2}(t)\,\mathscr{P}^{1/2}(s).$$

It is easily seen (compare with the arguments of §1.2) that the simplified integral equation

$$z(t) = \lambda \int_0^l \mathscr{L}(t,s)\,z(s)\,ds$$

has the characteristic numbers $\lambda_j = p_j^{-1}$ $(j = 1, 2, \cdots, m)$, where $p_1 \geqq p_2 \geqq \cdots \geqq p_m$ are the eigenvalues of the matrix

$$l\mathscr{P}_{\mathrm{av}} = \int_0^l \mathscr{P}(t)\,dt.$$

Using Theorem 3.2 and result 3 from §5, Chapter II, we obtain from (5.7) the bound

(5.8) $$\frac{1}{\lambda_1(\kappa)} \leqq \frac{l}{4\sin^2\pi\kappa}p_1;$$

in particular, for $\kappa = 1/2$ we will have

(5.9) $$1/\lambda_1 \leqq lp_1/4.$$

Thus we have derived the following result.

TEST III. *All solutions of equation* (5.1) *with a periodic H-matrix* $\mathscr{P}(t)$ $(\equiv \mathscr{P}(t+l))$ *of positive type are stably bounded as long as*[23)]

(5.10) $$\int_0^l \mathscr{P}(t)\,dt \leqq \frac{4}{l}\,I_m.$$

When this condition is satisfied, all the multiplicators of the first kind of equation (5.1) *lie on the arc* $\rho = \exp(i\theta)$ $(0 < \theta < \kappa)$, *and of the second kind—on the arc* $\rho = \exp(i\theta)$ $(-\kappa \leqq \theta < 0)$, *where* κ *is the smallest positive root of the equation*

(5.11) $$\sin\pi\kappa = \tfrac{1}{2}\sqrt{lp_1(\mathscr{P})}.$$

According to our earlier notation, $p_1(\mathscr{P})$ denotes the largest eigen-

---

23) We are using the notation according to which the relation $\mathscr{A} \leqq \mathscr{B}$ (or equivalently $\mathscr{B} \geqq \mathscr{A}$ or $\mathscr{B} - \mathscr{A} \geqq 0$) for two Hermitian matrices of order n means that the matrix $\mathscr{B} - \mathscr{A}$ is Hermitian-nonnegative.

value of the matrix $l\mathscr{P}_{\text{av}}$ (the left side of (5.10)). The condition (5.10) is equivalent to the condition

$$p_1(\mathscr{P}) < 4/l, \tag{5.12}$$

which, according to (5.9), guarantees the inequality $\lambda_1(\mathscr{P}) > 1$.

Hence the first assertion of Test III follows.

As regards the second assertion, on the basis of (5.8) it is obtained from the same considerations as was the second assertion of Test I_κ (§4).

Just as for every differential equation with periodic coefficients, a number ρ_0 is called a *multiplicator* of equation (5.1) if this equation has a nonnull solution $y_0(t)$ such that

$$y_0(t+l) \equiv \rho_0 y_0(t) \qquad (-\infty < t < \infty).$$

Obviously the multiplicators of equation (5.1) coincide with those of the corresponding canonical equation (5.4) (with $\mu = 1$). This circumstance enables us to carry over to equation (5.1) all the terminology concerning multiplicators from the theory of canonical equations. It is easily seen that in accordance with this a multiplicator ρ_0 of (5.1) will be a multiplicator of the first (second) kind if for all solutions y_0 corresponding to ρ_0 $(y_0(t+l) \equiv \rho_0 y_0(t),\ -\infty < t < \infty)$ the expression

$$\operatorname{Im}((dy^*/dt)\,y)$$

(which is independent of t) is positive (negative).

We remark that for a scalar equation (5.1) (the case $m = 1$) the test (5.10) coincides with the well-known criterion of A. M. Ljapunov from his famous dissertation (cf. [**1**], p. 203).

3. In a paper by M. G. Kreĭn [**7**] it was shown, as a generalization of a result obtained by A. M. Ljapunov in 1899 (see [**2**], pp. 404-406), that result D) carries over to the case in which the two conditions involved in the H-matrix $\mathscr{P}(t)$ being of positive type (§1.3) are replaced by two substantially weaker conditions, namely:

1) for any n-dimensional vector $\eta \neq 0$ the vector-function $\mathscr{P}(t)\eta$ is different from zero on a set of positive measure;

2) the matrix $\mathscr{P}_{\text{av}}$ is Hermitian-nonnegative: $\eta^*\mathscr{P}_{\text{av}}\eta \geqq 0$.

Let us represent the matrix-function $\mathscr{P}(t)$ as a difference $\mathscr{P}(t) = \mathscr{P}_1(t) - \mathscr{P}_2(t)$, where the $\mathscr{P}_k(t)$ $(k = 1, 2)$ are Hermitian-nonnegative matrix-functions.

If $\mathscr{P}(t)$ satisfies the conditions 1) and 2), the H-matrix $\mathscr{P}_1(t)$ in a decomposition of the indicated type will always be of positive type. To see this, it suffices to show that $\mathscr{P}_{1\text{av}}$ will be a Hermitian-positive matrix.

Let us for the moment assume the contrary, i.e. that for some n-dimensional vector η_0 $(\neq 0)$

$$l\eta_0^* \mathscr{P}_{1\text{av}} \eta_0 = \int_0^l \eta_0^* \mathscr{P}_1(t) \eta_0 dt = 0. \tag{5.13}$$

Then we will have

$$l\eta_0^* \mathscr{P}_{\text{av}} \eta_0 = - \int_0^l \eta_0^* \mathscr{P}_2(t) \eta_0 dt.$$

The left side of this equality, by virtue of condition 2), is nonnegative, and the right side, by definition of the decomposition $\mathscr{P} = \mathscr{P}_1 - \mathscr{P}_2$, is nonpositive; hence

$$\int_0^l \eta_0^* \mathscr{P}_2(t) \eta_0 dt = 0. \tag{5.14}$$

It follows from (5.13) and (5.14) that $\mathscr{P}_1(t)\eta = \mathscr{P}_2(t)\eta = 0$ almost everywhere, whence $\mathscr{P}(t)\eta = 0$ almost everywhere. We have arrived at a contradiction.

The reciprocal of the first positive eigenvalue $\lambda_1(\mathscr{P})$ of the boundary value problem (5.3) can always be obtained as the solution of a maximum problem:

$$\lambda_1^{-1}(\mathscr{P}) = \max_{y \in \mathfrak{A}} \frac{\int_0^l y^* \mathscr{P}(t) y dt}{\int_0^l \frac{dy^*}{dt} \frac{dy}{dt} dt}, \tag{5.15}$$

where $\mathfrak{A}$ is the set of all absolutely continuous n-dimensional vector-functions $y(t)$ $(0 \leqq t \leqq l)$ satisfying the condition $y(0) + y(l) = 0$.

Since the numerator in the right side of (5.15) can only increase if $\mathscr{P}(t)$ is replaced by $\mathscr{P}_1(t)$, we conclude that

$$\lambda_1(\mathscr{P}) \geqq \lambda_1(\mathscr{P}_1).$$

On the other hand, as was clarified in §5.1, the number $\lambda_1(\mathscr{P}_1)$ will be greater than 1 if

$$\int_0^l \mathscr{P}_1(t) dt < \frac{4}{l} I_m. \tag{5.16}$$

Thus, by virtue of the indicated generalization of result D), when conditions 1) and 2) are satisfied, all solutions of (5.1) will be stably bounded if (5.16) holds.

One can show that the second assertion of Test III remains valid for the case under consideration regarding the equation (5.1), if the number κ is here defined as the smallest positive root of the equation

$$\sin \pi\kappa = \tfrac{1}{2}\sqrt{l p_1(\mathscr{P}_1)}. \tag{5.17}$$

It turns out that if (5.16) (or the more general condition $\lambda_1(\mathscr{P}_1) > 1$) holds, then the arguments of the multiplicators of the first kind $\rho_{j1}^+ = \exp(i\theta_{j1}^+)$ $(0 < \theta_{11}^+ \leqq \theta_{21}^+ \leqq \cdots \leqq \theta_{m1}^+ < \pi)$ of the equation

$$d^2y/dt^2 + \mathscr{P}_1(t)y = 0 \tag{5.18}$$

will always be not less than those of the multiplicators of the first kind $\rho_j^+ = \exp(i\theta_j^+)$ $(0 < \theta_1^+ \leqq \theta_2^+ \leqq \cdots \leqq \theta_m^+ < \pi)$ of equation (5.1), i.e. $\theta_j^+ \leqq \theta_{j1}^+$ $(j = 1, 2, \cdots, m)$ and, of course, the corresponding assertion holds for the multiplicators of the second kind of the equations (5.1) and (5.18).

Obviously the bound $0 < \theta_j^+ \leqq \pi\kappa$ will possibly be more exact, and the test (5.16) for the stable boundedness of the solutions of (5.1) will possibly be stronger, if we carry out the "minimal" decomposition $\mathscr{P}(t) = \mathscr{P}_1(t) - \mathscr{P}_2(t)$,[24)] i.e. put $\mathscr{P}_1(t) = \mathscr{P}_+(t)$ in (5.16) and (5.17).

4. The construction of the H-matrix $\mathscr{P}_+(t)$ involves analytic difficulties which in practice frequently cannot be overcome. It is rather easy to obtain *some* decomposition $\mathscr{P}(t) = \mathscr{P}_1(t) - \mathscr{P}_2(t)$ of $\mathscr{P}(t)$ into Hermitian-nonnegative components, but for a random choice of this decomposition the test (5.16) may prove to be very crude.

It turns out, however, that when conditions 1) and 2) are satisfied there is an entirely different route for obtaining tests for the stable boundedness of the solutions of (5.1). The point is that when these conditions are satisfied, equation (5.1) can always be transformed (M. G. Kreĭn [7]) into a canonical equation, and indeed into one of positive type. In this way we obtain tests for the stable boundedness of the solutions of (5.1) which are substantially simpler than the test (5.16), and which have a number of essential advantages (in particular, when $\mathscr{P}_{\text{av}} = 0$), by virtue of which they usually "work" better than the test (5.16) (even if we put $\mathscr{P}_1(t) = \mathscr{P}_+(t)$ in the latter).

Given an l-periodic H-matrix $\mathscr{P}(t)$, we can always define an l-periodic H-matrix $\mathscr{Q}(t)$ by

---

24) If $\mathscr{A}$ is a Hermitian matrix, then $\mathscr{A}_+$ denotes the "smallest" Hermitian-nonnegative matrix among all those which "exceed" $\mathscr{A}$. The matrix $\mathscr{A}_+$ can also be characterized by the fact that in the decomposition $\mathscr{A} = \mathscr{A}_+ - \mathscr{A}_-$ both $\mathscr{A}_+$ and $\mathscr{A}_-$ are Hermitian-nonnegative and mutually orthogonal $(\mathscr{A}_+\mathscr{A}_- = \mathscr{A}_-\mathscr{A}_+ = 0)$.

(5.19) $$\mathscr{Q}(t) = \int_0^t (\mathscr{P}(s) - \mathscr{P}_{\text{av}})\, ds.$$

We put

(5.20) $$z = -\mathscr{Q} y - dy/dt.$$

Then for any solution y of equation (5.1) we will have

$$\frac{dz}{dt} = -\mathscr{Q}(t)\frac{dy}{dt} + \mathscr{P}_{\text{av}} y,$$

in virtue of which equation (5.1) is equivalent to the system

(5.21) $$\frac{dy}{dt} = -\mathscr{Q} z - z, \quad \frac{dz}{dt} = (\mathscr{Q}^2 + \mathscr{P}_{\text{av}}) y + \mathscr{Q} z.$$

Thus, for the direct sum $x = y \dot{+} z$ we again obtain a canonical equation with $\mathscr{Y} = \mathscr{Y}_m$, but with a new Hamiltonian

(5.22) $$\mathscr{A}(t) = \begin{pmatrix} \mathscr{Q}^2(t) + \mathscr{P}_{\text{av}} & \mathscr{Q}(t) \\ \mathscr{Q}(t) & I_m \end{pmatrix}.$$

It is not hard to show that when $\mathscr{P}(t)$ satisfies conditions 1) and 2) of §5.3, the matrix-function $\mathscr{A}(t)$ will be an H-matrix of positive type.

In fact, for any vector $\xi = \eta \dot{+} \zeta$ we have, by (5.22),

$$\xi^* \mathscr{A}(t) \xi = \eta^* \mathscr{P}_{\text{av}} \eta + |\mathscr{Q}(t)\eta - \zeta|^2 \geqq 0.$$

Therefore the relation

$$\int_0^l \xi^* \mathscr{A}(t) \xi\, dt = 0$$

is possible only if

$$\eta^* \mathscr{P}_{\text{av}} \eta = 0 \quad \text{and} \quad \mathscr{Q}(t)\eta - \zeta \equiv 0.$$

It follows from the first relation that $\mathscr{P}_{\text{av}}\eta = 0$, and from the second, by differentiation with respect to t, that $(\mathscr{P}(t) - \mathscr{P}_{\text{av}})\eta \equiv 0$, from which we obtain $\mathscr{P}(t)\eta \equiv 0$. But then, by condition 2), $\eta = 0$ and so $\zeta = \mathscr{Q}(t)\eta \equiv 0$.

The average of the H-matrix $\mathscr{A}(t)$ has the form

$$\mathscr{A}_{\text{av}} = \begin{pmatrix} (\mathscr{Q}^2)_{\text{av}} + \mathscr{P}_{\text{av}} & 0 \\ 0 & I_m \end{pmatrix},$$

and consequently the set of all eigenvalues of the matrix $il\mathscr{Y}_m\mathscr{A}_{\text{av}}$ coincides with the set of all roots of the equation (in q)

$$\det(l^2(\mathscr{Q}^2)_{\text{av}} + l^2\mathscr{P}_{\text{av}} - q^2 I_m) = 0.$$

Thus, if we denote by $q_1^2, q_2^2, \cdots, q_m^2$ $(q_1 \geqq q_2 \geqq \cdots \geqq q_m > 0)$ the sequence of eigenvalues of the matrix

$$l^2[(\mathscr{Q}^2)_{\text{av}} + \mathscr{P}_{\text{av}}], \tag{5.23}$$

we will have for the Hamiltonian $\mathscr{A}(t)$

$$\omega_j^+ = \omega_j^- = q_j \qquad (j = 1, 2, \cdots, m).$$

Thus, the inequality (4.7) from Test II assumes the form

$$\sum_{j=1}^{m} \frac{q_j}{2j-1} < \frac{\pi}{2}, \tag{5.24}$$

and we can formulate the following test.

TEST IV. *All solutions of equation* (5.1) *with an H-matrix $\mathscr{P}(t)$ which satisfies conditions* 1) *and* 2) *will be stably bounded as long as the inequality* (5.24) *is satisfied.*

When this inequality is satisfied, all the multiplicators of (5.1) *of the first kind lie on the arc $\rho = \exp(i\theta)$ $(0 < \theta < 2\pi\kappa_1)$, and of the second-kind—on the arc $\rho = \exp(i\theta)$ $(-2\pi\kappa_1 < \theta < 0)$, where κ_1 is the unique root in the interval $(0, \frac{1}{2})$ of the equation*

$$\sum_{j=1}^{m} q_j \left(\frac{1}{j+\kappa-1} + \frac{1}{j-\kappa} \right) = 2\pi.$$

For the case where $\mathscr{P}_{\text{av}} = 0$, the test (5.24) differs essentially from the test

$$p_1(\mathscr{P}_+) < 4/l,$$

which we discussed in §5.3.

Indeed, let us consider in the space $\mathbf{H}_n(l)$ of all l-periodic H-matrices of order n (see §3.2) the subspace $\mathbf{H}_n^0(l)$, consisting of all $\mathscr{P}(t) \in \mathbf{H}_n(l)$ for which $\mathscr{P}_{\text{av}} = 0$. It is easily seen that the functional $p_1(\mathscr{P}_+)$ is a continuous functional on $\mathbf{H}_n^0(l)$, while the left side of (5.24) is a continuous functional of the values of a *completely continuous* operator in $\mathbf{H}_n^0(l)$. Therefore it turns out that it is possible to find $\mathscr{P}(t) \in \mathbf{H}_n^0(l)$ for which (5.24) is satisfied, and $p_1(\mathscr{P}_+)$ will be arbitrarily large.

5. Let us consider in detail the case of a scalar equation

$$d^2y/dt^2 + p(t)\,y = 0 \tag{5.25}$$

with a periodic coefficient

$$p(t) \sim a_0 + \sum_{k=1}^{\infty} \left(a_k \cos\frac{2\pi kt}{l} + b_k \sin\frac{2\pi kt}{l} \right). \tag{5.26}$$

Since we have written $p(t)$ instead of $\mathscr{P}(t)$ (in order to emphasize the scalar nature of the function), we shall write $q(t)$ instead of $\mathscr{Q}(t)$. According to (5.26) and the definition (5.19) we have

$$q(t) = \frac{l}{2\pi} \sum_{k=1}^{\infty} \frac{1}{k} \left\{ a_k \sin\frac{2\pi kt}{l} - b_k \cos\frac{2\pi kt}{l} \right\}. \tag{5.27}$$

Test IV enables us to state the following.

If $a_0 = p_{\mathrm{av}} = 0$, then all solutions of equation (5.25) *will be stably bounded as long as*

$$l \int_0^l q^2(t)\,dt < \frac{\pi^2}{4}. \tag{5.28}$$

By (5.27) this condition is equivalent to

$$\sum_{k=1}^{\infty} \frac{a_k^2 + b_k^2}{k^2} < \frac{2\pi^4}{l^3}.$$

The test (5.28) was obtained by M. G. Kreĭn [**7**], who, however, did not suppose that it is exact. That it is was established by T. M. Karaseva [**1**], [**2**].[25)]

The transformation (5.20), when applied to equation (5.25), leads to the system (compare with (5.21))

$$\begin{aligned} \frac{dy}{dt} &= -\,q(t)\,y - z, \\ \frac{dz}{dt} &= (q^2(t) + a_0)\,y + q(t)\,z. \end{aligned} \tag{5.29}$$

We note that for this system, with $a_0 = p_{\mathrm{av}} \geqq 0$ (see (4.18) and (4.20)),

25) In the paper [**2**] by T. M. Karaseva an exact test for the stable boundedness of the solutions of the equation (5.25) was found in terms of the two parameters $a = p_{\mathrm{av}}$ and $b = l\int_0^l q^2(t)\,dt$. The investigation of the boundaries of the "regions of stability" in the (a, b) plane proved to be a very difficult problem. In the equations of these boundaries there unexpectedly appeared the periods of hyperelliptic integrals.

$$\omega = l^{1/2}\left(\int_0^l q^2(t)\,dt + a_0 l\right)^{1/2}, \quad \Delta = a_0^{1/2} l,$$

and the test (4.24) for the stable boundedness of the solutions of the system (5.29) assumes the form

$$l^{1/2}\left(\int_0^l q^2(t)\,dt + a_0 l\right)^{1/2} + \left(l\int_0^l q^2(t)\,dt\right)^{1/2} < \pi. \tag{5.30}$$

For $a_0 = p_{\text{av}} = 0$ this test goes over into the test (5.28). That this test is exact for $a_0 = 0$ is indicated by the fact that the general test (4.24) is exact for $\Delta = 0$.

From the investigations of T. M. Karaseva [**1**], [**2**] it also follows that for a given $a_0 > 0$ the test (5.30) for the equation (5.25) is never exact. However, it does not follow from this that for any fixed ω and $\Delta > 0$ the test (4.24) for a Hamiltonian system of second order is not exact. For example, as was already shown in §4.5, for $\omega = \Delta > 0$ the test (4.24) is always exact. The point is that not every Hamiltonian system of second order can be reduced to the form (5.29), where $q_{\text{av}} = 0$.

§6. Characteristics of the distribution density of the spectrum of a selfadjoint canonical boundary value problem

We continue the study, begun in §§1 and 2, of the general properties of the spectrum of a selfadjoint canonical boundary value problem:

$$\mathscr{Y}\frac{dx}{dt} - \mu\mathscr{H}(t)\,x = 0, \quad \mathscr{A}x(a) + \mathscr{B}x(b) = 0. \tag{6.1}$$

1. The spectrum of eigenvalues μ of this problem, taking account of multiplicity, can be arranged in a nondecreasing sequence $\{\mu_j\}$ $(j = \pm 1, \pm 2, \cdots)$:

$$\cdots \leqq \mu_{-2} \leqq \mu_{-1} < 0 \leqq \mu_1 \leqq \mu_2 \leqq \cdots \tag{6.2}$$

We denote by $n_+(r)$ and $n_-(r)$ the distribution functions of the positive and negative parts of the spectrum $\{\mu_j\}$.

Let us recall that by Lemma 2.1, for a skewperiodic problem (4.1) $(\mathscr{A} = \mathscr{B} = I_n)$ the spectrum $\{\mu_j\}$ coincides with the spectrum of the real component of the operator model A_M:

$$(A_M f)(t) = \int_t^b \mathscr{H}^{1/2}(t)\,\mathscr{Y}\,\mathscr{H}^{1/2}(s)\,f(s)\,ds.$$

Since the operator A_M has an n-dimensional imaginary component, Theorem III.2.4 is applicable to it. In particular, on the basis of this theorem we can assert the existence of a common limit

$$\lim_{r\to\infty}\frac{n_\pm(r)}{r}=\frac{h}{\pi}, \tag{6.3}$$

where

$$\left|\frac{1}{2}\int_a^b \operatorname{sp}(\mathscr{Y}\mathscr{H}(s))\,ds\right|\leqq h\leqq \Lambda_1\left(\frac{1}{2}\mathscr{Y}\int_a^b \mathscr{H}(s)\,ds\right). \tag{6.4}$$

We have used here Lemma V.3.1, according to which the spectrum of the imaginary component of the operator A_M coincides with the spectrum of the matrix

$$\frac{1}{2i}\mathscr{Y}\int_a^b \mathscr{H}(s)\,ds.$$

Let us recall that by the general definition of the symbol $\Lambda_p(X)$ (see §2.3), $\Lambda_1(X)$ denotes the sum of the moduli of all the eigenvalues of the square matrix X.

Recalling, further, Theorem 1.3, we see that the existence of a common limit (6.3) holds for any selfadjoint boundary value problem (6.1), and that the number h does not depend upon the choice of the matrices $\mathscr{A}$ and $\mathscr{B}$ in the boundary condition in (6.1). Thus the number h in (6.3) is completely determined by the signature matrix $\mathscr{Y}$ and the H-matrix $\mathscr{H}(t)$ (of positive type), and hence is a functional of $\mathscr{Y}$ and $\mathscr{H}(t)$.

We have the following theorem.

THEOREM 6.1 (M. G. KREĬN [8]).[26] *For any selfadjoint boundary value problem* (6.1),

$$\lim_{r\to\infty}\frac{n_\pm(r)}{r}=\frac{1}{2\pi}\int_a^b \Lambda_1(\mathscr{Y}\mathscr{H}(t))\,dt. \tag{6.5}$$

In particular, if $\mathscr{Y}=\pm iI_n$, *then*

$$\lim_{r\to\infty}\frac{n_\pm(r)}{r}=\frac{1}{2\pi}\int_a^b \operatorname{sp}\mathscr{H}(t)\,dt. \tag{6.6}$$

Let us note that the second equality (for the case $\mathscr{Y}=\pm iI_n$) is

[26] In the cited reference, this theorem is stated without proof.

an immediate consequence of the bound (6.4) (or simply of the fact that for a dissipative Volterra operator A the constant h is equal to the trace of its imaginary component); see formula (III.2.27).

To establish the general formula (6.5) is far from simple. It will be derived in Chapter VII simultaneously with the derivation of the general formula (III.2.26) for Volterra operators with nuclear imaginary component.

2. Let us consider the case where (6.1) is a Hamiltonian equation and $n = 2$. In this case (6.1) can be written as a system of two scalar equations:

$$\begin{aligned} \frac{dx_1}{dt} &= -\mu(b(t)\,x_1 + c(t)\,x_2), \\ \frac{dx_2}{dt} &= \mu(a(t)\,x_1 + b(t)\,x_2) \end{aligned} \qquad (a \leqq t \leqq b), \tag{6.7}$$

where the real coefficients $a(t)$, $b(t)$ and $c(t)$ $(\in L_1(a,b))$ satisfy the conditions (see §4)

$$a(t) \geqq 0,\ c(t) \geqq 0,\ a(t)c(t) - b^2(t) \geqq 0 \quad (a \leqq t \leqq b)$$

and

$$\int_a^b a(t)\,dt \int_a^b c(t)\,dt - \left(\int_a^b b(t)\,dt \right)^2 > 0.$$

The boundary conditions in (6.1) can be written in the form

$$\begin{cases} \alpha_{11}x_1(a) + \alpha_{12}x_2(a) + \beta_{11}x_1(b) + \beta_{12}x_2(b) = 0, \\ \alpha_{21}x_1(a) + \alpha_{22}x_2(a) + \beta_{21}x_1(b) + \beta_{22}x_2(b) = 0, \end{cases} \tag{6.8}$$

while for the case of real α, β the condition (1.19) can be written in the form

$$\alpha_{11}\alpha_{22} - \alpha_{12}\alpha_{21} = \beta_{11}\beta_{22} - \beta_{12}\beta_{21}. \tag{6.9}$$

If both sides of this relation equal zero, then we have to require that the rank of the matrix

$$\begin{Vmatrix} \alpha_{11} & \alpha_{12} & \beta_{11} & \beta_{12} \\ \alpha_{21} & \alpha_{22} & \beta_{21} & \beta_{22} \end{Vmatrix},$$

be equal to 2, i.e. that the boundary conditions (6.8) be linearly independent.

In obtaining (6.7) and (6.8) we used the fact that

$$\mathscr{Y} = \begin{pmatrix} 0 & 1 \\ -1 & 0 \end{pmatrix}, \quad \mathscr{A}(t) = \begin{pmatrix} a(t) & b(t) \\ b(t) & c(t) \end{pmatrix}.$$

Since in this case the eigenvalues of the matrix-function $\mathscr{Y}\mathscr{A}(t)$ are equal to $\pm i\sqrt{ac - b^2}$, Theorem 6.1 yields the following result.

The following asymptotic formula holds for the distribution functions $n_{\pm}(r)$ *of the positive and negative parts of the spectrum of the boundary value problem* (6.7), (6.8):

$$\lim_{r\to\infty} \frac{n_{\pm}(r)}{r} = \frac{1}{\pi}\int_a^b \sqrt{a(t)c(t) - b^2(t)}\,dt. \tag{6.5$'$}$$

This result was first stated in a paper by M. G. Kreĭn [1]. From this result it is rather simple to obtain an asymptotic formula for the spectrum of frequencies of a string with any mass distribution, as will be shown in §8.

3. We shall now give applications of other results concerning abstract Volterra operators which were discussed in the book [GK] (§§10 and 11 of Chapter IV). These results were established with the aid of the large machinery of the theory of entire functions.

For convenience we first give the statements of the results to be applied.

A) ([GK], THEOREM IV.10.1). *Let A be a Volterra operator with finite-dimensional imaginary component. If the negative characteristic numbers α_j of the operator $A_{\mathscr{R}}$ satisfy the condition*

$$\sum_{\alpha_j<0} \frac{1}{\sqrt{|\alpha_j|}} < \infty, \tag{6.10}$$

then the finite limit

$$\lim_{r\to\infty} \frac{n_+(r; A_{\mathscr{R}})}{\sqrt{r}} \tag{6.11}$$

exists for its positive characteristic numbers.

B) ([GK], COROLLARY IV.11.1). *Let A be a Volterra operator with a finite-dimensional imaginary component, and let $\{\lambda_j\}$ be the complete system of eigenvalues of its real component. Then for any p from the interval $\frac{1}{2} < p \leqq 1$ the two series*

$$\sum_{\lambda_j<0} |\lambda_j|^p, \quad \sum_{\lambda_j>0} \lambda_j^p$$

converge or diverge together.

C) ([GK], COROLLARY IV.11.2). *Let A be a Volterra operator with a finite-dimensional imaginary component, and let* $\lambda_1^+ \geqq \lambda_2^+ \geqq \cdots$ and $-\lambda_1^- \leqq -\lambda_2^- \leqq \cdots$ *be, respectively, the complete systems of positive and negative eigenvalues of the real component $A_{\mathscr{R}}$. Then for any p ($\frac{1}{2} < p < 1$) and any slowly varying function $L(r)$, the relation*

$$\lambda_n^- = O(n^{-1/p}L(n)) \quad (n \to \infty)$$

implies the relation

$$\lambda_n^+ = O(n^{-1/p}L(n)) \quad (n \to \infty).$$

These results imply the following theorem.

THEOREM 6.2. *For any selfadjoint boundary value problem* (6.1) *the following assertions hold*:

1) *if for some p ($\frac{1}{2} < p \leqq 1$) one of the sums*

$$\sum_{\mu_j > 0} \mu_j^{-p}, \quad \sum_{\mu_j < 0} |\mu_j|^{-p} \tag{6.12}$$

is finite, then so is the other;

2) *if for some p ($\frac{1}{2} < p < 1$) and some slowly varying function $L(r)$ one of the distribution functions $n_\pm(r)$ satisfies the relation*

$$n_\pm(r) = O(r^p L(r)), \tag{6.13}$$

then so does the other;

3) *if*

$$\sum_{\mu_j < 0} \frac{1}{\sqrt{|\mu_j|}} < \infty \quad \Bigl(\sum_{\mu_j > 0} \frac{1}{\sqrt{\mu_j}} < \infty\Bigr), \tag{6.14}$$

then the finite limit $\lim_{r \to \infty}(n_+(r)/\sqrt{r})$ $(\lim_{r \to \infty}(n_-(r)/\sqrt{r}))$ *exists.*

PROOF. By Theorem 1.3 it is sufficient to prove Theorem 6.2 for the spectrum of a selfadjoint skewperiodic problem (2.1), i.e. for the spectrum of the real component G of a model Volterra operator $A_M = G + iK$.

Since the imaginary component K of this operator is not more than n-dimensional (§2), the results A), B) and C) are applicable to the operators A_M and G.

Applying result A) to the operator A_M leads to assertion 3) for the problem (2.1), applying result B) leads to assertion 1), and applying result C) leads to assertion 2).

The theorem is proved.

We shall give (in §7) an example in which the spectrum of the problem (2.1) is bounded from below, and hence the condition (6.14) is satisfied; also an explicit expression will be obtained for $\lim(n_+(r)/\sqrt{r})$ which will show that it is positive. It follows, in particular, that the requirement $p > \frac{1}{2}$ in assertion 1) is essential (for $p = \frac{1}{2}$ assertion 1) is false). We remark that by virtue of (6.5), the series (6.12) always converge for $p > 1$.

§7. Characteristics of the distribution density of the spectrum of a second-order boundary value problem

1. A selfadjoint boundary value problem for a system of differential equations of any order can be transformed to an equivalent boundary value problem for a canonical system of differential equations. By virtue of this fact, the theorems and results of the preceding section lead to a number of corollaries for selfadjoint boundary value problems with ordinary differential operators of any order.

In this section we shall illustrate applications of the theorems of §6, using as examples boundary value problems with second-order differential operators.

A formally selfadjoint quasi-differential[27] second-order operator D, acting on m-dimensional vector-functions $y(t) = \{y_j(t)\}_1^m$ $(a \leq t \leq m)$, can be written in sufficiently general form as follows:

$$Dy = -\frac{d}{dt}\left(\mathscr{P}(t)\frac{dy}{dt} + \mathscr{Q}(t)y\right) + \mathscr{Q}^*(t)\frac{dy}{dt} + \mathscr{R}(t)y, \tag{7.1}$$

where $\mathscr{P}(t)$ and $\mathscr{R}(t)$ are H-matrices of order m on the interval $[a,b]$, and $\mathscr{Q}(t)$ is any complex matrix-function of order m on the interval $[a,b]$.

We shall be interested in selfadjoint boundary value problems for the equation

$$Dy - \lambda\mathscr{S}(t)y = 0, \tag{7.2}$$

where $\mathscr{S}(t)$ is an H-matrix of order m defined on $[a,b]$.

In the present case we mean, by a selfadjoint problem, one which reduces to a selfadjoint problem for a canonical differential equation.

---

[27] We wish to emphasize, by the term "quasi-differential", that we do not require that the coefficients $\mathscr{P}(t)$ and $\mathscr{Q}(t)$ be differentiable. The operator D is applicable to those absolutely continuous vector-functions y for which $\mathscr{P}(t)\,dy/dt + \mathscr{Q}(t)y$ is absolutely continuous.

In what follows we will assume that *all elements of the matrix-functions* $\mathscr{P}^{-1}$, $\mathscr{P}^{-1}\mathscr{Q}$, $\mathscr{Q}^*\mathscr{P}^{-1}\mathscr{Q}$, $\mathscr{R}$ *and* $\mathscr{S}$ *belong to* $L_1[a,b]$ *and that almost everywhere* $\det \mathscr{P}(t) \neq 0$.

With these assumptions the equation (7.2) can easily be transformed to a canonical equation. In fact, putting

$$\mathscr{P}(t)\frac{dy}{dt} + \mathscr{Q}(t)y = -z, \tag{7.3}$$

we obtain from (7.1) and (7.2)

$$\frac{dz}{dt} = -\mathscr{Q}^*\frac{dy}{dt} - \mathscr{R}y + \lambda\mathscr{S}y.$$

Determining dy/dt from (7.3) and inserting this expression in the last relation, we arrive at the system

$$\begin{aligned} \frac{dy}{dt} &= -\mathscr{P}^{-1}\mathscr{Q}y - \mathscr{P}^{-1}z, \\ \frac{dz}{dt} &= (\mathscr{Q}^*\mathscr{P}^{-1}\mathscr{Q} - \mathscr{R})y + \mathscr{Q}^*\mathscr{P}^{-1}z + \lambda\mathscr{S}y. \end{aligned} \tag{7.4}$$

Introducing the direct sum $u = y \dotplus z$ of the vector-functions y and z, we can write the system (7.4) in the form

$$du/dt = -\mathscr{Y}_m(\mathscr{H}_0(t) + \lambda\mathscr{H}_1(t))u, \tag{7.5}$$

where

$$\begin{gathered} \mathscr{Y}_m = \begin{pmatrix} 0 & I_m \\ -I_m & 0 \end{pmatrix}, \quad \mathscr{H}_0(t) = \begin{pmatrix} \mathscr{Q}^*\mathscr{P}^{-1}\mathscr{Q} - \mathscr{R} & \mathscr{Q}^*\mathscr{P}^{-1} \\ \mathscr{P}^{-1}\mathscr{Q} & \mathscr{P}^{-1} \end{pmatrix}, \\ \mathscr{H}_1(t) = \begin{pmatrix} \mathscr{S} & 0 \\ 0 & 0 \end{pmatrix}. \end{gathered} \tag{7.6}$$

If, according to the rule stated in §1.1, we put

$$u(t) = \mathscr{U}_0(t)x, \tag{7.7}$$

where $\mathscr{U}_0(t)$ is the matrizant of the equation (7.5) for $\lambda = 0$, i.e.

$$d\mathscr{U}_0/dt = -\mathscr{Y}_m\mathscr{H}_0(t)\mathscr{U}_0, \quad \mathscr{U}_0(0) = I_n \qquad (n = 2m),$$

then equation (7.5) is transformed to the equation

$$dx/dt = -\lambda\mathscr{Y}_m\mathscr{H}(t)x, \tag{7.8}$$

where

$$\mathscr{A}(t) = \mathscr{U}_0^*(t)\mathscr{A}_1(t)\mathscr{U}_0(t).$$

We now supplement our basic assumptions concerning the coefficients with the following assumption.

a_0) *The matrix-function $\mathscr{S}(t)$ $(a \leqq t \leqq b)$ is almost everywhere Hermitian-nonnegative, and for no solution $y_0(t)$ $(\not\equiv 0)$ of the equation $Dy = 0$ can the vector-function $\mathscr{S}(t)y_0(t)$ be almost everywhere equal to zero.*

In particular, this condition will be satisfied if the matrix-function $\mathscr{S}(t)$ is Hermitian-positive almost everywhere on $[a, b]$.

We shall show that with this additional assumption the H-matrix $\mathscr{A}(t)$ will be of positive type (see §1.3).

That the quadratic form $\xi^*\mathscr{A}(t)\xi$, $\xi \in E_n$, is nonnegative is obvious. It remains to verify that if for some $\xi \in E_n$ we have

$$\mathscr{A}(t)\xi = \mathscr{U}_0^*(t)\mathscr{A}_1(t)\mathscr{U}_0(t)\xi = 0 \quad \text{(almost everywhere)}, \tag{7.9}$$

then $\xi = 0$. Now the vector-function $x_0(t) = \mathscr{U}_0(t)\xi$ is a solution of equation (7.5) for $\lambda = 0$, and hence if it is written as a direct sum $x_0(t) = y_0(t) \dotplus z_0(t)$, then $y_0(t)$ will be a solution of the equation $Dy = 0$, and $z_0(t)$ will be represented in terms of $y_0(t)$ by (7.3). It follows from (7.9) that $\mathscr{A}_1(t)\mathscr{U}_0(t)\xi = 0$, and this says, by (7.6), that $\mathscr{S}(t)y_0(t) = 0$ almost everywhere. But then, by a_0), we will have $y_0(t) \equiv 0$, and hence $z_0(t) \equiv 0$, $\mathscr{U}_0(t)\xi \equiv 0$, $\xi = 0$.

Thus, under the stated assumptions the H-matrix $\mathscr{A}(t)$ is of positive type.

Since $\mathscr{U}_0(t)$ is a $\mathscr{Y}_m$-unitary matrix, we have $\mathscr{Y}_m\mathscr{U}_0^* = \mathscr{Y}_m\mathscr{U}_0^{-1}$, and therefore

$$\Lambda_1(\mathscr{Y}\mathscr{A}(t)) = \Lambda_1(\mathscr{U}_0^{-1}(t)\mathscr{Y}_m\mathscr{A}_1(t)\mathscr{U}_0(t)) = \Lambda_1(\mathscr{Y}_m\mathscr{A}_1(t)) \equiv 0 \quad (a \leqq t \leqq b), \tag{7.10}$$

since

$$\mathscr{Y}_m\mathscr{A}_1(t) = \begin{pmatrix} 0 & 0 \\ -\mathscr{S}(t) & 0 \end{pmatrix},$$

and all the eigenvalues of this matrix equal zero for all t.

Now let

$$\mathscr{A} = \begin{pmatrix} \mathscr{A}_{11} & \mathscr{A}_{12} \\ \mathscr{A}_{21} & \mathscr{A}_{22} \end{pmatrix} \quad \text{and} \quad \mathscr{B} = \begin{pmatrix} \mathscr{B}_{11} & \mathscr{B}_{12} \\ \mathscr{B}_{21} & \mathscr{B}_{22} \end{pmatrix}$$

be two square matrices of order $2m$ such that

$$\mathscr{A}\mathscr{Y}_m\mathscr{A}^* = \mathscr{B}\mathscr{Y}_m\mathscr{B}^*,$$

and such that the rank of the matrix $[\mathscr{A}, \mathscr{B}]$ equals $2m$.

Let us consider the boundary value problem

$$\frac{d}{dt}\left(\mathscr{P}\frac{dy}{dt} + \mathscr{Q}y\right) - \mathscr{Q}^*y - \mathscr{R}y + \lambda\mathscr{S}y = 0,$$

$$\left(\mathscr{A}_{j1}y + \mathscr{A}_{j2}\left(\mathscr{P}\frac{dy}{dt} + \mathscr{Q}y\right)\right)\Big|_{t=a} \tag{7.11}$$

$$+\left(\mathscr{B}_{j1}y + \mathscr{B}_{j2}\left(\mathscr{P}\frac{dy}{dt} + \mathscr{Q}y\right)\right)\Big|_{t=b} = 0 \quad (j = 1, 2).$$

By means of the substitution (7.3) and then (7.7) this problem is transformed into a selfadjoint boundary value problem for the equation (7.5). Hence, taking account of (7.10) and Theorem 6.1, we arrive at the following conclusion.

1. *The spectrum of the boundary value problem* (7.11) *consists of real numbers, and the distribution functions* $n_{\pm}(r)$ *of its positive and negative parts satisfy the condition*

$$\lim_{r\to\infty} \frac{n_{\pm}(r)}{r} = 0.$$

Naturally, all three assertions of Theorem 6.2 carry over to the spectrum of the problem (7.11).

2. For the boundary value problems (7.11) which one usually encounters, the spectrum is bounded from below, and for these problems, in particular, on the basis of Theorem 6.2 (assertion 3)) we can assert the existence of a finite limit $\lim_{r\to\infty}(n_+(r)/\sqrt{r})$. Thus, in this case the preceding equality is clarified from a more profound point of view.

Further on we shall obtain the exact value of $\lim_{r\to\infty}(n_+(r)/\sqrt{r})$ for a number of important cases.

For boundary value problems of the type (7.11), instead of the condition a_0) for the coefficient $\mathscr{S}$, the following condition for the two coefficients is usually satisfied.

a) *The matrix-functions* $\mathscr{P}^{-1}(t)$ *and* $\mathscr{S}(t)$ $(a \leqq t \leqq b)$ *are* H*-matrices of positive type with elements from* $L_1(a, b)$.

We shall begin our considerations with the case $\mathscr{Q}(t) \equiv 0$ and $\mathscr{R}(t) \equiv 0$. For this case it is quite simple to establish the following result.

2. *When condition* a) *is satisfied, the spectrum of the boundary value problem*

$$\frac{d}{dt}\left(\mathscr{P}(t)\frac{dy}{dt}\right)+\lambda\mathscr{S}(t)y=0,$$

(7.12)
$$\left(\mathscr{A}_{j1}y(a)+\mathscr{A}_{j2}\left(\mathscr{P}\frac{dy}{dt}\right)\right)\bigg|_{t=a}+\left(\mathscr{B}_{j1}y(b)+\mathscr{B}_{j2}\left(\mathscr{P}\frac{dy}{dt}\right)\right)\bigg|_{t=b}=0\qquad(j=1,2)$$

is bounded from below, and its distribution function $n(r)$ *satisfies the asymptotic formula*

(7.13)
$$\lim_{r\to\infty}\frac{n(r)}{\sqrt{r}}=\frac{1}{\pi}\int_a^b\Lambda_{1/2}(\mathscr{P}^{-1}(t)\mathscr{S}(t))\,dt.$$

Indeed, by Theorem 2.1 on the interlacing of the spectra it suffices to establish result 2 for a boundary value problem with rigid boundary conditions:

(7.14)
$$\frac{d}{dt}\left(\mathscr{P}(t)\frac{dy}{dt}\right)+\lambda\mathscr{S}(t)y=0,\qquad y(a)=0,\qquad y(b)=0.$$

Since (7.14) implies that

$$\int_a^b\frac{dy^*(t)}{dt}\mathscr{P}(t)\frac{dy(t)}{dt}\,dt=\lambda\int_a^b y^*(t)\mathscr{S}(t)y(t)\,dt,$$

we conclude from condition a) that the spectrum of the problem (7.14) is positive.

Putting

$$\lambda=\mu^2\quad\text{and}\quad\mathscr{P}(t)\,dy/dt=-\mu z$$

in (7.14), we transform the system (7.14) into a canonical system for the vector-function $x=y\dotplus z$:

(7.15)
$$dx/dt=-\mu\mathscr{Y}_m\mathscr{H}_2(t)x,\qquad y(a)=y(b)=0,$$

where

$$\mathscr{H}_2(t)=\begin{pmatrix}\mathscr{S}(t)&0\\0&\mathscr{P}^{-1}(t)\end{pmatrix}.$$

For the distribution function $n_2(r)$ of the spectrum of the problem (7.15) we will have (Theorem 6.1)

$$\lim_{r\to\infty}\frac{n_2(r)}{r}=\frac{1}{2\pi}\int_a^b \Lambda_1(\mathscr{Y}_m\mathscr{A}_2(t))\,dt. \tag{7.16}$$

Obviously the distribution function $n(r)$ for the problem (7.14) is related to the distribution function $n_2(r)$ by $n(r)=n_2(\sqrt{r})$. On the other hand, since

$$\det(\mathscr{Y}_m\mathscr{A}_2(t)-\zeta I_{2m})=\det\begin{pmatrix}-\zeta I_m & \mathscr{P}^{-1}(t)\\ -\mathscr{S}(t) & -\zeta I_m\end{pmatrix}$$
$$=\det(\mathscr{P}^{-1}(t)\mathscr{S}(t)+\zeta^2 I_m),$$

it follows that

$$\Lambda_1(\mathscr{Y}_m\mathscr{A}_2(t))=2\Lambda_{1/2}(\mathscr{P}^{-1}(t)\mathscr{S}(t)).$$

Therefore the formula (7.13) is a consequence of the formula (7.16).

3. We shall show that under very general assumptions concerning the matrix-functions $\mathscr{P}$, $\mathscr{Q}$, $\mathscr{R}$ and $\mathscr{S}$ the formula (7.13) remains valid for the general boundary value problem (7.11).

Since the question of the validity of (7.13) for the problem (7.11) is in no way connected with the choice of boundary conditions, we may choose the rigid boundary conditions

$$y(a)=y(b)=0.$$

We define the matrix-function $\mathscr{V}(t)$ $(a\leqq t\leqq b)$ as the solution of the system[28)]

$$\mathscr{P}\frac{d\mathscr{V}}{dt}+\mathscr{Q}\mathscr{V}=0,\qquad \mathscr{V}(0)=I_n \tag{7.17}$$

and then make the substitution

$$y=\mathscr{V}y_1 \tag{7.18}$$

in the differential equation (7.11). It is easily seen that

$$D(\mathscr{V}y_1)=-\frac{d}{dt}\left(\mathscr{P}\mathscr{V}\frac{dy_1}{dt}\right)+\mathscr{Q}^*\frac{d\mathscr{V}}{dt}y_1+\mathscr{Q}^*\mathscr{V}\frac{dy_1}{dt}+\mathscr{R}\mathscr{V}y_1,$$

and consequently

28) Recall that according to our basic assumptions (see p. 299), the elements of the matrix-functions $\mathscr{P}^{-1}$ and $\mathscr{P}^{-1}\mathscr{Q}$ belong to $L_1[a,b]$.

$$\mathscr{V}^* D(\mathscr{V} y_1) = -\frac{d}{dt}\left(\mathscr{V}^* \mathscr{P} \mathscr{V} \frac{dy_1}{dt}\right) + \frac{d\mathscr{V}^*}{dt} \mathscr{P} \mathscr{V} \frac{dy_1}{dt}$$
$$+ \mathscr{V}^* \mathscr{Q}^* \frac{d\mathscr{V}}{dt} y_1 + \mathscr{V}^* \mathscr{Q}^* \mathscr{V} \frac{dy_1}{dt} + \mathscr{V}^* \mathscr{R} \mathscr{V} y_1.$$

By (7.17)

$$\mathscr{V}^* \mathscr{Q}^* \frac{d\mathscr{V}}{dt} = -\mathscr{V}^* \mathscr{Q}^* \mathscr{P}^{-1} \mathscr{Q} \mathscr{V},$$

and taking Hermitian conjugates in (7.17) yields

$$\frac{d\mathscr{V}^*}{dt} \mathscr{P} + \mathscr{V}^* \mathscr{Q}^* = 0.$$

Therefore

$$\mathscr{V}^* D(\mathscr{V} y_1) = -\frac{d}{dt}\left(\mathscr{P}_1 \frac{dy_1}{dt}\right) + \mathscr{R}_1(t) y_1,$$

where

$$\text{(7.19)} \qquad \begin{aligned} \mathscr{P}_1(t) &= \mathscr{V}^*(t) \mathscr{P}(t) \mathscr{V}(t), \\ \mathscr{R}_1(t) &= \mathscr{V}^* \mathscr{R} \mathscr{V} - \mathscr{V}^* \mathscr{Q}^* \mathscr{P}^{-1} \mathscr{Q} \mathscr{V}. \end{aligned}$$

Therefore, making the substitution (7.18) in the boundary conditions and in the equation (7.11), after first multiplying it from the left by $\mathscr{V}^*$, leads to the boundary value problem

$$\text{(7.20)} \qquad \frac{d}{dt}\left(\mathscr{P}_1(t) \frac{dy}{dt}\right) + \mathscr{R}_1(t) y + \lambda \mathscr{S}_1(t) y = 0,$$
$$y(a) = y(b) = 0,$$

where $\mathscr{P}_1(t)$ and $\mathscr{R}_1(t)$ are defined in (7.19), and

$$\text{(7.21)} \qquad \mathscr{S}_1(t) = \mathscr{V}^*(t) \mathscr{S}(t) \mathscr{V}(t).$$

We remark that if condition a) is satisfied by the matrix-functions $\mathscr{P}$ and $\mathscr{Q}$, then it will be satisfied by the matrix-functions $\mathscr{P}_1$ and $\mathscr{Q}_1$, as the matrizant $\mathscr{V}(t)$ is absolutely continuous and $\det \mathscr{V}(t) \neq 0$ $(a \leq t \leq b)$.

Since

$$\Lambda_{1/2}(\mathscr{P}_1^{-1} \mathscr{S}_1(t)) = \Lambda_{1/2}(\mathscr{V}^{-1}(t) \mathscr{P}^{-1}(t) \mathscr{S}(t) \mathscr{V}(t)) = \Lambda_{1/2}(\mathscr{P}^{-1}(t) \mathscr{S}(t)),$$

formula (7.13) remains valid for the boundary value problem (7.20) for the case $\mathscr{R}_1(t) \equiv 0$.

We shall show that this conclusion is also valid for $\mathscr{R}_1(t) \not\equiv 0$, provided there exists a constant $\gamma > 0$ such that the matrix-functions

$$\mathscr{S}_1(t) \pm \gamma \mathscr{R}_1(t) \tag{7.22}$$

are Hermitian-nonnegative for all values of $t \in [a, b]$.

In fact, when this condition is satisfied we can assert that

$$| y^*(t) \mathscr{R}_1(t) y(t) | \leqq \gamma y^*(t) \mathscr{S}_1(t) y(t) \qquad (a \leqq t \leqq b)$$

for any continuous n-dimensional vector-function $y(t)$ $(a \leqq t \leqq b)$, and therefore

$$\left| \int_a^b y^*(t) \mathscr{R}_1(t) y(t) \, dt \right| \leqq \gamma \int_a^b y^*(t) \mathscr{S}_1(t) y(t) \, dt. \tag{7.23}$$

On the other hand, if for some λ $(\neq 0)$ the system (7.20) has a nontrivial solution $y(t)$ $(\not\equiv 0)$, then

$$\int_a^b \frac{dy^*}{dt} \mathscr{P}_1 \frac{dy}{dt} \, dt - \int_a^b y^*(t) \mathscr{R}_1 y(t) \, dt = \lambda \int_a^b y^*(t) \mathscr{S}_1 y(t) \, dt. \tag{7.24}$$

The coefficient of λ in the right side of (7.24) is obviously nonnegative. It cannot vanish, since in this case, by (7.23), the second and hence the first integral in the left side of (7.24) would vanish. But since, by condition a), the matrix-function $\mathscr{P}(t)$ is almost everywhere Hermitian-positive, it would follow that $dy/dt = 0$ almost everywhere, i.e. $y(t) \equiv \xi$, where ξ is a constant, different from zero. But then, by condition a) again, the right side of (7.24) is different from zero. We have arrived at a contradiction. Thus, the coefficient of λ in the right side of (7.24) is positive. Since the first of the integrals in the left side of (7.24) is nonnegative, we conclude from (7.23) that $\lambda \geqq -\gamma$.

We have thus shown that the spectrum of the problem (7.20) is bounded from below. Let

$$\lambda_1 \leqq \lambda_2 \leqq \cdots \tag{7.25}$$

be the complete sequence of eigenvalues of (7.20), and

$$\lambda_1' \leqq \lambda_2' \leqq \cdots \tag{7.26}$$

the complete sequence of eigenvalues of the same problem with $\mathscr{R}_1(t) \equiv 0$.

Formula (7.13) has already been proved for the distribution function of the sequence (7.26); it will be valid for the sequence (7.25) if we prove that

$$\lambda'_k - \gamma \leqq \lambda_k \leqq \lambda'_k + \gamma \qquad (k = 1, 2, \cdots). \tag{7.27}$$

These relations follow easily from the fact that the numbers λ_k $(k = 1, 2, \cdots)$ are the successive minimax of the ratio of quadratic forms

$$\int_a^b \left(\frac{dy^*}{dt}\mathscr{P}_1(t)\frac{dy}{dt} + y^*(t)\mathscr{R}_1(t)y(t)\right)dt \Big/ \int_a^b y^*(t)\mathscr{S}_1(t)y(t)\,dt, \tag{7.28}$$

considered on the set of all continuously differentiable vector-functions $y(t)$ $(a \leqq t \leqq b)$ which vanish for $t = a, b$, while the numbers λ'_k $(k = 1, 2, \cdots)$ are the successive minimax of the ratio of quadratic forms

$$\int_a^b \frac{dy^*}{dt}\mathscr{P}_1(t)\frac{dy}{dt}\,dt \Big/ \int_a^b y^*(t)\mathscr{S}_1(t)y(t)\,dt. \tag{7.29}$$

Since, by (7.23), the absolute value of the difference of the ratios (7.28) and (7.29) does not exceed γ, the inequalities (7.27) follow.

We now note that by virtue of (7.19) and (7.21), the condition that the matrix-functions (7.22) be Hermitian-nonnegative is equivalent to the following condition.

b) *There exists a positive* γ *such that the matrix-functions*

$$\gamma\mathscr{S}(t) \pm (-\mathscr{R}(t) + \mathscr{Q}^*(t)\mathscr{P}^{-1}(t)\mathscr{Q}(t))$$

are Hermitian-nonnegative.

This condition will be satisfied, for example, if the matrix-function $-\mathscr{R} + \mathscr{Q}^*\mathscr{P}^{-1}\mathscr{Q}$ is bounded, and the matrix-function $\mathscr{S}(t)$ is *uniformly Hermitian-positive*, i.e. there exists a $\delta > 0$ such that

$$\xi^*\mathscr{S}(t)\xi \geqq \delta\xi^*\xi \qquad (\xi \in E_n).$$

We have thus proven the following result.

3. *If conditions* a) *and* b) *are satisfied, then the spectrum of the boundary value problem* (7.11) *is bounded from below and its distribution function* $n(r)$ *satisfies the asymptotic formula*

$$\lim_{r\to\infty}\frac{n(r)}{\sqrt{r}} = \frac{4}{\pi}\int_a^b \Lambda_{1/2}(\mathscr{P}^{-1}(t)\mathscr{S}(t))\,dt. \tag{7.30}$$

By definition of the functional Λ_p,

$$\Lambda_{1/2}(\mathscr{P}^{-1}(t)\mathscr{S}(t)) = \sum_{j=1}^n \sqrt{\omega_j(t)} \qquad (\sqrt{\omega_j} \geqq 0;\ j = 1, 2, \cdots, n),$$

where $\omega_1(t) \geqq \omega_2(t) \geqq \cdots \geqq \omega_n(t)$ $(\geqq 0)$ are the roots of the secular equation $\det(\mathscr{S}(t) - \omega\mathscr{P}(t)) = 0$.

For the scalar case $(n = 1)$ (7.30) goes over into the formula

$$\lim_{r\to\infty} \frac{n(r)}{\sqrt{r}} = \frac{4}{\pi} \int_a^b \sqrt{\frac{\mathscr{S}(t)}{\mathscr{P}(t)}}\, dt,$$

which is well known from the theory of Sturm-Liouville equations, where it is usually established under somewhat more restrictive assumptions regarding the coefficients $\mathscr{P}$, $\mathscr{Q}$, $\mathscr{R}$ and $\mathscr{S}$ of the equation in the boundary value problem (7.11).

What is interesting about result 3 is the breadth of the conditions imposed upon the coefficients of the equation in (7.11)—the matrix-functions $\mathscr{P}$, $\mathscr{Q}$, $\mathscr{R}$, $\mathscr{S}$. When these coefficients are sufficiently smooth the method of Birkhoff-Tamarkin enables one to obtain a more precise asymptotic formula which yields, in accordance with the degree of smoothness of these functions, the series of successive terms of an asymptotic expansion of $n(r)$ for $r \to \infty$ (see M. A. Naĭmark [1] and I. M. Rapaport [2]).

We must emphasize, on the other hand, that in deriving result 3 we did not make full use of the possibilities as concerns the generality of our assumptions regarding the coefficients of the second order boundary value problems which can be handled thanks to Theorem 6.1. This circumstance will be illustrated with the scalar equation of a string, which is considered in §8.

§8. The spectrum of an inhomogeneous string with an arbitrary mass distribution

1. Suppose that an inhomogeneous string Σ is stretched between the fixed points $x = 0$ and $x = l$ by a unit force. We denote the mass of the segment $[0, x]$ by $\mathscr{M}(x)$. The function $\mathscr{M}(x)$ $(0 \leqq x \leqq l;\ \mathscr{M}(0) = 0)$ is obviously a nondecreasing right-continuous function.[29] Aside from this we shall assume that $\mathscr{M}(x)$ is a *completely arbitrary* nondecreasing function; it can have an absolutely continuous part $\mathscr{M}_a(x)$, a pure jump part $\mathscr{M}_d(x)$, and a singular part $\mathscr{M}_s(x)$:

$$\mathscr{M}(x) = \mathscr{M}_a(x) + \mathscr{M}_d(x) + \mathscr{M}_s(x) \tag{8.1}$$
$$(\mathscr{M}_a(0) = \mathscr{M}_d(0) = \mathscr{M}_s(0) = 0).$$

---

[29] Since the left end of the string Σ is by hypothesis fixed, we may assume that it does not carry any mass, i.e. that at the point $x = 0$ $\mathscr{M}(x)$ is continuous from the right: $\mathscr{M}(+0) = \mathscr{M}(0) = 0$.

The nondecreasing function $\mathscr{M}(x)$ has a derivative $\mathscr{M}'(x)$ almost everywhere, and

$$\mathscr{M}_a(x) = \int_0^x \mathscr{M}'(\xi)\,d\xi.$$

Let us consider a transverse harmonic vibration of the string, parallel to the y-axis,

$$y(x;t) = y(x)\sin(pt + \alpha).$$

Its amplitude function y and frequency p will be subject to the following set of conditions:

$$dy' + p^2 y\,d\mathscr{M} = 0, \qquad y(0) = y(l) = 0. \tag{8.2}$$

The above equation is to be understood as the relation

$$y'_{\pm}(x) = -p^2 \int_0^{x\pm 0} y(\xi)\,d\mathscr{M}(\xi) + y'_{+}(0), \tag{8.3}$$

where $y'_{+}(x)$ $(y'_{-}(x))$ denotes the right (left) derivative.

It follows from (8.3) and the condition $y(0) = 0$ that

$$y(x) = -\lambda \int_0^x (x - \xi)\,y(\xi)\,d\mathscr{M}(\xi) + xy'_{+}(0) \qquad (\lambda = p^2),$$

and if, moreover, we take into account that $y(l) = 0$, we arrive at a weighted integral equation

$$y(x) = \lambda \int_0^l G_0(x,\xi)\,y(\xi)\,d\mathscr{M}(\xi) \qquad (\lambda = p^2), \tag{8.4}$$

where $G_0(x,\xi)$ is the Green's function of the string Σ with fixed endpoints, i.e.

$$G_0(x,\xi) = \begin{cases} x(l-\xi)/l & (x \leqq \xi), \\ \xi(l-x)/l & (x \geqq \xi). \end{cases}$$

Thus, the squares of the successive frequencies[30)]

$$p_1 < p_2 < \cdots$$

of the harmonic oscillations of a string are the successive characteristic numbers of the integral equation (8.4).

[30)] As is known, the frequencies of a string with ordinary (Stürmian) constraints at the endpoints (in particular, with fixed endpoints) are always simple. For a study of the oscillation properties of a string Σ by means of the theory of weighted integral equations, see the book by F. R. Gantmaher and M. G. Kreĭn [1].

If the string Σ does not consist of a thread with a finite number of beads strung on it (i.e. if the function $\mathscr{M}(x)$ has an infinite number of points of increase), there will be infinitely many of these frequencies. For this case the following theorem holds.

THEOREM 8.1. *For the sequence of frequencies $\{p_n\}$ of the string Σ we have the asymptotic formula*

$$\lim_{n\to\infty}\frac{n}{p_n}=\frac{1}{\pi}\int_0^l\sqrt{\frac{d\mathscr{M}}{dx}}\,dx. \tag{8.5}$$

The validity of this formula for the general case was shown by M. G. Kreĭn [4]. For smooth functions $\mathscr{M}(x)$ formula (8.5) goes back to Stürm and Liouville (see Courant and Hilbert [1]).

PROOF. It will be simpler for us to start directly with the system (8.2) rather than the integral equation (8.4). We put

$$S=l+\mathscr{M}(l)$$

and map the interval $[0,l]$ monotonically into the interval $[0,S]$, putting

$$s=x+\mathscr{M}(x)\qquad(0\leqq x\leqq l).$$

We can now introduce the nondecreasing inverse function $x=x(s)$, which we can consider defined for all $s\in[0,S]$. To do this we only have to agree that for all s from any interval

$$x_0+\mathscr{M}(x_0-0)<s<x_0+\mathscr{M}(x_0+0), \tag{8.6}$$

where x_0 is a jump point of the function $\mathscr{M}(x)$ ($\mathscr{M}(x_0-0)<\mathscr{M}(x_0+0)$), the function $x(s)$ is set equal to x_0.

The substitution $x=x(s)$ transforms the functions $\mathscr{M}(x)$, $\mathscr{M}_a(x)$, $\mathscr{M}_d(x)$ and $\mathscr{M}_s(x)$ into nondecreasing functions which we denote by $\mathscr{M}[s]$, $\mathscr{M}_a[s]$, $\mathscr{M}_d[s]$ and $\mathscr{M}_s[s]$ respectively.

From $\Delta s=\Delta x+\Delta\mathscr{M}(x)$ follows

$$0\leqq\frac{\Delta x}{\Delta s}\leqq 1,\qquad 0\leqq\frac{\Delta\mathscr{M}[s]}{\Delta s}\leqq 1,$$

and thus the derivatives

$$c(s)=\frac{dx}{ds},\qquad a(s)=\frac{d\mathscr{M}[s]}{ds} \tag{8.7}$$

exist for almost all $s\in(0,S)$, and

$$0 \leqq c(s) \leqq 1, \quad 0 \leqq a(s) \leqq 1 \qquad (0 \leqq s \leqq S).$$

Putting $z = p^{-1}dy/dx$ and going over to the variable s, we can write the system (8.2) in the form

$$\begin{aligned} \frac{dy}{ds} &= -pc(s)z, \\ \frac{dz}{ds} &= pa(s)y, \qquad y(0) = y(S) = 0. \end{aligned} \tag{8.8}$$

The boundary value problem (8.8) is a special case of the selfadjoint boundary value problem (6.7), in which $\mu = p$, the coefficient $b(t)$ equals 0, and $a(t)$ and $c(t)$ have the values (8.7).

If we denote by $n(r; \Sigma) = n_+(r; \Sigma)$ the distribution function of the frequencies of the string Σ, then on the basis of formula (6.5′) applied to the boundary value problem (8.8) we can write

$$\lim_{r \to \infty} \frac{n(r; \Sigma)}{r} = \lim_{n \to \infty} \frac{n}{p_n} = \frac{1}{\pi} \int_0^S \sqrt{\frac{d\mathcal{M}}{ds} \cdot \frac{dx}{ds}}\, ds. \tag{8.9}$$

Carrying out the obvious formal transformation in the integral on the right side, we obtain (8.5).

We shall show that the right sides of (8.5) and (8.9) do indeed coincide. By (8.1)

$$\frac{d\mathcal{M}[s]}{ds} = \frac{d\mathcal{M}_a[s]}{ds} + \frac{d\mathcal{M}_d[s]}{ds} + \frac{d\mathcal{M}_s[s]}{ds}.$$

Let x_0 be a jump point of $\mathcal{M}(x)$. By definition the function $x(s)$ has a constant value throughout the interval (8.6), while the functions $\mathcal{M}[s]$ and $\mathcal{M}_d[s]$ vary linearly in this interval, so that

$$\frac{d\mathcal{M}[s]}{ds} = \frac{d\mathcal{M}_d[s]}{ds} = 1,$$

$$\frac{d\mathcal{M}_a[s]}{ds} = \frac{d\mathcal{M}_s[s]}{ds} = 0,$$

$$\frac{dx}{ds} = 0.$$

Therefore we have, in the interval (8.6),

$$\frac{d\mathcal{M}_d[s]}{ds} \cdot \frac{dx(s)}{ds} = 0. \tag{8.10}$$

Let $\{x_j\}$ be the set of jump points of the function $\mathscr{M}(x)$, and let

$$\mathscr{I}_j = (x_j + \mathscr{M}(x_j - 0), x_j + \mathscr{M}(x_j + 0)) \qquad (j = 1, 2, \cdots)$$

be the corresponding sequence of intervals. Then on the one hand

$$\mathscr{M}_d[S] = \int_0^S \frac{d\mathscr{M}_d}{ds}\, ds,$$

and on the other hand

$$\mathscr{M}_d[S] = \mathscr{M}_d(l) = \sum_j (\mathscr{M}(x_j + 0) - \mathscr{M}(x_j - 0))$$
$$= \sum_j \int_{\mathscr{I}_j} \frac{d\mathscr{M}_d}{ds}\, ds.$$

Thus $d\mathscr{M}_d/ds = 0$ almost everywhere outside the intervals $\mathscr{I}_j$ $(j = 1, 2, \cdots)$, and therefore (8.10) holds almost everywhere. Hence

$$(8.11) \quad \begin{aligned} \int_0^S \sqrt{\frac{d\mathscr{M}_a[s]}{ds}\frac{dx}{ds}}\, ds &\leqq \int_0^S \sqrt{\frac{d\mathscr{M}[s]}{ds}\frac{dx}{ds}}\, ds \\ &\leqq \int_0^S \sqrt{\frac{d\mathscr{M}_a[s]}{ds}\frac{dx}{ds}}\, ds + \int_0^S \sqrt{\frac{d\mathscr{M}_s[s]}{ds}\frac{dx}{ds}}\, ds. \end{aligned}$$

By virtue of the fact that the entire variation of the function $\mathscr{M}_s(x)$ takes place on a set of measure zero, for any $\epsilon > 0$ we can find an open set $\mathscr{O}_\epsilon \subset [0, l]$, consisting even of a finite number of intervals, the sum of whose lengths is less than ϵ, such that the variation of $\mathscr{M}_s(x)$ on this set will be greater than $\mathscr{M}_s(l) - \epsilon$. If we denote the image of $\mathscr{O}_\epsilon$ on the interval $[0, S]$ by $\mathscr{Q}_\epsilon$, then we can write

$$\int_{\mathscr{Q}_\epsilon} \frac{dx}{ds}\, ds < \epsilon, \qquad \int_{\mathscr{Q}_\epsilon} \frac{d\mathscr{M}_s[s]}{ds}\, ds > \mathscr{M}(l) - \epsilon.$$

Let us denote by F_ϵ the complement of $\mathscr{Q}_\epsilon$ with respect to the interval $[0, S]$. Then by virtue of the second inequality we will have

$$\int_{F_\epsilon} \frac{d\mathscr{M}_s}{ds}\, ds < \epsilon.$$

By the Cauchy-Bunjakovskiĭ inequality we have

$$\left(\int_{\mathscr{Q}_\epsilon} \sqrt{\frac{d\mathscr{M}_s}{ds}\frac{dx}{ds}}\, ds \right)^2 \leqq \left(\int_{\mathscr{Q}_\epsilon} \frac{d\mathscr{M}_s}{ds}\, ds \right)\left(\int_{\mathscr{Q}_\epsilon} \frac{dx}{ds}\, ds \right) < \epsilon^2 \mathscr{M}_s^2(l),$$

$$\left(\int_{F_\epsilon}\sqrt{\frac{d\mathscr{M}_s}{ds}\frac{dx}{ds}}\,ds\right)^2 \leqq \left(\int_{F_\epsilon}\frac{d\mathscr{M}_s}{ds}\,ds\right)^2\left(\int_{F_\epsilon}\frac{dx}{ds}\,ds\right)^2 < \epsilon^2 l^2.$$

Thus

$$\int_0^S \sqrt{\frac{d\mathscr{M}_s}{ds}\frac{dx}{ds}}\,ds < \epsilon(l + \mathscr{M}_s(l)).$$

Since $\epsilon > 0$ is arbitrary, we conclude that the integral on the left equals zero. Recalling (8.11), we conclude that

$$\int_0^S \sqrt{\frac{d\mathscr{M}}{ds}\frac{dx}{ds}}\,ds = \int_0^S \sqrt{\frac{d\mathscr{M}_a}{ds}\frac{dx}{ds}}\,ds = \int_0^S \sqrt{\frac{d\mathscr{M}_a}{dx}}\,dx.$$

We have thus finally established formula (8.5). Of course, (8.5) holds for the frequencies of a string under any other commonly used conditions regarding constraints on its endpoints (free sliding: $y' = 0$, or elastic constraints: $y' - hy = 0$ at the endpoints).

In particular, it follows from Theorem 8.1 that the relation

$$\lim_{n\to\infty}\frac{n}{p_n} = 0$$

holds for the frequencies of a string if and only if its mass distribution function $\mathscr{M}(x)$ is singular, i.e. $\mathscr{M}'(x) = 0$ almost everywhere.

In particular, this will be the case for a regular Stieltjes string, i.e. a thread weighted with an infinite sequence of beads (point masses) which converge to the right endpoint. The regularity of the string means that its length as well as the sum of the masses of all the beads is finite. We further remark that for a Stieltjes string the left endpoint may slide freely in a direction perpendicular to its equilibrium position, and can be weighted with a mass (see F. R. Gantmaher and M. G. Kreĭn [**1**]).

2. The methods which have been used enable us to obtain the following generalization of Theorem 8.1.

Let $\mathscr{M}(x) = \|\mathscr{M}_{jk}(x)\|_1^m$ $(0 \leqq x \leqq l)$ be a Hermitian matrix-function satisfying the following conditions:

1) for any $(0 \leqq) x_1 < x_2 (\leqq l)$ the matrix $\mathscr{M}(x_2) - \mathscr{M}(x_1)$ is Hermitian-nonnegative,

2) the matrix $\mathscr{M}(l) - \mathscr{M}(0)$ is Hermitian-positive.

We consider the following boundary value problem for continuous vector-functions $y(x) = \{y_j(x)\}_1^m$ $(0 \leqq x \leqq l)$:

$$dy' + \lambda d\mathscr{M}(x)y = 0, \quad y(0) = y(l) = 0. \tag{8.12}$$

Analogously to what was done in the scalar case, the problem (8.12) is to be regarded as equivalent to a weighted integral equation for the vector-function $y(x)$:

$$y(x) = \lambda \int_0^l G_0(x,\xi)y(\xi) \; d\mathscr{M}(\xi),$$

where $G_0(x,\xi)$ is the scalar kernel defined in §8.1.

We may suppose without loss of generality that $\mathscr{M}(0) = 0$. It is convenient to introduce a new variable s by means of

$$s = x + \operatorname{sp}\mathscr{M}(x) \qquad (0 \leqq x \leqq l).$$

Following this the inverse function $x = x(s)$ $(0 \leqq s \leqq S)$ is defined, where $S = l + \operatorname{sp}\mathscr{M}(l)$.

Under the transformation $x \to s$ the matrix-function $\mathscr{M}(x)$ goes over into the matrix-function $M[s]$, whose elements satisfy Lipschitz conditions:

$$| m_{jk}[s + \Delta s] - m_{jk}[s]| \leqq \Delta s \qquad (j, k = 1, 2, \cdots, m),$$

and the boundary value problem (8.11) can be transformed into a selfadjoint boundary value problem (8.8) for the $2m$-dimensional vector-function $y[s] \dotplus z[s]$, where $z[s] = -(dy/ds)/\lambda$. Applying Theorem 6.1 to the problem (8.8) after arguments of the same type as were used in the proof of Theorem 8.1, we obtain the following result.

THEOREM 8.2. *Let $n(r; \mathscr{M})$ be the distribution function of the spectrum of the boundary value problem* (8.12); *then*

$$\lim_{r\to\infty} \frac{n(r;\mathscr{M})}{\sqrt{r}} = \frac{1}{\pi}\int_0^l \Lambda_{1/2}\left(\frac{d\mathscr{M}}{dx}\right) dx.$$

This result, for a function $\mathscr{M}(s)$ with absolutely continuous elements, is a corollary of result 3 of §7 for the problem (7.11), with $\mathscr{P}(t) = I_m$, $\mathscr{Q}(t) = \mathscr{R}(t) = 0$, $\mathscr{S}(t) = d\mathscr{M}/ds$, $t = s$.

Theorem 8.2, as is Theorem 8.1, is interesting, in particular, in that it contains the following assertion: the leading term in the asymptotic behavior of $n(r; \mathscr{M})$ is determined solely by the absolutely continuous part of the function $\mathscr{M}(s)$.

CHAPTER VII

FUNDAMENTAL THEOREM ON THE DENSITY OF THE SPECTRUM OF THE REAL COMPONENT OF A VOLTERRA OPERATOR WITH NUCLEAR IMAGINARY COMPONENT

As we know from Theorem III.2.4 (assertion 2), the spectrum of the real component of a Volterra operator with nuclear imaginary component has a certain density. This density differs only by the factor $1/\pi$ from the absolute value of the trace of the imaginary component, if the latter is definite (nonnegative or nonpositive). As was already mentioned in §2 of Chapter III, for the general case of an arbitrary Volterra operator with nuclear imaginary component the density of the spectrum of the real component of the operator can be expressed in terms of its imaginary component in the form of a special integral, which is constructed with the help of a maximal eigenchain of the Volterra operator.

The establishment of this result (Theorem 1.1) as an important result of a lengthy chain of investigations, involving Volterra operators with nuclear imaginary component, which were carried out in Chapter III, constitutes the principal and most difficult problem of the present chapter. This problem is solved in §§1–3.

As we have already mentioned repeatedly, its solution is closely connected with the derivation of formula (6.5) of Chapter VI, which expresses the density of the spectrum of a canonical selfadjoint boundary value problem in terms of its H-matrix.

In the short concluding §5 we give an application of Theorem 1.1 to the theory of integral equations.

§1. Formulation of the basic theorem on the equality of the functionals $h(\cdot\,;\mathfrak{P})$ and $l(\cdot\,;\mathfrak{P})$

1. Let $\mathfrak{P}$ be a maximal chain of orthoprojectors. As before, we denote by $\hat{\mathfrak{S}}_1[\mathfrak{P}]$ the set of all selfadjoint nuclear operators H whose blocks on every gap (P_-, P_+) of the chain $\mathfrak{P}$ equal zero:

$$(P_+ - P_-)H(P_+ - P_-) = 0.$$

Thus if $\mathfrak{P}$ is a continuous chain, then $\hat{\mathfrak{S}}_1[\mathfrak{P}]$ coincides with the set of all selfadjoint operators from $\mathfrak{S}_1$.

In the general case $\hat{\mathfrak{S}}_1[\mathfrak{P}]$ is a closed subset of $\mathfrak{S}_1$ which is linear in the sense of multiplication by real numbers.

By Theorems I.5.1 and I.10.1 the integral

$$A = 2i \int_{\mathfrak{P}} PH\,dP$$

is meaningful for $H \in \hat{\mathfrak{S}}_1[\mathfrak{P}]$, and yields a Volterra operator A with imaginary component $A_{\mathscr{I}} = H$.

By Theorem III.2.4 we have, for its real component

$$G = A_{\mathscr{R}} \left(= i \int_{\mathfrak{P}} (PH\,dP - dP\,HP) = \mathscr{S}(H;\mathfrak{P}) \right),$$

the equality

$$\lim_{r\to\infty} \frac{n_+(r;G)}{r} = \lim_{r\to\infty} \frac{n_-(r;G)}{r} = \frac{h}{\pi}, \tag{1.1}$$

and hence for $n(r;G) = n_+(r;G) + n_-(r;G)$

$$\lim_{r\to\infty} \frac{n(r;G)}{r} = \lim_{r\to\infty} n s_n(G) = \frac{2h}{\pi}. \tag{1.2}$$

The quantity $h \geqq 0$ may thus be regarded as a functional on $\hat{\mathfrak{S}}_1[\mathfrak{P}]$, generated by $\mathfrak{P}$:

$$h = h(H;\mathfrak{P}) \qquad (H \in \hat{\mathfrak{S}}_1[\mathfrak{P}]).$$

Indeed, given $H \in \hat{\mathfrak{S}}_1[\mathfrak{P}]$ we can construct the operator G by the above formula, and from the spectrum of G we can determine the quantity h.

This functional is obviously nonnegative and even;

$$h(H;\mathfrak{P}) = h(-H;\mathfrak{P}),$$

and moreover

$$h(aH;\mathfrak{P}) = |a|\,h(H;\mathfrak{P}) \qquad (-\infty < a < \infty).$$

We shall see shortly that the functional $h(\cdot\,;\mathfrak{P})$ is subadditive on $\hat{\mathfrak{S}}_1[\mathfrak{P}]$, i.e. $h(H_1 + H_2;\mathfrak{P}) \leqq h(H_1;\mathfrak{P}) + h(H_2;\mathfrak{P})$. We shall first prove a weaker assertion.

LEMMA 1.1. *The square root of the functional h is subadditive on $\hat{\mathfrak{S}}_1[\mathfrak{P}]$, i.e.*

(1.3) $$\sqrt{h(H_1+H_2;\mathfrak{P})} \leqq \sqrt{h(H_1;\mathfrak{P})} + \sqrt{h(H_2;\mathfrak{P})} \quad (H_{1,2} \in \hat{\mathfrak{S}}_1[\mathfrak{P}]).$$

PROOF. Given $H_1, H_2 \in \hat{\mathfrak{S}}_1[\mathfrak{P}]$ and putting $H = H_1 + H_2$, we form the operators

$$G_k = i\int_{\mathfrak{P}} (PH_k dP - dPH_k P) \qquad (k = 1, 2)$$

and

$$G = G_1 + G_2 = i\int_{\mathfrak{P}} (PHdP - dPHP).$$

Since $G = G_1 + G_2$, by (II.1.2) we have

$$s_{n+m-1}(G) \leqq s_n(G_1) + s_m(G_2),$$

so that

$$(n+m)s_{n+m-1}(G) \leqq \left(1 + \frac{m}{n}\right) ns_n(G_1) + \left(1 + \frac{n}{m}\right) ms_m(G_2).$$

We now choose any ξ $(0 < \xi < 1)$ and pass to the limit $n, m \to \infty$ in the above inequality in such a way that $m/n \to \xi$; then by (1.2) we obtain

(1.4) $$h(H;\mathfrak{P}) \leqq (1+\xi)h(H_1;\mathfrak{P}) + (1+\xi^{-1})h(H_2;\mathfrak{P}).$$

The minimum of the right side, for $\xi \in (0,1)$, is attained for

$$\xi = \sqrt{h(H_2;\mathfrak{P})/h(H_1;\mathfrak{P})}.$$

Inserting this value for ξ in (1.4), we obtain (1.3). The lemma is proved.

Let us recall that by Theorem III.2.4

(1.5) $$|\operatorname{sp} H| \leqq h(H;\mathfrak{P}) \leqq |H|_1.$$

This inequality, in conjunction with the preceding lemma, enables us to make an important deduction.

LEMMA 1.2. *The functional* $h(\cdot;\mathfrak{P})$ *is continuous on* $\hat{\mathfrak{S}}_1[\mathfrak{P}]$ *in the metric of* $\mathfrak{S}_1$.

PROOF. In fact, the functional $\sqrt{h(\cdot;\mathfrak{P})}$ is continuous (and even uniformly so) on $\hat{\mathfrak{S}}_1[\mathfrak{P}]$ in the metric of $\mathfrak{S}_1$ since, by (1.3) and (1.5), for any $H_1, H_2 \in \hat{\mathfrak{S}}_1[\mathfrak{P}]$ we have

$$\left|\sqrt{h(H_1;\mathfrak{P})} - \sqrt{h(H_2;\mathfrak{P})}\right| \leqq \sqrt{h(H_1 - H_2;\mathfrak{P})} \leqq \sqrt{|H_2 - H_1|_1}.$$

Lemma 1.1 has played its role and will not be further used.

2. The functional h also has the important property of *block additivity*.

LEMMA 1.3. *Let* $H \in \hat{\mathfrak{S}}_1[\mathfrak{P}]$ *and* $P_1 \in \mathfrak{P}$. *Then*

$$h(H;\mathfrak{P}) = h(P_1HP_1;\mathfrak{P}) + h(Q_1HQ_1;\mathfrak{P}), \tag{1.6}$$

where $Q_1 = I - P_1$.

PROOF. We have

$$H = (P_1 + Q_1)H(P_1 + Q_1) = H_{11} + H_{12} + H_{12}^* + H_{22}, \tag{1.7}$$

where $H_{11} = P_1HP_1,\ H_{12} = P_1HQ_1,\ H_{22} = Q_1HQ_1$.

Applying the transformator $\mathscr{S}(\cdot;\mathfrak{P})$ to each term in (1.7), we obtain for $G = \mathscr{S}(H;\mathfrak{P})$ the decomposition

$$G = G_{11} + G_{12} + G_{12}^* + G_{22},$$

where

$$G_{11} = P_1GP_1,\ G_{12} = P_1GQ_1,\ G_{22} = Q_1GQ_1.$$

Since $H_{12} = P_1H(I - P_1)$, it is easily seen that

$$G_{12} = i\int_{\mathfrak{P}} (PH_{12}dP - dPH_{12}P) = i\int_{P_1}^{I} P_1HdP = iP_1H(I - P_1) = iH_{12},$$

where the last integral is taken over the segment of the chain $\mathfrak{P}$ from P_1 to I. Thus

$$G = (G_{11} + G_{22}) + i(H_{12} - H_{12}^*) = G' + G''.$$

Together with H, the operator H_{12} and hence the operator $G'' = i(H_{12} - H_{12}^*)$ is nuclear. Therefore

$$\lim_{n\to\infty} ns_n(G'') = 0.\text{[1)]}$$

But then, by result 1 of Chapter II §1,

$$\lim_{n\to\infty} ns_n(G) = \lim_{n\to\infty} ns_n(G'). \tag{1.8}$$

On the other hand, since the operators G_{11} and G_{22} are mutually orthogonal, the spectrum of their sum $G' = G_{11} + G_{22}$ is the union of the spectra of G_{11} and G_{22}, and therefore

$$\lim_{r\to\infty} \frac{n(r;G')}{r} = \lim_{r\to\infty} \frac{n(r;G_{11})}{r} + \lim_{r\to\infty} \frac{n(r;G_{22})}{r},$$

i.e., by (1.2)

[1)] We have here used the fact that for a convergent series $\sum_{n=1}^{\infty} a_n$ with nonincreasing terms one always has $\lim_{n\to\infty} na_n = 0$.

$$\lim_{n\to\infty} ns_n(G') = \lim_{n\to\infty} ns_n(G_{11}) + \lim_{n\to\infty} ns_n(G_{22}). \tag{1.9}$$

A comparison of (1.8) and (1.9) yields the desired relation (1.6). The lemma is proved.

We shall now show that the block additivity of the functional $h(\cdot\,;\mathfrak{P})$ enables us to obtain upper and lower bounds for it.

3. Let $\mathfrak{z} = \{P_k\}_0^n$ be a partition of the maximal chain $\mathfrak{P}$. On the basis of Lemma 1.3 we can assert that

$$h(H;\mathfrak{P}) = \sum_{k=1}^{n} h(\Delta P_k H \Delta P_k;\mathfrak{P}), \tag{1.10}$$

where

$$\Delta P_k = P_k - P_{k-1} \qquad (k = 1, 2, \cdots, n).$$

Using the bounds (1.5), we obtain

$$\sum_{k=1}^{n} |\operatorname{sp}(\Delta P_k H \Delta P_k)| \leqq h(H;\mathfrak{P}) \leqq \sum_{k=1}^{n} |\Delta P_k H \Delta P_k|_1. \tag{1.11}$$

For a given $H \in \hat{\mathfrak{S}}_1[\mathfrak{P}]$ the two bounding terms in this inequality are nonnegative functionals, defined on the set $\mathfrak{Z}$ of all partitions $\mathfrak{z}$ of the chain $\mathfrak{P}$. If we pass from a partition $\mathfrak{z}$ to a finer one $\mathfrak{z}' \supset \mathfrak{z}$, the value of the right side decreases (see [GK], Theorem III.8.7), and the value of the left side increases. Therefore

$$\int_{\mathfrak{P}} |\operatorname{sp}(dPHdP)| \leqq h(H;\mathfrak{P}) \leqq \int_{\mathfrak{P}} |dPHdP|_1, \tag{1.12}$$

where the integral on the right has the value

$$\inf_{\mathfrak{z}\in\mathfrak{Z}} \sum_{k=1}^{n} |\Delta P_k H \Delta P_k|_1 = \lim \sum_{k=1}^{n} |\Delta P_k H \Delta P_k|_1,$$

and that on the left, the value

$$\sup_{\mathfrak{z}\in\mathfrak{Z}} \sum_{k=1}^{n} |\operatorname{sp}(\Delta P_k H \Delta P_k)| = \lim \sum_{k=1}^{n} |\operatorname{sp}(\Delta P_k H \Delta P_k)|.$$

These limits are to be understood as the limits in the sense of Šatunovskiĭ along the set $\mathfrak{Z}$.

We remark that for a positive operator $H \in \hat{\mathfrak{S}}_1[\mathfrak{P}]$ the two bounding terms in (1.12) obviously coincide. One can also show that they coincide for any operator $H \in \hat{\mathfrak{S}}_1[\mathfrak{P}]$ if the chain $\mathfrak{P}$ has rank 1. In the general case the two bounding terms in (1.12) do not coincide. This follows, in particular, from Theorem 5.1 to be established further on.

Our aim is to establish the following fundamental theorem.

THEOREM 1.1. *For any maximal chain* $\mathfrak{P}$ *and any* $H \in \hat{\mathfrak{S}}_1[\mathfrak{P}]$ *we have*

$$h(H;\mathfrak{P}) = \int_{\mathfrak{P}} |dPHdP|_1. \tag{1.13}$$

Let us give an equivalent formulation of this theorem, which in a certain respect is more detailed.

THEOREM 1.1′. *Let* $A = G + iH$ *be a Volterra operator with nuclear imaginary component* H *and maximal eigenchain* $\mathfrak{P}$. *Then*

$$\lim_{r\to\infty} \frac{n_{\pm}(r;G)}{r} = \frac{1}{\pi}\int_{\mathfrak{P}} |dPHdP|_1.$$

In connection with the formulation of Theorem 1.1′ it remains only to recall that if a Volterra operator A has maximal eigenchain $\mathfrak{P}$, then the blocks of its imaginary component H on the gaps of the chain $\mathfrak{P}$ equal zero, and its real component $G(=\mathscr{S}(H;\mathfrak{P}))$ can be uniquely reconstructed.

The proof of Theorem 1.1—1.1′ requires a certain amount of preparation, and will be completed in §3.

4. We introduce, provisionally, the notation

$$l(H;\mathfrak{P}) = \int_{\mathfrak{P}} |dPHdP|_1 \tag{1.14}$$

for the integral in (1.13). We note the following properties of the nonnegative functional $l(H;\mathfrak{P})$ defined on $\hat{\mathfrak{S}}_1[\mathfrak{P}]$:

1) $l(aH;\mathfrak{P}) = |a|\, l(H;\mathfrak{P}) \quad (-\infty < a < \infty)$;
2) $l(H_1 + H_2;\mathfrak{P}) \leqq l(H_1;\mathfrak{P}) + l(H_2;\mathfrak{P})$;
3) $l(H;\mathfrak{P}) = \operatorname{sp} H \quad \text{if} \quad H \geqq 0$;
4) $l(H;\mathfrak{P}) \leqq |H|_1$.

The property 1) follows trivially from the definition of $l(H;\mathfrak{P})$. The property 2) follows from the corresponding property of each of the sums

$$\sum_{k=1}^{n} |\Delta P_k H \Delta P_k|_1.$$

Together, these two properties show that $l(\cdot;\mathfrak{P})$ is a quasi-norm.

The property 3) is easily obtained if we note that for $H \geqq 0$ we always have $\Delta PH\Delta P \geqq 0$, and consequently

$$|\Delta PH\Delta P|_1 = \mathrm{sp}(\Delta PH\Delta P).$$

To obtain property 4), it suffices to break up H into its nonnegative components and use properties 1), 2) and 3).

It follows from properties 2) and 4) that

5) *the functional* $l(\cdot;\mathfrak{P})$ *is uniformly continuous in the nuclear metric on* $\hat{\mathfrak{S}}_1[\mathfrak{P}]$:

$$|l(H_1;\mathfrak{P}) - l(H_2;\mathfrak{P})| \leqq |H_1 - H_2|_1.$$

It is also easily seen that, similarly to the functional $h(\cdot;\mathfrak{P})$, the functional l has the property of block additivity.

6) *Let* $H \in \hat{\mathfrak{S}}_1[\mathfrak{P}]$ *and let* $\mathfrak{z} = \{P_j\}_0^n$ *be a partition of the maximal chain* $\mathfrak{P}$. *Then*

$$l(H;\mathfrak{P}) = \sum_{j=1}^{n} l(\Delta P_j H \Delta P_j;\mathfrak{P}), \tag{1.15}$$

where $\Delta P_j = P_j - P_{j-1}$ $(j = 1, 2, \cdots, n)$.

5. Our proof of Theorem 1.1—1.1′ makes essential use of a number of reductions. We begin with a reduction which is based on the following two remarks.

I) The continuity of the functionals $l(\cdot;\mathfrak{P})$ and $h(\cdot;\mathfrak{P})$ on $\hat{\mathfrak{S}}_1[\mathfrak{P}]$ guarantees that they coincide *on all* of $\hat{\mathfrak{S}}_1[\mathfrak{P}]$ as long as they coincide on a subset which is dense in $\hat{\mathfrak{S}}_1[\mathfrak{P}]$.

II) On the other hand, the method presented in §1 of Chapter V (Theorem 1.2) of extending a chain to a continuous chain guarantees that the functionals $l(\cdot;\mathfrak{P})$ and $h(\cdot;\mathfrak{P})$ will coincide for *any* maximal chain $\mathfrak{P}$ as long as they coincide for *any continuous* maximal chain.

Therefore, recalling Theorem 2.2 of Chapter V, we conclude that

It is sufficient to prove Theorem 1.1 *for the case where* $\mathfrak{H}$ *is realized in the form of a space* $L_2^{(r)}(a,b)$, *the chain* $\mathfrak{P} = \hat{\mathfrak{P}}$ *is the maximal chain consisting of the truncation projectors, and the operator* H *is a selfadjoint finite-dimensional operator.*

§2. The functionals h and l as functionals of the H-matrix

1. In this section we shall denote by K a finite-dimensional selfadjoint operator in some space $L_2^{(r)}(a,b)$. According to the final deduction in §1, to prove Theorem 1.1 it remains to establish the equality

$$h(K;\hat{\mathfrak{P}}) = l(K;\hat{\mathfrak{P}}), \tag{2.1}$$

where $\hat{\mathfrak{P}} = \{P(t)\}_a^b$ is the chain of truncation projectors in the space $L_2^{(r)}(a,b)$.

The operator K (being finite-dimensional) is an integral operator

$$(Kf)(t) = \int_a^b \mathscr{K}(t,s)f(s)\,ds \qquad (f \in L_2^{(r)})$$

with kernel $\mathscr{K}(t,s)$ of the form

$$\mathscr{K}(t,s) = \sum_{j=1}^{n} \epsilon_j \psi_j(t)\psi_j^*(s) \qquad (a \leqq t, s \leqq b), \tag{2.2}$$

where n is the dimension of K, $\epsilon_j = \pm 1$ $(j = 1, 2, \cdots, n)$, and the $\psi_j(t)$ $(j = 1, 2, \cdots, n)$ are linearly independent vector-functions from $L_2^{(r)}(a, b)$. We recall that the vector-functions $\psi_j(t)$ are columns, and the $\psi_j^*(s)$ are rows, both with r elements.

The operator K is the imaginary component of the Volterra operator

$$(Af)(t) = 2i\int_t^b \mathscr{K}(t,s)f(s)\,ds.$$

The real component G of A is the integral operator

$$(Gf)(t) = -i\int_a^b \operatorname{sign}(t-s)\,\mathscr{K}(t,s)f(s)\,ds.$$

Obviously $G = \mathscr{S}(K; \hat{\mathfrak{P}})$, and by definition of the functional h (§1.1) we have

$$h(K; \hat{\mathfrak{P}}) = \frac{\pi}{2}\lim_{r\to\infty}\frac{n(r; G)}{r},$$

where $n(r; G)$ is the distribution function of the spectrum of characteristic numbers of the operator G, i.e. of the spectrum of the integral equation

$$\phi(t) = -i\mu\int_a^b \operatorname{sign}(t-s)\,\mathscr{K}(t,s)\phi(s)\,ds. \tag{2.3}$$

We associate with the operator K the matrix-function[2)]

$$\mathscr{H}(t) = \| h_{jk}(t)\|_1^n = 2\|\psi_j^*(t)\psi_k(t)\|_1^n \quad (a \leqq t \leqq b),$$

its *H-matrix*, and the constant matrix

$$\mathscr{Y} = i\|\epsilon_j\delta_{jk}\|_1^n,$$

its *signature matrix*.

[2)] In §3, Chapter V, the H-matrix $\mathscr{H}(t)$ was associated with a *Volterra* operator A with imaginary component $A_{\mathscr{J}} = K$.

The matrix $\mathscr{Y}$ is determined by the operator K up to the order of its diagonal elements. Once the matrix $\mathscr{Y}$ for the operator K has been chosen (see Chapter V, §4), the H-matrix $\mathscr{A}(t)$ will be determined up to a transformation $\mathscr{A}(t) \to S^*\mathscr{A}(t)S$, where S is a constant $\mathscr{Y}$-unitary matrix: $S\mathscr{Y}S^* = \mathscr{Y}$, and consequently the matrix-function $\mathscr{Y}\mathscr{A}(t)$ will be determined up to a similarity transformation:

$$\mathscr{Y}\mathscr{A}(t) \to S^{-1}\mathscr{Y}\mathscr{A}(t)S \quad (= \mathscr{Y}S^*\mathscr{A}(t)S).$$

We recall that the matrix-function $\mathscr{A}(t)$ is always an H-matrix of *positive type* (see the definition in Chapter V, §3), since

$$\frac{1}{2^n}\det \int_a^b \mathscr{A}(t)\,dt = \det \left\| \int_a^b \psi_j^*(t)\psi_k(t)\,dt \right\|_1^n$$
$$= \det \|(\psi_j, \psi_k)\|_1^n > 0.$$

2. We shall show that the functionals $h(K; \hat{\mathfrak{P}})$ and $l(K; \hat{\mathfrak{P}})$ can be regarded as functionals of the product $\mathscr{Y}\mathscr{A}$ associated with the operator K.

More precisely, we independently define two new functionals $h_1(\mathscr{Y}\mathscr{A})$ and $l_1(\mathscr{Y}\mathscr{A})$ as follows:

$$h_1(\mathscr{Y}\mathscr{A}) = h(K; \hat{\mathfrak{P}}) \qquad \text{and} \qquad l_1(\mathscr{Y}\mathscr{A}) = l(K; \hat{\mathfrak{P}})$$

whenever $\mathscr{Y}\mathscr{A}$ corresponds to the operator K.

As was already remarked, the correspondence $K \to \mathscr{Y}\mathscr{A}$ is not single-valued. The inverse mapping $\mathscr{Y}\mathscr{A} \to K$ is certainly not single-valued (in this mapping even the dimension r of the space $L_2^{(r)}$ in which K acts is not uniquely determined).

Following this the proof of (2.1) reduces to the proof of the equality

$$h_1(\mathscr{Y}\mathscr{A}) = l_1(\mathscr{Y}\mathscr{A}).$$

The functional $h(K; \hat{\mathfrak{P}})$ is determined by the spectrum of the equation (2.3).

We shall show that the spectrum of the equation (2.3) *coincides with the spectrum of the selfadjoint skewperiodic boundary value problem*[3)]

$$dx/dt + \mu\mathscr{Y}\mathscr{A}(t)x = 0, \qquad x(a) + x(b) = 0. \tag{2.4}$$

In fact, if we use the expression (2.2) for the kernel $\mathscr{K}(t,s)$, then equation (2.3) can be rewritten in the form of a system:

[3)] For the convenience of the reader we essentially repeat here the considerations of Chapter V, §3.

$$\phi(t) = \sum_{k=1}^{n} \xi_k(t)\psi_k(t),$$

$$\xi_j(t) = -i\mu\epsilon_j \int_a^b \operatorname{sign}(t-s)\psi_j^*(s)\phi(s)\,ds \qquad (2.5)$$

$$(j = 1, 2, \cdots, n).$$

Each of the last n equations is equivalent to an appropriate differential equation with a boundary condition:

$$d\xi_j/dt = -2i\mu\epsilon_j\psi_j^*(t)\phi(t), \quad \xi_j(a) + \xi_j(b) = 0$$
$$(j = 1, 2, \cdots, n).$$

Inserting in these equations the expression for ϕ from (2.5), we obtain for the vector-function $x(t) = \{\xi_j(t)\}_1^n$ the boundary value problem (2.4).

Thus $n(r; G) = n(r; \mathscr{Y}\mathscr{A})$, where $n(r; \mathscr{Y}\mathscr{A})$ is the distribution function of the problem (2.4), and hence

$$h(K; \hat{\mathfrak{P}}) = \frac{\pi}{2} \lim_{r\to\infty} \frac{n(r; \mathscr{Y}\mathscr{A})}{r}. \qquad (2.6)$$

We have thus found the functional $h_1(\mathscr{Y}\mathscr{A})$:

$$h_1(\mathscr{Y}\mathscr{A}) = \frac{\pi}{2} \lim_{r\to\infty} \frac{n(r; \mathscr{Y}\mathscr{A})}{r}.$$

3. We now turn to the construction of the functional l_1. We recall that for a square matrix X of order n, $\Lambda_1(X)$ denotes the sum of the moduli of all the eigenvalues of X:

$$\Lambda_1(X) = \sum_{j=1}^{n} |\lambda_j(X)|.$$

If S is a nonsingular square matrix of the same order as X, then obviously

$$\Lambda_1(S^{-1}XS) = \Lambda_1(X).$$

We shall first show that

The spectrum of nonzero eigenvalues of the operator K coincides with the spectrum of eigenvalues of the matrix

$$\frac{1}{2i}\mathscr{Y}\int_a^b \mathscr{A}(t)\,dt,$$

and consequently[4)]

$$|K|_1 = \frac{1}{2}\Lambda_1\left(\mathscr{Y}\int_a^b \mathscr{A}(t)\,dt\right) \tag{2.7}$$

In fact, using the expression (2.2) for the kernel $\mathscr{K}(t,s)$, we can rewrite the equation $K\phi = \lambda\phi$ in the form of an equivalent system:

$$\lambda\phi(t) = \sum_{k=1}^{n}\psi_k(t)\,\xi_k,$$

$$\xi_j = \epsilon_j\int_a^b \psi_j^*(s)\,\phi(s)\,ds \qquad (j = 1,2,\cdots,n).$$

Inserting the value of ϕ from the first equation into the second, we obtain the system of equations

$$\lambda\xi_j = \sum_{k=1}^{n}\int_a^b \epsilon_j\psi_j^*(s)\,\psi_k(s)\,ds\,\xi_k \quad (j = 1,2,\cdots,n),$$

which can be rewritten in the form

$$\lambda\xi = \frac{1}{2i}\mathscr{Y}\int_a^b \mathscr{A}(t)\,dt\,\xi \qquad (\xi = \{\xi_j\}_1^n).$$

The assertion is proved.

We now note that (2.7) implies the equality

$$|\Delta PK\Delta P|_1 = \frac{1}{2}\Lambda_1\left(\mathscr{Y}\int_\alpha^\beta \mathscr{A}(t)\,dt\right),$$

where $[\alpha,\beta]$ is any subinterval of $[a,b]$, and $\Delta P = P(\beta) - P(\alpha)$.

Consequently, recalling the definition (1.14) of the functional $l(K;\hat{\mathfrak{P}})$, we obtain

$$l(K;\hat{\mathfrak{P}}) = \int_{\hat{\mathfrak{P}}} |dPKdP|_1 = \frac{1}{2}\lim\sum_{k=1}^{N}\Lambda_1\left(\mathscr{Y}\int_{t_{k-1}}^{t_k}\mathscr{A}(t)\,dt\right), \tag{2.8}$$

where lim denotes the limit in the sense of Šatunovskiĭ along the set of all partitions

$$a = t_0 < t_1 < t_2 < \cdots < t_N = b$$

of the interval $[a,b]$.

Thus we have also found the functional $l_1(\mathscr{Y}\mathscr{A})$:

---

4) This result, in essence, coincides with Lemma V.3.1.

$$l_1(\mathscr{Y}\mathscr{A}) = \frac{1}{2}\lim\sum_{k=1}^{N}\Lambda_1\left(\mathscr{Y}\int_{t_{k-1}}^{t_k}\mathscr{A}(t)\,dt\right). \tag{2.9}$$

It follows from (2.6) and (2.8) that both sides of (2.1) are completely determined by the H-matrix $\mathscr{A}(t)$ and the signature matrix $\mathscr{Y}$ of the operator K. Therefore to prove (2.1) it is sufficient to prove it for any other operator K', even one acting in some other space $L_2^{(r')}(a,b)$, but having the same H-matrix $\mathscr{A}(t)$ and signature matrix $\mathscr{Y}$. For such an operator K' we will find it convenient to choose an n-dimensional model operator K_M, defined in $L_2^{(n)}(a,b)$ by the formula (see Chapter V, §3)

$$(K_Mf)(t) = \frac{1}{2i}\int_a^b \mathscr{A}^{1/2}(t)\mathscr{Y}\mathscr{A}^{1/2}(s)f(s)\,ds \qquad (f\in L_2^{(n)}(a,b)).$$

4. We have thus shown that it is sufficient to prove the basic Theorem 1.1—1.1′ for models K_M of selfadjoint operators of finite dimension n. We now show that one can carry out a further reduction, namely, that it suffices to prove Theorem 1.1—1.1′ for model operators having a *$\mathscr{Y}$-regular* H-matrix.

An H-matrix $\mathscr{A}(t)$ is by definition $\mathscr{Y}$-regular if the following conditions are satisfied:

a) all elements of $\mathscr{A}(t)$ are analytic functions of $t\in[a,b]$;

b) for every $t\in[a,b]$ the matrix $\mathscr{A}(t)$ is Hermitian-positive;

c) the multiplicities of the successive eigenvalues of the matrix $\mathscr{Y}\mathscr{A}(t)$ do not depend upon $t\in[a,b]$.

We begin with the following general remark. Let

$$K = \sum_{j=1}^{n}\epsilon_j(\cdot,\psi_j)\psi_j \quad\text{and}\quad K' = \sum_{j=1}^{n}\epsilon_j(\cdot,\psi_j')\psi_j'$$

be two operators acting in the same space; then

$$|K-K'|_1 \leqq \sum_{j=1}^{n}|\psi_j-\psi_j'|\,(|\psi_j|+|\psi_j'|). \tag{2.10}$$

We define the *smoothed* operator $K_\delta(\delta>0)$ by

$$(K_\delta f)(t) = \frac{1}{2i}\int_a^b \mathscr{N}_\delta(t)\mathscr{Y}\mathscr{N}_\delta(s)f(s)\,ds \qquad (f\in L_2^{(n)}(a,b)), \tag{2.11}$$

putting

$$\mathscr{N}_\delta(t) = \frac{1}{2\delta}\int_{t-\delta}^{t+\delta}\mathscr{A}^{1/2}(s)\,ds + \delta I_n;$$

we assume here that $\mathscr{A}^{1/2}(s) = 0$ for $s \notin [a,b]$. Obviously $\mathscr{N}_\delta(t)$ is a continuous matrix-function which assumes only Hermitian-positive values. As $\delta \to 0$ the columns of this matrix-function tend in the metric of $L_2^{(n)}(a,b)$ to the corresponding columns of the matrix-function $\mathscr{A}^{1/2}(t)$. Therefore, by the inequality (2.10), as $\delta \to 0$ the operator K_δ tends to the operator K_M in the metric of $\mathfrak{S}_1$.

Given $\epsilon > 0$ we now form a Hermitian matrix-function $\mathscr{N}_{\delta,\epsilon}(t)$ with analytic elements which differ in the uniform norm from the corresponding elements of the matrix-function $\mathscr{N}_\delta(t)$ by no more than ϵ. Obviously for sufficiently small ϵ $(0 < \epsilon < \epsilon(\delta))$ all the values of the matrix-function $\mathscr{N}_{\delta,\epsilon}(t)$ will be Hermitian-positive.

For such ϵ we define the operator $K_{\delta,\epsilon}$ by (2.11) with $\mathscr{N}_\delta(t)$ replaced by $\mathscr{N}_{\delta,\epsilon}(t)$. The H-matrix of this operator will be the matrix-function $\mathscr{A}_{\delta,\epsilon}(t) = \mathscr{N}^2_{\delta,\epsilon}(t)$, which obviously satisfies the requirements a) and b).

Thus, to prove (2.1) in the general case it suffices to prove it for an operator K having an H-matrix $\mathscr{A}(t)$ which satisfies the conditions a) and b).

Now let $\mathscr{A}(t)$ be an arbitrary H-matrix satisfying the conditions a) and b). By virtue of the Hermitian-positivity and the analyticity of the matrix-function $\mathscr{A}(t)$ all the eigenvalues of the matrix $i\mathscr{Y}\mathscr{A}(t)$ will be real analytic functions of $t \in [a,b]$.[5] Therefore there exists a finite number of points $(a <)\alpha_1 < \alpha_2 < \cdots < \alpha_{p-1}(< b)$ such that inside each of the intervals (α_j, α_{j+1}) $(j = 0, 1, \cdots, p-1; \alpha_0 = a, \alpha_p = b)$ the distinct successive eigenvalues of the matrix-function $i\mathscr{Y}\mathscr{A}(t)$ will have constant multiplicities. On each interval of the form $[\alpha_j + \epsilon, \alpha_{j+1} - \epsilon]$ $(\epsilon > 0)$ the H-matrix $\mathscr{A}(t)$ is now $\mathscr{Y}$-regular.

By means of the chain $\hat{\mathfrak{P}} = \{P(t)\}_a^b$ of truncation projectors in $L_2^{(n)}(a,b)$ we define the quantities $h(\alpha,\beta)$ and $l(\alpha,\beta)$ for any subinterval $(\alpha,\beta) \subset (a,b)$, putting

$$h(\alpha,\beta) = h(\Delta P K_M \Delta P; \hat{\mathfrak{P}}), \qquad l(\alpha,\beta) = l(\Delta P K_M \Delta P; \hat{\mathfrak{P}}),$$

where $\Delta P = P(\beta) - P(\alpha)$. By virtue of (2.7) and the bound 4) from §1.4,

$$h(\alpha,\beta) \leqq l(\alpha,\beta) \leqq |\Delta P K_M \Delta P|_1 = \frac{1}{2}\Lambda_1\left(\mathscr{Y}\int_\alpha^\beta \mathscr{A}(t)\,dt\right)$$

and thus $l(\alpha,\beta)$ and $h(\alpha,\beta) \to 0$ as $\beta - \alpha \to 0$. Using the block additivity of the functionals h and l, we write

---

[5] Since the eigenvalues are the roots of a secular equation with analytic coefficients.

$$h(a,b) = \sum_{j=0}^{p-1} h(\alpha_j, \alpha_{j+1}), \quad l(a,b) = \sum_{j=0}^{p-1} l(\alpha_j, \alpha_{j+1}). \tag{2.12}$$

On the other hand, since

$$h(\alpha_j, \alpha_{j+1}) = h(\alpha_j, \alpha_j + \epsilon) + h(\alpha_j + \epsilon, \alpha_{j+1} - \epsilon) + h(\alpha_{j+1} - \epsilon, \alpha_{j+1}),$$

it follows that

$$h(\alpha_j, \alpha_{j+1}) = \lim_{\epsilon \downarrow 0} h(\alpha_j + \epsilon, \alpha_{j+1} - \epsilon) \quad (j = 0, 1, 2, \cdots, p-1). \tag{2.13}$$

Similarly

$$l(\alpha_j, \alpha_{j+1}) = \lim_{\epsilon \downarrow 0} l(\alpha_j + \epsilon, \alpha_{j+1} - \epsilon) \quad (j = 0, 1, 2, \cdots, p-1). \tag{2.14}$$

If (2.1) were already established for operators K with a $\mathscr{Y}$-regular H-matrix, then for sufficiently small ϵ (> 0)

$$h(\alpha_j + \epsilon, \alpha_{j+1} - \epsilon) = l(\alpha_j + \epsilon, \alpha_{j+1} - \epsilon) \quad (j = 0, 1, 2, \cdots, p-1).$$

But then, by virtue of (2.12), (2.13) and (2.14), $h(a,b) = l(a,b)$, which yields (2.1) for any K_M with an H-matrix $\mathscr{A}(t)$ satisfying the conditions a) and b).

Thus, we have reduced the proof of Theorem 1.1—1.1′ to the case of a model operator K_M with a $\mathscr{Y}$-regular H-matrix.

5. In connection with the relation (2.8) the conjecture that

$$l(K; \hat{\mathfrak{P}}) = \frac{1}{2} \int_a^b \Lambda_1(\mathscr{Y}\mathscr{A}(t))\, dt \tag{2.15}$$

arises naturally. For the most general case (an H-matrix $\mathscr{A}(t)$ with summable elements) this equality will be established in §4. For our immediate purposes the following lemma is enough.

LEMMA 2.1. *If the H-matrix $\mathscr{A}(t)$ $(a \leqq t \leqq b)$ of the operator K is continuous, then* (2.15) *holds.*

PROOF. Let us denote by $\mathbf{H}_n$ the set of all Hermitian matrices X $(= X^*)$ of order n. Obviously $\mathbf{H}_n$ can be regarded as an n^2-dimensional real linear space. It is also obvious that $\Lambda_1(\mathscr{Y} X)$ $(X \in \mathbf{H}_n)$ is a continuous functional on this space.

Starting with the continuous H-matrix $\mathscr{A}(t)$ of the operator K, we form the set $C(\mathscr{A})$ of Hermitian matrices consisting of all matrices $\mathscr{A}_{\alpha,\beta}$ $(a \leqq \alpha < \beta \leqq b)$ of the form

$$\mathscr{A}_{\alpha,\beta} = \frac{1}{\beta - \alpha} \int_\alpha^\beta \mathscr{A}(t)\, dt$$

as well as their limits, which are the matrices $\mathscr{A}(t)$ $(a \leqq t \leqq b)$.

Since $C(\mathscr{A})$ is a compact (closed bounded) set in $\mathbf{H}_n$, the functional $\Lambda_1(\mathscr{Y} X)$ is uniformly continuous on $C(\mathscr{A})$, i.e. for any $\epsilon > 0$ there exists an $\eta = \eta_\epsilon > 0$ such that

$$|\Lambda_1(\mathscr{Y} X') - \Lambda_1(\mathscr{Y} X'')| < \epsilon$$

whenever $X', X'' \in C(\mathscr{A})$ are such that

$$|X' - X''| < \eta.^{6)}$$

On the other hand, since

$$|\mathscr{A}_{\alpha,\beta} - \mathscr{A}(\xi)| = \frac{1}{\beta - \alpha} \left| \int_\alpha^\beta [\mathscr{A}(t) - \mathscr{A}(\xi)] dt \right|$$
$$\leqq \frac{1}{\beta - \alpha} \int_\alpha^\beta |\mathscr{A}(t) - \mathscr{A}(\xi)|\, dt$$

and by hypothesis the matrix-function $\mathscr{A}(t)$ is continuous, for any $\eta > 0$ there exists a $\delta = \delta_\eta > 0$ such that $|\mathscr{A}_{\alpha,\beta} - \mathscr{A}(\xi)| < \eta$ for $\beta - \alpha < \delta$ and $\alpha \leqq \xi \leqq \beta$.

Thus, for any $\epsilon > 0$ there exists a $\delta = \delta_\epsilon > 0$ such that

$$|\Lambda_1(\mathscr{Y} \mathscr{A}_{\alpha,\beta}) - \Lambda_1(\mathscr{Y} \mathscr{A}(\xi))| < \epsilon \text{ for } \beta - \alpha < \delta_\epsilon \text{ and } \alpha \leqq \xi \leqq \beta.$$

Hence for any partition $a = t_0 < t_1 < t_2 < \cdots < t_N = b$ such that $t_j - t_{j-1} < \delta$ $(j = 1, 2, \cdots, N)$ and any $\xi_j \in (t_{j-1}, t_j)$ $(j = 1, 2, \cdots, N)$ we will have

$$\left| \sum_{k=1}^N \Lambda_1 \left(\mathscr{Y} \int_{t_{k-1}}^{t_k} \mathscr{A}(t)\, dt \right) - \sum_{k=1}^N \Lambda_1(\mathscr{Y} \mathscr{A}(\xi_k))(t_k - t_{k-1}) \right|$$
$$\leqq \sum_{k=1}^N |\Lambda_1(\mathscr{Y} \mathscr{A}_{t_{k-1},t_k}) - \Lambda_1(\mathscr{Y} \mathscr{A}(\xi_k)|\,(t_k - t_{k-1}) < \epsilon,$$

from which (2.15) follows.

6. Comparing (2.6) and (2.15) shows that the relation (2.1), at least

---

6) Where $|\cdot|$ is any fixed norm in $\mathbf{H}_n$.

for operators K with continuous H-matrices, is equivalent to the following:

$$\lim_{r\to\infty} \frac{n(r;\mathscr{Y}\mathscr{A})}{r} = \frac{1}{\pi}\int_a^b \Lambda_1(\mathscr{Y}\mathscr{A}(t))\,dt. \tag{2.16}$$

What is essential for us is that the reduction of the proof of Theorem 1.1, achieved in subsection 4 of this section, enables us to assert that *the proof of Theorem* 1.1 *will be completed as soon as we establish the preceding relation for any* $\mathscr{Y}$*-regular H-matrix* $\mathscr{A}(t)$.

§3. Conclusion of the proof of the fundamental theorem

1. Let the H-matrix $\mathscr{A}(t)$ $(a \leqq t \leqq b)$ satisfy the conditions of $\mathscr{Y}$-regularity a), b) and c) of §2.4. We denote by

$$\omega_1(t) < \omega_2(t) < \cdots < \omega_p(t) \ \ (<0<)\, \omega_{p+1}(t) < \cdots < \omega_q(t) \tag{3.1}$$

all distinct eigenvalues of the matrix-function $i\mathscr{Y}\mathscr{A}(t)$. By virtue of condition c) their number q does not depend upon $t \in [a,b]$, and they have constant multiplicities $\kappa_1, \kappa_2, \cdots, \kappa_q$. Obviously the sum $\kappa_1 + \cdots + \kappa_q$ equals the order n of the H-matrix $\mathscr{A}(t)$.[7)]

With each $\omega_j(t)$ there is associated, in the n-dimensional complex space E_n with the usual Euclidean metric, a κ_j-dimensional subspace $L_j(t)$, the corresponding eigenspace. The direct sum of the $L_j(t)$ yields E_n:

$$E_n = L_1(t) \dot{+} L_2(t) \dot{+} \cdots \dot{+} L_q(t).$$

This decomposition gives rise to a corresponding decomposition of the unit operator I_n into a sum of projectors

$$I_n = Q_1(t) + Q_2(t) + \cdots + Q_q(t), \tag{3.2}$$

which are pairwise orthogonal:

$$Q_j Q_k = \begin{cases} Q_j, & k = j, \\ 0, & k \neq j, \end{cases} \tag{3.3}$$

and map E_n onto $L_1(t), L_2(t), \cdots, L_q(t)$, respectively.

Obviously

$$i\mathscr{Y}\mathscr{A}(t) = \sum_{j=1}^{q} \omega_j(t)\, Q_j(t). \tag{3.4}$$

---

7) The sum $\kappa_1 + \cdots + \kappa_p$ of the first p numbers κ_j equals the number of negative ϵ_j $(=\pm 1)$ in the matrix $\mathscr{Y} = i\,\|\epsilon_j\delta_{jk}\|_1^n$.

We shall show that along with $\mathscr{A}(t)$, the projectors $Q_j(t)$ $(j = 1, 2, \cdots, q)$ are analytic matrix-functions. Indeed, let $t_0 \in [a, b]$ and let C_j be a circle with center at the point $\omega_j(t_0)$ $(j = 1, 2, \cdots, q)$ with radius so small that all the other points $\omega_k(t_0)$ $(k \neq j)$ lie outside it. Then for δ (> 0) sufficiently small the point $\omega_j(t)$ will lie inside C_j, and the other points $\omega_k(t)$ $(k \neq j)$ will lie outside C_j, for all $t \in (t_0 - \delta, t_0 + \delta)$. For these t we will have

$$Q_j(t) = -\frac{1}{2\pi i} \oint_{C_j} (\mathscr{A}(t) - \zeta I_n)^{-1} d\zeta \quad (t_0 - \delta < t < t_0 + \delta),$$

and since $\mathscr{A}(t)$ is an analytic matrix-function, it is clear that the matrix-function $Q_j(t)$ is analytic in the interval $(t_0 - \delta, t_0 + \delta)$. Since t_0 was an arbitrarily chosen point, we conclude that $Q_j(t)$ is an analytic matrix-function on the entire interval $[a, b]$.

We need the following lemma.[8)]

LEMMA 3.1 (JU. L. DALECKIĬ [1]). *Let the matrix-function $S(t)$ $(a \leqq t \leqq b)$ be the Cauchy solution of the differential equation*

$$\frac{dS}{dt} = \left(\sum_{j=1}^{q} \frac{dQ_j}{dt} Q_j \right) S, \quad S(a) = I_n. \tag{3.5}$$

Then

$$S(t)\, Q_k(a) = Q_k(t)\, S(t) \qquad (k = 1, 2, \cdots, q). \tag{3.6}$$

PROOF. Differentiating (3.3) with respect to t yields

$$Q_k' Q_j + Q_k Q_j' = \delta_{jk} Q_k' \qquad (j, k = 1, 2, \cdots, q).$$

It follows that

$$Q_k Q_j' Q_j = \begin{cases} 0 & (j = k), \\ -Q_k' Q_j & (j \neq k). \end{cases}$$

Using these relations and also (3.2), we multiply (3.5) from the left by Q_k, obtaining

$$Q_k \frac{dS}{dt} = -Q_k' \sum_{j \neq k} Q_j S = (Q_k' Q_k - Q_k') S.$$

Therefore

$$\frac{d}{dt}(Q_k S) = Q_k \frac{dS}{dt} + Q_k' S = Q_k' Q_k S,$$

8) For $q = 2$ this lemma was established earlier by Ju. L. Daleckiĭ and S. G. Kreĭn [1].

and consequently

$$\frac{d}{dt}(Q_k S) = \left(\sum_{j=1}^{q} Q_j' Q_j\right)(Q_k S) \qquad (k = 1, 2, \cdots, q).$$

Thus the matrix-function $Q_k(t)\,S(t)$ satisfies the same homogeneous differential equation as does the matrix-function $S(t)$. The same equation is satisfied by any matrix-function $S(t)\,C$, where C is a constant matrix. Putting $C = Q_k(a)$, we find that the matrices $Q_k(t)\,S(t)$ and $S(t)\,C$ coincide for $t = a$, and hence coincide for all $t \in [a, b]$, which yields (3.6). The lemma is proved.

We remark that along with the matrices $Q_j(t)$, the matrix-function $S(t)$ is analytic in t.

In proving the lemma we only used the fact that the $Q_j(t)$ $(j = 1, 2, \cdots, q)$ are differentiable matrix-functions which, for each $t \in [a, b]$, are pairwise orthogonal and have sum I_n.

If the $Q_j(t)$ $(j = 1, 2, \cdots, q)$ are taken from the decomposition (3.4), then they will have the additional property of being $\mathscr{Y}$*-Hermitian*, i.e.

$$\mathscr{Y}\,Q_j(t) = Q_j^*(t)\,\mathscr{Y} \qquad (j = 1, 2, \cdots, q;\ a \leqq t \leqq b). \tag{3.7}$$

In fact, taking the Hermitian conjugate of (3.4) yields

$$i\,\mathscr{A}(t)\,\mathscr{Y} = \sum_{j=1}^{q} \omega_j(t)\,Q_j^*(t).$$

Multiplying from the right and left by $\mathscr{Y}$ and taking into account that $\mathscr{Y}^2 = -I_n$, we obtain

$$i\mathscr{Y}\,\mathscr{A}(t) = -\sum_{j=1}^{q} \omega_j(t)\,\mathscr{Y}\,Q_j^*(t)\,\mathscr{Y}.$$

It is easily seen that the matrices $-\mathscr{Y}\,Q_j^*(t)\,\mathscr{Y}$ $(j = 1, 2, \cdots, q)$ are pairwise orthogonal projectors whose sum is I_n. Comparing the above decomposition with (3.4), we see that $-\mathscr{Y}\,Q_j^*(t)\,\mathscr{Y} = Q_j(t)$ $(j = 1, 2, \cdots, q)$, which is equivalent to (3.7).

For brevity we put

$$G(t) = \mathscr{Y} \sum_{j=1}^{q} Q_j'(t)\,Q_j(t).$$

Using $G(t)$, equation (3.5) can be rewritten in the form

$$dS/dt = -\mathscr{Y}\,G(t)\,S, \quad S(a) = I_n.$$

The relations (3.7) enable us to assert that

$$G(t) = \sum_{j=1}^{q} Q_j^{*\prime}(t)\mathscr{Y} Q_j(t) = \sum_{j=1}^{q} Q_j^{*\prime}(t) Q_j^{*}(t)\mathscr{Y} = - \left(\sum_{j=1}^{q} \mathscr{Y} Q_j(t) Q_j^{\prime}(t)\right)^{*},$$

and since

$$\sum_{j=1}^{q} [Q_j^{\prime}(t) Q_j(t) + Q_j(t) Q_j^{\prime}(t)] = \frac{d}{dt}\sum_{j=1}^{n} Q_j^{2}(t) = \frac{d}{dt}\sum_{j=1}^{n} Q_j(t) = 0,$$

the matrix-function $G(t)$ proves to be Hermitian.

Thus $S(t)$, being the matrizant of a canonical system, is a $\mathscr{Y}$-unitary matrix (see Chapter VI, §1), i.e.

$$S^{*}(t)\mathscr{Y} S(t) = \mathscr{Y}. \tag{3.8}$$

2. To complete the proof of Theorem 1.1 (see §2.6) it remains to establish the relation

$$\lim_{r\to\infty} \frac{n(r)}{r} = \frac{1}{\pi}\int_a^b \Lambda_1(\mathscr{Y}\mathscr{H}(t))\,dt \tag{3.9}$$

under the assumption that the H-matrix $\mathscr{H}(t)$ is $\mathscr{Y}$-regular (see conditions a), b), c) of §2.4).

Here $n(r)$ denotes the distribution function of the spectrum of the boundary value problem (2.4). By the theorem on the interlacing of spectra (Theorem VI.1.3) the limit in the left side of (3.9) will be the same if by $n(r)$ we understand the distribution function of the spectrum of the boundary value problem

$$dx/dt = -\mu\mathscr{Y}\mathscr{H}(t)x, \quad \mathscr{E}x(a) + x(b) = 0, \tag{3.10}$$

where $\mathscr{E}$ is any $\mathscr{Y}$-unitary matrix ($\mathscr{E}^{*}\mathscr{Y}\mathscr{E} = \mathscr{Y}$).

We remark that the spectrum of the boundary value problem (3.10) can be found from the equation

$$\det(\mathscr{U}(b;\mu) + \mathscr{E}) = 0, \tag{3.11}$$

where $\mathscr{U}(t;\mu)$ is the matrizant of (3.10), i.e.

$$d\mathscr{U}/dt = -\mu\mathscr{Y}\mathscr{H}(t)\mathscr{U}, \quad \mathscr{U}(a;\mu) = I_n. \tag{3.12}$$

Leaving the $\mathscr{Y}$-unitary matrix $\mathscr{E}$ unspecified for the moment (subsequent calculations will suggest the most suitable choice for $\mathscr{E}$), we transform (3.10), putting

$$x(t) = S(t)y(t). \tag{3.13}$$

Since by (3.4) and (3.6)

$$-\mathscr{Y}\mathscr{A}(t)S(t) = i\sum_{j=1}^{q}\omega_j(t)Q_j(t)S(t) = iS(t)\sum_{j=1}^{q}\omega_j(t)Q_j(a),$$

substituting (3.13) into (3.6) yields

$$\frac{dy}{dt} = i\mu\sum_{j=1}^{q}\omega_j(t)Q_j(a)y + \mathscr{Y}G_1(t)y,\ \mathscr{E}y(a) + S(b)y(b) = 0, \tag{3.14}$$

where

$$\mathscr{Y}G_1(t) = -S^{-1}(t)\frac{dS}{dt} = S^{-1}(t)\mathscr{Y}G(t)S(t).$$

Since $S(t)$ is a $\mathscr{Y}$-unitary matrix-function and hence $S^{-1}(t)\mathscr{Y} = \mathscr{Y}S^*(t)$, we have

$$G_1(t) = S^*(t)G(t)S(t),$$

and thus $G_1(t)$, along with $G(t)$, is an analytic Hermitian matrix-function.

3. We have to carry out a number of further transformations and obtain various bounds. It will be easier to follow the calculations if we first indicate their purpose. We are trying to prove that for an appropriate choice of the matrix $\mathscr{E}$, the spectrum of the boundary value problem (3.14) lies in some "neighborhood" of the spectrum of the simpler boundary value problem

$$\frac{dz}{dt} = i\mu\sum_{j=1}^{q}\omega_j(t)Q_j(a)z,\quad z(a) + z(b) = 0. \tag{3.15}$$

In particular, this means that in seeking the leading term in the asymptotic distribution of the spectrum the term $\mathscr{Y}G_1(t)y$ in (3.14), which does not contain the parameter μ, can be discarded.

The spectrum and the complete system of eigenvector-functions of the problem (3.15) can be found in explicit form.

In fact, multiplying the equation in (3.15) successively by the projectors $Q_k(a)$ $(k = 1, 2, \cdots, q)$ (and then changing the index k to j) shows that the problem (3.15) breaks up into q independent boundary value problems

$$dz_j/dt = i\mu\omega_j(t)z_j,\ z_j(a) + z_j(b) = 0 \qquad (j = 1, 2, \cdots, q), \tag{3.16}$$

where the vector-function $z_j(t)$ $(= Q_j(a)z(t))$ in the jth problem is subject to the condition

$$z_j(t) \in L_j(a) \ (= Q_j(a) E_n) \qquad (j = 1, 2, \cdots, q).$$

From the differential equation in (3.16) we find that

$$z_j(t) = \exp(i\mu\Omega_j(t)) z_j(a),$$

where

$$\Omega_j(t) = \int_a^t \omega_j(s)\, ds \qquad (j = 1, 2, \cdots, q),$$

and therefore the boundary condition $z_j(a) + z_j(b) = 0$ reduces to the equation

$$1 + \exp(i\mu\Omega_j(b)) = 0.$$

This equation has only simple roots, given by

$$\mu_{jk} = \frac{2k+1}{\Omega_j(b)} \qquad (k = 0, \pm 1, \pm 2, \cdots). \tag{3.17}$$

Giving j the values $1, 2, \cdots, q$, we obtain q series of numbers (some of these series may coincide termwise, and also one of these series may turn out to be a part of another one).[9] All these series together, taking multiplicities into account, form the spectrum of eigenvalues of the boundary value problem (3.15).

With each number $\mu = \mu_{jk}$ of the jth series we can associate an eigen-vector-function

$$z_j(t) = \exp(i\mu_{jk}\Omega_j(t))\xi_j,$$

where ξ_j is any vector of the lineal $Q_j(a) E_n = L_j(a)$ $(j = 1, 2, \cdots, q)$. Thus these eigenvector-functions form a κ_j-dimensional family (recall that κ_j, the dimension of $L_j(a)$, is equal to the multiplicity of $\omega_j(t)$ as an eigenvalue of the matrix $\mathscr{Y}\mathscr{A}(t)$).

Consequently, if we denote by $n^{(0)}(r)$ $(r > 0)$ the number of eigen-values, taken with their multiplicities from the boundary value problem (3.15), lying in the interval $[-r, r]$, and by $n_j(r)$ the corresponding number of those from the sequences (3.17), then

$$n^{(0)}(r) = \sum_{j=1}^{q} \kappa_j n_j(r). \tag{3.18}$$

---

[9] Since, by (3.1), $\Omega_1(b) < \Omega_2(b) < \cdots < \Omega_q(b)$, the first case can occur for the jth and lth series if $\Omega_j(b) = -\Omega_l(b)$, and the second case, when one of the ratios $|\Omega_j(b)| / |\Omega_l(b)|$, $|\Omega_l(b)| / |\Omega_j(b)|$ is an odd integer.

On the other hand, it follows from (3.17) that[10)]

$$n_j(r) = 2\left[\frac{|\Omega_j(b)|r}{2\pi} + \frac{1}{2}\right] = \frac{1}{\pi}|\Omega_j(b)|r + O(1). \tag{3.19}$$

From (3.18) and (3.19) we conclude that

$$\begin{aligned}\lim_{r\to\infty}\frac{n^{(0)}(r)}{r} &= \frac{1}{\pi}\sum_{j=1}^{q}|\Omega_j(b)|\kappa_j = \frac{1}{\pi}\sum_{j=1}^{q}\kappa_j\int_a^b|\omega_j(t)|\,dt \\ &= \frac{1}{\pi}\int_a^b\Lambda_1(\mathscr{Y}\mathscr{A}(t))\,dt\end{aligned} \tag{3.20}$$

We have thus proved that the distribution function $n^{(0)}(r)$ of the eigenvalues of the boundary value problem (3.15) satisfies the asymptotic relation (3.9).

4. We now turn to the investigation of the boundary value problem (3.14), which we shall regard as a perturbation of the problem (3.15).

We introduce the matrizant $\mathscr{V}(t;\mu)$ of (3.14):

$$\frac{d\mathscr{V}}{dt} = i\mu\sum_{j=1}^{q}\omega_j(t)Q_j(a)\mathscr{V} + \mathscr{Y}G_1(t)\mathscr{V},\quad \mathscr{V}(a;\mu) = I_n. \tag{3.21}$$

Let us consider, on the other hand, the matrizant $Z(t)$ corresponding to the block-diagonal part $\mathscr{Y}\hat{G}_1(t)$ of the matrix $\mathscr{Y}G_1$:

$$dZ/dt = \mathscr{Y}\hat{G}_1(t)Z,\quad Z(a) = I_n, \tag{3.22}$$

where

$$\mathscr{Y}\hat{G}_1(t) = \sum_{j=1}^{q}Q_j(a)\mathscr{Y}G_1(t)Q_j(a) = \mathscr{Y}\sum_{j=1}^{q}Q_j^*(a)G_1(t)Q_j(a).$$

Obviously $Z(t)$ is an analytic $\mathscr{Y}$-unitary matrix-function; moreover, it is a block-diagonal matrix-function,

$$Z(t) = \sum_{j=1}^{q}Z_j(t),$$

whose blocks

$$Z_j(t) = Q_j(a)Z(t) = Z(t)Q_j(a) \tag{3.23}$$

are determined from the equations

$$dZ_j/dt = \mathscr{Y}Q_j^*(a)G_1(t)Q_j(a)Z_j,\quad Z_j(a) = Q_j(a).$$

---

10) The symbol $[\xi]$ denotes the integer part of the number ξ.

If we carry out the substitution

$$\mathscr{V}(t;\mu) = Z(t)\mathscr{V}_0(t;\mu) \tag{3.24}$$

in (3.21), taking into account that $\sum_{j=1}^{q}\omega_j(t)Q_j(a)$ and $Z(t)$ commute, we obtain

$$\frac{d\mathscr{V}_0}{dt} = i\mu\sum_{j=1}^{q}\omega_j(t)Q_j(a)\mathscr{V}_0 + \mathscr{Y}G_0(t)\mathscr{V}_0, \tag{3.25}$$

where by virtue of (3.21)

$$\begin{aligned}\mathscr{Y}G_0(t) &= Z^{-1}(t)\mathscr{Y}G_1(t)Z(t) - Z^{-1}(t)\frac{dZ}{dt} \\ &= Z^{-1}(t)\mathscr{Y}\Big[G_1(t) - \sum_{j=1}^{q}Q_j^*(a)G_1(t)Q_j(a)\Big]Z(t).\end{aligned} \tag{3.26}$$

But since $Z^{-1}(t)\mathscr{Y} = \mathscr{Y}Z^*(t)$, we finally have

$$G_0(t) = Z^*(t)\left[G_1(t) - \sum_{j=1}^{q}Q_j^*(a)G_1(t)Q_j(a)\right]Z(t).$$

Therefore $G_0(t)$ is also a Hermitian matrix-function, as was to be expected (since all the substitutions were carried out by means of $\mathscr{Y}$-unitary matrix-functions). The equation (3.25), though it inherits the form of equation (3.21), nevertheless has an advantage which will be essential in further constructions: the diagonal blocks of its coefficient $\mathscr{Y}G_0$ are identically zero, i.e.

$$Q_j(a)\mathscr{Y}G_0(t)Q_j(a) = 0 \qquad (a \leq t \leq b;\ j = 1, 2, \cdots, q). \tag{3.27}$$

This follows at once from (3.23) and (3.26).

5. We now introduce the matrix-function

$$\mathscr{M}(t) = \sum_{j=1}^{q}\Omega_j(t)Q_j(a),$$

where as before

$$\Omega_j(t) = \int_a^t \omega_j(s)\,ds \qquad (j = 1, 2, \cdots, q)$$

are the matrix "exponents" of the matrizant

$$\exp(i\mu\mathscr{M}(t)) = \sum_{j=1}^{q}\exp(i\mu\Omega_j(t))Q_j(a) \tag{3.28}$$

of equation (3.15).

If we carry out the substitution

$$\mathscr{V}_0(t;\mu) = \exp(i\mu\mathscr{M}(t))\mathscr{W}(t;\lambda) \tag{3.29}$$

in (3.25), we obtain the following equation:

$$\frac{d\mathscr{W}}{dt} = \exp(-i\mu\mathscr{M}(t))\mathscr{Y}G_0(t)\exp(i\mu\mathscr{M}(t))\mathscr{W}, \quad \mathscr{W}(a;\mu) = I_n,$$

which can be transformed into an integral equation

$$\mathscr{W}(t;\mu) = I_n + \int_a^t \exp(-i\mu\mathscr{M}(s))\mathscr{Y}G_0(s)\exp(i\mu\mathscr{M}(s))\mathscr{W}(s;\mu)\,ds.$$

This Volterra integral equation enables us to obtain $\mathscr{W}(t;\mu)$ in the form of a series;

$$\mathscr{W}(t;\mu) = I_n + F_1(t;\mu) + F_2(t;\mu) + \cdots,$$

where

$$F_1(t;\mu) = \int_a^t \exp(-i\mu\mathscr{M}(s))\mathscr{Y}G_0(s)\exp(i\mu\mathscr{M}(s))\,ds \tag{3.30}$$

and

$$F_k(t;\mu) = \int_a^t \exp(-i\mu\mathscr{M}(s))\mathscr{Y}G_0(s)\exp(i\mu\mathscr{M}(s))F_{k-1}(s)\,ds \tag{3.31}$$

$$(k = 2, 3, \cdots).$$

We shall show that in the strip $\Pi = \{\mu; |\operatorname{Im}\mu| \leqq 1\}$ of the complex plane, the matrix $\mathscr{W}(b;\mu)$ admits the asymptotic representation

$$\mathscr{W}(b;\mu) = I_n + O(1/\mu) \quad (\operatorname{Re}\mu \to \pm\infty; \mu \in \Pi). \tag{3.32}$$

Putting

$$\mathscr{W}(t;\mu) = I + F(t;\mu); \quad F(t;\mu) = \sum_{k=1}^{\infty} F_k(t;\mu),$$

we can write the relation (3.32), which we wish to establish, in the form

$$|F(b;\mu)| = O(1/\mu) \quad (\operatorname{Re}\mu \to \pm\infty; \mu \in \Pi), \tag{3.33}$$

where $|\cdot|$ denotes a norm (arbitrarily defined) in the n^2-dimensional space of all complex matrices of order n. It is not hard to see from (3.28) that there exists a constant γ such that

$$|\exp(-i\mu\mathscr{M}(t))\mathscr{Y}G_0(t)\exp(i\mu\mathscr{M}(t))| \leqq \gamma \quad (\mu \in \Pi; a \leqq t \leqq b).$$

It follows at once from this bound and from (3.31) that

$$| F_k(t; \mu)| \leqq \gamma \int_a^t | F_{k-1}(s; \mu)| \, ds \quad (k = 2, 3, \cdots; \mu \in \Pi),$$

and so

$$| F_k(t; \mu)| \leqq \gamma^{k-1} \frac{(t-a)^{k-1}}{(k-1)!} \rho(\mu) \quad (k = 1, 2, \cdots; \mu \in \Pi),$$

where

$$\rho(\mu) = \max_{a \leqq t \leqq b} | F_1(t; \mu)| .$$

Therefore

$$| F(b; \mu)| \leqq (e^{\gamma(b-a)} - 1)\rho(\mu),$$

and thus the bound (3.33) will be established if we show that

$$\rho(\mu) = O(1/\mu) \quad (\mathrm{Re}\,\mu \to \pm \infty; \mu \in \Pi). \tag{3.34}$$

To this end we insert in (3.30) the expression for $\exp(\pm i\mu \mathscr{M}(t))$ from (3.28). This yields

$$F_1(t; \mu) = \sum_{j \neq k} \int_a^t \exp(-i\mu \Omega_{jk}(s)) T_{jk}(s) \, ds, \tag{3.35}$$

where

$$\Omega_{jk}(t) = \int_a^t [\omega_j(s) - \omega_k(s)] ds \qquad (j, k = 1, 2, \cdots, n),$$

and

$$T_{jk}(t) = Q_j(a) \mathscr{Y} G_0(t) Q_k(a) \qquad (j, k = 1, 2, \cdots, q).$$

Let us clarify that the summation in the right side of (3.35) need only be carried out over pairs of unequal j and k ($j \neq k$; $j, k = 1, 2, \cdots, q$), in view of the fact that according to (3.27) the diagonal blocks equal zero:

$$T_{jj}(t) = 0 \qquad (j = 1, 2, \cdots, q).$$ [11]

The fact that the function $\omega_j(t) - \omega_k(t)$ for $j \neq k$ does not change sign on the interval $[a, b]$ enables us to integrate by parts every term in the sum (3.35), which yields

[11] The necessity of the substitution (3.24), which eliminated the diagonal blocks, now becomes clear. The presence of diagonal blocks would have led to the appearance in (3.35) of integrals of nonoscillatory functions.

$$\int_a^t \exp(-i\mu\Omega_{jk}(s))\, T_{jk}(s)\, ds = -\frac{1}{i\mu}\left[\frac{\exp(-i\mu\Omega_{jk}(s))\, T_{jk}(s)}{\omega_j(s)-\omega_k(s)}\right]_a^t + \frac{1}{i\mu}\int_a^t \exp(-i\mu\Omega_{jk}(s))\frac{d}{ds}\left[\frac{T_{jk}(s)}{\omega_j(s)-\omega_k(s)}\right] ds$$

$$(j \neq k;\ j,k = 1,2,\cdots,q),$$

from which follows (3.34), and hence (3.33).

6. We now write down a formula which connects the matrizant $\mathscr{U}(t;\mu)$ of the original canonical equation (3.10) with the matrizant $\mathscr{W}(t;\mu)$. Since the equation (3.14) is obtained from (3.10) by means of the substitution $y = S(t)x$, it follows that

$$\mathscr{U}(t;\mu) = S(t)\mathscr{V}(t;\mu).$$

Now recalling the subsequent substitutions (3.24) and (3.29), we obtain

$$\mathscr{U}(t;\mu) = S(t)Z(t)\exp(i\mu\mathscr{M}(t))\mathscr{W}(t;\mu).$$

Therefore the equation (3.11), from which the spectrum of the problem (3.10) is determined, assumes the form

$$\det(S(b)Z(b)\exp(i\mu\mathscr{M}(b))\mathscr{W}(b;\mu) + \mathscr{E}) = 0. \tag{3.36}$$

Using the existing arbitrariness in the choice of the $\mathscr{Y}$-unitary matrix $\mathscr{E}$, we put

$$\mathscr{E} = S(b)Z(b).$$

As a result of this choice the equation (3.36) assumes a comparatively simple form:

$$(D(\mu) \equiv)\det(\exp(i\mu\mathscr{M}(b))\mathscr{W}(b;\mu) + I_n) = 0.$$

We put

$$D_0(\mu) \equiv \det(\exp(i\mu\mathscr{M}(b)) + I_n) = \det\left(\sum_{j=1}^{q}(1+\exp(i\mu\Omega_j(b)))\,Q_j(a)\right).$$

Obviously

$$D_0(\mu) = \text{const}\prod_{j=1}^{q}(1+\exp(i\mu\Omega_j(b))^{\kappa_j},$$

and the set of zeros of $D_0(\mu)$, counting their multiplicities, coincides with the union of the series (3.17), if each is considered the appropriate number of times.

Choosing for the moment an arbitrary δ, $0 < \delta < 1$, we consider the region Π_δ obtained from the strip Π by removing from the latter all discs μ_{jk} ($j = 1, 2, \cdots, q$; $k = 0, \pm 1, \pm 2, \cdots$) of radius δ with centers at the zeros of $D_0(\mu)$. We introduce the matrix-function $\mathscr{R}(\mu) = (\exp(-i\mu\mathscr{M}(b)) + I_n)^{-1}$. Since

$$\mathscr{R}(\mu) = \sum_{j=1}^{q} (1 + \exp(i\mu\Omega_j(b)))^{-1} Q_j(a),$$

there exists a constant C_δ such that

$$|\mathscr{R}(\mu)| = |(\exp(i\mu\mathscr{M}(b)) + I_n)^{-1}| < C_\delta \quad (\mu \in \Pi_\delta). \tag{3.37}$$

Using the equality $\mathscr{W}(b;\lambda) = I_n + F(\lambda)$ $(F(\lambda) = F(b;\lambda))$, we can write

$$\exp(i\lambda\mathscr{M}(b))\mathscr{W}(b;\lambda) + I_n = (\exp(i\lambda\mathscr{M}(b)) + I_n)(I_n + \mathscr{R}(\lambda)F(\lambda)).$$

Hence

$$D(\lambda)/D_0(\lambda) = \det(I_n + \mathscr{R}(\lambda)F(\lambda)),$$

where according to (3.33) and (3.37)

$$|\mathscr{R}(\lambda)F(\lambda)| = O(1/\lambda) \quad (\lambda \in \Pi_\delta, \operatorname{Re}\lambda \to \pm\infty).$$

Therefore there always exists an N_δ such that we will have

$$|D(\lambda)/D_0(\lambda) - 1| < 1 \quad \text{for } \lambda \in \Pi_\delta,\ |\operatorname{Re}\lambda| > N_\delta. \tag{3.38}$$

The steps in the progressions (3.17) are equal, respectively, to $2\pi/\Omega_j(b)$ $(j = 1, 2, \cdots, q)$. We choose δ, $0 < \delta < 1$, so that

$$\delta < \frac{1}{2q} \min\left(\frac{2\pi}{\Omega_1(b)}, \cdots, \frac{2\pi}{\Omega_q(b)}\right). \tag{3.39}$$

Let us denote by $G_\delta^{(\nu)}$ $(\nu = 0, \pm 1, \pm 2, \cdots)$ an *enumeration of the regions* ("clusters") complementary to Π_δ with respect to the strip Π $(|\operatorname{Im}\lambda| \leqq 1)$. It is easily seen that by virtue of the condition (3.39), each of the regions $G_\delta^{(\nu)}$ is the union of not more than q open discs of radius δ with centers from (3.17).

It follows from (3.38) that all the zeros of $D(\lambda)$ which do not lie in the interval $(-N_\delta, N_\delta)$ lie in the regions $G_\delta^{(\nu)}$. Moreover, since we have the inequality $|D(\lambda) - D_0(\lambda)| < D_0(\lambda)$ on the boundary of every region $G_\delta^{(\nu)}$ which does not intersect this interval, by Rouché's theorem the functions $D(\lambda)$ and $D_0(\lambda)$ have the same number of zeros, taking into account multiplicities, in each of these regions $G_\delta^{(\nu)}$.

Therefore if we denote by $n(r)$ and $n^{(0)}(r)$ the number of zeros of $D(\lambda)$ and $D_0(\lambda)$ in the interval $[-r,r]$, the difference $n(r)-n^{(0)}(r)$ will be a bounded function, and consequently

$$\lim_{r\uparrow\infty}\frac{n(r)}{r}=\lim_{r\uparrow\infty}\frac{n^{(0)}(r)}{r}.$$

Recalling (3.20), we see that formula (3.9) is valid for the case in question, in which $\mathscr{A}(t)$ is an analytic Hermitian-positive matrix-function and each of the distinct successive eigenvalues of the matrix $\mathscr{Y}\mathscr{A}(t)$ has the same multiplicity for all $t\in[a,b]$.

Theorem 1.1 is proved.

§4. Proof of the theorem on the asymptotic behavior of the spectrum of a selfadjoint canonical boundary value problem

In this section we shall prove Theorem VI.6.1, according to which one has, for the spectrum of any selfadjoint canonical boundary value problem

$$\mathscr{Y}\,dx/dt-\mu\mathscr{A}(t)x=0,\quad \mathscr{A}x(a)+\mathscr{B}x(b)=0, \tag{4.1}$$

the relation

$$\lim_{r\to\infty}\frac{n_\pm(r)}{r}=\frac{1}{2\pi}\int_a^b\Lambda_1(\mathscr{Y}\mathscr{A}(t))\,dt. \tag{4.2}$$

We already know, by virtue of Theorem VI.1.2 on the interlacing of spectra, that (4.2) will be established for an arbitrary selfadjoint boundary value problem as soon as it is established for a problem with skewperiodic boundary conditions:

$$\mathscr{Y}\,dx/dt-\mu\mathscr{A}(t)x=0,\quad x(a)+x(b)=0. \tag{4.3}$$

We shall again use the fact that the spectrum of this problem coincides with the spectrum of the real component G of a model Volterra operator $A_M=G+iK$ in $L_2^{(n)}(a,b)$. This component is defined by

$$(Gf)(t)=\frac{1}{2}\int_a^b\mathscr{A}^{1/2}(t)\mathscr{Y}\mathscr{A}^{1/2}(s)\operatorname{sign}(t-s)f(s)\,ds,$$

while the imaginary component K is defined by

$$(Kf)(t)=\frac{1}{2i}\int_a^b\mathscr{A}^{1/2}(t)\mathscr{Y}\mathscr{A}^{1/2}(s)f(s)\,ds. \tag{4.4}$$

The existence and equality of the two limits,

$$\lim_{r\to\infty}\frac{n_\pm(r)}{r}=\frac{1}{2}\lim_{r\to\infty}\frac{n(r)}{r}\quad (n(r)=n_+(r)+n_-(r)),\tag{4.5}$$

is already known from Theorem III.2.4.

According to the basic Theorem 1.1 we have

$$\lim_{r\to\infty}\frac{n_\pm(r)}{r}=\frac{1}{\pi}\int_{\hat{\mathfrak{P}}}|dPKdP|_1,\tag{4.6}$$

where $\hat{\mathfrak{P}}=\{P(t)\}_a^b$ is the chain of truncation projectors in $L_2^{(n)}(a,b)$.

On the other hand, on the basis of Lemma 2.1 for the case of a continuous H-matrix $\mathscr{A}(t)$ [12)]

$$\int_{\hat{\mathfrak{P}}}|dPKdP|_1=\frac{1}{2}\int_a^b\Lambda_1(\mathscr{Y}\mathscr{A}(t))\,dt.$$

Thus for a continuous H-matrix $\mathscr{A}(t)$ Theorem VI.6.1 is proved.

Starting with this fact and with the continuity, in the norm of $\mathfrak{S}_1$, of the right side of (4.6) [13)] as a functional of K, we shall prove the validity of (4.2) for the general case.

To this end, putting $\mathscr{A}^{1/2}(t)=0$ for $t\notin[a,b]$, we define for $\epsilon>0$ the smoothed H-matrix

$$Q_\epsilon(t)=\frac{1}{2\epsilon}\int_{t-\epsilon}^{t+\epsilon}\mathscr{A}^{1/2}(s)\,ds\quad(a\leq t\leq b),$$

and then the operator K_ϵ in $L_2^{(n)}(a,b)$, putting

$$(K_\epsilon f)(t)=\int_a^b Q_\epsilon(t)\mathscr{Y}Q_\epsilon(s)f(s)\,ds\quad(a\leqq t\leqq b).$$

To the operator K_ϵ there corresponds the H-matrix

$$\mathscr{A}_\epsilon(t)=Q_\epsilon^2(t)\quad(a\leqq t\leqq b),$$

---

12) In establishing this equality in §2 it was assumed that the matrix $\mathscr{Y}$ has the form $\mathscr{Y}=i\|\epsilon_j\delta_{jk}\|_1^n$; however, in fact only its properties $\mathscr{Y}=-\mathscr{Y}^*$, $\mathscr{Y}^2=-I_n$ were used. However, as was noted in §1, Chapter VI, for a suitable choice of the unitary matrix S the transformation $x=SX_1$ enables one to transform the problem (4.1) into a problem of the same kind with H-matrix $\mathscr{A}_1=S^*\mathscr{A}S$ and signature matrix $\mathscr{Y}_1=S^*\mathscr{Y}S$ which has the form indicated above, while the right and left side of (4.6) remain unaltered.

13) In the considerations which follow we could have started with the fact that the left side of (4.6) (which equals $(1/\pi)h(K;\hat{\mathfrak{P}})$) is a functional of K continuous in the norm of $\mathfrak{S}_1$, which is already known from Lemma 1.2.

and by virtue of its continuity we have

$$\frac{1}{\pi}\int_{\hat{\mathfrak{P}}} |\, dPK_\epsilon dP \,|_1 = \frac{1}{2\pi}\int_a^b \Lambda_1(\mathscr{Y}\mathscr{A}_\epsilon(t))\,dt. \tag{4.7}$$

Since for ϵ (> 0) sufficiently small the columns of the H-matrix $Q_\epsilon^2(t)$ differ arbitrarily little in the metric of $L_2^{(n)}$ from the columns of $\mathscr{A}(t)$, the operator K_ϵ will differ arbitrarily little in the metric of $\mathfrak{S}_1$ from K. Thus for sufficiently small ϵ the left side of (4.7) will differ arbitrarily little from the right side of (4.6).

We choose a decreasing sequence of positive numbers ϵ_n ($n = 1, 2, \cdots$) tending to zero such that almost everywhere $Q_{\epsilon_n}(t) \to \mathscr{A}^{1/2}(t)$, and hence $\mathscr{A}_{\epsilon_n}(t) \to \mathscr{A}(t)$, as $n \to \infty$.

Let us first assume that $\mathscr{A}(t)$ ($a \leqq t \leqq b$) is a bounded matrix-function. If all the elements of $\mathscr{A}(t)$ are bounded in modulus by the number M, then the elements of $\mathscr{A}^{1/2}(t)$, and along with them the elements of $Q_\epsilon(t)$, will be bounded in modulus by $\sqrt{M}$. But then all the elements of the H-matrix $\mathscr{A}_\epsilon(t)$ will be bounded in modulus by nM.

Therefore, if we put $\epsilon = \epsilon_n$ in (4.7) we can pass to the limit $n \to \infty$, obtaining

$$\int_{\hat{\mathfrak{P}}} |\, dPKdP \,|_1 = \frac{1}{2}\int_a^b \Lambda_1(\mathscr{Y}\mathscr{A}(t))\,dt. \tag{4.8}$$

It remains for us to consider the case in which the H-matrix $\mathscr{A}(t)$ is not bounded. For any $N > 0$ and $t \in [a, b]$ we put $\mathscr{A}_N(t) = \mathscr{A}(t)$ if all the elements of $\mathscr{A}(t)$ are bounded in modulus by N, and $\mathscr{A}_N(t) = 0$ otherwise. It is easily seen that for N sufficiently large, which we will henceforth assume, $\mathscr{A}_N(t)$ will be, along with $\mathscr{A}(t)$, an H-matrix of positive type.

We associate with $\mathscr{A}_N(t)$ the operator K_N by formula (4.4). Since $\mathscr{A}_N(t)$ is a bounded H-matrix, as was already proved we will have

$$\int_{\hat{\mathfrak{P}}} |\, dPK_N dP \,|_1 = \frac{1}{2}\int_a^b \Lambda_1(\mathscr{Y}\mathscr{A}_N(t))\,dt. \tag{4.9}$$

Since $\mathscr{Y}$ is a unitary matrix, we have

$$\Lambda_1(\mathscr{Y}\mathscr{A}_N(t)) \leqq \Lambda_1(\mathscr{Y}\mathscr{A}(t))$$
$$\leqq \sum_{j=1}^n s_j(\mathscr{Y}\mathscr{A}(t)) = \sum_{j=1}^n s_j(\mathscr{A}(t)) = \operatorname{sp}\mathscr{A}(t).$$

By hypothesis $\operatorname{sp}\mathscr{A}(t) \in L_1(a,b)$, and since $\Lambda_1(\mathscr{Y}\mathscr{A}_N(t)) \uparrow \Lambda_1(\mathscr{Y}\mathscr{A}(t))$ almost everywhere as $N \uparrow \infty$, the limit of the right side of (4.9) as $N \uparrow \infty$ will be the right side of (4.8). The same may be stated with regard to the left sides of (4.9) and (4.8), since the operator K_N tends to K in the norm of $\mathfrak{S}_1$.

Thus the relation (4.2), and with it Theorem VI.6.1, is proved for the general case.

REMARK 4.1. As has been noted, the existence and equality of the limits (4.5) for the spectrum $\{\mu_j\}$ of the problem (4.3) follows from Theorem III.2.4; moreover, it is also shown in the book [GK] (§8, Chapter IV) that the entire function

$$\mathscr{D}(\mu) = \prod_j \left(1 - \frac{\mu}{\mu_j}\right) \exp\left(\frac{\mu}{\mu_j}\right) \tag{4.10}$$

is of not greater than exponential type and satisfies the condition

$$\int_{-\infty}^{\infty} \frac{|\ln|\mathscr{D}(\mu)||}{1+\mu^2} d\mu < \infty. \tag{4.11}$$

The relation (4.2) enables us to state that the right side of (4.2) multiplied by π yields the exponential type of this function.

It follows from the results of a paper by M. G. Kreĭn [**1**] (see also his paper [**9**]) that for the spectrum $\{\mu_j\}$ of the general problem (4.1) the function $\mathscr{D}(\mu)$, defined by (4.10) (if some $\mu_j = 0$, then the factor in (4.10) which becomes infinite is to be replaced by μ), also satisfies the condition (4.11).

We can assert that

$$\mathscr{D}(\mu) = \text{const} \cdot \det(\mathscr{A} + \mathscr{B}\mathscr{U}(b;\mu))e^{-i\alpha\mu}, \tag{4.12}$$

where $\mathscr{U}(t;\mu)$ is the matrizant of (4.1), and α is some real number.

If $\mathscr{Y}$, $\mathscr{A}(t)$, $\mathscr{A}$ and $\mathscr{B}$ are real matrices, then obviously $\alpha = 0$. For the problem (4.3) ($\mathscr{A} = \mathscr{B} = I_n$)

$$\alpha = \operatorname{sp}\left(\frac{1}{i}\mathscr{Y}\int_a^b \mathscr{A}(t)\,\mathrm{dt}\right) \quad (= \operatorname{sp} K)$$

and in this case, on the basis of Theorem III.2.4, we can assert that

$$\lim_{r\to\infty} \sum_{|\mu_j| \leq r} \frac{1}{\mu_j} = 0,$$

which was already used in establishing Theorem VI.2.1.

§5. Asymptotic properties of the characteristic numbers for a class of integral equations

The theorems of the preceding sections enable us to obtain some new asymptotic formulas for the spectrum of eigenvalues of integral operators of a special class.

Let $L_2^{(r)}(a,b)$ be the Hilbert space of vector-functions $f(t) = \{f_j(t)\}_1^r$, and let $\mathscr{K}(t,s) = \| k_{pq}(t,s) \|_1^r$ $(a \leqq t, s \leqq b)$ be a Hilbert-Schmidt matrix kernel.

We recall (see [GK], Chapter III, §10.5) that if the operator K, defined in $L_2^{(r)}(a,b)$ by

$$(5.1) \qquad (Kf)(t) = \int_a^b \mathscr{K}(t,s) f(s)\, ds \quad (f \in L_2^{(r)}(a,b)),$$

is a nuclear operator, then its trace can be computed from the formula

$$(5.2) \qquad \operatorname{sp} K = \lim_{\delta \to 0} \int_a^b \operatorname{sp} \mathscr{K}_\delta(s,s)\, ds,$$

where

$$\mathscr{K}_\delta(t,s) = \frac{1}{4\delta^2} \int_{t-\delta}^{t+\delta} \int_{s-\delta}^{s+\delta} \mathscr{K}(u,v)\, du\, dv.$$

If the kernel $\mathscr{K}(t,s)$ is continuous, then

$$\operatorname{sp} K = \int_a^b \operatorname{sp} \mathscr{K}(s,s)\, ds.$$

We have the following theorem.

THEOREM 5.1. *Let $\mathscr{K}(t,s)$ $(a \leqq t, s \leqq b)$ be a Hermitian kernel which defines a nuclear operator in $L_2^{(r)}(a,b)$. Then the distribution functions $n_\pm(r; G)$ of the characteristic numbers of the operator G, defined by*

$$(5.3) \qquad (Gf)(t) = i \int_a^b \operatorname{sign}(t-s) \mathscr{K}(t,s) f(s)\, ds \quad (f \in L_2^{(r)}(a,b)),$$

satisfy the relations[14)]

$$(5.4) \qquad \lim_{r \to \infty} \frac{n_\pm(r; G)}{r} = \frac{1}{\pi} \lim_{\delta \to 0} \int_a^b \operatorname{sp} | \mathscr{K}_\delta(s,s) |\, ds,$$

[14)] We denote by $|\mathscr{C}|$ the matrix modulus of the matrix $\mathscr{C}$: $|\mathscr{C}| = (\mathscr{C}\mathscr{C}^*)^{1/2}$. If $\mathscr{C} = \mathscr{C}^*$, then $\operatorname{sp}|\mathscr{C}| = \Lambda_1(\mathscr{C})$.

where

$$\mathscr{K}_\delta(t,s) = \frac{1}{4\delta^2}\int_{t-\delta}^{t+\delta}\int_{s-\delta}^{s+\delta}\mathscr{K}(v,u)\,du\,dv.$$

If, moreover, $\mathscr{K}(t,s)$ *is continuous, then* (5.4) *can be replaced by*

$$\lim_{r\to\infty}\frac{n_\pm(r;G)}{r} = \frac{1}{\pi}\int_a^b \mathrm{sp}\,|\mathscr{K}(s,s)|\,ds. \tag{5.5}$$

PROOF. We first consider the case in which $\mathscr{K}(t,s)$ is a continuous finite-dimensional kernel:

$$\mathscr{K}(t,s) = \sum_{j=1}^n \epsilon_j\psi_j(t)\psi_j^*(s) \qquad (\epsilon_j = \pm 1).$$

Since the operator G is the real part $A_{\mathscr{R}}$ of the Volterra operator A,

$$(Af)(t) = 2i\int_t^b \mathscr{K}(t,s)f(s)\,ds,$$

which has the finite-dimensional imaginary component

$$(A_{\mathscr{I}}\,f)(t) = \int_a^b \mathscr{K}(t,s)f(s)\,ds, \tag{5.6}$$

it follows from what was proved in §§1—3 that

$$\lim_{r\to\infty}\frac{n_\pm(r;G)}{r} = \frac{1}{\pi}\,h(A_{\mathscr{I}}\,;\mathfrak{P}) = \frac{1}{2\pi}\int_a^b \Lambda_1(\mathscr{Y}\mathscr{H}(s))\,ds, \tag{5.7}$$

where $\mathscr{H}(t)$ and $\mathscr{Y}$ are the H-matrix and signature matrix corresponding to the operator $A_{\mathscr{I}}$.

The kernel $\mathscr{K}(t,s)$ can be written in the form

$$\mathscr{K}(t,s) = \mathscr{M}(t)\mathscr{N}^*(s), \tag{5.8}$$

where

$$\mathscr{M}(t) = \{\psi_1(t),\cdots,\psi_n(t)\},$$
$$\mathscr{N}(t) = \{\epsilon_1\psi_1(t),\cdots,\epsilon_n\psi_n(t)\}$$

are rectangular $r\times n$ matrix-functions with columns $\psi_j(t)$ and $\epsilon_j\psi_j(t)$ $(j = 1,2,\cdots,n)$ respectively. It is also easily seen that

$$\mathscr{Y}\mathscr{H}(t) = 2\mathscr{N}^*(t)\mathscr{M}(t). \tag{5.9}$$

It follows from (5.8) and (5.9) that the matrices $\mathscr{K}(t,t)$ and $\mathscr{Y}\mathscr{H}(t)/2$

$(a \leqq t \leqq b)$ have the same nonzero eigenvalues (taking into account their multiplicities).[15)]

Consequently

$$\tfrac{1}{2}\Lambda_1(\mathscr{Y}\mathscr{A}(t)) = \mathrm{sp}|\mathscr{K}(t,t)|,$$

from which, by (5.7), the equality (5.5) follows.

We now consider the second case, in which the kernel $\mathscr{K}(t,s)$ has the form

$$\mathscr{K}(t,s) = \sum_{j=1}^{\infty} \epsilon_j \psi_j(t)\psi_j^*(s), \tag{5.10}$$

where the $\psi_j(t)$ $(\in L_2^{(r)}(a,b))$ are continuous vector-functions, and the series converges absolutely and uniformly in the square $a \leqq t,\ s \leqq b$. We form the kernel

$$\mathscr{K}_N(t,s) = \sum_{j=1}^{N} \epsilon_j \psi_j(t)\psi_j^*(s)$$

and associate with it, by (5.3) and (5.6), the operators $G^{(N)}$ and $A_{\mathscr{J}}^{(N)}$. Then, as has been proved,

$$\lim_{r\to\infty} \frac{n_\pm(r;G^{(N)})}{r} = \frac{1}{\pi}\int_a^b \mathrm{sp}|\mathscr{K}_N(s,s)|\,ds. \tag{5.11}$$

The operators $A_{\mathscr{J}}^{(N)}$ tend in the norm of $\mathfrak{S}_1$ to the operator $A_{\mathscr{J}}$ as $N\to\infty$. By virtue of the continuity of the functional $h(\cdot;\hat{\mathfrak{P}})$, it follows from (5.11) that

$$\lim_{r\to\infty}\frac{n_\pm(r;G)}{r} = \frac{1}{\pi}\lim_{N\to\infty}\int_a^b \mathrm{sp}|\mathscr{K}_N(s,s)|\,ds = \frac{1}{\pi}\int_a^b \mathrm{sp}|\mathscr{K}(s,s)|\,ds.$$

We now consider the general case. The kernel $\mathscr{K}(t,s)$ can be expanded in a bilinear series

$$\mathscr{K}(t,s) = \sum_{j=1}^{\infty}\lambda_j\phi_j(t)\phi_j^*(s) \qquad (\phi_j\in L_2^{(r)}(a,b)),$$

which converges in the metric of $L_2^{(r)}((a,b)\times(a,b))$. Let δ (>0) be an arbitrary sufficiently small number; then

$$\mathscr{K}_\delta(t,s) = \frac{1}{4\delta^2}\int_{t-\delta}^{t+\delta}\int_{s-\delta}^{s+\delta}\mathscr{K}(u,v)\,du\,dv = \sum_{j=1}^{\infty}\lambda_j\,\psi_{j,\delta}(t)\psi_{j,\delta}^*(s), \tag{5.12}$$

---

15) We have used here the following elementary result: if A and B are $r\times n$ rectangular matrices, then the spectrum of nonzero eigenvalues of the matrices AB^* and B^*A coincide.

where

$$\psi_{j,\delta}(t) = \frac{1}{2\delta}\int_{t-\delta}^{t+\delta} \phi_j(s)\,ds = (S_\delta \phi_j)(t).$$

Let us associate with the kernel $\mathscr{K}_\delta(t,s)$, in familiar fashion, the operators $G^{(\delta)}$, $A_{\mathscr{I}}^{(\delta)}$ and $A^{(\delta)}$. Obviously

$$A_{\mathscr{I}}^{(\delta)} = S_\delta A_{\mathscr{I}} S_\delta \quad \text{and} \quad G^{(\delta)} = S_\delta G S_\delta.$$

Since $S_\delta \times S_\delta$ is a linear operator which maps the space $L_2^{(r)}(Q)$ ($Q = [a,b] \times [a,b]$) continuously into the space $C^{(r)}(Q)$ of continuous vector-functions, the series (5.12) converges uniformly and absolutely, and hence, as was proved,

$$\lim_{r\to\infty} \frac{n_\pm(r; G^{(\delta)})}{r} = \frac{1}{\pi}\int_a^b \mathrm{sp}\,|\mathscr{K}_\delta(s,s)|\,ds. \tag{5.13}$$

Taking into account that as $\delta \to 0$ the operator $A_{\mathscr{I}}^{(\delta)}$ tends in the norm of $\mathfrak{S}_1$ to the operator $A_{\mathscr{I}}$, and that hence

$$h(A_{\mathscr{I}}^{(\delta)}; \hat{\mathfrak{P}}) \to h(A_{\mathscr{I}}; \hat{\mathfrak{P}}),$$

we obtain (5.4) from (5.13).

For the case in which the kernel $\mathscr{K}(t,s)$ is continuous, the limit as $\delta \to 0$ of the right side of (5.4) is obviously equal to the right side of (5.5).

The theorem is proved.

For the case of a scalar single-pair kernel

$$\mathscr{K}(t,s) = \begin{cases} \psi(t)\,\chi(s) & (a \leqq t \leqq s \leqq b), \\ \psi(s)\,\chi(t) & (a \leqq s < t \leqq b) \end{cases}$$

with differentiable functions $\psi(t)$ and $\chi(t)$, the formula (5.4), with certain refinements, was obtained by P. D. Kalafati [1, 2].

We also mention that the asymptotic behavior of the characteristic numbers of integral equations with smooth kernels which have discontinuities on the diagonal was studied in a paper by G. M. Mordasova [1].

APPENDIX

UNICELLULAR OPERATORS AND RELATED ANALYTIC PROBLEMS

Until recently the theory of unicellular operators and the theory of the decomposition of a linear operator into a "sum" of unicellular operators have been beset by a paucity of results and a host of problems. The situation has changed markedly for the better in recent years. Certain of these recent results are so remarkable that we have devoted this Appendix to them. Unfortunately, instead of complete proofs (which would take us far afield) we are forced to restrict ourselves to brief clarifications and comments.

Let us recall that a bounded linear operator A acting in $\mathfrak{H}$ is said to be *unicellular* if for any two invariant subspaces $\mathfrak{L}_1$ and $\mathfrak{L}_2$ ($\mathfrak{L}_1 \neq \mathfrak{L}_2$) of A we have either $\mathfrak{L}_1 \subset \mathfrak{L}_2$ or $\mathfrak{L}_2 \subset \mathfrak{L}_1$. As was already noted in §9, Chapter I, by virtue of a well-known theorem of F. Riesz (see [GK], Chapter I) the spectrum of a unicellular operator is always a connected set. From a bound obtained by V. I. Macaev [2] for the resolvent of a linear operator with imaginary component of class $\mathfrak{S}_\omega$, and from earlier results of Ju. I. Ljubič and V. I. Macaev [1] (on the existence, for a closed linear operator whose resolvent has a certain behavior, of invariant subspaces which split the spectrum of the operator) we immediately have the following result.

A) *Every unicellular operator A of the form $A = H + T$, where $H = H^*$ and $T \in \mathfrak{S}_\omega$, has a one-point spectrum, and this point lies on the real axis.*[1)]

[1)] This assertion does not require that the selfadjoint operator H be bounded. More generally, for any (not necessarily unicellular) operator $A = H + T$, where $H = H^*$ (with H, generally speaking, unbounded) and $T \in \mathfrak{S}_\infty$, every *condensation point* of the spectrum of H (i.e. a limit point of the spectrum of H or an eigenvalue of infinite multiplicity) will be a condensation point of the spectrum of A, and conversely (see, for example, Gohberg-Kreĭn [1], §5). If $T \in \mathfrak{S}_\omega$ and the spectrum of A has at least two distinct points, then there exist two subspaces $\mathfrak{L}_1$ and $\mathfrak{L}_2$ whose intersection consists solely of the zero vector, which lie entirely in the domain $\mathfrak{D}(H)$ $(= \mathfrak{D}(A))$, and are invariant with respect to A. We mention in passing a remarkable result of V. I. Macaev.

If $T \in \mathfrak{S}_\omega$, and $H = H^*$ has a discrete spectrum (i.e. a spectrum without finite condensation points), then the system of root vectors of the operator $A = H + T$ is complete. If the completely continuous operator T does not belong to $\mathfrak{S}_\omega$, then there exists a self-

On the other hand, if an operator A $(\in \mathfrak{R})$ is unicellular, then obviously so is the operator $A - aI$, where a is an arbitrary number. Therefore in studying the question of the unicellularity of an operator A $(\in \mathfrak{R})$ with $A_{\mathscr{I}} \in \mathfrak{S}_\omega$, we can assume in advance, without loss of generality, that the entire spectrum of A consists of the point $\lambda = 0$. We note further (see [GK], Theorem I.5.2) that for every operator A with imaginary component $A_{\mathscr{I}} \in \mathfrak{S}_\infty$ the set of condensation points of its spectrum coincides with the set of condensation points of its real component $A_{\mathscr{R}}$. Hence if $\lambda = 0$ is the only spectral point of A, then A is completely continuous, i.e. it is a Volterra operator.

In all the sections of this Appendix, except for the third, we shall consider unicellular operators with $A_{\mathscr{I}} \in \mathfrak{S}_1$ (§§1, 2, 4, 5) or with $A_{\mathscr{I}} \in \mathfrak{S}_p$ (§6). Therefore without loss of generality we may restrict ourselves at once, in these considerations, to Volterra operators, and indeed to simple operators (obviously a nonsimple operator is never unicellular).

In §3 we study unicellular contractions T which "differ" from a unitary operator by a nuclear operator: $I - T^*T \in \mathfrak{S}_1$. The results of §3 enable us to interpret the unicellularity of dissipative operators with nuclear imaginary component in a new way.

Let us emphasize that problems of the theory of unicellular operator turn out to be connected with certain deep analytic problems of the theory of inverse problems for canonical differential equations. These problems are of independent interest (see §§4 and 5).

§1. The decomposition of simple dissipative Volterra operators into sums of unicellular operators

1. A linear operator acting in a complex n-dimensional space E_n is unicellular if and only if it has exactly n distinct (nonzero) invariant subspaces.

Taking over the terminology from infinite-dimensional Hilbert spaces to finite-dimensional spaces, we can call an operator A in E_n a *Volterra* operator if its spectrum consists of the single point $\lambda = 0$.

---

adjoint operator H with a discrete spectrum such that the operator $A = H + T$ has no spectrum (i.e. the operator A^{-1} is a Volterra operator).

The first assertion is contained in Theorem 3 of a note by V. I. Macaev [**2**], and the second assertion was given in his doctoral dissertation.

We should mention that for a general (not necessarily selfadjoint) operator A the condensed spectrum is defined here as the complement of the set $\tilde{\rho}(A)$ of normal points of A (for whose definition see [GK], Chapter I, §2.3).

Let $\mathfrak{L}_1 \subset \mathfrak{L}_2 \subset \cdots \subset \mathfrak{L}_n = E_n$ be the n invariant subspaces of some unicellular Volterra operator A; then $A\mathfrak{L}_n = \mathfrak{L}_{n-1}$, $A\mathfrak{L}_{n-1} = \mathfrak{L}_{n-2}, \cdots, A\mathfrak{L}_2 = \mathfrak{L}_1$. Choosing an arbitrary vector $e_1 \in \mathfrak{L}_n - \mathfrak{L}_{n-1}$ and putting $e_k = A^k e_1$ $(k = 1, 2, \cdots, n)$, we obtain a basis $\{e_k\}_1^n$ of E_n which is *cyclic* with respect to A:

$$e_2 = Ae_1, \; e_3 = Ae_2, \cdots, e_n = Ae_{n-1},$$

and such that $Ae_n = 0$. The matrix of A relative to the basis $\{e_1, e_2, \cdots, e_n\}$ will have all elements equal to zero, with the exception of those lying just below the principal diagonal, which will all equal unity, i.e. it will be a matrix consisting of one Jordan cell with zero principal diagonal.[2)]

The converse is also obvious; if an operator A in E_n has a spectrum consisting solely of zero, and A is cyclic (i.e. there exists a basis of E_n which is cyclic with respect to A), then A is unicellular.

We now recall that an operator $A \in \mathfrak{R}$, acting in a separable infinite-dimensional Hilbert space $\mathfrak{H}$, is said to be *cyclic* if there exists a vector $e \in \mathfrak{H}$ such that the linear hull of the sequence of vectors $\{A^k e\}_0^\infty$ is *dense* in $\mathfrak{H}$. It was proved in §9, Chapter I (Theorem 9.1) that *every unicellular operator* A $(\in \mathfrak{R})$ *is cyclic.*

As was remarked there (without substantiation), this result does not admit a converse. It turns out that in every class $\mathfrak{S}_p$ $(p > 0)$ there exist cyclic Volterra operators which, nonetheless, are not unicellular. In fact, let us consider in the space $\mathfrak{H} = L_2^{(2)}(0,1)$ of vector-functions $f = \{f_1(x), f_2(x)\}$ the Volterra operator A defined by

$$g = \{g_1(x), g_2(x)\} = Af = \left\{ \int_0^x f_1(t)\,dt, \; - \int_0^x f_2(t)\,dt \right\} \quad (0 \leqq x \leqq 1).$$

Let u denote the vector-function from $L_2^{(2)}(0,1)$ whose components are identically equal to unity. Then

$$u^{(n)} = A^n u = \{x^n/n!, (-x)^n/n!\} \qquad (n = 1, 2, \cdots).$$

Since the even powers x^{2n} $(n = 0, 1, \cdots)$, as well as the odd powers x^{2n+1} $(n = 0, 1, \cdots)$, form a complete system of functions in $L_2(0,1)$, it follows that the system of vector-functions $u^{(n)}$ is complete in $L_2^{(2)}(0,1)$. Thus A is a cyclic operator (with generating element u); nevertheless this operator is obviously not unicellular (the subspaces consisting of all elements with first or with second component identically equal to

[2)] This is the origin of the term "unicellular operator".

zero are invariant subspaces of A).[3)]

The operator A has a two-dimensional real component, and it belongs to the ideal $\mathfrak{S}_\Omega$, and hence to every ideal $\mathfrak{S}_p$ for $p > 1$. Each of its odd powers A^{2l+1} $(l = 0, 1, 2, \cdots)$ is also a Volterra operator, indeed cyclic with the same generating element u (by virtue of the fact that any system of powers x^{a+dn} $(n = 0, 1, \cdots)$ and, in particular, the systems with $a = 0,\ 2l+1,\ d = 2(2l+1)$ are dense in $L_2(0,1)$). Choosing the integer l sufficiently large, we obtain an operator which belongs to $\mathfrak{S}_p$ for any prescribed arbitrarily small $p > 0$.

Nevertheless, it may be that any cyclic *dissipative* Volterra operator is unicellular. At all events, this is the case for cyclic dissipative Volterra operators with nuclear imaginary component; i.e. the following result holds (compare with Theorem I.9.1).

THEOREM 1.1 (G. È. KISILEVSKIĬ [**5**]). *A simple dissipative Volterra operator with nuclear imaginary component is unicellular if and only if it is cyclic.*

2. As is known, every linear operator in a finite-dimensional space either is unicellular or can be broken up into the direct sum of unicellular operators. G. È. Kisilevskiĭ [**5**] has generalized this result to dissipative Volterra operators in a Hilbert space $\mathfrak{H}$.

Let $\{\mathfrak{L}_\nu\}_{\nu\in N}$ be a finite or countable set of subspaces of the space $\mathfrak{H}$. We shall say that $\mathfrak{H}$ *breaks up into the quasi-direct sum of the subspaces* $\mathfrak{L}_\nu$ $(\nu \in N)$, and write

$$\mathfrak{H} = \sum_{\nu\in N} \dot{+}\, \mathfrak{L}_\nu, \tag{1.1}$$

if the following two conditions are satisfied:

1) *The subspaces* $\mathfrak{L}_\nu$ $(\nu \in N)$ *are linearly independent* (i.e. for any finite collection of subspaces $\mathfrak{L}_1, \mathfrak{L}_2, \cdots, \mathfrak{L}_m$ from the system $\{\mathfrak{L}_\nu\}$ the equality $f_1 + f_2 + \cdots + f_m = 0$, where $f_k \in \mathfrak{L}_k$ $(k = 1, 2, \cdots, m)$, is possible only for $f_1 = f_2 = \cdots = f_m = 0$).

2) *The linear hull of the subspaces* $\mathfrak{L}_\nu$ $(\nu \in N)$ *is dense in* $\mathfrak{H}$.

If the space $\mathfrak{H}$ can be decomposed into a quasi-direct sum of invariant subspaces $\{\mathfrak{L}_\nu\}_{\nu\in N}$ of some operator A $(\in \mathfrak{R})$, and A_ν $(= A\,|\,\mathfrak{L}_\nu)$ is the restriction of A to $\mathfrak{L}_\nu$ $(\nu \in N)$, then we shall say that *the operator A breaks up into the quasi-direct sum of the operators A_ν* $(\nu \in N)$, and write

---

3) This example was communicated to us independently by L. A. Sahnovič and G. È. Kisilevskiĭ.

$$A = \sum_{\nu \in N} \dot{+} A_\nu. \tag{1.2}$$

If A is, say, a Volterra or dissipative operator, or has nuclear imaginary component, then obviously the same is true of each of the operators A_ν ($\nu \in N$) in the decomposition (1.2). If $A_{\mathscr{J}} \in \mathfrak{S}_1$, then it turns out that (1.2) implies

$$\operatorname{sp} A_{\mathscr{J}} = \sum_{\nu \in N} \operatorname{sp}(A_\nu)_{\mathscr{J}}.$$

THEOREM 1.2 (G. È. KISILEVSKIĬ [5]). *Every simple dissipative Volterra operator with nuclear imaginary component either is unicellular, or admits a decomposition into the quasi-direct sum of unicellular operators* A_k ($k = 1, 2, \cdots, \omega; \omega \leqq \infty$):

$$A = \sum_{k=1}^{\omega} \dot{+} A_k. \tag{1.3}$$

The cardinal number ω *and the positive numbers*

$$\sigma_k = \operatorname{sp}(A_k)_{\mathscr{J}} \qquad (k = 1, 2, \cdots, \omega)$$

are invariants of the operator A.[4]

If the imaginary component $A_{\mathscr{J}}$ *has finite dimension* n, *then* $\omega \leqq n$, *and the decomposition* (1.3) *can be chosen so that the imaginary component* $(A_k)_{\mathscr{J}}$ *of the operator* A_k *has dimension* $\leqq n - k + 1$ ($k = 1, 2, \cdots, \omega$).[5]

§2. A basic criterion for the unicellularity of a dissipative Volterra operator with nuclear imaginary component

1. The derivations of Theorems 1.1 and 1.2 make essential use of various facts from the theory of the characteristic function of a non-selfadjoint operator, and in particular use the criterion to be given below for the unicellularity of a dissipative Volterra operator, which is formulated in terms of its characteristic function.

Let us give the general definition of the characteristic operator-function $W_A(\lambda)$ of a nonselfadjoint operator A ($\in \mathfrak{R}$), in the form originally given by M. S. Livšic [2].

---

[4] That is, the number ω and the numbers σ_k ($k = 1, 2, \cdots, \omega$), up to their order, do not depend upon the choice of representation (1.3).

[5] The first and third assertions of Theorem 1.2 are stated and proved in a paper by G. E. Kisilevskiĭ [5]; the second assertion was communicated by him in a letter.

We denote by $\mathfrak{J}(A)$ the closure of the range of the imaginary component $A_{\mathscr{J}}$ of A:

$$\mathfrak{J}(A) = \overline{\mathfrak{R}(A_{\mathscr{J}})} = \overline{\mathfrak{R}(A - A^*)}.$$

The operator-function $W_A(\lambda)$ is defined on the set $\rho(A^*)$ of all regular points λ of A^* by[6)]

(2.1) $W_A(\lambda) = [I + 2i(\operatorname{sign} A_{\mathscr{J}})|A_{\mathscr{J}}|^{1/2}(A^* - \lambda I)^{-1}|A_{\mathscr{J}}|^{1/2}]|\mathfrak{J}(A),$

where the symbol $|\mathfrak{J}(A)$ denotes that the operator in the square brackets is restricted to the subspace $\mathfrak{J}(A)$. In order to clarify the formula (2.1), we note that

$$(A - \lambda I)(A^* - \lambda I)^{-1} = I + 2iA_{\mathscr{J}}(A^* - \lambda I)^{-1}.$$

On the other hand, we have

$$A_{\mathscr{J}} = |A_{\mathscr{J}}|^{1/2} \operatorname{sign} A_{\mathscr{J}}\, |A_{\mathscr{J}}|^{1/2};$$

therefore for the case in which $A_{\mathscr{J}}$ is invertible (and hence $\mathfrak{J}(A) = \mathfrak{H}$) formula (2.1) simplifies:

$$W_A(\lambda) = |A_{\mathscr{J}}|^{-1/2}(A - \lambda I)(A^* - \lambda I)^{-1}|A_{\mathscr{J}}|^{1/2}.$$

Let us also consider the case in which $A_{\mathscr{J}}$ is completely continuous and hence admits a spectral representation

$$A_{\mathscr{J}} = \sum_{j=1}^{r} \epsilon_j(\cdot, \psi_j)\psi_j, \tag{2.2}$$

where $\{\psi_j\}_1^r$ is some orthogonal (not normalized) system of elements,

6) If H is a selfadjoint operator and $H = \int_{-\infty}^{\infty} \lambda dE_\lambda$ is its spectral representation, then we put

$$\operatorname{sign} H = \int_{-\infty}^{\infty} \operatorname{sign} \lambda dE_\lambda = I - E(+0) - E(-0), \quad |H| = \int_{-\infty}^{\infty} |\lambda| dE_\lambda,$$

(in contrast with the norm we denote the "modulus" of the operator H by $|H|$). Obviously $|H| = H \operatorname{sign} H = (\operatorname{sign} H)H$. If, in particular, $H \in \mathfrak{S}_\infty$ and

$$H = \sum_j \lambda_j(\cdot, \phi_j)\phi_j \qquad ((\phi_j, \phi_k) = \delta_{jk},\ j, k = 1, 2, \cdots)$$

is the spectral representation of H, then

$$|H| = \sum_j |\lambda_j|(\cdot, \phi_j)\phi_j, \qquad \operatorname{sign} H = \sum_j \operatorname{sign} \lambda_j(\cdot, \phi_j)\phi_j.$$

$\epsilon_j = \pm 1$ $(j = 1, 2, \cdots, r)$, and r is the dimension of the range of $A_{\mathscr{J}}$.

The matrix of the operator $W_A(\lambda)$ relative to the basis $\{\phi_j\}_1^r$ $(\phi_j = \psi_j / |\psi_j|)$ of $\mathfrak{J}(A)$ is

$$\mathscr{W}(\lambda) = \| w_{jk}(\lambda) \|_1^r,$$

where

$$w_{jk}(\lambda) = (W_A(\lambda)\phi_k, \phi_j) \qquad (j, k = 1, 2, \cdots, r).$$

By (2.1) we have

$$\begin{aligned} w_{jk}(\lambda) &= \delta_{jk} + 2i((\operatorname{sign} A_{\mathscr{J}})|A_{\mathscr{J}}|^{1/2}(A^* - \lambda I)^{-1}|A_{\mathscr{J}}|^{1/2}\phi_k, \phi_j) \\ &= \delta_{jk} + 2i((A^* - \lambda I)^{-1}|A_{\mathscr{J}}|^{1/2}\phi_k, (\operatorname{sign} A_{\mathscr{J}})|A_{\mathscr{J}}|^{1/2}\phi_j), \end{aligned}$$

and since the representation (2.2) implies that

$$|A_{\mathscr{J}}|^{1/2}\phi_k = \psi_k, \qquad \operatorname{sign} A_{\mathscr{J}} |A_{\mathscr{J}}|^{1/2}\phi_j = \epsilon_j \psi_j,$$

we have

$$w_{jk}(\lambda) = \delta_{jk} + 2i\epsilon_j((A^* - \lambda I)^{-1}\psi_k, \psi_j) \qquad (j, k = 1, 2, \cdots, r).$$

We see that the matrix $\mathscr{W}(\lambda)$ is none other than the characteristic matrix-function of the operator A (see Chapter V, §4.3).[7]

In considering Volterra operators A it is convenient to go over from the characteristic operator-function $W_A(\lambda)$ to the *monodromy operator-function* $U_A(\lambda)$ (see Chapter V, §4.3), related to the former by

$$U_A(\mu) = W_A(1/\mu). \tag{2.3}$$

An independent definition of the function $U_A(\mu)$ is the following:

$$U_A(\mu) = [I - 2i\mu \operatorname{sign} A_{\mathscr{J}} |A_{\mathscr{J}}|^{1/2}(1 - \mu A^*)^{-1}|A_{\mathscr{J}}|^{1/2}]|\mathfrak{J}(A).$$

If A is a Volterra operator, then obviously $U_A(\mu)$ will be an entire operator-function of the complex variable μ.

For the case in which $A_{\mathscr{J}} \in \mathfrak{S}_1$ $(A \in \mathfrak{R})$, the determinant

$$\begin{aligned} \det U_A(\mu) &= \det(I - 2\mu i \operatorname{sign} A_{\mathscr{J}} |A_{\mathscr{J}}|^{1/2}(I - \mu A^*)^{-1}|A_{\mathscr{J}}|^{1/2}) \\ &= \det(I - 2\mu i(I - \mu A^*)^{-1}|A_{\mathscr{J}}|^{1/2} \operatorname{sign} A_{\mathscr{J}} |A_{\mathscr{J}}|^{1/2}) \\ &= \det(I - 2\mu i(I - \mu A^*)^{-1}A_{\mathscr{J}}) = \det((I - \mu A)(I - \mu A^*)^{-1}) \end{aligned}$$

[7] We remark that in the definition of the characteristic matrix-function $\mathscr{W}(\lambda) = \| w_{jk}(\lambda) \|_1^n$ given in Chapter V, §4.3, it was not required that the basis $\{\psi_j\}$ be orthogonal.

is meaningful.[8] Thus, by the notation from [GK], Chapter IV, §3.1,

$$\det U_A(\mu) = D_{A/A^*}(\mu). \tag{2.4}$$

For a Volterra operator A the perturbation determinant $D_{A/A^*}(\mu)$ has the quite simple value (see [GK], Chapter IV, §6.2)[9]

$$D_{A/A^*}(\mu) = \exp(-2i\mu \operatorname{sp} A_{\mathscr{J}}). \tag{2.5}$$

Thus

$$|\det U_A(\mu)| = \exp(2 \operatorname{Im} \mu \operatorname{sp} A_{\mathscr{J}}). \tag{2.6}$$

If A is dissipative, then the expressions for $W_A(\lambda)$ and $U_A(\mu)$ $(\lambda = 1/\mu)$ simplify:

$$W_A(\lambda) = [I + 2iA_{\mathscr{J}}^{1/2}(A^* - \lambda I)^{-1}A_{\mathscr{J}}^{1/2}] | \mathfrak{J}(A)$$

and

$$U_A(\mu) = [I - 2i\mu A_{\mathscr{J}}^{1/2}(I - \mu A^*)^{-1}A_{\mathscr{J}}^{1/2}] | \mathfrak{J}(A).$$

From the easily verified identity (see, for example, [GK], Chapter IV, §5)

$$W_A^*(\lambda) W_A(\lambda) = [I - 2 \operatorname{Im} \lambda A_{\mathscr{J}}^{1/2}(A - \bar{\lambda} I)^{-1}(A^* - \lambda I)^{-1} A_{\mathscr{J}}^{1/2}] | \mathfrak{J}(A) \tag{2.7}$$

it follows that

$$W_A^*(\lambda) W_A(\lambda) \leqq I \quad \text{for } \operatorname{Im} \lambda > 0, \tag{2.8}$$

$$W_A^*(\lambda) W_A(\lambda) \geqq I \quad \text{for } \operatorname{Im} \lambda < 0. \tag{2.9}$$

Let us denote by $s_j(\mu)$ $(j = 1, 2, \cdots)$ the sequence of s-numbers of the operator $U_A(\mu)$:

$$s_j^2(\mu) = \lambda_j(U_A^*(\mu) U_A(\mu)) \qquad (j = 1, 2, \cdots).$$

Assuming that $A_{\mathscr{J}} \in \mathfrak{S}_1$, we can write

---

[8] If $A_{\mathscr{J}} \in \mathfrak{S}_1$, then $|A_{\mathscr{J}}|^{1/2} \in \mathfrak{S}_2$, and consequently

$$C = -2\mu i \operatorname{sign} A_{\mathscr{J}} |A_{\mathscr{J}}|^{1/2} \in \mathfrak{S}_2, \; D = (I - \mu A^*)^{-1} |A_{\mathscr{J}}|^{1/2} \in \mathfrak{S}_2;$$

we have also used the fact that if $C, D \in \mathfrak{S}_2$, then $\det(I + CD) = \det(I + DC)$ (see [GK], Chapter IV, §§1 and 3).

[9] If we use a canonical model of a dissipative operator, then the equality $\det U_A(\mu) = \exp(-2i\mu \operatorname{sp} A_{\mathscr{J}})$ can be obtained as a corollary of the formula of Liouville-Ostrogradskiĭ. We have already used this equality in Chapter VI for the derivation of Theorem VI.2.2. We also note that formula (2.5) can be deduced from the general formula (3.21).

$$|\det U_A(\mu)|^2 = \det U_A^*(\mu) \det U_A(\mu) = \det(U_A^*(\mu) U_A(\mu)) = \prod_{j=1}^{\infty} s_j^2(\mu).$$

On the other hand, by (2.9) $s_j(\mu) \geqq 1$ $(j = 1, 2, \cdots)$ for $\operatorname{Im}\mu \geqq 0$. Thus

$$|\det U_A(\mu)| \geqq s_1(\mu) = |U_A(\mu)| \quad \text{for } \operatorname{Im}\mu \geqq 0.$$

Recalling (2.6), we obtain the bound

$$|U_A(\mu)| \leqq \exp(2 \operatorname{Im}\mu \operatorname{sp} A_{\mathscr{J}}) \qquad (\operatorname{Im}\mu \geqq 0). \tag{2.10}$$

As regards the lower halfplane $\operatorname{Im}\mu \leqq 0$, by (2.8) we have there

$$|U_A(\mu)| \leqq 1 \qquad (\operatorname{Im}\mu \leqq 0). \tag{2.11}$$

The bound (2.10) is familiar from the work of V. P. Potapov [1], and M. S. Brodskiĭ and M. S. Livšic [1].[10)]

2. We can now formulate the following fundamental criterion for the unicellularity of a Volterra operator of the class being considered.

THEOREM 2.1 (M. S. BRODSKIĬ AND G. È. KISILEVSKIĬ [1]). *In order that a simple dissipative Volterra operator A with nuclear imaginary component be unicellular, it is necessary and sufficient that*

$$\varlimsup_{\rho \to \infty} \left\{ \frac{1}{\rho} \ln |U_A(i\rho)| \right\} = 2 \operatorname{sp} A_{\mathscr{J}}. \tag{2.12}$$

The sufficiency of condition (2.12) had already been established by M. S. Brodskiĭ [1] in 1956. Then G. È. Kisilevskiĭ [2] proved the necessity of this condition for the case in which the imaginary component $A_{\mathscr{J}}$ is two-dimensional. For the general case ($A_{\mathscr{J}} \in \mathfrak{S}_1$) the necessity of (2.12) was established in a joint paper [1] of these authors.

Condition (2.12) is obviously a requirement that the bound (2.10) be exact. According to this bound, for the general case of a dissipative Volterra operator A with nuclear imaginary component we will have, for the quantity

$$\hbar(A) = \varlimsup_{\rho \to \infty} \left\{ \frac{1}{\rho} \ln |U_A(i\rho)| \right\},$$

[10)] The method of obtaining a bound for the norm of a matrix-function through its determinant was already used in the work of V. P. Potapov [1]. The bound (2.10), for the case of a finite-dimensional imaginary component $A_{\mathscr{J}}$, also follows almost immediately from the multiplicative representation (4.1). Since the latter can be extended to the case of dissipative Volterra operators with an infinite-dimensional component $A_{\mathscr{J}} \in \mathfrak{S}_1$ (see, in this connection, the end of §3), the bound (2.10) can be obtained by this route.

the inequality

$$\hbar(A) \leqq 2\,\mathrm{sp}\, A_{\mathscr{J}}.$$

If the strict inequality holds here, then the bound (2.10) can be improved, namely:

$$|U_A(\mu)| \leqq \exp(\hbar(A)\operatorname{Im}\mu) \quad \text{for } \operatorname{Im}\mu \geqq 0. \tag{2.13}$$

In fact, according to the bounds (2.10) and (2.11), for any unit vectors $f, g \in \mathfrak{H}$ the entire function $F(\mu) = F_{f,g}(\mu) = (U(\mu)f, g)$ satisfies the conditions

$$\sup_{-\infty<\mu<\infty} |F(\mu)| \leqq 1, \qquad \overline{\lim_{\rho\to\infty}} \left\{\frac{1}{\rho} \ln \max_{0\leqq\theta\leqq 2\pi} |F(\rho e^{i\theta})|\right\} < \infty.$$

But then, by a familiar theorem of the theory of entire functions (see B. Ja. Levin [1], Chapter 1, §14)

$$\ln|F(\mu)| \leqq h_F^+ \operatorname{Im}\mu \quad \text{for } \operatorname{Im}\mu \geqq 0,$$

where

$$h_F^+ = \overline{\lim_{\rho\to\infty}} \left\{\frac{1}{\rho} \ln|F(i\rho)|\right\}.$$

Since $h_F^+ \leqq \hbar(A)$ for any choice of the unit vectors $f, g \in \mathfrak{H}$, (2.13) follows.

On the basis of the foregoing and also of the relation (2.20) below, it is natural to call the functional $\hbar(A)$ the *exponential type* of the dissipative Volterra operator A.

3. We shall show that the criterion for unicellularity can be reformulated in terms of the Fredholm resolvent $A(\mu) = A(I-\mu A)^{-1}$ of the operator A as follows.

THEOREM 2.2. *In order that a simple dissipative Volterra operator A with nuclear imaginary component be unicellular, it is necessary and sufficient that*

$$\overline{\lim_{\rho\to\infty}} \left\{\frac{1}{\rho} \ln|A(-i\rho)|\right\} = 2\,\mathrm{sp}\, A_{\mathscr{J}}. \tag{2.14}$$

To establish the equivalence of conditions (2.14) and (2.12), we shall show that

$$\hbar(A) = \overline{\lim_{\rho\to\infty}} \left\{\frac{1}{\rho} \ln|A(-i\rho)|\right\}. \tag{2.15}$$

It follows from the identity (2.7) that

$$|W_A(\bar{\lambda})|^2 = 1 + 2\operatorname{Im}\lambda |T(\lambda)|^2 \quad \text{for } \operatorname{Im}\lambda \geqq 0,$$

where

$$T(\lambda) = (A^* - \bar{\lambda}I)^{-1}A_{\mathscr{J}}^{1/2}.$$

On the other hand,

$$|T(\lambda)|^2 = |T^*(\lambda)|^2 = |T(\lambda)T^*(\lambda)| = |(A^* - \bar{\lambda}I)^{-1}A_{\mathscr{J}}(A - \lambda I)^{-1}|,$$

and therefore

$$|W_A(\bar{\lambda})|^2 = |I + 2\operatorname{Im}\lambda(A^* - \bar{\lambda}I)^{-1}A_{\mathscr{J}}(A - \lambda I)^{-1}|.$$

Representing $2A_{\mathscr{J}}$ in the form

$$2A_{\mathscr{J}} = \frac{1}{i}[(A - \lambda I) - (A^* - \bar{\lambda}I) + 2\operatorname{Im}\lambda I],$$

we find that

$$\begin{aligned} \max(|S(\lambda)|^2, (\operatorname{Im}\lambda)^2|(A - \lambda I)^{-1}|^2) &\leqq |W_A(\bar{\lambda})|^2 \\ &\leqq |S(\lambda)|^2 + (\operatorname{Im}\lambda)^2|(A - \lambda I)^{-1}|^2 \qquad (\operatorname{Im}\lambda \geqq 0), \end{aligned} \tag{2.16}$$

where

$$S(\lambda) = I + i\operatorname{Im}\lambda(A - \lambda I)^{-1} = (A - \operatorname{Re}\lambda I)(A - \lambda I)^{-1}.$$

For $\lambda = -1/i\rho$ $(\rho > 0)$ we have

$$\begin{aligned} S(\lambda) &= -i\rho A(I + i\rho A)^{-1} = -i\rho A(-i\rho), \\ (\operatorname{Im}\lambda)^2(A^* - \bar{\lambda}I)^{-1}(A - \lambda I)^{-1} &= (I - i\rho A^*)^{-1}(I + i\rho A)^{-1} \\ &= (I - i\rho A(-i\rho))^*(I - i\rho A(-i\rho)). \end{aligned}$$

Thus, the substitution $\lambda = -1/i\rho$ $(\rho > 0)$ in (2.16) yields

$$|\rho A(-i\rho)|^2 \leqq |U_A(i\rho)|^2 \leqq |\rho A(-i\rho)|^2 + |I - i\rho A(-i\rho)|^2. \tag{2.17}$$

In particular, it follows that

$$|\rho A(-i\rho)| \leqq |U_A(i\rho)| \leqq 1 + \sqrt{2}\rho|A(-i\rho)|,$$

and so (2.15) holds.

REMARK 2.1. If an operator A $(\in \mathfrak{R})$ is dissipative, then

$$|(A - \lambda I)^{-1}| \leqq 1/|\operatorname{Im}\lambda| \qquad (\operatorname{Im}\lambda < 0).$$

Putting $\lambda = 1/\mu$, we find that the Fredholm resolvent $A(\mu)$ of A has the bound

$$(2.18)\qquad |A(\mu)| \leqq |A||(I-\mu A)^{-1}| \leqq |A||\mu|/\mathrm{Im}\,\mu \quad \text{for } \mathrm{Im}\,\mu > 0.$$

Thus $A(\mu)$ is bounded on each interior ray $\mu = \rho\exp(i\theta)$ $(0 < \theta < \pi)$ of the upper halfplane. Conversely, for a dissipative Volterra operator A with $A_{\mathscr{J}} \in \mathfrak{S}_1$ the resolvent $A(\mu)$ has exponential growth on each interior ray $\mu = \rho\exp(i\theta)$ $(-\pi < \theta < 0)$ of the lower halfplane, namely

$$(2.19)\qquad \overline{\lim_{\rho\to\infty}} \left\{\frac{1}{\rho}\ln|A(\rho e^{i\theta})|\right\} = \hbar(A)|\sin\theta| \quad (-\pi < \theta < 0).$$

This is easily obtained from relations (2.13) and (2.16).

We can also assert that the relation (2.16) is satisfied uniformly in θ, and therefore it is also valid for $\theta = 0, \pi$.

It will be shown below (see (2.26)) that the resolvent $A(\mu)$ admits the representation

$$(2.20)\qquad A(\mu) = i\int_0^{\hbar(A)} e^{i\mu t}T_t\,dt,$$

where T_t $(0 \leqq t < \infty)$ is a strongly continuous semigroup of contractions.

It follows at once from this representation that

$$|A(\mu)| \leqq \begin{cases} \hbar(A) & \text{for } \mathrm{Im}\,\mu \geqq 0,\\ \hbar(A)e^{|\mathrm{Im}\mu|\hbar(A)} & \text{for } \mathrm{Im}\,\mu \leqq 0.\end{cases}$$

The representation (2.20) enables us to show that the lim sup in (2.12), (2.14), (2.15) and (2.19) can be replaced by lim. Thus, for a dissipative Volterra operator A with nuclear imaginary component $A_{\mathscr{J}}$ the quantity $\hbar(A)$, defined by (2.15), is the *exponential type of the entire function* $A(\mu)$.

In conclusion we make a few comparisons with the finite-dimensional case.

If A is a Volterra operator in an n-dimensional space E_n, then $A^n = 0$ and

$$A(\mu) = A(I-\mu A)^{-1} = A + \mu A^2 + \cdots + \mu^{n-2}A^{n-1}.$$

The operator A will be unicellular if and only if $A^{n-1} \neq 0$. Thus for A to be unicellular it is necessary and sufficient that the resolvent $A(\mu)$ be a polynomial of degree precisely $n-2$.

If the degree of $A(\mu)$ (and hence, the order of growth of A) is lower, then A ceases to be unicellular.

The condition (2.14) is also a requirement of the maximality of the growth of the Fredholm determinant $A(\mu)$ of A as an entire operator-function, in other words, the condition of the maximality of its exponential type (or in the terminology of S. N. Bernšteĭn, of its *degree*).

This is the analogy with the finite-dimensional case. However, this analogy was established for a simple dissipative Volterra operator, *for which there is no analog* among the Volterra operators in E_n!

4. As we know (see §1.1), a Volterra operator A in E_n is unicellular if and only if the smallest value of p for which $A^p = 0$ is n.

It turns out that one can formulate a criterion for the unicellularity of a dissipative Volterra operator in $\mathfrak{H}$, which has points of contact with this elementary criterion.

To do this we need certain new concepts.

An operator B with domain $\mathfrak{D}(B)$ dense in $\mathfrak{H}$ is said to be *dissipative*, if

$$\operatorname{Im}(Bf,f) \geqq 0 \quad \text{for } f \in \mathfrak{D}(B). \tag{2.21}$$

A dissipative operator B is said to be *maximal* if it has no dissipative extensions.

A maximal dissipative operator is always closed. It is characterized by the property that for at least one λ (and then, for all λ) from the open lower halfplane $\operatorname{Im}\lambda < 0$ it has a bounded resolvent $(B - \lambda I)^{-1} \in \mathfrak{R}$ (see R. S. Phillips [1]).

The dissipative operators A $(\in \mathfrak{R})$ considered above, being everywhere defined, are obviously maximal dissipative operators.

Just as for bounded dissipative operators, one has for maximal dissipative operators the bound

$$|(B - \lambda I)^{-1}| \leqq 1/|\operatorname{Im}\lambda| \quad \text{for } \operatorname{Im}\lambda < 0.$$

If $\lambda = 0$ is not an eigenvalue of the dissipative operator A $(\in \mathfrak{R})$ (and hence of A^*), then the range $\mathfrak{R}(A)$ is dense in $\mathfrak{H}$. In this case, as is easily seen, the operator $B = -A^{-1}$ (which, generally speaking, is unbounded) will be maximal dissipative and

$$(B - \lambda I)^{-1} = -A(-\lambda). \tag{2.22}$$

Now let T_t $(0 \leqq t < \infty)$ be a one-parameter strongly continuous semigroup of uniformly bounded operators in $\mathfrak{H}$.[11] Then by a well-known theorem of K. Yosida (see E. Hille and R. S. Phillips [1]) this semigroup has an infinitesimal generating operator Γ, i.e. a closed linear operator Γ such that the set of all $f \in \mathfrak{H}$ for which the strong limit

---

[11] That is, 1) $T_0 = I$; 2) $T_{t+s} = T_t T_s$ $(0 \leqq s, t < \infty)$; 3) $\text{s-lim}_{t \downarrow 0} T_t = I$ and 4) $\sup |T_t| < \infty$. We refer the reader to the book by E. Hille and R. S. Phillips [1] for all facts regarding the theory of semigroups.

$$\text{s-}\lim_{t \downarrow 0} \{t^{-1}(T_t - I)f\}$$

exists coincides with $\mathfrak{D}(\Gamma)$, and this strong limit (s-lim) coincides with Γf. One writes $T_t = \exp(t\Gamma)$. Among the many justifications of this notation we point out that the entire left halfplane $\operatorname{Re} \zeta < 0$ consists of regular points of Γ, and for the resolvent $(\Gamma - \zeta I)^{-1}$ one has the formula

$$-(\Gamma - \zeta I)^{-1} f = \int_0^\infty e^{-\zeta t} T_t f \, dt. \tag{2.23}$$

It turns out (a theorem of Phillips [**1**]) that every maximal dissipative operator B, when multiplied by i, is the infinitesimal generator of some strongly continuous one-parameter semigroup of contractions T_t $(0 \leqq t < \infty)$, and conversely every such semigroup has infinitesimal generator Γ of the form $\Gamma = iB$, where B is a maximal dissipative operator. We shall say that a strongly continuous semigroup of contractions T_t $(0 \leqq t < \infty)$ is *nilpotent*, if for some value $t = a$ (and hence for all $t \geqq a$) we have $T_t = 0$. The smallest value of a for which $T_a = 0$ is called the *exponent of nilpotence* of the semigroup T_t.

The following result holds.

THEOREM 2.3. *Let A be a simple dissipative Volterra operator with nuclear imaginary component. Then the operator $-iA^{-1}$ is the generator of some nilpotent semigroup of contractions $T_t = \exp(-itA^{-1})$ with exponent of nilpotence s_A equal to $\hbar(A)$ $(\leqq 2 \operatorname{sp} A_{\mathcal{J}})$.*

Hence if we make use of Theorem 2.1 (or Theorem 2.2) we immediately obtain the following corollary.

COROLLARY. *The operator A of the preceding theorem will be unicellular if and only if the exponent of nilpotence s_A of the semigroup $\exp(-itA^{-1})$ is equal to $2 \operatorname{sp} A_{\mathcal{J}}$.*

Let us turn to the proof of the theorem.

For the semigroup $T_t = \exp(-itA^{-1})$ the relations (2.22) and (2.23) (for $\Gamma = iB$, $B = -A^{-1}$, $\lambda = -\mu$, $\zeta = -i\mu$) yield

$$A(\mu) f = i \int_0^\infty e^{i\mu t} T_t f \, dt. \tag{2.24}$$

Hence for $f, g \in \mathfrak{H}$

$$(A(\mu) f, g) = i \int_0^\infty e^{i\mu t} (T_t f, g) \, dt. \tag{2.25}$$

Since

$$|(A(\mu)f,g)| \leqq |A(\mu)| \, |f| \, |g|,$$

the left side of (2.25) is an entire function of exponential type, and its type does not exceed $\hbar(A)$. By the well-known Paley-Wiener theorem (see, for example, N. I. Ahiezer [1]) it follows that $(T_t f,g) = 0$ for $t \geqq \hbar(A)$. Since f and g are arbitrary vectors in $\mathfrak{H}$, we conclude that $T_t = 0$ for $t \geqq \hbar(A)$, so that $s_A \leqq \hbar(A)$.

On the other hand, if $T_t = 0$ for $t \geqq a$ $(= s_A)$, then by (2.24)

$$A(\mu)f = i\int_0^a e^{i\mu t} T_t f dt, \tag{2.26}$$

and so

$$|A(\mu)| \leqq \int_0^a |e^{i\mu t}| \, |T_t| \, dt \leqq a e^{|\mathrm{Im}\,\mu| a} \quad (\mathrm{Im}\,\mu < 0).$$

Recalling (2.15), we conclude that $\hbar(A) \leqq a = s_A$.

Thus the theorem is proved. Let us now comment on this theorem.

Choosing an arbitrary positive integer n, we put $T = \exp(-i\tau A^{-1})$, where $\tau = ([2\,\mathrm{sp}\,A_{\mathscr{J}}] + 1)/n$. The spectrum of A is concentrated at zero, that of A^{-1}—at infinity, and that of T—again at zero, since T is nilpotent: $T^n = 0$. Considering the powers T^p $(= \exp(-ip\tau A^{-1}))$ with an arbitrary positive exponent p, we can assert that $p = \hbar(A)/\tau$ is the smallest exponent of nilpotency of the operator T. This exponent does not exceed $2\,\mathrm{sp}\,A_{\mathscr{J}}/\tau$. For the case in which it achieves this value, and only for this case, is A unicellular. Nevertheless the operator T itself is not unicellular. Indeed, given any vector $f \in \mathfrak{H}$ we obtain an invariant subspace $\mathfrak{L}_f$, spanned by the vectors $f, Tf, \cdots, T^{n-1}f$, of dimension not greater than n. Obviously we can always find vectors $f,g \in \mathfrak{H}$ such that neither of the subspaces $\mathfrak{L}_f$ and $\mathfrak{L}_g$ contains the other.

One can show that the operators $T_t = \exp(-itA^{-1})$ are not Volterra operators.

To conclude this section we make a few bibliographical remarks.

Under the assumption that $A_{\mathscr{J}}$ is finite-dimensional, the nilpotence of the semigroup $T_t = \exp(-itA^{-1})$, where A is a simple dissipative Volterra operator, was established by B. Sz.-Nagy and C. Foiaş [**10**]. These authors showed (see [**10**], p. 322) that the exponent of nilpotency s_A of this semigroup does not exceed a_T, where a_T is a number for whose computation the authors indicated a procedure in terms of the operator $T = (iI + A)(iI - A)^{-1}$, and the exponent s_A is equal to a_T if and only if A is unicellular. We now know that the number a_T coincides with $2\,\mathrm{sp}\,A_{\mathscr{J}}$ (see (3.6)).

Sz.-Nagy and Foiaş obtained their results from an entirely different circle of ideas (see §3.4).

Following these authors, we consider by way of example a unicellular dissipative Volterra operator A with a one-dimensional imaginary component. This operator, the simplest of the class being considered, when normalized so that $\mathrm{sp}\,A_{\mathscr{J}} = \frac{1}{2}$ is unitarily equivalent to the operator of integration $\frac{1}{2}J$ in $L_2(0,1)$ (see Theorem I.8.1):

$$(\tfrac{1}{2}Jf)(x) = i\int_x^1 f(s)\,ds.$$

The domain of the operator $\Gamma = -i(\frac{1}{2}J)^{-1}$ is the set of all absolutely continuous functions $g(t) \in L_2(0,1)$ for which $g'(t) \in L_2(0,1)$ and $g(1) = 0$, and $\Gamma g = g'$. It is not hard to show that the operator Γ generates the semigroup $\exp(s\Gamma)$ of contractions defined by

$$(\exp(t\Gamma)f)(x) = \begin{cases} f(x+t) & \text{for } x+t \in [0,1], \\ 0 & \text{for } x+t \notin [0,1] \end{cases} \qquad (0 \leq t \leq 1).$$

The exponent s_A of nilpotence of this semigroup obviously equals 1, which is in complete accord with the corollary to Theorem 2.3.

§3. Unicellular contractions

1. Recall that an operator T ($\in \mathfrak{R}$) is called a *contraction* if $|T| \leq 1$. A contraction T is said to be *simple* if it has no invariant subspace on which its restriction is unitary. By a theorem of Langer—Sz.-Nagy—Foiaş (see [GK], Theorem V.3.1) a contraction can be split into the orthogonal sum of a simple contraction and a unitary operator. Obviously only a simple contraction can be unicellular.

From these results and the arguments of Ju. I. Ljubič and V. I. Macaev [1] and V. I. Macaev [2], from which the result A) formulated in the introduction was obtained, we also have the following result.

THEOREM 3.1. *An invertible*[12] *unicellular contraction T for which*

---

[12] The requirement that T be invertible in this theorem (as well as in other theorems where it appears) could have been replaced by a formally more general requirement, namely, that at least one point ζ ($|\zeta| < 1$) not be in the spectrum of T. If this condition is satisfied and $I - T^*T \in \mathfrak{S}_\infty$, then T can have only a discrete spectrum inside the unit disc, consisting of normal eigenvalues (see [GK], Chapter I). If the contraction T is unicellular, this spectrum must be absent, and hence T is invertible.

We also remark that Theorem 3.1 enables one to strengthen result A) (see the relation (3.3)).

$I - T^*T \in \mathfrak{S}_\omega$ has a one-point spectrum, and this point lies on the unit circle.

If T is a contraction with one-point spectrum α ($|\alpha| = 1$), then $T_\alpha = \bar{\alpha}T$ will be a contraction with one-point spectrum $\lambda = 1$. Therefore wherever it is convenient, we may assume without loss of generality that T has the one-point spectrum $\lambda = 1$.

2. If A ($\in \mathfrak{R}$) is a dissipative operator, then its Cayley transform

$$T = (iI - A)(iI + A)^{-1}, \tag{3.1}$$

as is easily seen, will be a contraction ($|T| \leqq 1$), for which the point $\lambda = -1$ is regular.

If, in turn, T is a contraction for which the point $\lambda = -1$ is regular, then the transformation inverse to (3.1),

$$A = i(I - T)(I + T)^{-1} \tag{3.2}$$

yields a bounded dissipative operator A.

From the relations

$$I - T^*T = 4(-iI + A^*)^{-1}A_{\mathcal{J}}(iI + A)^{-1}, \tag{3.3}$$

$$A_{\mathcal{J}} = (I + T^*)^{-1}(I - T^*T)(I + T)^{-1} \tag{3.4}$$

it follows that if $A_{\mathcal{J}}$ belongs to some two-sided ideal $\mathfrak{S}$ ($\subset \mathfrak{S}_\infty$), then $I - T^*T$ belongs to the same ideal, and conversely. If, moreover, we consider the relations

$$I - T = 2A(iI + A)^{-1} = 2iA(i),$$

$$A = iS(2I - S)^{-1} = 2iS(\tfrac{1}{2}) \quad (S = I - T),$$

where $S(\mu)$ is the Fredholm resolvent of the operator S, then we can conclude that the dissipative operator A will be a Volterra operator if and only if $S = I - T$ is a Volterra operator.

Thus the transformation (3.1) maps the class of dissipative operators with nuclear imaginary component, being considered above, onto the class of all contractions T having the following properties:

1) *$T - I$ is a Volterra operator;*

2) *$I - T^*T$ is a nuclear operator.*

Henceforth for convenience we shall call the operator $I - T^*T$ the deviation of T from unitarity or, briefly, the *deviation operator*.

We remark that property 1) (regarded as a condition) is more restrictive than the following condition:

1′) *T has a one-point spectrum $\lambda = 1$.*

Nevertheless, as is not hard to show, in the class of all operators T for which the deviation operator $I - T^*T$ is completely continuous (and a fortiori in the class of those operators for which the deviation operator $I - T^*T$ is nuclear), the properties 1) and 1′) are equivalent. Therefore the criterion established above for the unicellularity of a dissipative Volterra operator A with $A_{\mathscr{J}} \in \mathfrak{S}_1$ can be transformed into a criterion for unicellularity in the indicated class of contractions T. For example, Theorem 2.2 (and the inequality $\hbar(A) \leqq 2\,\mathrm{sp}\,A_{\mathscr{J}}$) enables us to state the following result.

THEOREM 3.2. *Let T be a simple contraction whose spectrum consists of the single point $\lambda = 1$, and whose deviation operator $I - T^*T$ is nuclear. Then*

$$\overline{\lim_{\rho \uparrow 1}}\{(1-\rho)\ln|(T-\rho I)^{-1}|\} \leqq -\ln\det T^*T. \tag{3.5}$$

Equality holds here if and only if T is unicellular.

Since, by (3.1),

$$(1-\rho)(T-\rho I)^{-1} = (I - iA)(I + irA)^{-1} = I + i(r-1)A(ir)$$
$$(r = (1+\rho)/(1-\rho)),$$

the left side of (3.5) coincides with twice the left side of (2.14). We shall show that the right sides of (3.5) and (2.14) are similarly related, i.e.

$$-\ln\det T^*T = 4\,\mathrm{sp}\,A_{\mathscr{J}}. \tag{3.6}$$

To do so, we put $\mu = -i$ in (2.5); then we find that

$$\det((I + iA)(I + iA^*)^{-1}) = \exp(-2\,\mathrm{sp}\,A_{\mathscr{J}}). \tag{3.7}$$

Since the right side is real, this equality will persist if we take the conjugate of the operator in the determinant, i.e.

$$\det((I - iA)^{-1}(I - iA^*)) = \exp(-2\,\mathrm{sp}\,A_{\mathscr{J}}). \tag{3.8}$$

Let us compute the product of the determinants in the left sides of (3.7) and (3.8). Obvious calculations yield[13)]

$$\det((I - iA^*)(I - iA)^{-1}(I + iA)(I + iA^*)^{-1})$$
$$= \det((iI - A^*)^{-1}(iI + A^*)(iI + A)^{-1}(iI - A)) = \det(T^*T),$$

from which (3.6) follows.

We further note that for a contraction T satisfying the hypotheses of Theorem 3.2 one has

---

[13)] We have used here the fact that if $X, Y, X^{-1} \in \mathfrak{R}$ and $XY - I \in \mathfrak{S}_1$, then $\det XY = \det YX$ (see [GK], Chapter IV, §1).

$$\overline{\lim_{\rho\uparrow 1}}(1-\rho)\ln M_T(\rho) = \overline{\lim_{\rho\uparrow 1}}(1-\rho)\ln|(T-\rho I)^{-1}|, \tag{3.9}$$

where

$$M_T(\rho) = \max_{0\leqq\theta\leqq 2\pi}|(T-\rho e^{i\theta}I)^{-1}| \quad (0<\rho<1). \tag{3.10}$$

In fact, for $A = i(I-T)(I+T)^{-1}$ we have

$$|(I-\mu A)^{-1}| = |I+\mu A(\mu)| \leqq 1 + \hbar(A)|\mu|\exp(\hbar(A)|\operatorname{Im}\mu|) \quad (\operatorname{Im}\mu \leqq 0).$$

Hence for the operator T with $\mu = i(1+\zeta)/(1-\zeta)$ we obtain

$$\begin{aligned}|(T-\zeta I)^{-1}| &= \left|\frac{2}{1-\zeta}(I+T)^{-1}(I-\mu A^{-1})\right| \\ &\leqq \frac{\text{const}}{1-\rho}\left(1+\frac{\hbar(A)}{1-\rho}\exp\left(\frac{2\hbar(A)}{1-\rho}\right)\right) \quad (\rho = |\zeta| < 1).\end{aligned} \tag{3.11}$$

We have here used the inequality $|\operatorname{Im}\mu| \leqq |\mu| \leqq 2(1-\rho)^{-1}$.

From (3.11) we can conclude that the left side of (3.9) does not exceed $2\hbar(A)$; on the other hand it is not less than the right side, which equals $2\hbar(A)$.

For completeness we also note that when the point ζ approaches the unit circle from the outside, the growth of the norm of the resolvent of any contraction T satisfies the bound

$$|(T-\zeta I)^{-1}| = \left|\sum_{k=0}^{\infty}\frac{1}{\zeta^{k+1}}T^k\right| \leqq \sum_{k=0}^{\infty}\frac{1}{|\zeta|^{k+1}} = \frac{1}{|\zeta|-1}. \tag{3.12}$$

3. We shall omit formulating the result for contractions which corresponds to Theorem 2.1. To formulate this result we would have to introduce the concept of the characteristic function of a contraction (see M. S. Livšic and V. P. Potapov [1], Ju. L. Šmul'jan [1, 2], B. Sz.-Nagy and C. Foiaş [6] and V. T. Poljackiĭ [1]), which we have not used up till now.

Instead of this we present a criterion for the unicellularity of a contraction T which in certain connections seems to be more complete and sounds more effective than Theorem 3.2.

THEOREM 3.3. *Let T be a simple invertible contraction with nuclear deviation operator ($I - T^*T \in \mathfrak{S}_1$). In order that T be unicellular, it is necessary and sufficient that*

1) *T have no eigenvalues inside the unit disc;*
2) *the relation*

$$\overline{\lim_{\rho \uparrow 1}} \{ (1-\rho) \ln M_T(\rho) \} = -\ln \det T^*T \tag{3.13}$$

hold, where

$$M_T(\rho) = \max_{|\zeta|=\rho} |(T - \zeta I)^{-1}| \quad (\rho < 1).$$

If the first condition is satisfied and T *is not unicellular, then instead of the equal sign in* (3.13) *one has* $<$.

We shall prove this theorem, omitting certain details. We first present certain facts (which are of independent interest) whose derivation is not very difficult.

First of all, if for some invertible operator T $(\in \mathfrak{R})$ one has $I - T^*T \in \mathfrak{S}_1$, then

$$T^{*-1} - T = T^{*-1}(I - T^*T) \in \mathfrak{S}_1.$$

Hence in this case the perturbation determinant

$$\begin{aligned} D_{T/T^{*-1}}(\zeta) &= \det((T - \zeta I)(T^{*-1} - \zeta I)^{-1}) \\ &= \det(T^*(T - \zeta I)(I - \zeta T^*)^{-1}) \end{aligned} \tag{3.14}$$

exists (see [GK], Chapter IV). Here ζ belongs to $\rho(T^*)$, the resolvent set of T^*. For brevity we put $\Delta_T(\zeta) = D_{T/T^{*-1}}(\zeta)$.

It will be more convenient for us to deal with the normalized function

$$d_T(\zeta) = \Delta_T(\zeta) / \Delta_T^{1/2}(0) \tag{3.15}$$

in place of $\Delta_T(\zeta)$. We can write this function in another way. Since by hypothesis $T^*T = I - H_1$, where $H_1 \in \mathfrak{S}_1$, it follows that also $(T^*T)^{1/2} = I - H$, where $H \in \mathfrak{S}_1$. We have

$$\Delta_T^{1/2}(0) = [\det(T^*T)]^{1/2} = \det((T^*T)^{1/2}) \quad (= \det(I - H)).$$

From the polar representation $T = U(T^*T)^{1/2}$, where U is some unitary operator, we have $T^* = (T^*T)^{1/2}U^*$ and, by virtue of (3.14),

$$\Delta_T(\zeta) = \det(T^*T)^{1/2} \det(U^*(T - \zeta I)(I - \zeta T^*)^{-1}).$$

Thus[14] we obtain a new way of writing $d_T(\zeta)$:

---

[14] The relation (3.16) is important, besides everything else, in that it enables us to define the function $d_T(\zeta)$ for the case in which $\det T^*T = 0$ (i.e., when an inverse T^{-1} $(\in \mathfrak{R})$ does not exist, and hence T^{*-1} does not exist). In order to define the function $d_T(\zeta)$ by (3.16) it is essential that besides the condition $I - T^*T \in \mathfrak{S}_1$ the condition which we spoke of in footnote 12 be satisfied (the operator T should have at least one regular point inside the unit disc). One can show that when this condition is satisfied the function $d_T(\zeta)$ does not depend upon the choice of the unitary operator U in the polar representation $T = U(T^*T)^{1/2}$ (this choice is nonunique when $\det T^*T = 0$).

$$d_T(\zeta) = \det(U^*(T - \zeta I)(I - \zeta T^*)^{-1}). \tag{3.16}$$

We shall show, secondly, that for every contraction T with nuclear deviation operator the function $d_T(\zeta)$ satisfies the inequality

$$|d_T(\zeta)| \leqq 1 \qquad (|\zeta| < 1). \tag{3.17}$$

In fact, we have

$$\overline{d_T(\zeta)} = \det((I - \bar{\zeta}T)^{-1}(T^* - \bar{\zeta}I)U)$$

and

$$\overline{d_T(\zeta)}d_T(\zeta) = \det((I - \bar{\zeta}T)^{-1}(T^* - \bar{\zeta}I)(T - \zeta I)(I - \zeta T^*)^{-1}).$$

It can be shown that the value of the determinant on the right does not change if we interchange the first two factors. Thus

$$\begin{aligned} |d_T(\zeta)|^2 &= \det((T^* - \bar{\zeta}I)(I - \bar{\zeta}T)^{-1}(T - \zeta I)(I - \zeta T^*)^{-1}) \\ &= \det((I - \zeta T^*)^{-1}(T^* - \bar{\zeta}I)(I - \bar{\zeta}T)^{-1}(T - \zeta I)), \end{aligned}$$

i.e.

$$|d_T(\zeta)|^2 = \det(T_\zeta^* T_\zeta), \tag{3.18}$$

where

$$T_\zeta = (T - \zeta I)(I - \bar{\zeta}T)^{-1}. \tag{3.19}$$

From the easily verified identity

$$I - T_\zeta^* T_\zeta = (1 - |\zeta|^2)(I - \zeta T^*)^{-1}(I - T^*T)(I - \bar{\zeta}T)^{-1}$$

it follows that for every contraction T the operator T_ζ will also be a contraction for any ζ with $|\zeta| < 1$.

This proves the inequality (3.17).

We note, thirdly, that by a well-known theorem of the theory of functions (see I. I. Privalov [1], Chapter II, §6.2), inequality (3.17) and the inequality

$$d_T(0) = (\det(T^*T))^{1/2} > 0 \tag{3.20}$$

imply that $d_T(\zeta)$ can be represented in the following form:

$$d_T(\zeta) = B_T(\zeta)\exp\left(-\int_0^{2\pi} \frac{e^{it} + \zeta}{e^{it} - \zeta}\, d\sigma_T(t)\right), \tag{3.21}$$

where

$$B_T(\zeta) = \prod_j \frac{\alpha_j - \zeta}{1 - \bar{\alpha}_j\zeta} \frac{\bar{\alpha}_j}{|\alpha_j|} \tag{3.22}$$

is the Blaschke product formed from all the zeros of the function $d_T(\zeta)$, and $\sigma_T(t)$ $(0 \leq t \leq 2\pi)$ is some nondecreasing function (for definiteness we normalize this function by the conditions $\sigma_T(0) = 0$, $\sigma_T(t) = \sigma_T(t-0)$ for $0 < t \leq 2\pi$). We now proceed directly to the proof of the theorem. We begin with the second assertion.

Recalling that the function $d_T(\zeta)$ differs from the perturbation determinant $\Delta_T(\zeta)$ only by a constant factor, we conclude (see [GK], Chapter IV, §2) that the sequence $\{\alpha_j\}$ in the representation (3.22) yields precisely the complete sequence of all eigenvalues of the contraction T lying inside the unit disc (if T is a simple contraction, that it cannot have any other eigenvalues).[15]

We note a relation which is obtained from (3.20) and (3.21) for $\zeta = 0$:[16]

$$(\det(T^*T))^{1/2} = \prod_j |\alpha_j| \exp\left(-\int_0^{2\pi} d\sigma_T(t)\right). \tag{3.23}$$

If the contraction T has no eigenvalues inside the unit disc (i.e. $B_T(\zeta) = 1$), this relation transforms into

$$\int_0^{2\pi} d\sigma_T(t) = -\tfrac{1}{2}\ln\det(T^*T). \tag{3.24}$$

Let us denote this expression by a_T:

$$a_T = -\tfrac{1}{2}\ln\det(T^*T).$$

We remark that this functional in fact coincides with the functional a_T introduced by B. Sz.-Nagy and C. Foiaş (see the end of §2).

We now rewrite relation (3.18) in the following form:

$$\det(T_\zeta^{-1}T_\zeta^{*-1}) = |d_T^{-1}(\zeta)|^2$$

and use it to estimate the resolvent $(T-\zeta I)^{-1}$. From this relation there follows (compare the arguments on pp. 356-357)

$$|T_\zeta^{-1}| \leq |d_T^{-1}(\zeta)|. \tag{3.25}$$

Further, according to (3.21) and (3.24), if $B_T(\zeta) \equiv 1$, then

---

[15] In the report by M. G. Kreĭn [**25**] there is given a spectral interpretation, obtained by the authors, of the function $\sigma_T(t)$ in terms of a chain of invariant subspaces of T.

[16] If the system of root vectors of a simple invertible contraction T is complete in $\mathfrak{H}$, then and only then do we have $\det T^*T = \prod|\alpha_j|$ (see [GK], Theorem V.3.2), i.e., by (3.23) in this case and only in this case do we have $d_T(\zeta) = B_T(\zeta)$. This deduction can be obtained immediately by applying the indicated criterion of completeness to the operator T_ζ.

$$\ln|d_T^{-1}(\zeta)| = \operatorname{Re}\int_0^{2\pi}\frac{e^{it}+\zeta}{e^{it}-\zeta}\,d\sigma_T(t) \leqq \frac{2a_T}{1-|\zeta|}. \tag{3.26}$$

Taking into account now the bounds

$$\begin{aligned} |T_\zeta^{-1}| &= |(I-\bar{\zeta}T)(T-\zeta I)^{-1}| \leqq 2|(T-\zeta I)^{-1}|, \\ |(T-\zeta I)^{-1}| &= |(I-\bar{\zeta}T)^{-1}T_\zeta^{-1}| \leqq \frac{1}{1-|\zeta|}|T_\zeta^{-1}|, \end{aligned} \qquad (|\zeta|<1)$$

and the bounds (3.25) and (3.26), we find that

$$\ln|(T-\zeta I)^{-1}| \leqq \ln\frac{1}{1-|\zeta|} + \frac{2a_T}{1-|\zeta|} \quad (|\zeta|<1). \tag{3.27}$$

From this bound on the growth of the resolvent it follows that

$$\begin{aligned} \varlimsup_{\rho\uparrow 1}\{(1-\rho)\ln M_T(\rho)\} &= \varlimsup_{|\zeta|\uparrow 1}\{(1-|\zeta|)\ln|T_\zeta^{-1}|\} \\ &\leqq \varlimsup_{|\zeta|\uparrow 1}\left\{(1-|\zeta|)\operatorname{Re}\int_0^{2\pi}\frac{e^{it}+\zeta}{e^{it}-\zeta}\,d\sigma_T(t)\right\}. \end{aligned} \tag{3.28}$$

It now follows from (3.26) that always

$$\varlimsup_{\rho\uparrow 1}\{(1-\rho)\ln M_T(\rho)\} \leqq 2a_T. \tag{3.29}$$

Thus the second assertion of Theorem 3.3 is proved.

To prove the first (basic) assertion of Theorem 3.3 we have to use the preceding Theorems 3.1 and 3.2.

The necessity of the first condition of the theorem is obvious, and the necessity of the second condition (equality (3.13)) follows from Theorems 3.1 and 3.2.

It remains to clarify the sufficiency of these conditions. When they are fulfilled, then by (3.28) and (3.29), which we now have to take with the equal signs, we have

$$\varlimsup_{|\zeta|\uparrow 1}\left\{(1-|\zeta|)\operatorname{Re}\int_0^{2\pi}\frac{e^{it}+\zeta}{e^{it}-\zeta}\,d\sigma_T(t)\right\} = 2\int_0^{2\pi}d\sigma_T(t) = 2a_T.$$

It is not hard to show that this relation is valid if and only if the function $\sigma_T(t)$ has only one point of increase. If this point is $t=t_0$, then the entire spectrum of the simple contraction T reduces to the point $\alpha=\exp(it_0)$.[17]

[17] One can show that in the general case the spectrum of a simple contraction T consists of the sequence of points $\{\alpha_j\}$, the limit points of this sequence, and the image points $\exp(it)$ on the unit circle of all points of increase t of the function $\sigma_T(t)$.

Without loss of generality we may suppose that $\alpha = 1$. But then the operator T will satisfy the conditions of Theorem 3.2, and (3.9) will hold for T. Hence, by (3.13) equality will hold in (3.5).

Thus Theorem 3.3 is proved.

4. As was already mentioned in §2.4, the investigations of B. Sz.-Nagy and C. Foiaş [**5**—**11**] have made it possible to obtain a criterion for the unicellularity of a contraction from another circle of ideas.

Let us denote by $H^{(\infty)}$ the class of all functions $u(\zeta)$, defined and holomorphic in the open disc $|\zeta| < 1$. If $u(\zeta) \in H^{(\infty)}$ and T is a contraction, then, as the aforementioned authors have shown [**7**], there is a natural way to give meaning to the expression $u(T)$. Let

$$u(\zeta) = \sum_{k=0}^{\infty} c_k \zeta^k \quad (|\zeta| < 1)$$

be the Maclaurin series of the function $u(\zeta)$; if $\| T \| < 1$, then in accordance with the usual definition of a function of an operator we put

$$u(T) = \sum_{k=0}^{\infty} c_k T^k;$$

in the general case $u(T)$ is defined as the following strong limit:

$$u(T) = \text{s-}\lim_{r \uparrow 1} u(rT).$$

It turns out that this strong limit always exists for any function $u(\zeta) \in H^{(\infty)}$ and any contraction T $(\| T \| \leq 1)$.

Let us denote by $\mathbf{C}_0$ the class of all contractions T, for each of which there exists a function $d(\zeta) \in H^{(\infty)}$, $d(\zeta) \not\equiv 0$, which annihilates this contraction, i.e. such that $d(T) = 0$.

The following important result (B. Sz.-Nagy and C. Foiaş [**5**]) holds.

1. *Let $T \in \mathbf{C}_0$. Then there exists a unique, up to a constant factor c $(|c| = 1)$, function $m_T(\zeta)$ such that*

1) *$m_T(\zeta)$ is an inner function;*[18)]

2) *$m_T(T) = 0$;*

3) *$m_T(\zeta)$ is a divisor in the ring $H^{(\infty)}$ of every function $u(\zeta) \in H^{(\infty)}$ for which $u(T) = 0$.*

The function $m_T(\zeta)$, whose existence is asserted by 1) for $T \in \mathbf{C}_0$, is called the *minimal function* of T.

[18)] A function $u(\zeta) \in H^{(\infty)}$ is called an *inner function* if $\lim_{\rho \uparrow 1} |u(\rho e^{i\theta})| = 1$ for almost every $\theta \in [0, 2]$ or, what is equivalent, if the function $u(\zeta)$ admits a representation of the type (3.21) in which the nondecreasing function $\sigma(t)$ is singular (i.e., has a derivative which equals zero almost everywhere).

B. Sz.-Nagy and C. Foiaş [6] have shown that if $T \in \mathbf{C}_0$, then

$$\text{s-}\lim_{n\to\infty} T^n = \text{s-}\lim_{n\to\infty} T^{*n} = 0. \tag{3.30}$$

It turns out that this condition is also necessary for T to belong to the class $\mathbf{C}_0$ for the important case, already considered above, of a contraction with nuclear deviation operator. Namely:

2) *Let T be a simple invertible contraction for which $I - T^*T \in \mathfrak{S}_1$. Then the condition* (3.30) *is necessary and sufficient for T to belong to the class $\mathbf{C}_0$. When this condition is satisfied, the above constructed function $d_T(\zeta)$ can be taken as an annihilating function, i.e. $d_T(T) = 0$.*

This result can be regarded as an analog of the Hamilton-Cayley theorem for matrices.

We have the following result.

THEOREM 3.4. *Let T be a simple invertible contraction with $I - T^*T \in \mathfrak{S}_1$. In order that T be unicellular, it is necessary and sufficient that the following two conditions be satisfied:*

1) *the spectrum of T consists of only one point, which lies on the unit circle;*
2) $m_T(\zeta) \equiv \text{const}\, d_T(\zeta)$.

For this theorem, as for Theorem 3.3, it is essential that the Hilbert space $\mathfrak{H}$ in which T acts be infinite-dimensional.

It is useful to make a parallel with what can be asserted for simple contractions T acting in n-dimensional space E_n.

A contraction T acting in E_n will be simple if and only if its entire spectrum lies inside the unit circle. Therefore in the finite-dimensional case condition 1) is ***absurd*** where it requires that the single spectral point shall lie on the unit circle. However the first part of condition 1) (the existence of a one-point spectrum) is a necessary condition for unicellularity in the finite-dimensional case.

For a simple invertible contraction T in E_n the function $d_T(\zeta)$ is a rational function

$$d_T(\zeta) = \text{const}\frac{\det(T - \zeta I)}{\det(I - \zeta T^*)},$$

whose numerator is the characteristic polynomial, and whose denominator is the mirror image of the numerator; these polynomials are mutually prime. Therefore the relation $d_T(T) = 0$ will always be satisfied on the basis of the familiar Hamilton-Cayley theorem, which is in complete accord with result 2.

For a simple contraction T in E_n the function $d_T(\zeta)$ can be used in place of the characteristic polynomial, and the function $m_T(\zeta)$ can be used in place of the minimal polynomial. Therefore condition 2) together with the first part of condition 1) is, for the case being considered, a necessary and sufficient condition for the unicellularity of T.

Let us remark also that in the formulation of condition 2) it appeared that it was first necessary to require that $T \in \mathbf{C}_0$. However, the operator T, as can be shown, will always belong to the class $\mathbf{C}_0$ whenever the preceding condition of the theorem is satisfied. If we denote by α ($|\alpha| = 1$) the unique spectral point of T, whose existence is asserted in condition 1), then condition 2) could have been formulated as follows:

2′) *the minimal function* $m_T(\zeta)$ *of* T *has the form*

$$m_T(\zeta) = \exp\left(a_T \frac{\alpha + \zeta}{\alpha - \zeta}\right).$$

This follows from the general representation (3.21) of the function $d_T(\zeta)$, in which under the hypotheses of the theorem and condition 1) the Blaschke product $B_T(\zeta)$ is identically equal to 1, and the function $\sigma_T(t)$ has a single point of increase $t_0 = \arg\alpha$ $(0 \leq t_0 < 2\pi)$.

Result 2 and Theorem 3.4 were established by Sz.-Nagy and Foiaş, but with a narrower formulation: in place of the *nuclearity* of the operator $I - T^*T$ these authors required that it be finite-dimensional. The validity of these results in the extended formulation given here was established by Ju. P. Ginzburg.[19)]

We shall show that Theorem 3.4 follows (formally) very simply from Theorem 2.3.

By virtue of our previous results (for example, the general Theorem 3.1), we may suppose at once that the contraction T being considered $(I - T^*T \in \mathfrak{S}_1)$ has the one-point spectrum $\lambda = 1$. It remains to show that condition 2) of Theorem 3.4 is a necessary and sufficient condition for the unicellularity of T.

Under the assumptions which have been made,

$$d_T(\zeta) = \exp(a_T(1 + \zeta)(1 - \zeta)^{-1}),$$

the operator $A = i(I - T)(I + T)^{-1}$ is a dissipative Volterra operator, and finally $a_T = 2\,\mathrm{sp}\,A_{\mathscr{I}}$. By Theorem 2.3

$$d_T(T) = \exp(a_T(I + T)(I - T)^{-1}) = \exp(-2i(\mathrm{sp}\,A_{\mathscr{I}})A^{-1}) = 0.$$

---

19) In a report in a seminar by M. G. Kreĭn. A recent note by Sz.-Nagy and Foiaş [12] supports the assertion that they have also been able to obtain result 2 and Theorem 3.4 in the extended form given here.

Consequently, T has a minimal function $m_T(\zeta)$. It must be a divisor of the function $d_T(\zeta)$, and therefore $m_T(\zeta) = c\exp(a(1+\zeta)(1-\zeta)^{-1})$ $(|c| = 1)$, where a is the smallest positive number such that $m_T(T) = 0$. Since, then, $m_T(T) = c\exp(-iaA^{-1})$, by Theorem 2.3 we have $a \leqq 2\,\mathrm{sp}\,A_{\mathscr{J}} = a_T$, and equality holds if and only if $a = a_T$, i.e. $m_T(\zeta) = cd_T(\zeta)$.

We also mention that in the work of Sz.-Nagy and Foiaş [**6**—**10**] the function $d_T(\zeta)$ is defined not by (3.16), but by means of the characteristic operator-function $\theta_T(\zeta)$ of the contraction T. This function, in turn, is defined by

$$\theta_T(\zeta) = [-T + \zeta(I - TT^*)^{1/2}(T - \zeta I)(I - \zeta T^*)^{-1}(I - T^*T)^{1/2}] | \mathfrak{D}_T \qquad (|\zeta| < 1),$$

where $\mathfrak{D}_T$ is the closure of the range of the operator $I - T^*T$. If $I - T^*T$ is finite-dimensional (n-dimensional), then $\theta_T(\zeta)$ will be n-dimensional. It is easily shown that in this case $\det\theta_T(\zeta) = \epsilon d_T(\zeta)$ $(|\epsilon| = 1)$.

If the contraction T is related to a dissipative operator A by (3.1), then, as can be shown, the function $\theta_T(\zeta)$ coincides up to constant unitary factors (on the right and left)—which map $\mathfrak{J}(A)$, the closure of the range of $A_{\mathscr{J}}$, onto $\mathfrak{D}_T$—with the function $W_A(z)$, where $\zeta = (i-z)/(i+z)$.

Similarly to the way in which Theorems 2.1 and 2.2 can be obtained one from the other by replacing $|U(i\rho)|$ by $|A(-i\rho)|$, Theorems 3.2 and 3.3 also have duals, which are obtained from these theorems by replacing the resolvent $(T - \zeta I)^{-1}$ by the function $\theta_T(\zeta)$. In many cases this is very convenient, for example when $I - T^*T$ is finite-dimensional.

§4. The problem of unicellularity and inverse problems for a canonical equation

1. The criterion for unicellularity of M. S. Brodskiĭ and G. È. Kisilevskiĭ (Theorem 2.1) together with certain other results of the theory of characteristic operator-functions leads to a number of important deductions in the theory of canonical differential equations.

In his investigations in the theory of analytic matrix-functions V. P. Potapov [**1**] obtained (among other more general results) the following result.

THEOREM 4.1 (V. P. POTAPOV [**1**]).[20] *Let $\mathscr{Y}$ be a signature matrix*

---

[20] Let us recall (Chapter V, §3) that an H-matrix $\mathscr{A}(t)$ $(0 \leqq t \leqq l)$ (i.e. a Hermitian matrix-function with elements from $L_1(0,l)$) is called an H-matrix of positive type if

of order n ($\mathcal{Y}^* = -\mathcal{Y}$, $\mathcal{Y}^2 = -I_n$), *and let* $\mathcal{U}(\mu) = \| u_{jk}(\mu) \|_1^n$ *be an entire matrix-function. In order that* $\mathcal{U}(\mu)$ *admit a representation in terms of a multiplicative integral*

$$(4.1) \qquad \mathcal{U}(\mu) = \overset{\curvearrowleft}{\int_0^l} e^{-\mu \mathcal{Y} \mathcal{H}(t)\, dt},$$

where $\mathcal{H}(t) = \| h_{jk}(t) \|_1^n$ $(0 \leqq t \leqq l)$ *is an* H*-matrix of positive type, it is necessary and sufficient that the matrix-function* $\mathcal{U}(\mu)$ *have the following properties*:

1) $\mathcal{U}(0) = I_n$, *and* $\mathcal{Y}\mathcal{U}'(0)$ *is a positive-definite matrix;*
2) *for every real* μ *the matrix* $\mathcal{U}(\mu)$ *is* $\mathcal{Y}$*-unitary, i.e.* $\mathcal{U}^*(\mu)\mathcal{Y}\mathcal{U}(\mu) = \mathcal{Y}$;
3) *for* $\operatorname{Im} \mu > 0$ *the matrix* $\mathcal{U}(\mu)$ *is* $(1/i)\mathcal{Y}$*-contractive, i.e.*

$$(4.2) \qquad \frac{1}{i}(\mathcal{Y} - \mathcal{U}^*(\mu)\mathcal{Y}\mathcal{U}(\mu)) \geqq 0 \quad (\operatorname{Im} \mu > 0).$$

Without going into other possibilities for defining a multiplicative integral, let us clarify that for a matrix $\mathcal{A}(t) = \| a_{jk}(t) \|_1^n$ with elements from $L_1(a,b)$ *the multiplicative integral*

$$\Pi(t) = \overset{\curvearrowleft}{\int_a^t} e^{\mathcal{A}(s)\, ds} \quad (a \leqq t \leqq b)$$

furnishes the solution $\Pi(t)$ $(a \leqq t \leqq b)$ of the matrix differential equation

$$d\Pi/dt = \mathcal{A}(t)\,\Pi(t)$$

which is singled out by the initial condition $\Pi(0) = I_n$.

To put it another way, $\Pi(t)$ $(a \leqq t \leqq b)$ is the solution of the integral equation

$$\Pi(t) = I_n + \int_a^t \mathcal{A}(s)\,\Pi(s)\, ds.$$

---

the following two conditions are satisfied:

a) the matrix-function $\mathcal{H}(t)$ is Hermitian-nonnegative (for almost all t);

b) its average, $\mathcal{H}_{\text{av}}$, is a Hermitian-positive matrix.

Potapov's theorem remains valid for more general representations than (4.1); these representations require that the H-matrix $\mathcal{H}(t)$ be only Hermitian-nonnegative (and the requirement b) is omitted); for this only the requirement, in condition 1), that $\mathcal{U}(0) = I_n$ is maintained for $\mathcal{U}(\mu)$. It was in precisely this general form that Theorem 4.1 was proved by Potapov [**1**, **2**].

Among various properties of a multiplicative integral we note a fundamental property which clarifies its name:

$$\overset{b}{\underset{a}{\curvearrowleft\!\!\int}} e^{\mathscr{A}(s)\,ds} = \overset{b}{\underset{c}{\curvearrowleft\!\!\int}} e^{\mathscr{A}(s)\,ds} \overset{c}{\underset{a}{\curvearrowleft\!\!\int}} e^{\mathscr{A}(s)\,ds};$$

here c is any number from the interval $a < c < b$.

In particular, (4.1) simply says that the matrix-function $\mathscr{U}(\mu)$ is the monodromy matrix of the corresponding canonical differential equation, i.e. that $\mathscr{U}(\mu) = \mathscr{U}(l; \mu)$, where the matrizant $\mathscr{U}(t; \mu)$ $(0 \leq t \leq l)$ is in turn defined as the solution of the Cauchy problem

$$\mathscr{Y}\frac{d\mathscr{U}}{dt} = \mu\mathscr{A}(t)\mathscr{U}, \quad \mathscr{U}(0; \mu) = I_n. \tag{4.3}$$

Thus, Potapov's theorem completely solves the problem, *what analytic properties characterize the monodromy matrix-function of a canonical equation with a given signature matrix* $\mathscr{Y}$?

Let us now indicate that the necessity of condition 1) follows trivially from the expansion (1.6) which was given in Chapter VI, §1, and according to which

$$\mathscr{U}(t; \mu) = I_n - \mu\mathscr{Y}\int_0^t \mathscr{A}(s)\,ds + O(\mu^2) \quad (\mu \to 0).$$

Indeed, by this expansion

$$\mathscr{Y}\mathscr{U}'(0) = \mathscr{Y}\frac{d\mathscr{U}(l; \mu)}{d\mu}\bigg|_{\mu=0} = \int_0^l \mathscr{A}(s)\,ds.$$

Property 2) of the monodromy matrix $\mathscr{U}(\mu)$ was already noted in Chapter VI, §1.2. An immediate generalization of the simple argument given there enables us to obtain property 3) at the same time as property 2). Indeed, let us form the matrix-function

$$\Omega(t; \mu) = \frac{1}{i}\left(\mathscr{Y} - \mathscr{U}^*(t; \mu)\mathscr{Y}\mathscr{U}(t; \mu)\right)$$

and differentiate it with respect to t, recalling (4.3).

This yields

$$\frac{d\Omega(t; \mu)}{dt} = \frac{1}{i}(\mu - \bar{\mu})\mathscr{U}^*(t; \mu)\mathscr{A}(t)\mathscr{U}(t; \mu),$$

and since $\Omega(0; \mu) = 0$ we obtain

$$\Omega(l;\mu) = 2\,\mathrm{Im}\,\mu \int_0^l \mathscr{U}^*(t;\mu)\mathscr{H}(t)\mathscr{U}(t;\mu)\,dt. \tag{4.4}$$

This equality immediately implies both properties 2) and 3).[21)]

What is deep and difficult about this result is the proof of the sufficiency of conditions 1), 2) and 3). V. P. Potapov obtained this result by a long chain of analytic constructions.

The sufficiency of these conditions can at the present time be obtained from general theorems on the multiplicative representation of generalized characteristic operator-functions (see M. S. Brodskiĭ [7]).

2. Without loss of generality the H-matrix $\mathscr{H}(t)$ in the representation (4.1) may be supposed normalized,[22)] i.e.

$$\mathrm{sp}\,\mathscr{H}(t) \equiv 1 \qquad (0 \leqq t \leqq l). \tag{4.5}$$

Indeed, this can always be accomplished by making the change of parameter $s = \mathrm{sp}\mathscr{H}(t)$ $(0 \leqq t \leqq l)$.

We will call a representation (4.1) with a normalized H-matrix $\mathscr{H}(t)$ $(0 \leqq t \leqq l)$ of positive type a *$\mathscr{Y}$-canonical representation.*

The question naturally arises,

Under what conditions will an entire matrix-function $\mathscr{U}(\mu)$ have a unique $\mathscr{Y}$-canonical representation?

In other words, under what conditions is the normalized H-matrix $\mathscr{H}(t)$ $(0 \leqq t \leqq l)$ in a $\mathscr{Y}$-canonical representation (4.1) (with a previously prescribed signature matrix $\mathscr{Y}$) defined uniquely almost everywhere?

A nontrivial answer to this question can first be given in the framework of the theory of unicellular operators.

From the results of M. S. Livšic [2] and M. S. Brodskiĭ [6] (see also the joint paper [1] of these authors) there follows an important result.

THEOREM 4.2. *Suppose that the matrix-function $\mathscr{U}(\mu)$ admits a canonical representation* (4.1). *In order that this representation be unique, it is necessary and sufficient that the simple part $A_{\mathscr{U}}$*[23)] *of the model operator*

---

21) Moreover, we can conclude from (4.4) that the matrix $-i(\mathscr{Y} - \mathscr{U}^*(\mu)\mathscr{Y}\mathscr{U}(\mu))$ is positive (negative) definite for $\mathrm{Im}\,\mu > 0$ (< 0).

22) This remark was already made in the theory of models (Chapter V, §3, Remark 3.2).

23) The notation $A_{\mathscr{U}}$ can be justified by the fact that this operator is determined by the function $\mathscr{U}(\mu)$ (having the properties enumerated in Theorem 4.1) up to a unitary transformation. This operator is characterized by the property that $\mathscr{U}(\mu)$ is its characteristic matrix-function (in the sense of the definition given in Chapter V, §4).

$$(A_M f)(t) = \mathscr{Y} \int_t^l \mathscr{A}^{1/2}(t) \mathscr{A}^{1/2}(s) f(s)\, ds \quad (f \in L_2^{(n)}(0,l)) \tag{4.6}$$

be unicellular.

3. We shall be interested below in the *definite* case, in which $\mathscr{Y} = iI_n$, and hence the representation (4.1) has the form

$$\mathscr{U}(\mu) = \overset{\curvearrowleft}{\int_0^l} e^{-i\mu \mathscr{A}(t)\, dt}, \tag{4.7}$$

where $\mathscr{A}(t)$ $(0 \leqq t \leqq l)$ is a normalized H-matrix of positive type.

We will call a representation (4.7) a *definite canonical representation*. By Potapov's theorem an entire matrix-function $\mathscr{U}(\mu)$ which admits a definite canonical representation (4.7) is characterized by the following properties:

I) $\mathscr{U}(0) = I_n$, $i\mathscr{U}'(0)$ *is a positive-definite matrix*;[24)]

II) *for real* μ *the matrix* $\mathscr{U}(\mu)$ *is unitary*;

III) *for* $\operatorname{Im}\mu < 0$ *the matrix* $\mathscr{U}(\mu)$ *is contractive*:

$$I_n - \mathscr{U}^*(\mu)\mathscr{U}(\mu) \geqq 0.$$

This result should be regarded as a generalization of a well-known result of the theory of functions, according to which the function $u(\mu) = \exp(-ih\mu)$ $(h > 0)$ is singled out among all other entire functions by the following properties:

1) $u(0) = 1$, $iu'(0) > 0$; 2) $|u(\mu)| = 1$ for $\operatorname{Im}\mu = 0$, and 3) $|u(\mu)| < 1$ for $\operatorname{Im}\mu < 0$.

Comparing Theorems 2.1 and 4.2 leads to the following:

THEOREM 4.3 (M. S. BRODSKIĬ AND G. È. KISILEVSKIĬ [1]).[25)] *In order that the matrix-function* $\mathscr{U}(\mu)$ *(having the properties* I), II), III)) *admit a unique definite canonical representation* (4.6), *it is necessary and sufficient that the condition*

---

24) It can be shown that the relation $\mathscr{U}(0) = I_n$ and conditions II) and III) for an entire matrix-function $\mathscr{U}(\mu)$ imply that the matrix $i\mathscr{U}'(0)$ is nonnegative. If the matrix $\mathscr{U}'(0)$ were singular, the space E_n could be broken up into the orthogonal sum of invariant subspaces of $\mathscr{U}(\mu)$, $E_n = Z \oplus N$, where Z is the set of all vectors annihilated by $\mathscr{U}'(0)$, and then $\mathscr{U}(\mu)$ would induce the identity operator on Z, and an operator-function $\hat{\mathscr{U}}(\mu)$ with the properties I), II) and III) on N.

25) The sufficiency of condition (4.8) had already been obtained by M. S. Brodskiĭ [**1, 6**].

(4.8) $$\overline{\lim_{\rho\to\infty}}\left\{\frac{1}{\rho}\ln|\mathscr{U}(i\rho)|\right\} = \operatorname{sp}(i\mathscr{U}'(0))$$

be satisfied.

Let us clarify that by (4.5) and (4.6)

$$l = \int_0^l \operatorname{sp}\mathscr{A}(t)\,dt = 2\operatorname{sp}(A_M)_{\mathscr{I}} = 2\operatorname{sp}(A_{\mathscr{U}})_{\mathscr{I}} = \operatorname{sp}(i\mathscr{U}'(0)).$$

Theorem 4.3 has an interesting corollary; to obtain this corollary we shall use the expansion of the function $\mathscr{U}(\mu)$ in a power series in μ (compare with Chapter VI, §1.2):

(4.9) $$\mathscr{U}(\mu) = I_n - i\mu\mathscr{A}_1(l) + (-i\mu)^2\mathscr{A}_2(l) + \cdots,$$

where

$$\mathscr{A}_1(t) = \int_0^t \mathscr{A}(s)\,ds, \quad \mathscr{A}_n(t) = \int_0^t \mathscr{A}(s)\mathscr{A}_{n-1}(s)\,ds \quad (n = 2, 3, \cdots).$$

Let us denote by $h(t)$ the largest eigenvalue of the Hermitian matrix $\mathscr{A}(t)$ $(0 \leqq t \leqq l)$. Obviously

(4.10) $$0 < h(t) \leq \operatorname{sp}\mathscr{A}(t) \quad (\equiv 1).$$

If by the norm of an nth order matrix we understand the norm of the linear operator corresponding to it in the n-dimensional Hilbert space E_n, then we have $h(t) = |\mathscr{A}(t)|$. Therefore

$$|\mathscr{A}_1(t)| \leqq \int_0^t h(s)\,ds,$$

and arguing by induction we obtain the bounds

$$|\mathscr{A}_n(t)| \leqq \frac{1}{n!}\left(\int_0^t h(s)\,ds\right)^n \qquad (n = 1, 2, \cdots).$$

Thus it follows from the expansion (4.9) that

$$|\mathscr{U}(\mu)| \leqq \exp\left(|\mu|\int_0^l h(s)\,ds\right).$$

If (4.8) holds, then a comparison with the preceding bound yields

$$l \leqq \int_0^l h(s)\,ds.$$

Taking (4.10) into account, we conclude that $h(t) \equiv 1 \equiv \operatorname{sp}\mathscr{A}(t)$.

Thus, all the eigenvalues of the matrix $\mathscr{H}(t)$, besides $h(t)$, equal zero.

COROLLARY 4.1. *If the matrix-function $\mathscr{U}(\mu)$ admits a unique definite canonical representation* (4.7), *then the H-matrix $\mathscr{H}(t)$ in this representation has rank equal to unity almost everywhere.*

4. Let us denote by $r(t)$ the rank of the H-matrix $\mathscr{H}(t)$ $(0 \leqq t \leqq l)$. We see that the relation $r(t) = 1$ (almost everywhere) is a necessary condition for the representation (4.7) to be unique. However this condition is *by no means* sufficient, as can be concluded from the following result.

THEOREM 4.4. *For any H-matrix $\mathscr{H}(t)$ $(0 \leqq t \leqq l)$ of positive type there exists an H-matrix $\mathscr{H}_0(t)$ $(0 \leqq t \leqq l)$ of positive type and with rank* 1 $(r_0(t) \equiv 1)$ *such that*[26)]

$$(4.11) \qquad \overset{l}{\widehat{\int_0}} \exp(-i\mu\mathscr{H}_0(t)\,dt) = \overset{l}{\widehat{\int_0}} \exp(-i\mu\mathscr{H}(t)\,dt).$$

This theorem is a simple consequence of the results noted above. In fact, let us form the simple dissipative Volterra operator A (with n-dimensional imaginary component) which has $\mathscr{H}(t)$ $(0 \leqq t \leqq l)$ as its H-matrix. This operator can obviously be defined, for example, as the simple part of the model operator

$$(A_M f)(t) = i\int_t^l \mathscr{H}^{1/2}(t)\mathscr{H}^{1/2}(s)f(s)\,ds \quad (f \in L_2^{(n)}(0,l)).$$

Then the right side of (4.11) will give us the monodromy matrix of the operator A (i.e. $\mathscr{U}(1/\lambda)$ will be the characteristic matrix-function of A).

On the other hand, according to Remark 3.1 of Chapter V, following Theorem V.3.1, on models of a dissipative operator A, there exists at least one canonical model with an H-matrix $\mathscr{H}_1(t)$ of rank $r(t) \equiv 1$. The multiplicative integral corresponding to $\mathscr{H}_1(t)$ will yield the characteristic matrix-function $\mathscr{W}_A(\lambda)$ $(\lambda = 1/\mu)$ in some orthonormal basis and, consequently, will be equal to $\mathscr{E}\mathscr{U}_A(\mu)\mathscr{E}^{-1}$, where $\mathscr{E}$ is some constant unitary matrix. Therefore if we put $\mathscr{H}_0(t) = \mathscr{E}\mathscr{H}_1(t)\mathscr{E}^{-1}$, the H-matrix $\mathscr{H}_0(t)$ will now have all the necessary properties.

---

26) Thus Corollary 4.1 is trivially contained in Theorem 4.4. We formulated it separately, since it enables us to prove Theorem I.7.3 (of G. È. Kisilevskiĭ), which will be discussed below, and on which the derivation of Theorem 4.4 is based.

Remark V.3.1 to Theorem V.3.1 on models was obtained on the basis of Theorem I.7.3, which states that *every dissipative Volterra operator A with nuclear imaginary component has a maximal eigenchain of rank $r = 1$.*

We can now clarify how Theorem I.7.3 is obtained.

Let us first assume that A is a unicellular operator.[27] Then in the representation (4.7) for $\mathscr{U}(\mu) = \mathscr{U}_A(\mu)$ the normalized H-matrix $\mathscr{A}(t)$ of positive type will be determined uniquely almost everywhere and has rank $r(t) \equiv 1$ (almost everywhere). Hence, we will have

$$\mathscr{A}(t) = \xi(t)\xi^*(t) \qquad (0 \leqq t \leqq l), \tag{4.12}$$

where $\xi(t)$ $(0 \leqq t \leqq l)$ is some vector-function from $L_2^{(n)}(0,l)$.

We form the Hermitian-nonnegative scalar kernel

$$\mathscr{K}(t,s) = \xi^*(s)\xi(t) \qquad (0 \leqq s,t \leqq l) \tag{4.13}$$

and the corresponding model operator in $L_2(0,l)$:

$$(A_{\mathscr{K}} f)(t) = i\int_t^l \mathscr{K}(t,s)f(s)\,ds \quad (f \in L_2(0,l)).$$

The operator A, being unitarily equivalent to the simple part of A_M, will be unitarily equivalent to the simple part of $A_{\mathscr{K}}$. Taking into account that $A_{\mathscr{K}}$ has a complete eigenchain consisting of the truncation operators $\hat{P}(t)$ $(0 \leqq t \leqq l)$, which chain has rank 1, we conclude that the operator A has a maximal eigenchain $\mathfrak{P} = \{P\}$ of rank 1.

We now consider the case in which A is not unicellular. By Theorem 2.1, it can be decomposed into the quasidirect sum

$$A = \sum_{k=1}^{\omega} \oplus A_k \qquad (\omega \leqq n)$$

of unicellular operators A_k $(k = 1, 2, \cdots, \omega)$. Since, as was just proved, each of the operators A_k has a maximal eigenchain $\mathfrak{P}_k$ $(k = 1, 2, \cdots, \omega)$ of rank 1, the same will be true of the operator A.

Thus Theorem I.7.3 is proved.

We remark that to every ordering of the chains $\mathfrak{P}_k$ $(k = 1, 2, \cdots, \omega)$ there corresponds a maximal eigenchain $\mathfrak{P}$ of rank 1 of A. Therefore the operator A will have at least $\omega!$ distinct maximal eigenchains of rank 1.

5. The property of unicellularity for a dissipative Volterra operator is very delicate. To support this assertion we give the following two results.

---

[27] For this case the theorem was proved in a paper by M. S. Brodskiĭ and G. È. Kisilevskiĭ [1].

A) (M. S. Brodskiĭ [1]). *Suppose that the matrix-function* $\mathscr{U}(\mu)$ *admits a definite canonical representation* (4.7) *with a piecewise constant* H*-matrix* $\mathscr{A}(t)$ *of rank* 1, *i.e. an* H*-matrix of the form* $\mathscr{A}(t) = \xi(t)\xi^*(t)$ $(0 \leqq t \leqq l)$, *where* $\xi(t)$ $(0 \leqq t \leqq l)$ *is a piecewise constant vector-function with values in* E_n. *Then the representation* (4.7) *will be unique if and only if no two vectors* $\xi(t-0)$ *and* $\xi(t+0)$ $(0 < t < l)$ *are mutually orthogonal*:

$$\xi^*(t-0)\xi(t+0) \neq 0 \qquad (0 < t < l). \tag{4.14}$$

This result can be proved by directly verifying that for the special H-matrix $\mathscr{A}(t)$ being considered the conditions (4.8) and (4.14) are equivalent. As this verification is instructive, we shall carry it out.

In fact, suppose that the H-matrix $\mathscr{A}(t)$ has the constant value $\mathscr{A}_0 = \xi\xi^*$ $(\xi \in E_n,\ \xi^*\xi = 1)$ on some interval (a,b); then

$$\begin{aligned}\overset{\curvearrowleft}{\int_a^b} \exp(-i\mu\mathscr{A}(t)\,dt) &= \exp(-i\mu(b-a)\mathscr{A}_0) \\ &= I_n + (\exp(-i\mu(b-a)) - 1)\mathscr{A}_0 \\ &= I_n + (\exp(-i\mu(b-a)) - 1)\xi\xi^*.\end{aligned}$$

We have used here the relation $\mathscr{A}_0^n = \mathscr{A}_0$ $(n = 1, 2, \cdots)$. Therefore, if the interval $[0,l]$ can be broken up into subintervals

$$[t_{j-1}, t_j] \quad (j = 1, 2, \cdots, p;\ 0 = t_0 < t_1 < \cdots < t_p = l)$$

such that

$$\mathscr{A}(t) = \xi_j\xi_j^* \quad (t_{j-1} < t < t_j;\ j = 1, 2, \cdots, p),$$

where the ξ_j are constant unit vectors from E_n $(\xi_j^*\xi_j = 1;\ j = 1, 2, \cdots, p)$, then

$$\begin{aligned}\mathscr{U}(\mu) = \overset{\curvearrowleft}{\int_0^l} e^{-i\mu\mathscr{A}(t)\,dt} &= \overset{\curvearrowleft}{\int_{t_{p-1}}^l} e^{-i\mu\mathscr{A}(t)\,dt} \cdots \overset{\curvearrowleft}{\int_0^{t_1}} e^{-i\mu\mathscr{A}(t)\,dt} \\ &= \overset{\curvearrowleft}{\prod_{j=1}^p} (I_n + (\exp(-i\mu l_j) - 1)\xi_j\xi_j^*).\end{aligned} \tag{4.15}$$

Here $l_j = t_j - t_{j-1}$ $(j = 1, 2, \cdots, p)$, and the curved arrow indicates that the successive matrix factors are written from right to left.

Carrying out the multiplication in (4.15), we see that $\mathscr{U}(\mu)$ has the following form:

$$\mathscr{U}(\mu) = I_n + \sum_{k=1}^{p} C_k \exp(-i\mu a_k),$$

where

$$0 < a_1 < a_2 < \cdots < a_p = \sum_{j=1}^{p} l_j = l.$$

The leading coefficient C_p is obviously found from the formula

$$C_p = \gamma \xi_1 \xi_n^*, \quad \text{where } \gamma = \prod_{j=1}^{p-1} \xi_{j+1}^* \xi_j. \tag{4.16}$$

Thus $C_p \neq 0$ if and only if $\gamma \neq 0$, i.e. when the condition (4.14) is fulfilled.

On the other hand, in the case being considered the limit

$$\lim_{\rho \to \infty} \left\{ \frac{1}{\rho} \ln | \mathscr{U}(i\rho) | \right\} \tag{4.17}$$

will always exist and, obviously, will be equal to l if and only if $C_p \neq 0$.

6. The limit superior in (4.8) can be regarded as a functional $\Phi(\mathscr{H})$ defined on the convex cone $\mathfrak{K}_{n\times n}$ of H-matrices $\mathscr{H}(t) = \| h_{jk}(t) \|_1^n$ $(0 \leq t \leq l)$ of positive type. The cone $\mathfrak{K}_{n\times n}$ can be regarded in a natural way as a cone in the real Banach space $\mathfrak{B}_{n\times n}$ of all H-matrices $\mathscr{H}(t) = \| h_{jk}(t) \|_1^n$ of order n with elements from $L_1(0, l)$ (l is some fixed positive number) with the norm defined by

$$\| \mathscr{H}(t) \| = \int_0^l | \mathscr{H}(t) | \, dt.$$

Result A) shows that *the functional* $\Phi(\mathscr{H})$ *is discontinuous*. In fact, let the H-matrix $\mathscr{H}(t)$ have the form (4.12) with a piecewise constant function $\xi(t)$ and with the quantity γ, defined in (4.16), equal to zero. For this H-matrix the limit (4.17) will be $< l$. On the other hand, an arbitrary small change in the value of the vectors ξ_j $(j = 1, 2, \cdots, p)$ can lead to $\gamma \neq 0$, and then the limit (4.17) for the corresponding H-matrix will equal l.

Following what has been said, the following result assumes special interest.

B) (G. È. Kisilevskiĭ [**5**]).[28] *If the matrix function* $\mathscr{U}(\mu)$ *admits a*

---

[28] This result is not explicitly formulated in Kisilevskiĭ's paper [**5**], although it is essentially proved if one takes note of the remark used below on the continuity of the vector-function $\xi(t)$ in the representation (4.12) of the H-matrix $\mathscr{H}(t)$ under the hypothesis of result B).

definite canonical representation (4.7) *with a continuous H-matrix* $\mathscr{A}(t)$ *of rank* 1 (*i.e.* $r(t)\equiv 1$), *then this representation is the unique definite canonical representation of the function* $\mathscr{U}(\mu)$.

We omit the proof of this result. We shall only prove that it is implied by the following result.[29)]

C) (G. È. Kisilevskiĭ [**5**]). *Let* $\mathscr{K}(t,s)$ $(0\leqq t,s\leqq l)$ *be a continuous Hermitian-nonnegative kernel, with* $\mathscr{K}(t,t)>0$ $(0\leqq t\leqq l)$. *Then the simple part of the integral operator*

$$(Kf)(t)=i\int_t^l \mathscr{K}(t,s)f(s)\,ds$$

is unicellular.

In fact, if the H-matrix $\mathscr{A}(t)$ in the representation (4.7) has rank $r(t)\equiv 1$ and is continuous, then, as is not hard to show, it will admit a representation (4.12) with a continuous vector-function $\xi(t)$ $(0\leqq t\leqq l)$. Therefore the kernel $\mathscr{K}(t,s)$, defined by (4.13), will be a continuous Hermitian-nonnegative kernel with $\mathscr{K}(t,t)=\xi^*(t)\xi(t)=\mathrm{sp}\,\mathscr{A}(t)=1$. Thus result B) is a consequence of result C) and Theorem 4.2.

In conclusion we point out that Theorem 4.1 of V. P. Potapov was generalized by Ju. P. Ginzburg [**1, 2**] to the case of infinite-dimensional normalized H-matrices $\mathscr{A}(t)$ of positive type.[30)] Theorems 4.2, 4.3 and 4.4 were proved by their authors for this more general case.

§5. Real unicellular operators and related analytic problems

1. Suppose that there is defined in the separable Hilbert space $\mathfrak{H}$ an involution S, which enables one to introduce the concepts of S-real elements, S-real operators, etc. (see Chapter V, §5).

An S-real operator A is said to be *S-unicellular* if, for any two of its invariant subspaces $\mathfrak{L}_1,\mathfrak{L}_2$ which have S-real kernels, at least one of the relations $\mathfrak{L}_1\subset\mathfrak{L}_2$, $\mathfrak{L}_2\subset\mathfrak{L}_1$ holds. Obviously every S-unicellular operator is simple.

In order to distinguish S-unicellularity from the concept of unicellularity introduced in §1, we shall refer to unicellularity in the sense of §1 as *complex* unicellularity.

---

29) In essence, result B) is equivalent to result C) for the case of degenerate kernels.

30) The first generalization of Theorem 4.1 to this case was obtained by M. S. Livšic [1] by operator methods (in all completeness only for $\mathscr{Y}=\pm iI$). Later an elegant derivation of Theorem 4.1 by means of the abstract triangular integral was given by M. S. Brodskiĭ [7].

If an S-real operator is *complex-unicellular*, then it is a fortiori *S-unicellular*. On the other hand, even in the finite-dimensional case an operator can be S-unicellular without being complex-unicellular. In a finite-dimensional space this occurs if and only if the S-real operator has, in all, only two distinct nonreal (complex conjugate) eigenvalues, to each of which there corresponds just one Jordan cell.

It is highly noteworthy that for Volterra operators in an infinite-dimensional space one observes situations for which there is no analog in a finite-dimensional space.

As is easily seen, an S-real Volterra operator in a finite-dimensional space will be S-unicellular if and only if it is complex-unicellular.

In contrast to this, in an infinite-dimensional Hilbert space $\mathfrak{H}$ there exist Volterra operators which are not complex-unicellular, but which turn out to be S-unicellular for a suitable choice of the involution S. We give an example of such an operator in §5.4.

Just as was the case for dissipative operators (see §4), the question of the unicellularity of an S-real operator with a finite-dimensional skewsymmetric component reduces to the delicate question of uniqueness for a certain inverse problem in the theory of differential equations. The solution of this problem requires, however, considerably more tools from the theory of entire functions than did the preceding problem. At the same time it is interesting in certain respects, since it concerns differential equations which are frequently encountered in problems of mathematical physics.

The questions treated in this section have a very direct relation to the theory of spectral functions of Sturm-Liouville differential operators and others, and also to the theory of the extrapolation of stationary random processes. However, due to lack of space these relations will not be revealed here.

2. According to Theorem V.5.3, every S-real simple Volterra operator A with a $2m$-dimensional skewsymmetric component A_α is S-unitarily equivalent[31] to the simple part of a certain model operator A_M, defined in $L_2^{(2m)}(0,l)$ by

$$(A_M f)(t) = \mathscr{A}^{1/2}(t)\mathscr{Y}_m \int_t^l \mathscr{A}^{1/2}(s) f(s)\, ds. \tag{5.1}$$

In this model $\mathscr{Y}_m$ is the Hamiltonian signature matrix of order $2m$:

---

31) An operator A ($\in \mathfrak{R}$) is *S-unitarily equivalent* to an operator A' acting in $L_2^{(m)}(0,l)$, if $UAU^{-1} = A'$, where U is a unitary mapping of $\mathfrak{H}$ onto $L_2^{(m)}$ for which $USf = \overline{Uf}$ (the bar indicates complex conjugation).

$$\mathcal{Y}_m = \begin{pmatrix} 0 & I_m \\ -I_m & 0 \end{pmatrix},$$

and $\mathscr{A}(t)$ $(0 \leqq t \leqq l)$ is a Hamiltonian (a real H-matrix) of positive type. Without loss of generality we may suppose that this Hamiltonian is normalized:

$$\text{sp}\,\mathscr{A}(t) = 1 \qquad (0 \leqq t \leqq l). \tag{5.2}$$

In §5 of Chapter V (see also §4, Chapter V) we also clarified that the multiplicative integral

$$\mathscr{U}(\mu) = \overset{l}{\underset{0}{\overset{\curvearrowleft}{\int}}} \exp(-\mu \mathcal{Y}_m \mathscr{A}(t)\,dt) \tag{5.3}$$

will be a monodromy matrix of the operator A (and hence $\mathscr{U}(1/\lambda)$ will be its characteristic matrix-function).

The representation of an entire matrix-function $\mathscr{U}(\mu)$ in the form of a multiplicative integral (5.3) will be called a *real canonical representation* (normalized, if the condition (5.2) is satisfied).

A function $\mathscr{U}(\mu)$ which has such a representation is obviously a real entire function, i.e. admits an everywhere convergent expansion

$$\mathscr{U}(\mu) = I + \mu C_1 + \mu^2 C_2 + \cdots$$

with real matrix coefficients C_k $(k = 1, 2, \cdots)$.

The theorem of V. P. Potapov (Theorem 4.1) can be supplemented by the following assertion: a real entire matrix-function $\mathscr{U}(\mu)$ which satisfies the hypotheses of that theorem for $\mathcal{Y} = \mathcal{Y}_m$, admits a real canonical representation.

Thus we can formulate the following result.

THEOREM 5.1. *In order that a matrix-function* $\mathscr{U}(\mu) = \| u_{jk}(\mu) \|_1^{2m}$ *of a complex variable admit a real canonical representation, it is necessary and sufficient that it be a real entire function and that it satisfy the following conditions*:

1) $\mathscr{U}(0) = I_{2m}$, *and* $\mathcal{Y}_m \mathscr{U}'(0)$ *is positive definite*;

2) $\mathscr{U}(\mu)$ *is a symplectic matrix,*[32] i.e.

$$\mathscr{U}^{\tau}(\mu)\, \mathcal{Y}_m \mathscr{U}(\mu) = \mathcal{Y}_m; \tag{5.4}$$

3) *for* $\operatorname{Im}\mu > 0$ *the matrix*

[32] We recall that $\mathscr{U}^{\tau}$ denotes the transpose of $\mathscr{U}$.

$$i(\mathscr{U}^*(\mu)\mathscr{Y}_m\mathscr{U}(\mu) - \mathscr{Y}_m) \tag{5.5}$$

is Hermitian-positive.

Let us clarify that since, for real μ, $\mathscr{U}(\mu)$ is a real matrix, for real μ the condition (5.4) coincides with the condition of the $\mathscr{Y}_m$-unitarity of $\mathscr{U}(\mu)$ (i.e. condition 2) of Theorem 4.1). On the other hand, if (5.4) holds for real μ, then by virtue of the analyticity of $\mathscr{U}(\mu)$ it will hold for all complex μ. By (5.4),

$$\det(\mathscr{U}^{\tau}(\mu)\mathscr{U}(\mu)) = (\det\mathscr{U}(\mu))^2 = 1.$$

Recalling that $\mathscr{U}(0) = I_{2m}$, we conclude that

$$\det\mathscr{U}(\mu) = 1. \tag{5.6}$$

For the case being considered, Theorem 4.2 can similarly be supplemented (compare with M. S. Brodskiĭ [8]).

THEOREM 5.2 *Suppose that the entire matrix-function* $\mathscr{U}(\mu) = \|u_{jk}(\mu)\|_1^{2m}$ *admits a real normalized canonical representation* (5.3) *with a Hamiltonian* $\mathscr{A}(t)$ $(0 \leq t \leq l)$. *In order that this Hamiltonian be the unique one*[33] *yielding the required representation of* $\mathscr{U}(\mu)$, *it is necessary and sufficient that the simple part of the model operator* (5.1) *corresponding to* $\mathscr{A}(t)$ $(0 \leq t \leq l)$ *be a real-unicellular operator in* $L_2^{(2m)}(0,1)$.

Let us clarify that in speaking of the real-unicellularity of an operator in $L_2^{(2m)}(0,1)$ we have in mind its S_0-unicellularity with respect to the involution S_0 of ordinary complex conjugation: $S_0 f = \bar{f}$.

3. In the language of the theory of differential equations, the relation (5.3) says that $\mathscr{U}(\mu)$ is the monodromy matrix of the Hamiltonian equation

$$\mathscr{Y}_m dx/dt - \mu\mathscr{A}(t)x - 0 \tag{5.7}$$

of phase dimension $2m$.

An important special case of equation (5.7) is that in which the Hamiltonian $\mathscr{A}(t)$ is quasi-diagonal, i.e.

$$\mathscr{A}(t) = \begin{pmatrix} \mathscr{A}_1(t) & 0 \\ 0 & \mathscr{A}_2(t) \end{pmatrix}, \tag{5.8}$$

---

[33] Up to its values on a set of measure zero.

where $\mathscr{H}_1(t)$ and $\mathscr{H}_2(t)$ $(0 \leqq t \leqq l)$ are real H-matrices of positive type.

Putting $x = y \dot{+} z$, where y and z are m-dimensional vector-functions, we can write (5.7) in the form of a system:

$$dy/dt = -\mu\mathscr{H}_2(t)z, \qquad dz/dt = \mu\mathscr{H}_1(t)y. \tag{5.9}$$

In particular, the second-order equation

$$d^2y/dt^2 + \mu^2\mathscr{P}(t)y = 0 \tag{5.10}$$

can be transformed (see Chapter VI, §5.1) to such a system. In this case $\mathscr{H}_2(t) = I_m$, $\mathscr{H}_1(t) = \mathscr{P}(t)$.

We further recall that the differential equation of an arbitrary weighted string (see Chapter VI, §8) reduces to a system of the form (5.9) (with $m = 1$).

In view of all this, the following result is of interest.

THEOREM 5.3. *In order that a matrix-function $\mathscr{U}(\mu)$ of a complex variable μ admit a real canonical representation* (5.3) *with a quasi-diagonal Hamiltonian $\mathscr{H}(t)$ $(0 \leqq t \leqq l)$, it is necessary and sufficient that it be a real entire function, that it satisfy conditions* 1), 2) *and* 3) *of Theorem* 5.1, *and that when it is written in the form of four square blocks*

$$\mathscr{U}(\mu) = \begin{pmatrix} \mathscr{U}_{11}(\mu) & \mathscr{U}_{12}(\mu) \\ \mathscr{U}_{21}(\mu) & \mathscr{U}_{22}(\mu) \end{pmatrix} \tag{5.11}$$

the matrix-functions $\mathscr{U}_{11}(\mu)$ and $\mathscr{U}_{22}(\mu)$ be even, and the matrix-functions $\mathscr{U}_{12}(\mu)$ and $\mathscr{U}_{21}(\mu)$ be odd.

Since for the general case of the representation (5.3) there is no uniqueness theorem, Theorem 5.3 cannot be obtained from Theorem 5.1; it has to be deduced independently (of course, this remark also applies to the relation of Theorem 5.1 to Theorem 4.1).

In various problems of mathematical physics and the theory of spectral functions of differential equations (see M. G. Kreĭn [**9, 23**] one encounters differential equations of the form

$$\mathscr{Y}_m dy/dt - \mathscr{G}(t)y - \mu y = 0 \qquad (0 \leqq t \leqq l), \tag{5.12}$$

where $\mathscr{G}(t)$ is a real symmetric matrix-function with elements from $L_1(0,l)$.[34]

---

[34] In the papers [**9, 23**] by Kreĭn the coefficient $\mathscr{G}(t)$ also satisfies the anticommutation condition $\mathscr{Y}_m\mathscr{G}(t) = -\mathscr{G}(t)\mathscr{Y}_m$. This can always be accomplished by transforming equation (5.12) by means of the substitution $y \to \mathscr{V}(t)y$, where $\mathscr{V}(t)$ is the unitary matrizant orthogonal to $\mathscr{Y}_m$: $d\mathscr{V}(t)/dt = -\frac{1}{2}(\mathscr{Y}_m\mathscr{G} + \mathscr{G}\mathscr{Y}_m)\mathscr{V}$, $\mathscr{V}(0) = I_{2m}$.

In §8, Chapter VII we showed that under certain conditions on the Hamiltonian $\mathscr{H}(t)$ equation (5.7) can be transformed to equation (5.12). If one starts with equation (5.12), it can be transformed to equation (5.7) in the following way.

We denote by $\mathscr{S}(t)$ the matrizant of equation (5.12) for $\mu = 0$, i.e. the solution of the matrix differential equation

$$\mathscr{Y}_m d\mathscr{S}/dt - \mathscr{G}(t)\mathscr{S} = 0, \tag{5.13}$$

singled out by the initial condition $\mathscr{S}(0) = I_{2m}$.

The substitution

$$y = \mathscr{S}(t)x \tag{5.14}$$

transforms equation (5.12) to equation (5.7), where

$$\mathscr{H}(t) = \mathscr{Y}_m \mathscr{S}^{-1}(t)\mathscr{Y}_m^{-1}\mathscr{S}(t) = \mathscr{S}^{\tau}(t)\mathscr{S}(t) \qquad (0 \leqq t \leqq l). \tag{5.15}$$

We have used here the fact that the solution $\mathscr{S}(t)$ of the Cauchy problem for the Hamiltonian equation (5.7) is always a symplectic matrix-function: $\mathscr{S}^{\tau}(t)\mathscr{Y}_m\mathscr{S}(t) = \mathscr{Y}_m$ $(0 \leqq t \leqq l)$ (see §1, Chapter VI).

Thus equation (5.12) can be transformed to a Hamiltonian equation (5.7) with a Hamiltonian $\mathscr{H}(t)$ of positive type, and having the additional properties:

1) *the elements of* $\mathscr{H}(t)$ *are absolutely continuous, and,* above all,
2) *the Hamiltonian* $\mathscr{H}(t)$ *is a symplectic matrix-function, i.e.*

$$\mathscr{H}(t)\mathscr{Y}_m\mathscr{H}(t) = \mathscr{Y}_m \quad (0 \leqq t \leqq l); \tag{5.16}$$

from which follows $\det \mathscr{H}(t) = 1$, so that the Hamiltonian $\mathscr{H}(t)$ is positive-definite.

We shall show that the preceding computations are reversible.

It is not hard to see that for a positive-definite real matrix $\mathscr{H}$ of order $2m$ the following three assertions are equivalent: a) the matrix $\mathscr{H}$ is symplectic, b) the spectrum of the matrix $\mathscr{Y}_m\mathscr{H}$ consists of the two points $\pm i$, and c) the matrix $\mathscr{H}$ can be represented in the form $\mathscr{H} = \mathscr{S}^{\tau}\mathscr{S}$, where $\mathscr{S}$ is a real symplectic matrix.

If now $\mathscr{H} = \mathscr{H}(t)$ $(0 \leqq t \leqq l)$ is a matrix-function with certain smoothness properties (if its elements are absolutely continuous or, say, p times continuously differentiable, etc.) and for each $t \in [0, l]$ one of the conditions a), b), c) is fulfilled, then there exists a representation (5.15) of the Hamiltonian $\mathcal{A}(t)$ in which the real symplectic matrix-function $\mathcal{S}(t)$ $(0 \leq t \leq l)$ will have the same smoothness properties as the Hamiltonian itself. The substitution $x = \mathcal{S}^{-1}(t)y$, effected with this matrix-function $\mathcal{S}(t)$, transforms the Hamiltonian equation (5.7) to equation (5.12).

Thus, properties 1) and 2) above of a Hamiltonian $\mathcal{A}(t)$ are characteristic for Hamiltonian equations (5.7) which can be transformed to equation (5.12) with a real symmetric matrix coefficient $\mathcal{G}(t)$, whose elements belong to the space $L_1(0,l)$.

It turns out that if the Hamiltonian $\mathcal{A}(t)$ *in a Hamiltonian equation is symplectic, then it can be uniquely reconstructed from the monodromy matrix* $\mathcal{U}(\mu)$ *of this equation.*[35)]

However, a stronger result holds. In order to formulate it in the language of the theory of multiplicative integrals, we introduce the following definition.

A real canonical representation (5.3) of a function $\mathcal{U}(\mu)$ will be called *totally symplectic* if the Hamiltonian $\mathcal{A}(t)$ is *symplectic* for almost all $t \in [0,l]$.

We will call this representation *densely symplectic* if there is a nowhere dense set in the interval $[0,l]$ such that the Hamiltonian $\mathcal{A}(t)$ is almost everywhere symplectic on its complement.

THEOREM 5.3. *If the entire matrix-function* $\mathcal{U}(\mu) = \|u_{jk}(\mu)\|_1^{2m}$ *admits a dense symplectic representation* (5.3), *then this representation is unique.*

Moreover, in this case $\mathcal{U}(\mu)$ *admits a unique normalized representation* (5.3).

This result was obtained rather long ago by M. G. Kreĭn (for $m = 1$, in 1951) as a corollary of his investigations on entire Hermitian operators[36)] and the problem of extending Hermitian-positive functions and matrix-functions, and also on inverse problems of the theory of spectral functions of differential operators (see his papers [**12–23**]). Theorem 5.3 is given here for the first time.

---

35) In particular, from the mondromy matrix of equation (5.12) one can always reconstruct uniquely the real symmetric matrix-function $\mathcal{G}(t)$ which anticommutes with $\mathcal{Y}_{lm}$ (compare with the results in the papers by L. A. Sahnovic [**1, 2**]).

36) Unfortunately, we cannot dwell here even on the definition of an *entire* operator.

4. For the case $m = 1$ we can formulate a more definitive result. In the fourth memoir of his outstanding studies in the theory of Hilbert spaces of entire functions, L. de Branges[37] (see [**5**], Theorem VI) established, in somewhat different form, the following result.

THEOREM 5.4. *If the entire matrix-function* $\mathscr{U}(\mu) = \|u_{jk}(\mu)\|_1^2$ *of second order admits a real canoncial representation*

$$\mathscr{U}(\mu) = \overset{l}{\overset{\frown}{\int_0}} \exp(-\mu \mathscr{Y}_1 \mathscr{H}(t)\,dt), \tag{5.17}$$

then it admits a unique representation of this kind with a normalized Hamiltonian.

We shall write the second-order Hamiltonian $\mathscr{H}(t)$ in question in the same form as in §§5 and 6 of Chapter VI:

$$\mathscr{H}(t) = \begin{pmatrix} a(t) & b(t) \\ b(t) & c(t) \end{pmatrix}. \tag{5.18}$$

That $\mathscr{H}(t)$ is a Hamiltonian of positive type means that the real functions $a(t)$, $b(t)$ and $c(t)$ (from $L_1(0,l)$) satisfy the conditions

$$\begin{gathered} a(t) \geqq 0, \quad c(t) \geqq 0, \quad a(t)c(t) - b^2(t) \geqq 0; \\ \int_0^l a(t)\,dt \int_0^l c(t)\,dt - \left(\int_0^l b(t)\,dt\right)^2 > 0. \end{gathered} \tag{5.19}$$

Normalization means that

$$a(t) + c(t) = 1 \quad (0 \leqq t \leqq l).$$

It is elementary to verify the relation

$$\mathscr{H}(t)\mathscr{Y}_1\mathscr{H}(t) = \det \mathscr{H}(t)\mathscr{Y}_1. \tag{5.20}$$

Thus, for the case being considered, that $\mathscr{H}(t)$ is symplectic is equivalent to its being unimodular: $\det \mathscr{H}(t) = 1$.

Therefore, if the condition

α) $\det \mathscr{H}(t)$ $(\equiv a(t)c(t) - b^2(t)) \neq 0$ *almost everywhere in* $[0, l]$ is satisfied, then the representation (5.17) can be transformed to a totally symplectic representation by means of a simple change of parameter.

---

[37] We mention that, apparently, the work of M. G. Kreĭn on the theory of entire Hermitian operators and more generally the work in his entire series [**12-24**], to which de Branges' studies have direct relationship, remained unknown to de Branges (who in his investigations repeated many results from this series).

In fact, the substitution

$$s = \int_0^t \sqrt{\det \mathscr{A}(t)}\, dt \tag{5.21}$$

in the representation (5.17) yields

$$\mathscr{U}(\mu) = \overset{\hat{l}}{\underset{0}{\overset{\curvearrowleft}{\int}}} \exp(-\mu \mathscr{Y}_1 \hat{\mathscr{A}}(s)\, ds),$$

where

$$\hat{\mathscr{A}}(s) = \mathscr{A}(t)/\sqrt{\det \mathscr{A}(t)}$$

$$\left(0 \leqq s \leqq \hat{l};\ \ \hat{l} = \int_0^l \sqrt{\det \mathscr{A}(t)}\, dt\right)$$

Obviously $\det \hat{\mathscr{A}}(s) = 1 \quad (0 \leqq s \leqq \hat{l})$.

It is now not difficult to see that the representation (5.17) can be transformed to a dense symplectic representation if (and only if) the following condition is satisfied.

β) *The interval* $[0, l]$ *contains no interval* Δ *on which* $\det \mathscr{A}(t) = 0$ *almost everywhere.*

In this case the substitution (5.21) should be replaced by the substitution

$$s = \int_0^t p(\tau)\, d\tau \quad (0 \leqq t \leqq l),$$

where $p(t)$ is any positive function from $L_1(0, l)$ which satisfies the condition

$$p(t) = (\det \mathscr{A}(t))^{1/2} \quad \text{if} \quad \det \mathscr{A}(t) \neq 0.$$

5. Comparison of Theorems 5.2 and 5.4 leads to the following result.[38)]

THEOREM 5.5. *Every simple S-real Volterra operator A with a two-dimensional skewsymmetric component is S-unicellular.*

In fact, such an operator is S-unitarily equivalent to the simple part of the real model operator A_M,

$$(A_M f)(t) = \mathscr{A}^{1/2}(t)\, \mathscr{Y}_1 \int_t^l \mathscr{A}^{1/2}(s) f(s)\, ds. \tag{5.22}$$

---

38) This result, as a conjecture by M. G. Kreĭn, was stated in the joint report by M. S. Brodskiĭ, I. C. Gohberg, M. G. Kreĭn and V. I. Macaev [1], and also in a paper by M. S. Brodskiĭ [8]. M. G. Kreĭn was led to this conjecture several years earlier by comparing his earlier conjecture (Theorem 5.4) with the investigations of M. S. Livšic and M. S. Brodskiĭ. L. de Branges [7] also noted that Theorem 5.5 can be obtained on the basis of Theorems 5.2 and 5.4.

Here $\mathscr{A}(t)$ $(0 \leqq t \leqq l)$ is a normalized Hamiltonian of positive type of order 2 ($l = |A_\alpha|_1$; see p. 243).

On the other hand, on the basis of Theorems 5.2 and 5.3 the simple part of A_M is real-unicellular.

Theorem 5.5 implies the following corollary.

COROLLARY 5.1. *A real model operator* A_M *in* $L_2^{(2)}(0,l)$ *is real-unicellular if and only if*

$$\det \mathscr{A}(s) \neq 0 \quad \textit{almost everywhere in} \quad [0,l]. \tag{5.23}$$

In fact, it remains for us to show that (5.23) is a necessary and sufficient condition for A_M to be a simple operator.[39] If $\det \mathscr{A}(t) = 0$ on some set $\mathscr{E}$ of positive measure, then there exists a vector-function $f_0(t)$ $(\in L_2^{(2)}(0,l))$, different from zero on $\mathscr{E}$, such that $\mathscr{A}^{1/2}(t) f_0(t) = 0$ everywhere. Then it follows from (5.23) and from the expression for the adjoint,

$$(A_M^* f)(t) = -\mathscr{A}^{1/2}(t)\mathscr{Y}_1 \int_0^t \mathscr{A}^{1/2}(s) f(s)\, ds,$$

that $A_M^* f_0 = A_M f_0 = 0$, while $f_0 \neq 0$:

$$(f_0, f_0) = \int_0^l f_0^*(t) f_0(t)\, dt = \int_{\mathscr{E}} f_0^*(t) f_0(t)\, dt > 0 \quad (\operatorname{mes}\mathscr{E} > 0).$$

Thus A_M is not a simple operator.

It is easy to verify that the condition (5.23) is also sufficient for $A_M f = A_M^* f = 0$ to imply $f = 0$.

According to Corollary 5.1, the model operator with $\mathscr{A}(t) \equiv I_2$, i.e. the operator defined by

$$(J_{(2)} f)(t) = \mathscr{Y}_1 \int_t^l f(s)\, ds \quad (f \in L_2^{(2)}(0,l)),$$

will be real-unicellular. From this example we can see that *an operator which is not complex-unicellular can turn out to be real-unicellular.*

In fact, the operator $J_{(2)}$ has as invariant subspaces $\mathscr{L}_+$ and $\mathscr{L}_-$, consisting respectively of all vector-functions of the form

$$\{f(t), if(t)\} \in L_2^{(2)}(0,l) \quad \text{and} \quad \{f(t), -if(t)\} \in L_2^{(2)}(0,l);$$

at the same time, these subspaces are mutually orthogonal.

Thus the operator $J_{(2)}$ is not complex-unicellular.

---

[39] This assertion is of a general nature: it concerns all model operators A_M. The trivial subspace of the operator A_M (the set of common zeros of A_M and A_M^*) consists of all $f \in L_2^{(n)}(0,l)$ for which $\mathscr{A}(t) f(t) = 0$ almost everywhere.

6. Theorem 4.1 for $n=2$ yields necessary and sufficient conditions for a given matrix-function $\mathscr{U}(\mu)=\|u_{jk}(\mu)\|_1^2$ to have a representation (5.20). These conditions can be formulated in a new interesting form, which we will obtain with the help of the following elementary lemma.

LEMMA 5.1.[40] *In order that the fractional-linear function*

$$w=(\alpha z+\beta)(\gamma z+\delta)^{-1}, \tag{5.24}$$

corresponding to the complex unimodular matrix

$$\mathscr{A}=\begin{pmatrix}\alpha & \beta\\ \gamma & \delta\end{pmatrix} \qquad (\alpha\delta-\beta\gamma=1),$$

map the upper halfplane Π_+ ($\operatorname{Im} z\geqq 0$) *onto a disc lying inside* Π_+, *it is necessary and sufficient that the matrix*

$$i(\mathscr{A}^*\mathscr{Y}_1\mathscr{A}-\mathscr{Y}_1) \tag{5.25}$$

be Hermitian-positive.

For the linear transformation

$$w_1=\alpha z_1+\beta z_2, \quad w_2=\gamma z_1+\delta z_2$$

the Hermitian-positivity of the matrix (5.25) means that for any z_1 and z_2, not simultaneously zero,

$$\frac{1}{i}(w_1\bar w_2-w_2\bar w_1)>\frac{1}{i}(z_1\bar z_2-z_2\bar z_1).$$

Hence for $z_1=1$, $z_2=z$ ($\operatorname{Im} z<0$) it follows that $w_2\neq 0$ and

$$\operatorname{Im}(w_1/w_2)>\operatorname{Im} z|w_2|^2,$$

which proves the sufficiency of the conditions of the lemma.

We note that the unimodularity of the matrix $\mathscr{A}$ was not used anywhere. However it is essential for the proof of the necessity of the conditions. We leave this proof to the reader (it is somewhat complicated, but still elementary).

We note that if the fractional-linear transformation (5.24) maps Π_+ onto a disc lying inside Π_+, then the same will be true of the fractional-linear transformation obtained from (5.24) by interchanging α and δ or by replacing β and γ by $-\gamma$ and $-\beta$.

We now introduce the following definition.

---

[40] This lemma was recently generalized by M. G. Kreĭn and Ju. L. Šmul′jan [1] to the case in which w, z, α, β, γ and δ are matrices of arbitrary order (or even bounded linear operators, acting in a Hilbert space).

We will call an entire matrix-function

$$\mathcal{U}(\mu) = \begin{pmatrix} u_{11}(\mu) & u_{12}(\mu) \\ u_{21}(\mu) & u_{22}(\mu) \end{pmatrix}$$

an *entire* ρ-*matrix*[41)] if its elements $u_{jk}(\mu)$ $(j, k = 1, 2)$ are real entire functions satisfying the following conditions:

a) $\mathcal{U}(0) = I_2$;

b) $\det\mathcal{U}(\mu) = 1$;

c) *for any* μ *from the upper halfplane* $(\operatorname{Im}\mu > 0)$ *the fractional-linear transformation*

$$w = \frac{u_{11}(\mu)z + u_{12}(\mu)}{u_{21}(\mu)z + u_{22}(\mu)}$$

maps the upper halfplane Π_+ *onto a disc lying inside* Π_+.

We note that an entire ρ-matrix-function remains such if we interchange the elements u_{11} and u_{22} or replace u_{12} and u_{21} by $-u_{21}$ and $-u_{12}$. We can now state the following result.

THEOREM 5.6. *In order that a matrix-function* $\mathcal{U}(\mu)$ *of order two admit a real canonical representation* (5.17), *it is necessary and sufficient that it be an entire* ρ-*matrix*.

This theorem is a corollary of Theorem 5.1 and Lemma 5.1. To show this, it remains to verify that conditions a), b) and c) are equivalent to conditions 1), 2) and 3) from Theorem 5.1 for $m = 1$. Conditions a) and 1) simply coincide. Since for $m = 1$

$$\mathcal{U}^{\tau}(\mu)\mathcal{Y}_1\mathcal{U}(\mu) = \det\mathcal{U}(\mu)\mathcal{Y}_1,$$

condition b) is equivalent to condition 2).

Finally, on the basis of Lemma 5.1 condition c) is equivalent to condition 3).

Recalling the connection between the equation of a string and a canonical equation (Chapter VI, §8), it is natural to conjecture that Theorems 5.5 and 5.6 should imply various results concerning inverse

---

41) In M. G. Kreĭn's paper [19] these matrix-functions were called *special* matrix-functions. Various special cases of ρ-matrix-functions have been studied by Stieltjes and Hamburger in moment problems and by H. Weyl in the theory of singular differential equations (see Ju. M. Berezanskiĭ [1] and N. I. Ahiezer [2]; in [2] they are called Nevanlinna matrices). To every entire Hermitian operator there corresponds an entire ρ-matrix (M. G. Kreĭn [13]) with the property $u_{11}(ir)/u_{21}(ir) = O(1/r)$ for $r \to +\infty$, and conversely. In all of these investigations, entire ρ-matrices are used to describe the spectral functions and resolvents of certain Hermitian operators with deficiency indices (1, 1). This circumstance clarifies the nomenclature "ρ-matrix".

problems for a string. In particular, from these theorems one can obtain a theorem stating that the mass distribution along a string with a given tension is completely determined by two of its frequency spectra: the spectrum obtained when both ends are fixed, and the spectrum obtained when one end is fixed and the other is free (M. G. Kreĭn **[3, 18, 24]**).

§6. Examples of unicellular and nonunicellular operators

1. The familiar operators from analysis J_α $(0 < \alpha < \infty)$ of fractional integration are instructive examples of unicellular operators; they are defined in the space $L_2(0,1)$ by

$$(J_\alpha f)(t) = \frac{1}{\Gamma(\alpha)} \int_0^t (t-s)^{\alpha-1} f(s)\, ds.$$

The operator J_α for $\alpha = 1$ is related to the operator

$$(Jf)(t) = 2i \int_t^1 f(s)\, ds,$$

considered in §§8 and 9 of Chapter I, by

$$J_1 = (i/2) J^*.$$

Since by Theorem I.9.1 the operator J is unicellular, it follows that J_1 is unicellular.

We have the following more complete result.

THEOREM 6.1. *The operators* J_α $(0 < \alpha < \infty)$ *are unicellular Volterra operators which are pairwise nonsimilar.*

We shall give below a purely analytic proof of the unicellularity of the operators J_α, based on a well-known theorem of Titchmarsh (see Theorem I.9.3).[42)]

We will precede this proof by a simple lemma. To simplify its statement, we associate with every function $f \in L_2(0,1)$ a number l_f $(\geqq 0)$, defined by

$$\int_0^{l_f} |f(t)|^2 dt = 0, \qquad \int_{l_f}^{l_f+\epsilon} |f(t)|^2 dt > 0 \quad (\epsilon > 0).$$

---

42) In §9, Chapter I it was shown that Titchmarsh's theorem is a consequence of Theorem I.9.1 on the unicellularity of the operator J. Now, in particular, it will be proved that conversely Titchmarsh's theorem implies Theorem I.9.1 (thus, these theorems are equivalent).

LEMMA 6.1. *Let $\mathfrak{L}$ be a closed subspace of $L_2(0,1)$. If for every $f \in \mathfrak{L}$ we have $l_f > 0$, then*

$$\inf_{f \in \mathfrak{L}} l_f > 0.$$

PROOF. Let us assume the contrary; then we can choose a sequence $f_n \in \mathfrak{L}$ $(n = 1, 2, \cdots)$ for which the sequence $l_n = l_{f_n}$ $(n = 1, 2, \cdots)$ is decreasing and tends to zero. Obviously

$$\int_{l_{n+1}}^{l_n} | f_{n+1}(t) |^2 dt > 0 \qquad (n = 1, 2, \cdots).$$

Let us consider the sequence of functions $\psi_n = b_n f_n$ $(n = 1, 2, \cdots)$, where $b_1 = 1$, and the further b_n are successively defined so that

$$\int_0^1 | \psi_{n+1}(t) |^2 dt < \frac{1}{3^2} \int_{l_n}^{l_{n-1}} | \psi_n(t) |^2 dt \qquad (n = 1, 2, \cdots).$$

For the functions ψ_n thus defined we obviously have

$$\int_0^1 | \psi_{n+k}(t) |^2 dt \leqq \frac{1}{3^{2k}} \int_{l_n}^{l_{n-1}} | \psi_n(t) |^2 dt \qquad (n = 1, 2, \cdots).$$

Form the function

$$\psi(t) = \sum_{n=1}^{\infty} \psi_n(t).$$

Obviously $\psi \in \mathfrak{L}$, and moreover

$$\begin{aligned} &\left(\int_{l_{n+1}}^{l_n} | \psi(t) |^2 dt \right)^{1/2} \\ &\qquad \geqq \left(\int_{l_{n+1}}^{l_n} | \psi_{n+1}(t) |^2 dt \right)^{1/2} - \sum_{j=n+2}^{\infty} \left(\int_{l_{n+1}}^{l_n} | \psi_j(t) |^2 dt \right)^{1/2} \\ &\qquad \geqq \left(\int_{l_{n+1}}^{l_n} | \psi(t) |^2 dt \right)^{1/2} - \sum_{j=n+2}^{\infty} \left(\int_0^1 | \psi_{n+1}(t) |^2 dt \right)^{1/2}. \end{aligned}$$

Hence

$$\int_{l_{n+1}}^{l_n} | \psi(t) |^2 dt \geqq \frac{1}{2} \int_{l_{n+1}}^{l_n} | \psi_{n+1}(t) |^2 dt > 0 \qquad (n = 1, 2, \cdots).$$

This shows that $l_\psi = 0$, which contradicts the hypothesis of the lemma. The lemma is proved.

PROOF OF THEOREM 6.1. We will first prove that the Volterra operator J_α $(0 < \alpha < \infty)$ is unicellular. To do this we shall show that a vector $f \in L_2(0,1)$ will be a generating vector of the operator J_α[43] if and only if $l_f = 0$.

In fact, it is obvious that $l_{J_\alpha f} \geqq l_f$, and so for $l_f > 0$ the vector f cannot be a generating vector.

Now let $l_f = 0$. We shall show that any vector $g \in L_2(0,1)$ which is orthogonal to all of the vectors $(J_\alpha)^n f$ $(n = 0, 1, \cdots)$ is equal to zero. Indeed, for such a vector

$$\int_0^1 \overline{g(t)} \int_0^t (t-s)^{n\alpha - 1} f(s)\, ds dt = 0 \quad (n = 1, 2, \cdots).$$

Replacing $t - s$ by s, we obtain

$$\int_0^1 \overline{g(t)} \int_0^t s^{n\alpha - 1} f(t-s)\, ds dt = 0 \quad (n = 1, 2, \cdots)$$

or

$$\int_0^1 \left[\int_s^1 \overline{g(t)} f(t-s)\, dt \right] s^{n\alpha - 1} ds = 0 \quad (n = 1, 2, \cdots).$$

Let n_0 be a positive integer for which $n_0\alpha - 1 > 0$. Denoting the function

$$\left[\int_s^1 \overline{g(t)} f(t-s)\, dt \right] s^{n_0\alpha - 1}$$

by $h(s)$, we obtain

$$\int_0^1 h(s) s^{n\alpha} ds = 0 \quad (n = 1, 2, \cdots). \tag{6.1}$$

Since the system of functions $\{t^{n\alpha}\}_{n=1}^{\infty}$ is complete in $L_2(0,1)$, it follows from (6.1) that $h(t) \equiv 0$, and so

$$\int_t^1 \overline{g(s)} f(s-t)\, ds = 0 \quad (0 \leqq t \leqq 1),$$

or

$$\int_0^t \overline{g(1-s)} f(t-s)\, ds = 0 \quad (0 \leqq t \leqq 1).$$

---

[43] That is, the linear hull of the system of vectors $(J_\alpha)^n f$ $(n = 0, 1, \cdots)$ is dense in $\mathfrak{H}$.

Since $l_f = 0$, it follows from Titchmarsh's theorem (Theorem I.9.3) that $g(t) \equiv 0$ $(0 \leqq t \leqq 1)$. Thus if $l_f = 0$, then f is a generating vector.

Let us now denote by $\mathfrak{H}_l$ $(0 \leqq l \leqq 1)$ the subspace of $L_2(0,1)$, consisting of all functions $f \in L_2(0,1)$ for which $l_f \geqq l$. It is easily seen that each of the subspaces $\mathfrak{H}_l$ $(0 < l < 1)$ is invariant with respect to the operators J_α $(0 < \alpha < \infty)$.

We shall show that any proper invariant subspace $\mathfrak{L}$ of an operator J_α coincides with one of the subspaces $\mathfrak{H}_l$ $(0 < l < 1)$. In fact, no vector f from $\mathfrak{L}$ is a generating vector for an operator J_α; hence $l_f > 0$. By Lemma 6.1

$$l = \min_{f \in \mathfrak{L}} l_f > 0,$$

so that $\mathfrak{L} \subseteq \mathfrak{H}_l$. Let us denote by $f_0 \in \mathfrak{L}$ a vector for which $l_{f_0} = l$.[44] From what was proved above, it is easy to deduce that the sequence $f_1, J_\alpha f_1, \cdots, J_\alpha^n f_1, \cdots$, where $f_1(t) = f_0(t+l)$, is complete in the space $L_2(0, 1-l)$.

It follows at once that the linear hull of the sequence $\{J_\alpha^n f_0\}_{n=0}^\infty$ is dense in $\mathfrak{H}_l$, and so $\mathfrak{L} = \mathfrak{H}_l$.

Thus, all the operators J_α $(0 < \alpha < \infty)$ are unicellular.

To complete the proof it remains to show that the operators J_α are pairwise nonsimilar. We make a general remark: if two operators A_1 and A_2 $(\in \mathfrak{S}_\infty)$ are similar, then

$$\sup_n \frac{s_n(A_1)}{s_n(A_2)} < \infty \quad \text{and} \quad \sup_n \frac{s_n(A_2)}{s_n(A_1)} < \infty.$$

Indeed, if $A_2 = BA_1B^{-1}$, and hence $A_1 = B^{-1}A_2B$, then

$$s_n(A_2) \leqq |B|\,|B^{-1}|\,s_n(A_1) \quad (n = 1, 2, \cdots),$$
$$s_n(A_1) \leqq |B|\,|B^{-1}|\,s_n(A_2) \quad (n = 1, 2, \cdots).$$

Let us study the asymptotic behavior of the s-numbers of the operator J_α. The numbers $\sigma_n = s_n^{-2}(J_\alpha)$ $(n = 1, 2, \cdots)$ are the characteristic numbers of the integral equation $\phi = \sigma J_\alpha J_\alpha^* \phi$ which, written out in detail, has the form

$$\phi(t) = \frac{\sigma}{\Gamma^2(\alpha)} \int_0^t (t-u)^{\alpha-1} \int_u^1 (s-u)^{\alpha-1} \phi(s)\, ds\, du. \tag{6.2}$$

Let us consider the case of integer α. In this case equation (6.2) is equivalent to the boundary value problem

[44] It is easily seen that such a vector exists (see the proof of Lemma 6.1).

(6.3) $$\phi^{(2\alpha)} - \sigma\phi = 0, \quad \phi^{(k)}(0) = \phi^{(\alpha+k)}(1) = 0 \quad (k = 0, 1, \cdots, \alpha - 1).$$

The following boundary value problem is simpler:

(6.4) $$\psi^{(2\alpha)} - \tau\psi = 0, \quad \psi^{(k)}(0) = \psi^{(k)}(1) = 0 \quad (k = 0, 1, \cdots, \alpha - 1).$$

Its fundamental functions and characteristic numbers are known:

$$\psi_n(t) = C_n \sin \pi n t; \quad \tau_n = (n\pi)^{2\alpha} \quad (n = 1, 2, \cdots).$$

Since the changed boundary conditions in the boundary value problem (6.4) cannot change the asymptotic behavior of the eigenvalues (see Theorem VI.1.3), we can assert that $\sigma_n \sim (n\pi)^{2\alpha}$; hence

(6.5) $$s_n(J_\alpha) \sim 1/(n\pi)^\alpha.$$

We will omit the proof that the asymptotic relation (6.5) also holds for noninteger α. This relation, together with the general remark made earlier regarding the s-numbers of similar operators, shows that the operators J_α are pairwise nonsimilar.

The theorem is proved.

Theorem 6.1 makes possible the following deduction.

In contrast with a finite-dimensional space, where all unicellular operators are similar to each other, *in an infinite-dimensional Hilbert space there exists at least a continuum of unicellular Volterra operators which are pairwise nonsimilar.*

We also note that all the unicellular operators considered in the preceding sections, with the exception of §5, were dissipative Volterra operators with nuclear imaginary components, and consequently their order $p(A)$[45] was always equal to unity. It follows from (6.5) that $p(J_\alpha) = 1/\alpha$. Hence *there exist unicellular Volterra operators of any finite order.*

One can show (see the more general assertion in §6.2) that for $0 < \alpha \leqq 1$ the values of the quadratic form $(J_\alpha f, f)$ $(f \in \mathfrak{H})$ lie in the sector $|\arg \lambda| < \pi\alpha/2$. Thus *the operators* iJ_α $(0 < \alpha \leqq 1)$ *are dissipative Volterra operators of order* $1/\alpha$.

For $1 < \alpha < \infty$ the values of the quadratic form $(J_\alpha f, f)$ $(f \in \mathfrak{H})$ fill up the entire complex plane. The point is that if the order of a Volterra operator is < 1, then this operator cannot be dissipative (see Theorem III.2.4).

[45] We recall that the *order* $p(A)$ of an operator $A \in \mathfrak{S}_\infty$ is the infimum of all numbers p for which $A \in \mathfrak{S}_p$.

2. The unicellularity of the operators J_α for the narrower range $0 < \alpha < 1$ can be obtained as a corollary of Theorem I.9.1 and a general result. To formulate the latter we need the concept of the fractional powers of a dissipative operator (see V. I. Macaev and Ju. A. Palant [1], T. Kato [1], B. Sz.-Nagy and C. Foiaş [4]).

If A $(\in \mathfrak{R})$ is a dissipative operator, then the operator A^ν $(0 < \nu < 1)$ is defined by

$$A^\nu = -\frac{\sin \pi\nu}{\pi} e^{\pi i\nu/2} \int_0^\infty \lambda^{\nu-1} A(A + i\lambda I)^{-1} d\lambda, \tag{6.6}$$

where the integral converges in norm.[46] Putting $A^{n+\nu} = A^n A^\nu$ $(n = 0, 1, \cdots;$ $0 < \nu < 1)$, we obtain the semigroup A^α $(0 < \alpha < \infty)$ of *powers* of the operator A.

1. (JU. A. PALANT).[47] *If A is a dissipative unicellular Volterra operator, then all of its powers A^α $(0 < \alpha < 1)$ are also unicellular Volterra operators.*

The operator IJ_1 is a dissipative unicellular Volterra operator. Substituting $A = iJ_1$ in (6.6) enables us to deduce that $(iJ_1)^\alpha = \exp(\pi i\alpha/2) J_\alpha$ (i.e. $J_1^\alpha = J_\alpha$). Consequently, the operators J_α $(0 < \alpha < 1)$ are Volterra and unicellular.

3. A *formal* analog of a finite-dimensional unicellular operator in an infinite-dimensional space is the operator S, whose matrix relative to some orthogonal basis $\{e_j\}_0^\infty$ of the space $\mathfrak{H}$ has the form

$$\left\| \begin{matrix} 0 & 0 & 0 & \dots \\ 1 & 0 & 0 & \dots \\ 0 & 1 & 0 & \dots \\ \cdot & \cdot & \cdot & \cdot\ \cdot \end{matrix} \right\|$$

Obviously S is defined by

$$Se_j = e_{j+1} \quad (j = 0, 1, 2, \cdots),$$

---

[46] For such a definition of the powers A^ν $(0 < \nu \leqq 1)$ of a dissipative operator A, all the values of the quadratic form $(A_\nu f, f)$ $(f \in \mathfrak{H})$ lie in the sector $0 \leqq \arg \lambda \leqq \pi\nu$ (see V. I. Macaev and Ju. A. Palant [1]).

It turns out that if the operator A is dissipative and invertible, then the operator $B = A^{1/n}$, where n is any integer, is the unique operator for which

$$B^n = A \quad \text{and} \quad 0 \leqq \arg(Bf, f) \leqq \pi/n.$$

A generalization of this result to the case of a maximal dissipative unbounded operator was obtained by H. Langer [1].

[47] Oral communication.

and is therefore called the (*right*) *shift* operator.

However, S is not only not a Volterra operator; it is not even completely continuous. Indeed, S is isometric and maps the space $\mathfrak{H}$ onto the subspace orthogonal to the line $\{\lambda e_0\}$.

This operator has nothing in common with a unicellular operator. Besides the obvious invariant subspaces (the closed linear hulls of the vectors $\{e_j\}_n^\infty$, $n = 1, 2, \cdots$), it has a "vast" set of other invariant subspaces which form a complicated structure. Nevertheless, this set admits a complete description, which we shall now give.

It is more convenient to describe all the invariant subspaces of the operator S after realizing the space $\mathfrak{H}$ in the form of the Hardy space $\boldsymbol{H}^2$. The space $\boldsymbol{H}^2$ consists of all analytic functions

$$f(z) = \sum_{j=0}^{\infty} c_j(f) z^j$$

for which

$$\sum_{j=0}^{\infty} |c_j|^2 < \infty .$$

Every such function automatically turns out to be holomorphic in the unit disc $|z| < 1$. The scalar product in $\boldsymbol{H}^2$ is defined in a natural way by

$$(f, g) = \sum_{j=0}^{\infty} c_j(f) \overline{c_j(g)} .$$

To map the space $\mathfrak{H}$ unitarily onto the space $\boldsymbol{H}^2$, we associate with every vector

$$f = \sum_{j=0}^{\infty} f_j e_j$$

the function

$$f(z) = \sum_{j=0}^{\infty} f_j z^j .$$

Clearly, under this mapping the operator S goes over into the operator of multiplication by the independent variable z.

With every inner function[48] $m(z)$ we associate the subspace $\mathfrak{L}_m$ $(\subset \boldsymbol{H}^2)$, consisting of all functions of the form $m(z) f(z)$ $(f(z) \in \boldsymbol{H}^2)$.

We shall say that an inner function is normalized, if the first nonzero coefficient in its Maclaurin series expansion is positive.

---

[48] For the definition of an inner function, see footnote 18), above.

THEOREM 6.2 (A. BEURLING [1]). *The subspaces $\mathfrak{L}_m$, where $m(z)$ is an arbitrary inner function, exhaust all the invariant subspaces of the operator of multiplication by z in $\boldsymbol{H}^2$. To distinct inner functions there correspond distinct subspaces.*

It follows from Theorem 6.2 that the restriction of the (right) shift operator to any one of its invariant subspaces is again a (right) shift operator which is unitarily equivalent to the original one.

4. Let us consider the operator S_Λ, to which there corresponds in some orthonormal basis $\{e_j\}_0^\infty$ the matrix

$$\left\| \begin{matrix} 0 & 0 & 0 & \dots \\ \lambda_1 & 0 & 0 & \dots \\ 0 & \lambda_2 & 0 & \dots \\ . & . & . & . \end{matrix} \right\|, \tag{6.7}$$

where $\{\lambda_j\}$ is a sequence of complex numbers.

A finite $(n \times n)$ matrix of the form (6.7) generates a unicellular operator in the space E_n if and only if all of the λ_j are different from zero. This condition is necessary for the unicellularity of the operator S_Λ in the infinite-dimensional case. Indeed, if $\lambda_k = 0$ for some k, then S_Λ has two nonintersecting invariant subspaces: one having basis $\{e_j\}_{j=0}^k$, and the other with basis $\{e_j\}_{j=k+1}^\infty$.

If $\lim \lambda_j = 0$, then, as is easily seen, S_Λ is a Volterra operator. However, the two conditions

$$\lambda_j \neq 0 \ (j = 1, 2, \cdots) \quad \text{and} \quad \lim_{j \to \infty} \lambda_j = 0$$

taken together still turn out to be insufficient for the unicellularity of S_Λ. N. K. Nikol'skiĭ [1] has shown that the operator S_Λ with

$$\lambda_j = \begin{cases} 2^{-j} & \text{for } j \neq 2^p + 1, \ p = 1, 2, \cdots, \\ 2^{-2(p+1)^2} & \text{for } j = 2^p + 1, \ p = 1, 2, \cdots \end{cases}$$

is not unicellular. However we have the following result (which is contained in more general results by N. K. Nikol'skiĭ [1]).

2. (N. K. NIKOL'SKIĬ [1]). *Let $\Lambda = \{\lambda_j\}$ be an arbitrary nonincreasing sequence of positive numbers for which the series $\sum_{j=1}^\infty \lambda_j^p$ converges for some p $(0 < p < \infty)$. Then the operator S_Λ is unicellular.*

We note that the condition $\sum_{j=1}^\infty \lambda_j^p < \infty$ means that S_Λ belongs to the class $\mathfrak{S}_p$.

For the special case $\lambda_j = 2^{-j}$ $(j = 1, 2, \cdots)$ the unicellularity of the corresponding operator S_Λ was established earlier by W. F. Donoghue [1].

BIBLIOGRAPHY

N. I. AHIEZER

1. *Lectures in the theory of approximation,* 2nd rev. ed., "Nauka", Moscow, 1965; English transl., Ungar, New York, 1956. MR **20** # 1872; MR **32** # 6108.
2. *The classical moment problem and some related questions in analysis,* Fizmatgiz, Moscow, 1961; English transl., Hafner, New York, 1965. MR **27** # 4028; MR **32** # 1518.

N. I. AHIEZER AND I. M. GLAZMAN

1. *The theory of linear operators in Hilbert space,* GITTL, Moscow, 1950; 2nd rev. ed., "Nauka", Moscow, 1966; English transl., Ungar, New York, 1961. MR **13**, 358; MR **34** # 6527.

V. M. ALEKSANDROV

1. *To the solution of certain contact problems of elasticity theory,* Prikl. Mat. Meh. **27** (1963), 970-972 = J. Appl. Math. Mech. **27** (1963), 1490-1494.

N. ARONSZAJN AND K. T. SMITH

1. *Invariant subspaces of completely continuous operators,* Ann. of Math. (2) **60** (1954), 345-350. MR **16**, 488.

N. H. ARUTJUNJAN

1. *A planar contact problem of the theory of creep,* Prikl. Mat. Meh. **23** (1959), 901-924 = J. Appl. Math. Mech. **23** (1959), 1283-1313.

N. H. ARUTJUNJAN AND M. M. MANUKJAN

1. *A contact problem in the theory of creep taking into account frictional force,* Prikl. Mat. Meh. **27** (1963), 813-820 = J. Appl. Math. Mech. **27** (1963), 1244-1254.

F. V. ATKINSON

1. *Some aspects of Baxter's functional equation,* J. Math. Anal. Appl. **7** (1963), 1-30. MR **27** # 5135.
2. *Discrete and continuous boundary problems,* Academic Press, New York, 1964. MR **31** # 416.

N. K. BARI

1. *Trigonometric series,* Fizmatgiz, Moscow, 1961; English transl., Macmillan, New York and Pergamon Press, Oxford, 1964. MR **23** # A3411; MR **30** # 1347.

M. A. BARKAR′ AND I. C. GOHBERG

1. *Factorization of operators with respect to a discrete chain of projectors in a Banach space,* Mat. Issled. **1** (1966), no. 1, 32-54; English transl., Amer. Math. Soc. Transl. (2) **90** (1970) (to appear). MR **34** # 6529.
2. *Factorization of operators in a Banach space,* Mat. Issled. **1** (1966), no. 2, 98-129; English transl., Amer. Math. Soc. Transl. (2) **90** (1970) (to appear). MR **35** # 780.

G. BAXTER

1. *An operator identity,* Pacific J. Math. **8** (1958), 649-663. MR **21** # 2298.

JU. M. BEREZANSKIĬ

1. *Expansions in eigenfunctions of selfadjoint operators,* "Naukova Dumka", Kiev, 1965; English transl., Transl. Math. Monographs, vol. 17, Amer. Math. Soc., Providence, R. I., 1968. MR **36** # 5768.

A. BEURLING

1. *On two problems concerning linear transformations in Hilbert space,* Acta Math. **81** (1948), 239-255. MR **10**, 381.

M. Š. Birman and M. Z. Solomjak

1. *On double Stieltjes operator integrals,* Dokl. Akad. Nauk SSSR **165** (1965), 1223-1226 = Soviet Math. Dokl. **6** (1965), 1567-1571. MR **32** # 6234.
2. *Double Stieltjes operator integrals,* Probl. Math. Phys., no. 1, Leningrad. Univ., Leningrad, 1966, pp. 33-67 = Topics in Math. Phys., Vol. I, Consultants Bureau, New York, 1967, pp. 25-54. MR **35** # 767b.
3. *Double Stieltjes operator integrals and problems on multipliers,* Dokl. Akad. Nauk SSSR **171** (1966), 1251-1254 = Soviet Math. Dokl. **7** (1966), 1618-1621. MR **35** # 767a.
4. *Double Stieltjes operator integrals.* II, Probl. Math. Phys., no. 2, Leningrad. Univ., Leningrad, 1967, pp. 26-60. (Russian) MR **38**, 2621.

E. Bishop

1. *A duality theorem for an arbitrary operator,* Pacific J. Math. **9** (1959), 379-397. MR **22** # 8339.

M. S. Brodskiĭ

1. *On Jordan cells of infinite-dimensional operators,* Dokl. Akad. Nauk SSSR **111** (1956), 926-929. (Russian) MR **19**, 48.
2. *On a problem of I. M. Gel'fand,* Uspehi Mat. Nauk **12** (1957), no. 2 (74), 129-132. (Russian) MR **20** # 1229.
3. *Integral representations of bounded non-selfadjoint operators with a real spectrum,* Dokl. Akad. Nauk SSSR **126** (1959), 1166-1169. (Russian) MR **21** # 7438.
4. *On the integral representation of dissipative completely continuous operators,* Naučn. Zap. Kafedr Mat. Fiz. Estest. Odessk. Gos. Ped. Inst. **24** (1959), 3-7. (Russian)
5. *On the triangular representation of completely continuous operators with one-point spectra,* Uspehi Mat. Nauk **16** (1961), no. 1 (97), 135-141: English transl., Amer. Math. Soc. Transl. (2) **47** (1965), 59-65. MR **24** # A426.
6. *Unicellularity criteria for Volterra operators,* Dokl. Akad. Nauk SSSR **138** (1961), 512-514 = Soviet Math. Dokl. **2** (1961), 637-639. MR **24** # A1015.
7. *A multiplicative representation of certain analytic operator-functions,* Dokl. Akad. Nauk SSSR **138** (1961), 751-754 = Soviet Math. Dokl. **2** (1961), 695-698. MR **24** # A1031.
8. *Unicellularity of real Volterra operators,* Dokl. Akad. Nauk SSSR **147** (1962), 1010-1012 = Soviet Math. Dokl. **3** (1962), 1749-1751. MR **26** # 6779.
9. *On the unicellularity of the integration operator and a theorem of Titchmarsh,* Uspehi Mat. Nauk **20** (1965), no. 5 (125), 189-192. (Russian) MR **32** # 8055.

M. S. Brodskiĭ and G. È. Kisilevskiĭ

1. *Criterion for the unicellularity of dissipative Volterra operators with nuclear imaginary components,* Izv. Akad. Nauk SSSR. Ser. Mat. **30** (1966), 1213-1228; English transl., Amer. Math. Soc. Transl. (2) **65** (1967), 282-296. MR **34** # 3310.

M. S. Brodskiĭ and M. S. Livšic

1. *Spectral analysis of non-selfadjoint operators and intermediate systems,* Uspehi Mat. Nauk **13** (1958), no. 1 (79), 3-85; English transl., Amer. Math. Soc. Transl. (2) **13** (1960), 265-346. MR **20** # 7221; MR **22** # 3982.

M. S. Brodskiĭ and Ju. L. Šmul'jan

1. *Invariant subspaces of a linear operator and divisors of its characteristic function,* Uspehi Mat. Nauk **19** (1964), no. 1 (115), 143-149. (Russian) MR **29** # 2645.

M. S. Brodskiĭ, I. C. Gohberg, M. G. Kreĭn and V. I. Macaev

1. *Some investigations in the theory of non-selfadjoint operators,* Proc. Fourth All-Union Math. Congress (Leningrad, 1961), vol. II, Sectional Lectures, "Nauka", Leningrad, 1964, pp. 261-271; English transl., Amer. Math. Soc. Transl. (2) **65** (1967), 237-251.

V. M. Brodskiĭ

1. *On the integral with respect to a chain of orthogonal projectors,* Bul. Akad. Štiince RSS Moldoven, **1966**, no. 4, 59-64. (Russian) MR **36** # 1997.
2. *Multiplicative representation of the characteristic functions of contraction operators,* Dokl. Akad. Nauk SSSR **173** (1967), 256-259 = Soviet Math. Dokl. **8** (1967), 362-366. MR **36** # 3154.

M. S. Budjanu

1. *A certain theorem on the factorization of operator functions,* Bul. Akad. Štiince RRS Moldoven, **1965**, no. 7, 22-32. (Russian) MR **33** # 7872.

M. S. Budjanu and I. C. Gohberg

1. *A general theorem on factorization of matrix functions,* Studies in Algebra and Math. Anal., Izdat. "Kaetja Moldovenjaske", Kishinev, 1965, pp. 116-121. (Russian) MR **36** # 726.

A. P. Calderon and A. Zygmund

1. *A note on the interpolation of sublinear operations,* Amer. Math. J. **78** (1956), 282-288. MR **18**, 586.

T. Carleman

1. *Zur Theorie der linearen Integralgleichungen,* Math. Z. **9** (1921), 196-217.
2. *Über die Abelsche Integralgleichung mit konstanten Integrationsgrenzen,* Math. Z. **15** (1922), 111-120.

H. Cartan

1. *Sur les matrices holomorphes de n variables complexes,* J. Math. Pures Appl. **19** (1940), 1-26. MR **1**, 312.

È. R. Cekanovskiĭ

1. *The real and imaginary parts of an unbounded operator,* Dokl. Akad. Nauk SSSR **139** (1961), 48-51 = Soviet Math. Dokl. **2** (1961), 881-884. MR **24** # A1619.
2. *Model elements of non-selfadjoint operators,* Dokl. Akad. Nauk SSSR **142** (1962), 1043-1046 = Soviet Math. Dokl. **3** (1962), 256-260. MR **25** # 455.
3. *On the description of the invariant subspaces and of the unicellularity of the operator of integration in the space* $W_2^{(r)}$, Uspehi Mat. Nauk **20** (1965), no. 6 (126), 169-172. (Russian) MR **32** # 8156.

M. Cotlar

1. *A unified theory of Hilbert transforms and ergodic theorems,* Revista Mat. Cuyana **1** (1955), 105-167. MR **18**, 893.

R. Courant and D. Hilbert

1. *Methoden der mathematischen Physik.* I, II, Interscience, New York, 1943, 1953; Russian transl., GITTL, Moscow, 1951. MR **5**, 97; MR **13**, 800; MR **16**, 426.

Ju. L. Daleckiĭ

1. *On continuous rotation of subspaces in a Banach space,* Uspehi Mat. Nauk **12** (1957), no. 3 (75), 147-154. (Russian) MR **20** # 1199.

Ju. L. Daleckiĭ and S. G. Kreĭn

1. *Properties of operators depending on a parameter,* Dokl. Akad. Nauk USSR (1960), 433-436. (Ukrainian)

L. De Branges

1. *Some Hilbert spaces of entire functions,* Proc. Amer. Math. Soc. **10** (1959), 840-846. MR **22** # 4833.
2. *Some Hilbert spaces of entire functions,* Trans. Amer. Math. Soc. **96** (1960), 259-295. MR **24** # A3289a.
3. *Some Hilbert spaces of entire functions.* II, Trans. Amer. Math. Soc. **99** (**1961**), 118-152. MR **24** # A3289b.

4. *Some Hilbert spaces of entire functions.* III, Trans. Amer. Math. Soc. **100** (1961), 73-115. MR **24** # A3289c.

5. *Some Hilbert spaces of entire functions.* IV, Trans. Amer. Math. Soc. **105** (1962), 43-83. MR **26** # 583.

6. *Some applications of spaces of entire functions,* Canad. J. Math. **15** (1963), 563-583. MR **27** # 3801.

7. *Invariant subspaces of nonselfadjoint transformations,* Bull. Amer. Math. Soc. **69** (1963), 587-590. MR **26** # 6780.

8. *Some Hilbert spaces of analytic functions.* I, Trans. Amer. Math. Soc. **106** (1963), 445-468. MR **26** # 2866.

9. *New and old problems for entire functions,* Bull. Amer. Math. Soc. **70** (1964), 214-223. MR **28** # 1305.

10. *Some Hilbert spaces of analytic functions.* II, J. Math. Anal. Appl. **11** (1965), 44-72. MR **35** # 778.

11. *Some Hilbert spaces of analytic functions.* III, J. Math. Anal. Appl. **12** (1965), 149-186. MR **35** # 779.

W. F. DONOGHUE

1. *The lattice of invariant subspaces of a completely continuous quasi-nilpotent transformation,* Pacific J. Math. **7** (1957), 1031-1035. MR **19**, 1066.

N. DUNFORD

1. *A survey of the theory of spectral operators,* Bull Amer. Math. Soc. **64** (1958), 217-274; Russian transl., Matematika **4** (1960), no. 1, 53. MR **21** # 3616.

N. DUNFORD AND J. T. SCHWARTZ

1. *Linear operators.* Part I: *General theory,* Pure and Appl. Math., vol. 7, Interscience, New York and London, 1958; Russian transl., IL, Moscow, 1962. MR **22** # 8302.

2. *Linear operators.* Part II: *Spectral theory, Selfadjoint operators in Hilbert space,* Interscience, New York, 1963; Russian transl., "Mir", Moscow, 1966. MR **32** # 6181.

C. L. DOLPH

1. *Recent developments in some non-self-adjoint problems of mathematical physics,* Bull. Amer. Math. Soc. **67** (1961), 1-69. MR **25** # 5612.

V. M. ENI

1. *On the multiplicity of an eigenvector and the multiplicity of a characteristic value of a matrix pencil,* Bul. Akad. Štiince RSS Moldoven. 1965, no. 7, 94-97. (Russian) MR **33** # 5642.

D. K. FADDEEV AND V. N. FADDEEVA

1. *Computational methods of linear algebra,* Fizmatgiz, Moscow, 1960; 2nd ed., 1963; English transl., Freeman, San Francisco, Calif., 1963. MR **28** # 4659; MR **28** # 1742.

F. R. GANTMAHER AND M. G. KREĬN

1. *Oscillation matrices and kernels and small oscillations of mechanical systems,* 2nd ed., GITTL, Moscow, 1950; German transl., Akademie-Verlag, Berlin, 1960; English transl., U. S. Atomic Energy Commission, AEC-tr-4481, 1961. MR **14**, 178; MR **22** # 5161.

I. M. GEL'FAND AND V. B. LIDSKIĬ

1. *On the structure of the regions of stability of linear canonical systems of differential equations with periodic coefficients,* Uspehi Mat. Nauk **10** (1955), no. 1 (63), 3-40; English transl., Amer. Math. Soc. Transl. (2) **8** (1958), 143-181. MR **17**, 482; MR **19**, 960.

I. M. GEL'FAND, D. A. RAĬKOV AND G. E. ŠILOV

1. *Commutative normed rings,* Fizmatgiz, Moscow, 1960; English transl., Chelsea, New York, 1964. MR **34** # 4940.

JU. P. GINZBURG

1. *On J-contractive operator-functions,* Dokl. Akad. Nauk SSSR **117** (1957), 171-173. (Russian) MR **20** # 1203.
2. *The factorization of analytic matrix functions,* Dokl. Akad. Nauk SSSR **159** (1964), 489-492 = Soviet Math. Dokl. **5** (1964), 1510-1514. MR **30** # 3228.
3. *Multiplicative representations of bounded analytic operator-functions,* Dokl. Akad. Nauk SSSR **170** (1966), 23-26 = Soviet Math. Dokl. **7** (1966), 1125-1128. MR **34** # 611.

I. C. GOHBERG

1. *On linear operators depending analytically upon a parameter,* Dokl. Akad. Nauk SSSR **78** (1951), 629-632. (Russian) MR **13**, 46.
2. *Relations between the spectra of the Hermitian components of nilpotent matrices and on the integral of triangular truncation,* Bul. Akad. Štiince RSS Moldoven. **1963**, no. 1, 27-37. (Russian) MR **35** # 2168.
3. *The factorization problem in normed rings, functions of isometric and symmetric operators and singular integral equations,* Uspehi Mat. Nauk **19** (1964) no. 1 (115), 71-124 = Russian Math. Surveys **19** (1964), no. 1, 63-114. MR **29** # 487.

I. C. GOHBERG AND M. G. KREĬN

1. *The basic propositions on defect numbers, root numbers and indices of linear operators,* Uspehi Mat. Nauk **12** (1957), no. 2 (74), 43-118; English transl., Amer. Math. Soc. Transl. (2) **13** (1960), 185-264. MR **20** # 3459; MR **22** # 3984.
2. *Systems of integral equations on a half line with kernels depending on the difference of arguments,* Uspehi Mat. Nauk **13** (1958), no. 2 (80), 3-72; English transl., Amer. Math. Soc. Transl. (2) **14** (1960), 217-287. MR **21** # 1506; MR **22** # 3954.
3. *Completely continuous operators with a spectrum concentrated at zero,* Dokl. Akad. Nauk SSSR **128** (1959), 227-230. (Russian) MR **24** # A1022.
4. *On the theory of triangular representations of nonselfadjoint operators,* Dokl. Akad. Nauk SSSR **137** (1961), 1034-1037 = Soviet Math. Dokl. **2** (1961), 392-396. MR **25** # 3370.
5. *Volterra operators with imaginary component in one class or another,* Dokl. Akad. Nauk SSSR **139** (1961), 779-782 = Soviet Math. Dokl. **2** (1961), 938-985. MR **25** # 3371.
6. *The effect of some transformations of kernels of integral equations upon the equations' spectra,* Ukrain. Mat. Ž. **13** (1961), no. 3, 12-38; English transl., Amer. Math. Soc. Transl. (2) **35** (1964), 263-295. MR **27** # 1788.
7. *On the problem of factoring operators in a Hilbert space,* Dokl. Akad. Nauk SSSR **147** (1962), 279-282 = Soviet Math. Dokl. **3** (1962), 1578-1582. MR **26** # 6777.
8. *Factorization of operators in Hilbert space,* Acta Sci. Math. Szeged **25** (1964), 90-123; English transl., Amer. Math. Soc. Transl. (2) **51** (1966), 155-188. MR **29** # 6313.
9. *Criteria for completeness of the system of root vectors of a contraction,* Ukrain. Mat. Ž. **16** (1964), 78-82; English transl., Amer. Math. Soc. Transl. (2) **54** (1966), 119-124. MR **29** # 2651.
10. *Introduction to the theory of linear nonselfadjoint operators,* "Nauka", Moscow, 1965; English transl., Transl. Math. Monographs, vol. 18, Amer. Math. Soc., Providence, R.I., 1969. MR **36** # 3137.
11. *On the multiplicative representation of the characteristic functions of operators close to unitary ones,* Dokl. Akad. Nauk SSSR **164** (1965), 732-735 = Soviet Math. Dokl. **6** (1965), 1279-1283. MR **33** # 571.

I. C. GOHBERG AND N. JA. KRUPNIK

1. *On the norm of the Hilbert transform in L_p spaces,* Funkcional. Anal. i Priložen. **2** (1968), no. 2, 91-92 = Functional Anal. Appl **2** (1968), 180-181.

V. P. GURARIĬ

1. *The Fourier transform in $L^2(-\infty, \infty)$ with a weight,* Mat. Sb. **58(100)** (1962), 439-452. (Russian) MR **26** # 2810.

H. HELSON

1. *Lectures on invariant subspaces,* Academic Press, New York and London, 1964. MR **30** # 1409.

E. HILLE AND R. S. PHILLIPS

1. *Functional analysis and semi-groups,* Amer. Math. Soc. Colloq. Publ., vol. 31, Amer. Math. Soc., Providence, R.I., 1957. MR **19**, 664.

K. HOFFMAN

1. *Banach spaces of analytic functions,* Prentice-Hall, Englewood Cliffs, N. J., 1962; Russian transl., IL, Moscow, 1963. MR **24** # A2844.

I. S. IOHVIDOV AND M. G. KREĬN

1. *Spectral theory of operators in a space with an indefinite metric.* I, Trudy Moskov. Mat. Obšč. **5** (1956), 367-432; English transl., Amer. Math. Soc. Transl. (2) **13** (1960), 105-175. MR **18**, 320; MR **22** # 3983.

V. A. JAKUBOVIČ

1. *Construction of the group of symplectic matrices and the structure of the set of non-stable canonical systems of differential equations with periodic coefficients,* Mat. Sb. **44 (86)** (1958), 313-352. (Russian) MR **20**, 3325.
2. *Arguments on the group of symplectic matrices,* Mat. Sb. **55 (97)** (1961), 255-280. (Russian) MR **26** # 5072.
3. *The structure of the functional space of complex canonical equations with periodic coefficients,* Dokl. Akad. Nauk SSSR **139** (1961), 54-57 = Soviet Math. Dokl. **2** (1961), 887-890. MR **25** # 5232.

P. D. KALAFATI

1. *On a new orthogonal system of functions,* Dokl. Akad. Nauk SSSR **105** (1955), 631-633. (Russian) MR **18**, 310.
2. *Integral equations with homogeneous skewsymmetric kernels,* Trudy Nikolaevskogo Korablestroitel. Inst. **8** (1956), 246-255. (Russian)

G. K. KALISCH

1. *A functional analysis proof of Titchmarsh's theorem on convolution,* J. Math. Anal. Appl. **5** (1962), 176-183. MR **25** # 4307.

T. M. KARASEVA

1. *On a test for the boundedness of the solutions of Hill's differential equations,* Prikl. Mat. Meh. **20** (1956), 549-551. (Russian) MR **18**, 306.
2. *On the rate of growth and boundedness of solutions of second order differential equations with periodic coefficients,* Izv. Akad. Nauk SSSR. Ser. Mat. **29** (1965), 41-70; English transl., Amer. Math. Soc. Transl. (2) **79** (1969), 181-214. MR **31** # 3664.

T. KATO

1. *Fractional powers of dissipative operators,* J. Math. Soc. Japan **13** (1961), 246-274. MR **25** # 1453.

M. V. KELDYŠ AND V. B. LIDSKIĬ

1. *On the spectral theory of non-selfadjoint operators,* Proc. Fourth All-Union Math. Congress (Leningrad, 1961), vol. I, Izdat. Akad. Nauk SSSR, Leningrad, 1963, pp. 101-120. (Russian) MR **30** # 1414.

G. È. KISILEVSKIĬ

1. *On the unicellularity of dissipative Volterra operators,* Ukrain. Mat. Ž. **16** (1964), 690-696. (Russian) MR **30** # 419.

2. *Conditions for unicellularity of dissipative Volterra operators with finite-dimensional imaginary component,* Dokl. Akad. Nauk SSSR **159** (1964), 505-508 = Soviet Math. Dokl. **5** (1964), 1527-1531. MR **30** # 5262.
3. *On the reduction of a certain class of infinite-dimensional operators to Jordan normal form,* First Republ. Math. Conf. of Young Researchers, Part I, Akad. Nauk Ukrain. SSR Inst. Mat., Kiev, 1965, pp. 332-341. (Russian) MR **33** # 3101.
4. *On the order of the characteristic matrix functions of dissipative Volterra operators,* Dokl. Akad. Nauk SSSR **159** (1964), 730-733 = Soviet Math. Dokl. **5** (1964), 1579-1582. MR **30** # 5163.
5. *Invariant subspaces of dissipative Volterra operators with nuclear imaginary components,* Izv. Akad. Nauk SSSR Ser. Mat. **32** (1968), 3-23 = Math. USSR Izv. **2** (1968), 1-20. MR **36** # 4375.

A. N. Kolmogorov
1. *Sur les fonctions harmoniques conjuguées et les séries de Fourier,* Fund. Math. **7** (1925), 23-28.

M. G. Kreĭn
1. *On the theory of entire matrix functions of exponential type,* Ukrain. Mat. Ž. **3** (1951), 164-173. (Russian) MR **14**, 981.
2. *On certain problems on the maximum and minimum of characteristic values and on the Ljapunov zones of stability,* Prikl. Mat. Meh. **15** (1951), 323-348; English transl., Amer. Math. Soc. Transl. (2) **1** (1955), 163-187. MR **13**, 348; MR **17**, 484.
3. *On inverse problems for a nonhomogeneous cord,* Dokl. Akad. Nauk SSSR **82** (1952), 669-672. (Russian) MR **14**, 649.
4. *On a generalization of investigations of Stieltjes,* Dokl. Akad. Nauk SSSR **87** (1952), 881-884. (Russian) MR **14**, 868.
5. *On integral equations generating differential equations of second-order,* Dokl. Akad. Nauk SSSR **97** (1954), 21-24. (Russian) MR **16**, 372; MR **17**, 1436.
6. *On a new method of solving linear integral equations of the first and second kind,* Dokl. Akad. Nauk SSSR **100** (1955), 413-416. (Russian) MR **17**, 375.
7. *On criteria of stable boundedness of solutions of periodic canonical systems,* Prikl. Mat. Meh. **19** (1955), 641-680. (Russian) MR **17**, 1088.
8. "The basic propositions of the theory of λ-zone of stability of a canonical system of linear differential equations with periodic coefficients", in *Memorial volume to A. A. Andronov,* Izdat. Akad. Nauk SSSR, Moscow, 1955, pp. 413-498. (Russian) MR **17**, 738.
9. *On the theory of the accelerants and S-matrices of canonical differential systems,* Dokl. Akad. Nauk SSSR **111** (1956), 1167-1170. (Russian) MR **19**, 277.
10. *Integral equations on the half-line with a kernel depending on the difference of the arguments,* Uspehi Mat. Nauk **13** (1958), no. 5 (83), 3-120; English transl., Amer. Math. Soc. Transl. (2) **22** (1962), 163-288. MR **21** # 1507.
11. *A contribution to the theory of linear nonselfadjoint operators,* Dokl. Akad. Nauk SSSR **130** (1960), 254-256 = Soviet Math. Dokl. **1** (1960), 38-40. MR **24** # A1024.
12. *Sur le problème du prolongement des fonctions Hermitiennes, positives et continues,* Dokl. Akad. Nauk SSSR **26** (1940), 17-22. MR **2**, 361.
13. *On a remarkable class of Hermitian operators,* C. R. (Dokl.) Acad. Sci. URSS **44** (1944), 175-179. MR **6**, 269.
14. *On the logarithm of an infinitely decomposable Hermite-positive function,* C. R. (Dokl.) Acad. Sci. URSS **45** (1944), 91-94. MR **6**, 269.
15. *On the problem of continuation of helical arcs in Hilbert space,* C. R. (Dokl.) Acad. Sci. URSS **45** (1944), 139-142. MR **6**, 269.

16. *The fundamental propositions of the theory of representations of Hermitian operators with deficiency index* (m, m), Ukrain. Mat. Ž. **1** (1949), no. 2, 3-66. (Russian) MR **14**, 56.
17. *Solution of the inverse Stürm-Liouville problem*, Dokl. Akad. Nauk SSSR **76** (1951), 21-24. (Russian) MR **12**, 613.
18. *Determination of the density of a nonhomogeneous symmetric cord by its frequency spectrum*, Dokl. Akad. Nauk SSSR **76** (1951), 345-348. (Russian) MR **13**, 43.
19. *On the indeterminate case of the Stürm-Liouville boundary problem in the interval* $(0, \infty)$, Izv. Akad. Nauk SSSR Ser. Mat. **16** (1952), 293-324. (Russian) MR **14**, 558.
20. *On some cases of effective determination of the density of an inhomogeneous cord from its spectral function*, Dokl. Akad. Nauk SSSR **93** (1953), 617-620. (Russian) MR **15**, 796.
21. *On a basic approximation problem of the theory of extrapolation and filtration of stationary random processes*, Dokl. Akad. Nauk SSSR **94** (1954), 13-16. (Russian) MR **16**, 53.
22. *On a method of effective solution of an inverse boundary problem*, Dokl. Akad. Nauk SSSR **94** (1954), 987-990. (Russian) MR **16**, 38.
23. *Continuous analogs of propositions on polynomials orthogonal on the unit circle*, Dokl. Akad. Nauk SSSR **105** (1955), 637-640. (Russian) MR **18**, 291.
24. *On some new problems of the theory of oscillations of Stürmian systems*, Prikl. Mat. Meh. **16** (1952), 555-568. (Russian) MR **14**, 650.
25. *Analytic problems and results in the theory of linear operators in Hilbert space*, Internat. Congr. Math. (Moscow, 1966), pp. 189-216; English transl., Amer. Math. Soc. Transl. (2) **90** (1970) (to appear).

M. G. KREĬN AND V. A. JAKUBOVIČ

1. *Hamiltonian systems of linear differential equations with periodic coefficients*, Proc. Internat. Sympos. Nonlinear Vibrations, vol. I, 1961, Izdat. Akad. Nauk. Ukrain. SSR, Kiev, 1963. (Russian) MR **28** # 312.

M. G. KREĬN AND G. JA. LJUBARSKIĬ

1. *Analytic properties of the multipliers of periodic canonical differential systems of positive type*, Izv. Akad. Nauk SSSR. Ser. Mat. **26** (1962), 549-572; English transl., Amer. Math. Soc. Transl. (2) **89** (1970) (to appear). MR **26** # 400.

M. G. KREĬN AND JU. L. SMUL'JAN

1. *On fractional-linear transformations with operator coefficients*, Mat. Issled. **2** (1967), no. 3, 64-96. (Russian)

S. G. KREĬN AND JU. I. PETUNIN

1. *Scales of Banach spaces*, Uspehi Mat. Nauk **21** (1966), no. 2 (128), 89-168 = Russian Math. Surveys **21** (1966), no. 2, 85-159. MR **33** # 1719.

H. K. LANGER

1. *Über die Wurzeln eines maximalen dissipativen Operators*, Acta Math. Acad. Sci. Hungar. **13** (1962), 415-424. MR **26** # 1757.
2. *Ein Zerspaltungssatz für Operatoren in Hilbertraum*, Acta Math. Acad. Sci. Hungar. **12** (1961), 441-445. MR **25** # 3378.

P. D. LAX AND R. S. PHILLIPS

1. *Scattering theory*, Pure and Appl. Math., vol. 26, Academic Press, New York, 1967. MR **36** # 530.

B. JA. LEVIN

1. *Distribution of zeros of entire functions*, GITTL, Moscow, 1956; English transl., Transl. Math. Monographs, vol. 5, Amer. Math. Soc., Providence, R. I., 1964. MR **19**, 402; MR **28** # 217.

V. B. LIDSKIĬ

1. *Conditions for completeness of the system of root subspaces for non-selfadjoint operators with discrete spectra,* Trudy Moskov. Mat. Obšč. **8** (1959), 83-120; English transl., Amer. Math. Soc. Transl. (2) **34** (1963), 241-281. MR **21** # 6539.

O. N. LITVINENKO

1. *A method for the synthesis of non-homogeneous lines,* Radiotehn. i Èlektron. **11** (1966), 1869-1870. (Russian)

M. S. LIVŠIC

1. *On a class of linear operators in a Hilbert space,* Mat. Sb. **19 (61)** (1946), 239-262; English transl., Amer. Math. Soc. Transl. (2) **13** (1960), 61-83. MR **8**, 588; MR **22** # 3981a.
2. *On the spectral resolution of linear non-selfadjoint operators,* Mat. Sb. **34 (76)** (1954), 145-199; English transl., Amer. Math. Soc. Transl. (2) **5** (1957), 67-114. MR **16**, 48; MR **18**, 748.
3. *Theory of non-selfadjoint operators and its applications,* Trudy Third All-Union Math. Congress, vol. 3, 1956, pp. 269-276. (Russian)
4. *Operators, oscillations, waves,* "Nauka", Moscow, 1966. (Russian)

M. S. LIVŠIC AND V. P. POTAPOV

1. *A theorem on the multiplication of characteristic matrix functions,* Dokl. Akad. Nauk SSSR **72** (1950), 625-628. (Russian) MR **11**, 669.

A. M. LJAPUNOV

1. *General problem of stability of motion,* GITTL, Moscow, 1950; French transl., Ann. of Math. Studies no. 17, Princeton Univ. Press, Princeton, N. J., and Oxford Univ. Press, London, 1947. MR **9**, 34; MR **12**, 612; MR **15**, 1139.
2. *On a series, related to the theory of linear differential equations with periodic coefficients,* Collected works, Vol. II, Izdat. Akad. Nauk SSSR, 1956, pp. 387-390.

JU. I. LJUBIČ AND V. I. MACAEV

1. *To the spectral theory of linear operators in Banach space,* Dokl. Akad. Nauk SSSR **131** (1960), 21-23 = Soviet Math. Dokl. **1** (1960), 184-186.
2. *Operators with separable spectrum,* Mat. Sb. **56 (98)** (1962), 433-468; English transl., Amer. Math. Soc. Transl. (2) **47** (1965), 89-129. MR **25** # 2450.

V. I. MACAEV

1. *On the growth of entire functions that admit a certain estimate from below,* Dokl. Akad. Nauk SSSR **132** (1960), 283-286 = Soviet Math. Dokl. **1** (1960), 548-552. MR **23** # A328.
2. *A class of completely continuous operators,* Dokl. Akad. Nauk SSSR **139** (1961), 548-551 = Soviet Math. Dokl. **2** (1961), 972-975. MR **24** # A1617.
3. *Volterra operators obtained from self-adjoint operators by perturbation,* Dokl. Akad. Nauk SSSR **139** (1961), 810-813 = Soviet Math. Dokl. **2** (1961), 1013-1016. MR **25** # 457.
4. *A method of estimation for the resolvents of non-selfadjoint operators,* Dokl. Akad. Nauk SSSR **154** (1964), 1034-1037 = Soviet Math. Dokl. **5** (1964), 236-241. MR **28** # 1495.
5. *Several theorems on completeness of root subspaces of completely continuous operators,* Dokl. Akad. Nauk SSSR **155** (1964), 273-276 = Soviet Math. Dokl. **5** (1964), 396-399. MR **28** # 2444.

V. I. MACAEV AND JU. A. PALANT

1. *On the powers of a bounded dissipative operator,* Ukrain. Mat. Ž. **14** (1962), 329-337. (Russian) MR **26** # 4184.

E. MAGENES

1. *Spazi di interpolazione ed equazioni a derivate parziali,* Atti del VII Congresso dell'Unione Matematica Italiana, Genova, 1963, Edizioni Cremonese, Rome, 1965, pp. 134-197; Russian transl., Uspehi Mat. Nauk **21** (1966), no. 2 (128), 169-218. MR **35** # 5925; **34** # 3074.

G. F. MANDŽAVIDZE

1. *Approximate solution of boundary problems of the theory of analytic functions,* Investigations on modern problems of the theory of functions of a complex variable, Fizmatgiz, Moscow, 1960, pp. 365-370. (Russian) MR **22** # 6890.

G. F. MANDŽAVIDZE AND B. V. HVEDELIDZE

1. *On the Riemann-Privalov problem with continuous coefficients,* Dokl. Akad. Nauk SSSR **123** (1958), 791-794. (Russian) MR **22** # 3806.

P. MASANI

1. *The Laurent factorization of operator-valued functions,* Proc. London Math. Soc. **6** (1956), 59-69. MR **18,** 138.

J. MIKUSINSKI

1. *Une simple démonstration du théorème de Titchmarsh sur la convolution,* Bull. Acad. Polon. Sci. Sér. Sci. Math. Astr. Phys. **7** (1959), 715-717. MR **22** # 2854.

B. S. MITJAGIN

1. *An interpolation theorem for modular spaces,* Mat. Sb. **66 (108)** (1965), 473-482. (Russian) MR **31** # 1562.

G. M. MORDASOVA

1. *Asymptotic behavior of the eigenvalues of a linear integral equation with discontinuous kernel,* Dokl. Akad. Nauk SSSR **142** (1962), 1022-1025 = Soviet Math. Dokl. **3** (1962), 234-238. MR **26** # 4134.

B. SZ.-NAGY

1. *Sur les contractions de l'espace de Hilbert,* Acta Sci. Math. Szeged **15** (1953), 87-92. MR **15,** 326.
2. *Sur les contractions de l'espaces de Hilbert.* II, Acta Sci. Math. Szeged **18** (1957), 1-14. MR **19,** 873.

B. SZ.-NAGY AND C. FOIAŞ

1. *Sur les contractions de l'espace de Hilbert.* III, Acta Sci. Math. Szeged **19** (1958), 25-45. MR **21** # 2188.
2. *Sur les contractions de l'espace de Hilbert.* IV, Acta Sci. Math. Szeged **21** (1960), 251-259. MR **23** # A3445.
3. *Sur les contractions de l'espace de Hilbert.* V: *Translations bilatérales,* Acta Sci. Math. Szeged **23** (1962), 106-129. MR **26** # 2890.
4. *Sur les contractions de l'espace de Hilbert.* VI: *Calcul fonctionnel,* Acta Sci. Math. Szeged **23** (1962), 130-167. MR **26** # 2891.
5. *Sur les contractions de l'espace de Hilbert.* VII: *Triangulations canoniques. Fonction minimum,* Acta Sci. Math. Szeged **25** (1964), 12-37. MR **30** # 455.
6. *Sur les contractions de l'espace de Hilbert.* VIII: *Fonctions caractéristiques. Modèles fonctionnels,* Acta Sci. Math. Szeged **25** (1964), 38-71. MR **30** # 2348.
7. *Sur les contractions de l'espace de Hilbert.* IX: *Factorisations de la fonction caractéristique. Sous-espaces invariants,* Acta Sci. Math. Szeged **25** (1964), 283-316. MR **31** # 5101.
8. *Sur les contractions de l'espace de Hilbert.* X: *Contractions similaires à des transformations unitaires,* Acta Sci. Math. Szeged **26** (1965), 79-91. MR **34** # 1856.
9. *Corrections et compléments aux Contractions.* IX, Acta Sci. Math. Szeged **26** (1965), 193-196. MR **33** # 4689.

10. *Sur les contractions de l'espace de Hilbert.* XI: *Transformations unicellulaires,* Acta Sci. Math. Szeged **26** (1965), 301-324. MR **32** # 8176.

11. *Décomposition spectrale des contractions presque unitaires,* C. R. Acad. Sci. Paris Ser. A-B **262** (1966), A440-A442. MR **33** # 3111.

12. *Sur les contractions de l'espace de Hilbert.* XII: *Fonctions intérieures admettant des facteurs extérieurs,* Acta Sci. Math. Szeged **27** (1966), 27-33. MR **33** # 6414.

M. A. Naĭmark

1. *Linear differential operators,* GITTL, Moscow, 1954; English transl., Ungar, New York, 1968; German transl., Akademie-Verlag, Berlin, 1960. MR **16**, 702.

N. K. Nikol'skiĭ

1. *Unicellularity and non-unicellularity of weighted shift operators,* Dokl. Akad. Nauk SSSR **172** (1967), 287-290 = Soviet Math Dokl. **8** (1967), 91-94. MR **36** # 728.

R. S. Phillips

1. *Dissipative operators and hyperbolic systems of partial differential equations,* Trans. Amer. Math. Soc. **90** (1959), 193-254; Russian transl., Matematika **6** (1962), no. 4, 11-70. MR **21** # 3669.

A. I. Plesner

1. *Spectral theory of linear operators,* "Nauka", Moscow, 1965. (Russian) MR **33** # 3106.

A. I. Plesner and V. A. Rohlin

1. *Spectral theory of linear operators,* Uspehi Mat. Nauk **1** (1946), no. 1 (11), 71-191; English transl., Amer. Math. Soc. Transl. (2) **62** (1967), 29-175. MR **9**, 43.

V. T. Poljackiĭ

1. *On the reduction of quasi-unitary operators to a triangular form,* Dokl. Akad. Nauk SSSR **113** (1957), 756-759. (Russian) MR **19**, 873.

G. Ja. Popov

1. *On a method of solution of the axisymmetric contact problem of the theory of elasticity,* Prikl. Mat. Meh. **25** (1961), 76-85 = J. Appl. Math. Mech. **25** (1961), 105-118. MR **27** # 3124.

2. *On an approximate method for solving certain planar contact problems of elasticity theory,* Izv. Akad. Nauk Armjan. SSR Ser. Fiz.-Mat. Nauk **14** (1961), no. 3, 81-96. (Russian)

3. *Certain properties of classical polynomials and their application to contact problems,* Prikl. Mat. Meh. **28** (1964), 442-451 = J. Appl. Math. Mech. **28** (1964), 544-555. MR **31** # 3647.

4. *Concerning a planar contact problem for an elastic half-strip,* Izv. Akad. Nauk Armjan. SSR Ser. Meh. **4** (1965), 153-157. (Russian)

V. P. Potapov

1. *The multiplicative structure of J-contractive matrix functions,* Trudy Moskov. Mat. Obšč. **4** (1955), 125-236; English transl., Amer. Math. Soc. Transl. (2) **15** (1960), 131-243. MR **17**, 958; MR **22** # 5733.

2. *On holomorphic matrix functions bounded in the unit circle,* Dokl. Akad. Nauk SSSR **72** (1950), 849-852. (Russian) MR **13**, 736.

I. I. Privalov

1. *Boundary properties of analytic functions,* GITTL, Moscow, 1950; German transl., Hochschulbücher für Math., Bd. 25, VEB Deutscher Verlag, Berlin, 1956. MR **13**, 926; MR **18**, 727.

I. E. Prokopovič

1. *On the solution of the plane contact problem taking account of creep,* Prikl. Mat. Meh. **20** (1956), 680-687. (Russian) MR **19**, 342.

I. M. RAPOPORT

1. *On a variational problem in the theory of ordinary differential equations with boundary conditions,* Dokl. Akad. Nauk SSSR **73** (1950), 889-890. (Russian) MR **12**, 183.
2. *On asymptotic methods in the theory of differential equations,* Izdat. Akad. Nauk Ukrain. SSR, Kiev, 1954. (Russian) MR **17**, 734.

F. RIESZ AND B. SZ.-NAGY

1. *Leçons d'analyse fonctionelle,* Akad. Kiadó, Budapest, 1952; 2nd ed., 1953; 3rd ed., Akad. Kiadó, Budapest and Gauthier-Villars, Paris, 1955; English transl., Ungar, New York, 1955. MR **14**, 286; MR **15**, 132; MR **16**, 837; MR **17**, 175.

M. RIESZ

1. *Sur les maxima des formes bilinéaires et sur les fonctionelles linéaires,* Acta Math. **49** (1926), 465-497.
2. *Sur les fonctions conjuguées,* Math. Z. **27** (1937), 213-244.

N. A. ROSTOVCEV

1. *On the solution of a plane contact problem,* Prikl. Mat. Meh. **17** (1953), 99-106. (Russian) MR **14**, 1146.

L. A. SAHNOVIČ

1. *On the reduction of Volterra operators to the simplest form and on inverse problems,* Izv. Akad. Nauk SSSR. Ser. Mat. **21** (1957), 235-262. (Russian) MR **19**, 970.
2. *Spectral analysis of Volterra operators and some inverse problems,* Dokl. Akad. Nauk SSSR **115** (1957), 666-669. (Russian) MR **19**, 866.
3. *Spectral analysis of operators of the form $Kf = \int_0^x f(t)k(x-t)\,dt$,* Izv. Akad. Nauk SSSR. Ser. Mat. **22** (1958), 299-308. (Russian) MR **20** #5409.
4. *The reduction of non-selfadjoint operators to triangular form,* Izv. Vysš. Učebn. Zaved. Matematika **1959**, no. 1 (8), 180-186. (Russian) MR **25** #460.
5. *A study of the "triangular form" of non-selfadjoint operators,* Izv. Vysš. Učebn. Zaved. Matematika **1959**, no. 4 (11), 141-149; English transl., Amer. Math. Soc. Transl. (2) **54** (1966), 75-84. MR **25** #461.

S. O. ŠATUNOVSKIǏ

1. *Introduction to analysis,* "Mathesis", Odessa, 1923.

R. SCHATTEN

1. *Norm ideals of completely continuous operators,* Ergebnisse der Math. und ihrer Grenzgebiete, Heft 27, Springer-Verlag, Berlin, 1960. MR **22** #9878.

I. SCHUR

1. *Über die characteristischen Wurzeln einer linearen Substitution mit einer Anwendung auf die Theorie der Integralgleichungen,* Math. Ann. **66** (1909), 488-510.

E. M. SEMENOV

1. *Interpolation of linear operators in symmetric spaces,* Dokl. Akad. Nauk SSSR **164** (1965), 746-749 = Soviet Math. Dokl. **6** (1965), 1294-1298. MR **32** #8139.

N. P. SERGEEV

1. *On a method for solving linear integral equations,* Leningrad. Gos. Ped. Inst. Učen. Zap. **33** (1956), 149-153. (Russian)

I. B. SIMONENKO

1. *Riemann's boundary value problem with a continuous coefficient,* Dokl. Akad. Nauk SSSR **124** (1959), 278-281. (Russian) MR **22** #3805.

JU. L. ŠMUL'JAN

1. *Isometric operators with infinite deficiency indices and their orthogonal extensions,* Dokl. Akad. Nauk SSSR **87** (1952), 11-14. (Russian) MR **14**, 882.
2. *Operators with degenerate characteristic functions,* Dokl. Akad. Nauk SSSR **93** (1953), 985-988. (Russian) MR **15**, 803.

S. L. SOBOLEV

1. *Remarks on the numerical solution of integral equations,* Izv. Akad. Nauk SSSR. Ser. Mat. **20** (1956), 431; English transl., Acad. R. P. Romîne. An. Romîno-Soviet. Ser. Mat.-Fiz. (3) **11** (1957), no. 4 (23), 5. MR **18**, 322; MR **21** #439.

I. JA. ŠTAERMAN

1. *The contact problem of the theory of elasticity,* GITTL, Moscow, 1949. (Russian) MR **11**, 756.

A. V. ŠTRAUS

1. *Characteristic functions of linear operators,* Izv. Akad. Nauk SSSR. Ser. Mat. **24** (1960), 43-74; English transl., Amer. Math. Soc. Transl. (2) **40** (1964), 1-37. MR **25** #4363.

E. C. TITCHMARSH

1. *The theory of functions,* 2nd ed., Oxford Univ. Press, Oxford, 1939; Russian transl., GITTL, Moscow, 1951. MR **33** #5850.
2. *Reciprocal formulae involving series and integrals,* Math. Z. **25** (1926), 321-381.

J. WERMER

1. *The existence of invariant subspaces,* Duke Math. J. **19** (1952), 615-622. MR **14**, 384.

N. WIENER AND P. MASANI

1. *The prediction theory of multivariate stochastic processes.* II: *The linear predictor.* Acta Math. **99** (1958), 93-137. MR **20** #4325.

F. WOLF

1. *Operators in Banach space which admit a generalized spectral decomposition,* Nederl. Akad. Wetensch. Proc. Ser. A **60** = Indag. Math. **19** (1957), 302-311. MR **19**, 869.

N. E. ŽUKOVSKIĬ

1. *Conditions for the finiteness of the integrals of the equation* $d^2y/dx^2 + py = 0$, Selected papers, Vol. 1, OGIZ, Moscow, 1948, pp. 246-253. (Russian) MR **11**, 574.

A. ZYGMUND

1. *Trigonometric series,* Vols. I, II, 2nd rev. ed., Cambridge Univ. Press, New York, 1959; Russian transl., "Mir", Moscow, 1965. MR **21** #6498; MR **31** #2554.

INDEX